Solarizing
Your Present Home
Practical Solar Heating Systems You Can Build

Edited by Joe Carter

Assistant Editor: John Blackford
Editorial Assistant: Margaret J. Balitas
Copy Editor: Dolores Plikaitis
Technical Editor: Herb Wade
Art Director: Karen A. Schell
Book Design: Kim E. Morrow
Photography: Rodale Press
Photography
Department

Rodale Press, Emmaus, Pa.

To all the people in the solar community whose good works made this book possible. The seeds they have planted will make future solar harvests all the more bountiful.

The plans and projects presented in *Solarizing Your Present Home* were screened from hundreds of possible entries. They were designed and built by knowledgeable people, and we publish them here as working examples of solar retrofit options. But the success of any project can be affected by local conditions, materials, and personal skills. For this reason, neither the people who contributed their plans and projects nor Rodale Press can assume any responsibility for any damages or losses incurred or injuries suffered as a result of construction of these designs. We recommend that before you begin any project, you study all directions, plans and graphics very carefully and consult with a knowledgeable person who has had experience with such construction.

Printed in the United States of America on recycled paper, containing a high percentage of de-inked fiber.

Library of Congress Cataloging in Publication Data
Main entry under title:

Solarizing your present home.

Includes bibliographies and index.
1. Solar heating—Amateurs' manuals. I. Carter, Joe.
TH7413.S64   643'.1   81-13855
ISBN 0-87857-367-4 hardcover   AACR2

4 6 8 10 9 7 5   hardcover

Contents

Introduction

Energy: A National Problem
with Some Household Remedies

We're all feeling it in one way or another—homeowners, store owners, manufacturers, farmers, truckers—everyone is having close encounters with energy. Rising prices and the threat of shortages have put us all on notice that it's time to be more careful with the fuels and power we consume. The problem is clear: Domestic energy resources that were once easily and cheaply recoverable have dwindled. It's costing more and more to get fossil fuels (oil, natural gas, coal) out of the ground, and our dependence on foreign sources for over 40 percent of our crude oil supply has hurt us deeply. Prices for imported oil are out of control, up by 300 percent in the last seven years and, as we've seen, foreign supply is by no means a sure supply.

The energy problem is not a seasonal one, nor does it follow on the heels of national or international events. There have been peaks when problem became crisis, but the troubles that seemed to have their beginnings in 1973, with a "recurrence" in 1978, have in fact been long abuilding. In the 1970s "energy," the Whole Big Thing that it is, finally became a highly visible issue that was having serious political, economic, social and global impact. In the 1980s the problem is still with us, steadily intensifying.

A few words can give but the simplest summary of what is really a huge and complex problem. It is the stuff that thick volumes are made of. How did we get caught by the energy problem? Where do things stand today? How do we get out of this mess? Questions fly about almost as thickly as do conflicting answers. Everyone has something to say about energy.

One thing is certain: The problem won't be diminished by any sort of benign neglect. The classic 20th century complaint reminds us that the surest things in life are death and taxes, but we can be quite sure that energy will increasingly affect our lives and our livelihoods. It can help us as much as it can hurt us, and whether the effects are for better or worse will depend on the solutions this country and we as individuals choose to pursue. In no uncertain terms we need answers, and results, soon.

In the search for answers, the complexity and magnitude of the energy problem are the dominant factors. Energy isn't a monolithic problem that awaits the sweep of a Big Techno-Fix to make things right again. It is rather a web of problems that lies over the entire fabric of our society. Until recent years there has been little if any need for this society to concern itself with energy supply, demand and cost. All the major sectors —industrial, commercial, transportation and residential—have been designed for and have grown on an unrestrained access to cheap energy. No blame, though. Cheap energy, for all it meant, was the way it was. But energy has gotten so uncheap so quickly that our patterns of consumption have had little time to adapt. Few people could have predicted that energy costs would rise 5 or 10 or 15 percentage points faster than inflation, but now we're caught between a rock and a hard place with little space in between: We've got to pay out to the spiraling cost of energy to keep the supply going and to buy the time we need to upgrade and restructure this country's myriad energy production, distribution and consumption systems.

Compounding this glum state of affairs is the fact that there is still great pressure for energy growth. Demand increases, and more power plants have to be built. This also makes energy cost more because "new energy" production is increasingly expensive. Every new power plant costs more per unit of energy produced than the one built before it. Every new oil or gas well produces costlier energy because all the easy-access wells have been tapped already, and we must dig deeper to get at new resources and transport them farther to point of use.

In just about any way it's examined the energy problem is a ponderous thing. It is multifaceted and very complicated, and so are the apparent solutions, especially when they're considered on a large scale. Some solutions are, of course, more desirable than others, and all have their backers and their detractors. For the sake of simplifying the argument, we can collect the bundle of possible energy solutions into two basic options:

Option: We can strive to increase dramatically domestic energy production, both to shed our dependence on foreign oil and to meet, and thereby to promote, a "normal" rate of energy growth.

Option: We can, as individuals and as a nation, work toward getting off the foreign oil-new production hook by decreasing the demand for fossil fuels and electric power, by increasing the efficiency of our energy use and by developing supplies from the solar-renewable family of energy sources—wood and other "biomass," wind, falling water and, of course, the sun.

The ultimate solution will most certainly involve a mix of these two approaches. But the critical question is this: Which of these two options, which are fundamentally so different, will become the dominant approach as we create our energy future? Both can be the solution to the problem, but in terms of the quality of our society and of the environment it inhabits, they will not lead to the same result.

Business As Usual

The first option can be characterized as a "more of the same" course of action. According to its supporters, this path follows the traditional, proven routes of our recent energy history: Increases in demand must be met with increases in production. Energy growth is vital to general economic growth; it guarantees jobs and helps to preserve the high standard of living that most Americans enjoy. But in real terms, sustained energy growth today essentially means building more coal-fired and nuclear-powered generating plants. It means seeking out and extracting every last barrel of crude oil and cubic foot of gas that lie under our lands and waters.

Nuclear power, which once looked like *the* answer for our energy future, is hardly a choice anymore. Twenty-five years after the development of the "peaceful atom" was begun, nuclear power technologies are still technically unreliable, and without the huge federal subsidies the industry has enjoyed, it could never be economically feasible. There are dangers at virtually every point in the nuclear fuel cycle, all of which revolve around the deadly effects of nuclear radiation. Even the "normal" processing and handling of nuclear fuel materials increase the amount of radiation present in the environment. Nuclear power plants generate tons of both low- and high-level radioactive wastes, and while the inventory of these wastes steadily increases, neither the government nor industry has come up with a trustworthy means of disposal.

We know that exposure to high levels of radiation causes sickness and death, but we do not know about the long-term effects of low-level radiation exposure. Yet we are supposed to accept as truth statements from the nuclear industry that increases in "background" radiation are not a threat to our health. Uranium is a nonrenewable source, and it is not the fuel we need to create a self-sustaining energy economy.

There are billions of tons of coal in the ground, enough some say to supply our energy needs through this and the next century and beyond. Our use of coal will undoubtedly have a tremendous increase in the years ahead, as it replaces foreign oil and takes the place of the nuclear industry. Newer technologies can create cleaner-burning liquid, gaseous and solid fuels from coal, which can help to reduce some of the environmental risks associated with coal combustion. But until these technologies are fully developed, we're faced with the prospect of daily throwing tons of particulates into the atmosphere. Acid rain, a result of sulfurous emissions from fossil fuel combustion combining with oxygen and hydrogen in the atmosphere, is killing freshwater lakes and eroding our buildings.

The greatest impact of increased coal production is in what will happen to the land it lies under. Despite the claims of those that support reclamation, strip mining is not benign in its effects on environmental quality. Land that is reclaimed is not the same land. Coal combustion also heavily impacts water supplies. Tremendous volumes of water are needed as a cooling agent in coal-fired plants. In the semiarid West where most of the coal in the United States is buried, this will mean robbing Peter to pay Paul, for there is little enough water to go around as it is.

If we rush to deplete our final reserves of oil and gas, we will lose valuable resources that we need for purposes other than energy production. Lubricants, plastics and synthetic fibers are all oil-derived, as are many kinds of medicines and other chemicals. Gas, too, has uses in other nonenergy production processes. And finding more of either resource will not necessarily curb the rapid price inflation of recent years. Price always increases with demand, and because these resources are becoming harder and harder to find, they will invariably cost more, demand notwithstanding.

Less Will Do More

If we focus our efforts instead on the second option—reducing demand and developing renewable energy sources—we will solve many more problems than we create. Conservation is the primary means by which the demand for energy can be reduced. In the book, *Energy Future,* the Energy Project at the Harvard Business School went so far as labeling conservation as a "key energy *source*" for this country (italics added). Indeed it is: Reducing consumption creates more slack in existing energy production capacity. "Efficiency growth" can create more energy supplies more quickly and economically than energy growth. The same Harvard study noted that the conservation energy source has the potential to reduce national energy consumption by fully 30 percent.

In 1978 this country consumed about 78 quadrillion Btu (78,000,000,000,-000,000) of energy, 13 quads of which were imported crude oil, with another 5½ quads imported in other forms of energy. Total imports in that year were less than 25 percent of total consumption, less energy than we stand to "create" with conservation. This is not to say that conservation can replace all of our import demand, nor can we expect conservation to save 20, 15 or even 5 quads of energy in the very near future, but in conservation we have at hand an enormous potential to relieve external and internal energy problems and to reduce substantially the pressure for energy growth.

There is a multitude of proven applications for solar energy in all its forms. Small-scale hydroelectric systems are being introduced in creeks and rivers, and small, once-abandoned dams and generating stations are being pressed into service, profitably. Windmills, once a fixture on the American landscape, are being returned to the tasks of pumping water and generating electricity because they too can compete favorably with other power sources. Solar energy is being used on the farm to keep livestock warm and healthy, to dry grain, nuts and fruits. Direct solar heat is even being used to pump water. In industry solar systems can produce fluid temperatures of 100 to 1000+°F for a wide variety of process heat applications.

The use of wood fuel has increased tremendously, and today the interest in producing alcohol fuels from both agricultural produce and waste is running high. These applications show great potential, but there is reason to be a little bearish toward the former and very bearish toward the latter. More woodburning means more atmospheric pollution. In some towns where woodstoves are the primary home heating system, air quality has been compromised. It is to be hoped that the research now underway will soon produce ways of cleaning the exhaust from woodstoves before it enters the atmosphere. There is also the question of resource availability and proper forestland management. Wood is indeed renewable, but at its own pace. Our forestlands are vast, but we must be careful not to deplete the wood resource with hasty and improper harvesting routines.

The technology for making alcohol from biomass is available, the economics favorable, but great care will have to be taken in the way land is managed for food and energy. If silage is put into stills instead of back into the ground, soil quality can be greatly compromised. Things are bad enough with soil these days: Every year three million acres of farmland are taken out of agricultural production and the equivalent of another three million acres is lost each year from topsoil erosion. While we have plenty of food in this country now, that may not always be the case if statistics like these continue to repeat. Using land for energy production will of course put more pressure on agricultural lands. We must not trade soil for energy.

Household Remedies

There are indeed some questions and conflicts concerning some solar-related energy technologies, but when it comes to household remedies for the energy problem, there is no argument. Conservation and solar improvements can work wonders on our home energy systems. We, the people, use a lot of energy to stay warm, to stay cool, to wash and cook. Like the other energy-consuming sectors, our demand for energy has been consistently on the rise. From 1960 to 1973, the population of the United States increased by 11 percent, yet residential energy consumption increased by 50 percent. Today the residential sector consumes 20 percent of all the energy production of the United States, a figure that does not include the consumption of gasoline. Half of that consumption is for space heating; 15 percent of residential energy goes for domestic water heating. Air conditioning, lighting and refrigeration each account for 6 to 7 percent of residential consumption, and the remainder includes cooking and other appliance loads.

The top three energy users—space heating and cooling and domestic water heating—all present attractive opportunities for conservation and solar energy. A study of energy consumption in Sweden showed that while Swedes enjoy a standard of living much like our own, their per capita energy consumption is about one-half that of United States' consumption levels. "Per capita consumption" applies to a country's total use of energy, not just residential use. But the point is that we shouldn't look at conservation as deprivation.

Recent studies of United States' residential energy use have shown that with the most common and most cost-effective conservation improvements, home heating energy use can easily be reduced by 30 percent. With additional improvements space heating loads have been cut from 50 to 75 percent. Domestic water heating energy use can easily be cut by 25 percent, and a 50 percent reduction has been demonstrated time and time again.

With electricity being the most expensive form of energy, air conditioning for space cooling can in some climates be extremely costly. It is also the most burdensome load for many electric utilities because summer is when peak demand occurs, and new power plants often have to be built to cover these peaks. Yet by simply shading windows through a hot summer's day, the heat gained by a house can be reduced by as much as 30 percent. Beyond that there are many techniques for *natural cooling* that can put air conditioners on long vacations.

Most conservation improvements have zero to very little impact on day-to-day living. Some do require a little habit modification, but by and large conservation is a matter of physically upgrading the energy system: your house. You can go a long way with conservation, and the biggest changes will be nice ones: lower bills and increased comfort.

The range of solar applications for homes gives virtually every single-family dwelling some degree of solar heating potential; ergo, this book. Once conservation improvements have "turned off" much of a home's energy use, solar improvements can reduce the remaining space and water heating loads by another 50 percent or more. In the next few years, photovoltaics (conversion of sunlight directly to electricity) will also become a cost-effective option for independent home energy production. This member of the solar family may well grow to dwarf its heat-collecting cousins in terms of the energy contribution it will make.

What conservation and solar energy sources have in common is that they are highly distributed, rather than highly centralized; they are spread quite democratically across the land, available to all. They promote personal self-reliance and they help to guarantee national energy security. Today it is, of course, the first option that dominates the energy scene; the second option is a lesser, but growing element in the mix of solutions. Where does it go from here? I think it can go where we want it to go. We should not see ourselves as being separate from this country's huge energy system, nor power-less to affect its future. We are in fact directly linked to it, and we can cast a vote, an effective vote for the direction of our individual and collective energy futures by making our homes less dependent on "imported" (through pipes and wires) energy, by making them more efficient and tuned-in to the sun.

Joe Carter

Working with This Book

If you've made a quick pass through this book, you can see that there's quite a lot going on. We've packed it. But there's a method to this melange of words, drawings and photographs, and it needs just a little explanation to help you make the most effective use of this "toolbox" of retrofit ideas.

Words

There are four basic manuscript elements mixed into this book in various combinations: projects and essays and our beloved boxes and mini-essays. The projects are, of course, the heart of this collection. With the help of some 40-odd expert contributors we assembled project ideas from all over the country, which solved one problem: making the book comprehensive; but created another: maintaining consistency from one project to the next. To do that we established some basic guidelines as to what each project should cover.

Overall, the projects are designed to provide enough information to take you from start to finish on your own project. But when is "enough" enough? That depends on the nature of the particular project and on your own level of experience. The simpler projects, such as a solar water heater and a window insulation panel, are described in total piece-by-piece detail. The more complex undertakings are broken down into their basic parts but, because of space limitations, they don't quite have the "blueprint" presentation of the simpler projects. That's where your own expertise fits in. We've assumed that if you're tackling a tough project, you've already got a good handle on the basis of construction, and you don't need to see or be told where every nail, nut and bolt goes. Let your background be your guide and, if you're starting small, it won't take long to work up to a major retrofit for your house.

When you've found the perfect project, here's what you can expect it to include: There is some description of the "real world" example from which the project is drawn. This usually includes some discussion of *why* the owners did what they did in terms of solving an energy problem and a design problem, and whatever other issues there were that had to be resolved. Maybe some of their answers will help your own decision making.

Then there is discussion, where appropriate, of construction, operation, maintenance, cost and benefit, ways to improve the performance or appearance of the finished product, and what might be involved in transposing a project from one climate to another. This body of information is drawn from either the owner's or the builder's or the writer's experience (many times they're the same person), and sometimes from the knowledge of "innocent bystanders." These sections give the kinds of insights that wouldn't show up on the blueprint: Do's and don't's, construction tips, mistakes that you shouldn't repeat, general wisdom from people who know.

The final part of the project format is the construction steps, which begin with the materials checklist. The steps themselves break down the job(s) at hand into a logical construction sequence, another sort of checklist to follow. A list of other references is sometimes included at the close of a project to help you pick up on more design or related technical information. At the end of most of the projects, you'll find a list of Hardware Focus items that notes manufacturers and/or sources of supply for certain specialized hardware and materials that are involved in the project.

Essays are an important companion to the projects. They generally provide overviews of large subject areas that can't be dealt with as projects. As you go through the book, you'll find essays on subjects like what solar glazing materials are all about, on design and construction parameters for attached solar greenhouses, on ways of using reflectors to increase solar energy gain and on ways to keep cool in the summer without switching on that costly air conditioner. Although they don't describe actual projects, several of the essays are full of project ideas, seeds that you might decide to germinate for your own retrofit plan. For those inclined to know more, essays will usually close with a few bibliographic entries, and occasionally they'll also contain some Hardware Focus entries.

The boxes in this book are little flashes of information and illustration that highlight a variety of subject areas: unique Hardware Focus items that warrant more coverage, little project ideas that are just too tiny to fit the project format, little bits of essay-type subject matter. That might be a fun way to skim the book—read the boxes. They're quick, to the point and downright pithy!

The mini-essays look like essays but, you guessed it, they're short. In the sections, "Solar Additions" and "The Whole House," we use mini-essays along with illustrations and photographs to present portraits of some rather elegant retrofit work. These sections are a sort of break in the action, a collection of imagination boosters that you can peruse in your armchair while you're piecing together your own retrofit puzzle.

Pictures

Even the pictures are part of a program. They, too, are broken down into four categories: idea illustrations, schematics, construction views, and 3-dimensional cutaway or "X-ray" views. *Idea illustrations* generally accompany essays, where they serve to illustrate some of the basic concepts that the essays present. *Schematics* are just what the name implies: detailed renderings of the plumbing and electrical systems that are part of some projects, mostly domestic water heating and air heating collector systems. They do not and realistically could not include every possible pipe fitting and wire nut, but they do present all the important components in the system design. And except for specialized solar controls, none of those components are beyond the standard inventories of your local plumbing and electrical suppliers. As for the solar controls, you will be able to acquire them from local solar businesses or from some of the manufacturers and mail-

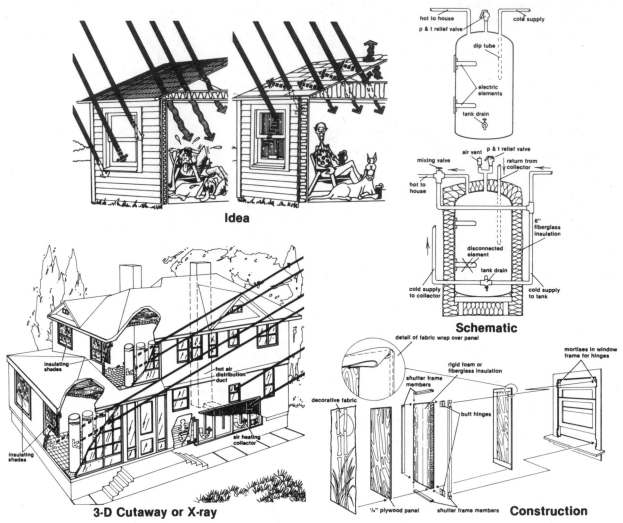

Idea

Schematic

3-D Cutaway or X-ray

Construction

The illustrations serve very different purposes, as these examples indicate. The idea illustrations inform, and sometimes entertain, with relatively simple line drawings. The other types are generally much more detailed, and they warrant your close attention, especially with the construction drawings where we've endeavored to show all the important pieces without turning the book into a pile of blueprints.

order sources that are listed in the Hardware Focus sections.

   Construction views appear in the projects in a couple of ways. Simpler build-it projects will have exploded views that show every piece of the finished project in relation to the other pieces it's connected with. Other views show the cross section of a completed assembly such as a glazing or a framing detail. More complex construction views look at a project from a perspective or isometric cutaway view so that they can show most or all of the many parts of larger retrofit projects. With illustrations like these you're urged to study them closely and carefully, for they do indeed present a lot of information. Our top-of-the-line illustrations are the *3-dimensional cutaway or X-ray views* of major retrofit projects such as building additions and whole-house retrofits, where at least three solar improvements have been built into one house. These views are intended to let you see into the space created by an addition or to show you how multiple retrofits are integrated by design and function.

Hardware Focus: Your Access to the Right Stuff

As you read into this book, you will (undoubtedly) be inspired to retrofit your home with a solar heating system. Along with all the standard materials these systems use, some of them require specialized hardware controls. But where do they come from, and where do you go to get them? To save time and effort with your solar shopping, we've borrowed a technique perfected by Ma Bell. Instead of dashing madly around your town, county or state trying to find products suitable for your solar installation, let your fingers do the walking through what we call the Hardware Focus sections of this book. These sections were researched and compiled by Margaret J. Balitas, and you'll find them at the ends of many projects and essays. All told they list literally hundreds of useful products, each with a brief description and the manufacturer's name, address, and phone number. The Hardware Focus sections run the gamut of off-the-shelf items, from special fasteners to complete packages like greenhouse kits and solar water heaters. No matter what item you need for your solar retrofit, you'll probably find a source for it in this book. In short, we've done your solar shopping for you.

Manufacturers can direct you to nearby distributors, or if you have a local solar contractor or solar products store (which is becoming more common; check Ma Bell's Yellow Pages under "Solar") you can shop there with confidence because you'll know about what you want.

Read This Book!

This book grows on itself, and it gets smarter as it goes along. So do you, if you read the whole thing. If you just read up to the project you're interested in, you'll probably have found your tree, but you may have missed the forest. If you read the whole thing, you'll have gained the benefit of a fuller knowledge of retrofitting, which could benefit the outcome of that first project that caught your fancy or which could affect your whole game plan. This book traverses the entire solar landscape, from energy conservation to domestic water heating to space heating to natural cooling. We'd like you to consider all these options, because they can all be of value to you and your home. Look at it this way: You are the homeowner, and this is your new owner's manual. If you bought a new car, would you just read halfway into the manual and then go rolling off into the sunset? Read this book! We hope you enjoy the trip.

GETTING STARTED

I

Photo I-1: Solar homes didn't just start in the '70s, '60s or '50s. There have always been homes that were consciously or intuitively designed with an eye for the path of the sun. Look around your town, and you'll probably find some nice looking examples, old and new.

"Every home is a solar home."

The quote belongs to Denis Hayes. When he said it, it was Sun Day, 1978, this country's first celebration of solar energy, and as a prominent solar activist, Hayes had led the creation of that day. Now, after a stint as the director of the federal Solar Energy Research Institute, Hayes continues his activist role as one of the foremost experts on the often complex social, political and economic aspects of widespread solar utilization. Coming from him, that remark was by no means hype or facetiousness. It was loaded with truth. Simply put, if the Great Orb washes its energy over your roofs and walls, you live in a solar home. There is of course a liberal dose of poetic license in that viewpoint, for in the strict thermodynamic sense your house may now gain only a minor percentage of its heat from the sun. But perhaps in Hayes' mind's eye was a vision of 80-odd million houses, all of them standing in the sunlight, all of them sharing the vast energy source. If we do the right things to them, our homes can greatly increase their use of that resource by becoming more finely tuned to the sun's comings and goings. They can become solar heated by design (or redesign) instead of by accident.

This book will help you do that. It will help you to better understand your place in the sun, and it will help you decide on the right ways to make your solar home as "solar" as it can be. The ways are many, but they are easily understood. To begin, you'll have to understand just what it is you're dealing with, this solar resource, and you'll have to become better acquainted with the house you're aiming to solarize, which is what "Getting Started" is all about. If you're still a little skeptical of Denis Hayes' vision, we hope that the things you're going to see and read will make you a true believer.

Photo I-2: Every house has a unique relationship with the sun, and finding out about "your place in the sun" is the first step toward knowing the value of your solar resource.

Your Place in the Sun

You can diagnose your home's heat losses and find the solar cures

What is a solar home? There seems to be a widespread notion that solar homes are the costly creations of an architectural elite serving an equally effete well-to-do clientele; they are ultra-ultra-modern and bear little resemblance to any architectural traditions; they are "not for me." The fact is, though, that intentional solar design has been a part of some architectural traditions for a long time, for centuries, and nowadays new solar homes are being built to compete in cost and consumer-attractiveness with conventional housing. Solar housing is becoming conventional!

But our concern in this book is about what "solar home" means for houses that are already here. How do you deal with the house you own in order to bring it into the solar age? For starters, know that "solar" is synonymous with "efficiency," and before you do anything to solarize, you've got to do everything you can, within economic reason, to conserve. This means making the house tighter against the elements, adding insulation, reducing the amount of fossil fueled energy needed to heat your water. These measures come first because they are cheaper, in terms of energy saved per dollar invested, and they make the solar potential of your house all the greater.

When your house has reached its maximum energy conservation efficiency, it's ready to enter higher realms of efficiency with solar improvements. The goal is still the same—to reduce even further your home's use of fossil fuel and electric power—but the means have changed. Now you invest in a different energy source, one that doesn't experience chronic price hikes or supply shortages. You may have to make some pretty fundamental alterations in your house in order to capture this resource, but with good design and execution, solar retrofits can blend very nicely with existing buildings. And it's as likely as not that you can combine solar improvements with some of the home improvements you're planning to get the double benefits of home enhancement and heat from the sun. On the energy benefits alone you can be assured that the accrued value of the fossil energy savings will equal the cost of most solar retrofits in five to ten years. Plus, you'll be reducing your monthly operating costs and freeing up more cash for saving or spending, and you'll be raising your home's selling value in an increasingly energy-conscious home-buying market.

There are so many options for solarizing: You can build heat collectors and put them on your roof, your walls, even out in the yard. You can cut holes in the right places and use glazing to admit the light and trap the heat of the sun's rays. You can use glazing over masonry walls and create heat storage from what was once just a heat sink. You can put a little sun into your domestic water heating system. This is all going to take a good bit of planning, so before you begin your energy upgrading, you've got to take stock of your holdings—your house, the grounds around it and the other houses around your grounds. Step One: Get acquainted with your place in the sun.

Photo I-3: From the back side a passive solar house looks pretty much like any other on the block. There are usually fewer north, east and west windows to help minimize winter heat loss and summer heat gain, but otherwise the differences are in things you can't see: super-tight construction, lots of insulation and a relatively open floor plan to promote even distribution of solar heat. . . .

Introducing: Your House

Is your house a mystery to you? Do you really know what's going on in there, what it's made of and how it works? In many ways a house is an organism, with some rather human characteristics: It has a system for heating and cooling itself, it breathes, it sweats, it responds to the climate. A house gets out of shape: When it's not properly cared for, its skin and bones deteriorate and the heart (central heating) goes bad. An ailing house can be rejuvenated with the right doses of structural or thermal or cosmetic medicine. And a house can, of course, grow when its people want more space. Houses live.

Fortunately you don't have to be a doctor of anything to gain the knowledge or the ability that's needed to give your house a checkup, prescribe some preventive medicine or perform minor or major surgery. But you do have to know what's going on in there: Things like how the thigh bone is connected to the hip bone, where the heat

Photo I-4: . . . but, seen from the sun side, you see the essence of passive solar space heating: lots of fixed glazing combined with operable windows, vents and window insulation for night use. And the whole thing is packaged in a fairly conventional house design. This is where solar retrofitting had some of its beginnings. Once architects and contractors figured out how to use solar design in new construction, it was a short jump to use it in existing houses. It was inevitable and supremely necessary, because in order for this country to make it into the solar age the existing housing stock, representing 20 percent of our energy use, must be energy-upgraded with conservation and solar improvements.

comes from and where it goes, and whether or not its "thermal envelope," that skin that fends off the elements, is all that it should be. As your own "house doctor" you can use this research to diagnose accurately your home's energy ills and strengths, and with a little more study you can render a realistic prognosis of just how much more energy-efficient the old place can be.

By the time you get through this introductory section, you'll have found out a lot about your house, and about what you can do in terms of conservation and solar improvements. You'll be finding out how much energy it uses to keep you warm, and you'll also find out how much more insulation to add to its thermal envelope so you can stay just as warm or warmer with less energy. You'll be finding out the solar potential of your house and how much heat it stands to gain from the sun. You'll be finding out how to create a master plan for maximizing your home's conservation efficiency and

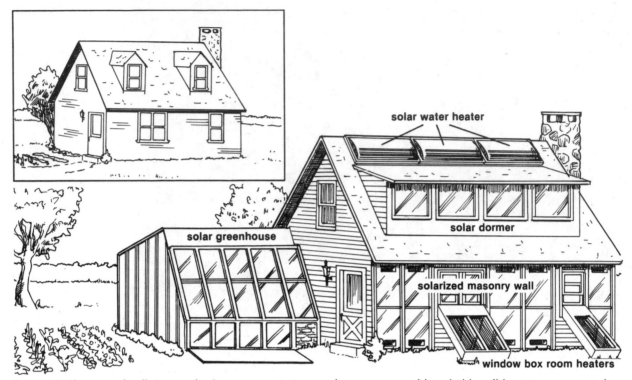

Figure I-1: If you made all these solar improvements to your house, you could probably sell heat to your next door neighbor since there would be such an excess of it. The point is that houses will generally present several opportunities for retrofits, all of which can improve thermal performance and looks.

solar potential. And when it comes to actually doing something with a hammer, you can get some "free advice," if you want it, on good work habits, job safety and tool buying.

But first, know thy house, as is—its structure and layout, its heat production and distribution system and what's there that's keeping the heat in, or letting it escape. Why are hard-to-heat rooms so hard to heat, and why are overheated rooms so overheated or so hard to cool off? Overall, what are your house's thermal or visual or spatial pluses and what are its minuses? You can also start looking at your house in terms of its solar heating and natural cooling potential, which essentially entails seeing where the sun's rays come in during the cold months and seeing where they sneak in, unwanted, during hot weather. Then you can start to imagine the ways of getting more of the former and less of the latter.

The more you know, the more control you'll have over your home's energy future. Houses are bigger than people only in size, and there's no need to be intimidated by what seem to be potentially enormous undertakings. Rigid as they are, houses are also flexible, and in home improvement and energy upgrading you can, within reason, do anything you want: push walls around, take them down, put them up, cut big and little holes in them, raise roofs, expand this way or that. Give yourself that sort of attitude and you'll free your imagination to come up with the right solution to problems and with some effective schemes for making your home a better place to live.

A Home Energy Tour

What is your house made of? You certainly know if it's a wood-frame or masonry structure, and you may know more about things like insulation levels and weathertightness and the age, type and condition of your heating system. But if you'll allow for the assumption that you and your house have yet to be, well, intimately introduced (no offense intended for them that knows more), you can embark on a little energy tour that will take you all through the house. With pencil and paper in hand to keep track of all the big and little things you have to look for, you may be amazed, for better or worse, at what your house has to show you. Along the way you'll also be finding out about the dynamics of heat. In order to better understand your home's thermal behavior you need to know a little about the basic processes of heat exchange, and when they come up, we'll give them names and relate them both to ordinary heating systems and to the ways they work in solar heating systems.

Let's start from the top and work down. You've either got a finished or unfinished attic, an attic crawl space or a flat roof that lies directly over heated living space. First, check the condition of the roof itself. Beware of brittle, curled-up composition shingles that break at the slightest touch or get "carried away" on a windy day. If your wood shakes or shingles are splitting or showing signs of rot, they may be ready for the kindling box. You may be blessed with a slate or tile roof that's in good condition, or burdened with one that needs a lot of help. Inside, if the roof pitches are unfinished and uninsulated, you'll be able to find evidence of leaks. Better yet, go up there in a downpour and look for drips and drops. Finished roof pitches and rooms under flat roofs may show stains from seepage. Whatever the case, start with a good roof. It's not insulation, and it's not a solar window, but a bad roof can wreak havoc upon your energy improvement plans.

A tight roof will protect your insulation, but how much is up there? Depending on the roof design, it'll either be up in the rafters or down in between the attic floor joists. In finished attics you may have to poke a little hole somewhere to see what's been put in. You should also check to make sure the attic (insulated floor) or roof (insulated rafters) is ventilated so that moisture doesn't build up in the insulation, which can greatly undercut its heat-resistance value.

As long as you're up there, you can do a little solar investigation. If you have a southeast- to southwest-facing roof pitch, feel the inside surface on a cold, sunny day. If it's uninsulated, it will be invitingly warm. Make a note of that in another list, which will be a catalog of your home's "solar surfaces."

What you're feeling in that warm surface is the result of a basic heat transfer mechanism: *conduction.* Heat is conducted through solid materials, and that can be good or bad in terms of your home's energy efficiency. Conductive heat loss, heat

Energy Tax Credits: Uncle Sam Wants You

Although consumers are enthusiastic about the federal energy tax credits for residential conservation and solar improvements, there has been some confusion about exactly what qualifies and how much the credits are worth. For one thing the law has changed a couple times, but even when different regulations are established, the Internal Revenue Service (IRS) hasn't been all too clear in its interpretation. Recently, however, the IRS eliminated some of the confusion by issuing its final regulations on the subject.

According to the new regulations, a solar retrofit must include five components in order to qualify for a tax credit: a solar collection area, an absorber, a storage mass, a heat distribution method and heat regulation devices. However, credits aren't allowed for any devices that have a dual purpose, e.g., structural as well as heating. A Trombe wall does not qualify if the wall supports part of the house, but the glazing and any other related hardware added to the wall qualify. So would air ducts and blowers that were installed to circulate warm air from the wall.

There are actually two kinds of tax credits, one for energy conservation, and one for renewable energy. The energy conservation credit, which is worth 15 percent of investments up to $2000 (a maximum credit of $300), includes such things as insulation for the home or the water heater, caulking and weather stripping, storm doors and windows, fuel-conserving furnaces or burners, flue restrictors and furnace ignition systems that replace gas pilot lights. The renewable energy credit, worth 40 percent of any amount up to $10,000 (a maximum credit of $4000), applies to solar heating, geothermal and wind-powered equipment. The energy conservation credit covers the cost of new materials and labor, purchased after April 1977, and must be designed to last at least three years. The renewable energy credit is good for new materials and labor purchased after December 21, 1979, and must be designed to last at least five years.

These credits are nonrefundable, meaning they can only be applied to taxes you owe. If the credit exceeds your tax liability in the particular tax year, you may apply the balance of the credit against future taxes. You may apply year

working its way through walls and windows to the outside, is what you want to minimize with insulation. Conductive heat gain, when the summer sun blasts heat from the outside to the inside, is also reduced with insulation. A little later you'll see how conduction works with other heat transfer mechanisms, when you find more solar surfaces.

Next, descend to the heated living spaces (or stay where you are if it's a finished attic or flat-roofed house). First off, is the door or hatch to the unheated attic weatherstripped and insulated? It should be because it's part of the thermal envelope, and wherever possible that envelope should be treated to slow down the flow of heat in both directions, inside and out.

by year until you reach the maximum, but once you do, you can't apply for any more credits until you move to a new home. You may have to check with your local IRS office for the latest regulations and for hair-splitting interpretations of your particular situation, preferably before you start work. Or, don't worry about anything and just decide for yourself what's right for your situation.

Tax credits are better than tax deductions because they are subtracted directly from the amount you owe the government. A $500 credit reduces the tax you owe by $500, while a deduction only reduces it by the tax rate assigned to your tax bracket. For example, if you are in the 20 percent tax bracket, a $500 deduction reduces your taxes by $500 × 20 percent, or $100.

You are eligible for the credits whether you own or rent your home, as long as you pay for the equipment and materials and use them in your principal residence. Vacation or second homes don't qualify. Landlords are not eligible, although they may apply for business investment energy tax credits. You may also apply, if you are the owner of a condominium or a stockholder in a cooperative housing corporation, for credit on eligible systems bought jointly with other members of the organization. Your share of the expenses is credited according to the same limits applying to any homeowner or renter. Where two or more homeowners own equipment jointly, each owner may file individually for the maximum credit for their share of the investment of up to $2000 for energy conservation and $10,000 for renewable energy.

State and even city or county governments sometimes offer income tax credits or property tax abatements for solar improvements, so it pays to check on current legislation in your area. There are also grants and low-interest loans available from the newly formed, federally sponsored Solar Bank, which makes money available, primarily on the basis of income, to people who want solar improvements. If you do take one of these federally subsidized energy loans or grants, however, you may *not* apply for a tax credit. To get the most from these loans and credits, estimate your costs in advance, then determine which combination is most beneficial to you. But don't delay. The 40 percent federal solar tax credit is slated to expire in the middle of this decade, and if you miss out, you'll lose the chance to solarize at a bargain price.

In heated living spaces, no matter what floor you're on, you've got to examine the tightness of windows and exterior doors and the insulation levels in the exterior walls. Unweather-stripped doors and windows are a major source of heat loss by *infiltration*. Infiltration is a two-way street: Warm air can escape through things like leaky window channels, and cold air can get in the same way. The latter effect is especially dramatic on cold windy days ("why are the curtains fluttering . . . ?"). Look for other points of possible air exchange: wire and plumbing passing through exterior walls, bathroom and kitchen vents that go outside, light fixtures that recess up into unheated spaces. These are items that rate first-priority attention (see "The Turned-Off House" in this section).

Energy Arrows

This is probably a good time to explain the meaning of the funny looking arrows that you'll be seeing running every which way through all the illustrations. Some are straight and bold and black; some are curved or squiggly or chopped into dashes or cross stripes; there's a reason for every one of them. The arrows represent energy that has been lanced earthward by the sun, and when it reaches a building, that energy behaves in several different ways. The following symbols thus represent the dynamics of radiant solar energy (heat) exchange as it passes from the atmosphere through glazing and into and around interior space.

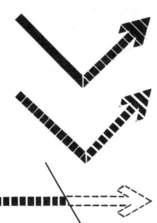

These arrows show the path of direct solar radiation through the atmosphere.

When the radiation passes through glazing or strikes a reflector, it loses a fraction of its energy value.

If the radiant energy strikes an interior reflector, it loses a bit more of its energy, but it's still useful for heating and daylighting.

When radiant solar energy strikes an object, some of its energy is absorbed as heat. This is an important heat exchange activity for it means that heat can be stored for later use in thermal mass (water in containers and masonry walls and floors).

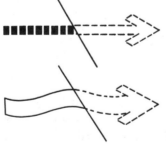

When that thermal mass decides it's time to give back heat to the living space (when the inside air begins to cool), it *reradiates* what it has stored. Radiant heat feels good.

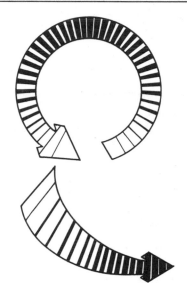

With all this heat charging around, it's got to start heating up the inside air, and when a given volume of air gets warmer, it gets lighter and wants to travel. It rises (by *natural convection*) and is replaced by cooler air, thus initiating a *convective flow* or *loop*. Like reradiation, this is a nice thing because heat can thereby get to places that the sun's rays can't.

But sometimes heated air has to be pushed with a fan or blower so that it will go somewhere that natural convection can't take it, like down, and pronto. *Forced-air* flow (forced convection)—down, up or over—can sometimes be a more efficient mode of heat transfer than natural convection, even though you have to use a little energy (electricity) to move a lot of energy (heat).

This one's easy; these arrows denote the direction of water flow. Water flows in a pipe in one of three directions: this way, that way or the other way.

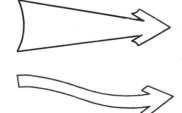

Finally, wind. It can be a strong chilling gust or a gentle cooling breeze.

Figure I-2: Thoroughness is the key when you make your home energy tour. If you can diligently track down the good, the bad and the ugly, you'll be off to a good start in your energy improvement odyssey.

In frame houses, checking for wall insulation will probably require some more hole-poking. If so, poke at a high level near the ceiling to see if the stuff has settled over the years. Then poke in the middle and near the floor to see if it's there at all. (You might also be able to check for insulation by pulling a light switch or socket plate from an exterior wall.) At this time you'll also be finding out about the actual wall construction, if it's a standard 2×4 stud wall, an old timber frame or newer building with 2×6 studding, and hopefully insulation to match.

Next, you want to check out the heat distribution system. Does the placement of forced-air registers or hot water radiators or electric baseboard units make sense? Or are there rooms that have always seemed either too hot or too cold during the heating season? Those conditions may be the fault of the design of the heat distribution system, or they could point back to problems with the house structure itself: lots of air leaks or inadequate insulation. If you're heating with solid fuels (wood, coal), can warm air reach the upstairs or rooms furthest from the stove? Once again there may be some problems with distribution, which could be solved with the addition of small vent openings in walls and floors. There may also be rooms that routinely overheat in summer. Are they perhaps lacking in adequate ventilation or window shading? Simple improvements in those two areas can reduce unwanted heat gain by as much as 50 percent.

As you go through the house, try to pinpoint any specific heating or cooling problems there seem to be, along with what you think the reasons are. Some of these problems may have conservation or solar solutions, or they might simply require some modifications in the heat and distribution system. In the latter case you may need the services of a heating contractor; try to find one who makes conservation part of his business. Heating and cooling problems may also be caused in part by the way the

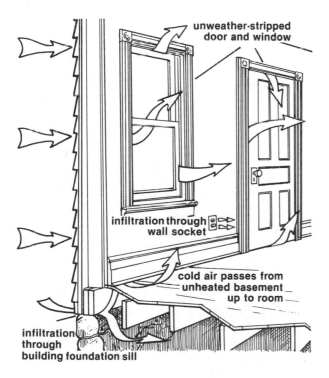

Figure I-3: Infiltration is a sneaky thing. Whenever there's a way for cold air to get in where it's warm, it will. The pressure exerted on a house by wind can significantly increase the rate of cold air infiltration.

unweather-stripped
door and window

infiltration through
wall socket

cold air passes from
unheated basement
up to room

infiltration
through
building foundation sill

interior space is divided up. If you're thinking about changing room sizes by moving partitions around, there may be ways to do that *and* to improve heat distribution.

Natural convection is another basic mode of heat transfer, and it can have a great deal to do with interior heat distribution. It's a simple event: As air heats it expands, which means that the same number of air molecules occupies a larger volume. Thus, warm air is less dense, and because of that it rises and is replaced by cooler, denser air. When you're heating the house, you can feel the effect of convection by climbing a ladder to the ceiling where the air will feel somewhat warmer than it does down on the floor. One goal of increasing your house's efficiency is to minimize that temperature difference. For example, with warmer air at the ceiling the rate of heat loss through the roof is increased. The way out of that problem is to use circulating fans or blowers to pull the warmer air from the ceiling and redistribute it back down to lower levels. This is heat transfer by *forced convection.* Convection also works against you if your fireplace damper isn't shut tight because the chimney will rapidly convect room air up, up and away.

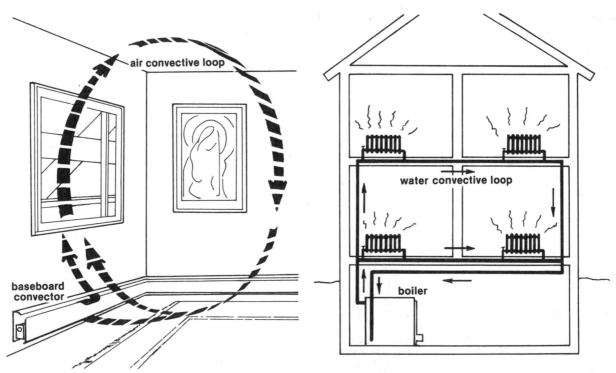

Figure I-4: The panel at left shows an air convective loop created by a baseboard "convector" heater. Heated air rises and collects near the ceiling, while cooler air "feeds" the baseboard unit. A water convective loop is typified by the operation of an old-style boiler system that circulates hot water without a pump. Hot water rises and flows through radiators that pass heat into the room. The cooler water or steam returns to the boiler. This flow, called a thermosiphon flow, is created by the temperature difference between the supply and return water.

But convection has its benefits too. If you have hot water or electric baseboard heating, you're being warmed by a convective flow of air created by the heating element. Because heated air rises out of the top of the baseboard, cooler air is drawn in through the bottom slot, and a *convective loop* is initiated. As you'll see later, convection, both natural and forced, has an important role in heat distribution in solar heating systems.

The energy tour continues down to the basement where again you should check for air leaks and insulation levels. If the basement isn't heated, a worthy improvement would be to add insulation under the first floor, which will make the heated side of the floor feel warmer. Finished, heated basements need insulation in the basement walls.

The Heating System

If you don't heat with electric baseboards or a solid fuel burner (e.g., a woodstove), you must have some kind of central heating system. If it heats water, it's a boiler, and if it heats air, it's a furnace. Take a long look at the contraption and, if you don't already know, try to figure out how it works. Find where the fuel or power goes in, where the hot water or air goes out, and where the return water or air comes back. Does the system use a water pump or a blower? If it doesn't use either, it's an oldie that runs by natural convection. (Remember? Heated air or water will rise; in this case, through ducts and pipes.)

Along with the improvements that may be needed for the house's thermal envelope, be prepared to upgrade the heating system, for this too will usually be more cost-effective than most solar improvements. Heating system improvements will range from basic tune-ups to burner replacements to add-on components (stack dampers, stack heat reclaimers, dual setback thermostats) to total replacement of your antique rattletrap for a new-generation sports car with all the latest in high-efficiency features. Again, you should try to find a heating tradesman who has done his conservation homework and who can help you sort out what the best moves will be.

Radiation: the Good Kind

While we're on the subject of heating systems, we can talk about yet another form of heat transfer some systems use to keep you warm: *radiation*. Radiation is essentially an energy exchange from a warm or hot object to one that is cooler. Radiative heat distribution is most easily identified with standard hot water or steam radiators and wood- and coal-burning stoves. The way these heaters work shows that radiation has some attractive characteristics. For example, when the radiator gets hot with 180 to 200°F water and even hotter steam, it literally sends out through space rays of energy that don't become heat until they strike something, like your body. The same thing happens with solid fuel stoves, only all the more intensely because of their much higher operating temperatures. This is a pretty efficient way to heat, by directly heating objects instead

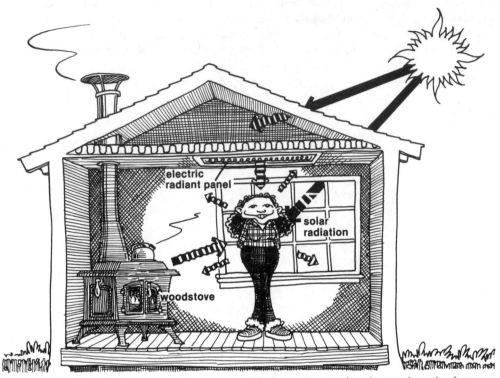

Figure I-5: Radiant heat can come from several sources: stoves, electric panels and, of course, the sun. A radiant source beams energy omnidirectionally, which doesn't become heat until it's intercepted by an object. It's a nice feeling when your body is the object.

of heating air (by convection) to surround objects with warmth. For this reason, radiant heating is becoming more popular in industrial and residential applications. In the wide open spaces of industry, electric or gas radiant heaters can be placed to beam energy directly onto a group of people. Great savings are realized over heating large volumes of air. In houses, electric radiant panels installed in walls and ceilings do the same thing. Even concrete slab floors can be equipped with hot water pipes (which can be solar heated) or with electric elements to provide radiant comfort.

Radiant heat transfer has a lot to do with your comfort even at lower temperatures. For example, when it gets cold outside, the surface temperature of walls and windows drops accordingly. If you begin to feel chilly, it's because with your 98.6°F body temperature, *you* are the radiator, the higher temperature source feeding heat to colder surfaces. That is why adding insulation increases your comfort: The wall surface temperature goes up because of a reduction in the rate of conductive heat loss. And because the temperature difference between you and the wall is reduced, you're beaming away less body heat. As we'll see shortly, radiation is an essential solar heat transfer mechanism.

But first we ought to do a little review. What does your tour log tell you? Are you looking at a string of impending conservation improvements, or is the house tight and well-insulated and ready for some solar action? Do you suspect an energy liability in your heating system? Is heat being properly distributed or are there cold spots? Do the room divisions make sense for your household or would a little partition-shifting improve the available space and possibly improve heat distribution? These are just some of the cues for determining the thermal status of your house, and since this isn't an in-depth conservation book (it's an in-depth solar book!), it neither asks all of the questions nor gives all the answers. But you can refer to the conservation books listed at the end of "The Turned-Off House," in this section, to get the most up-to-date information on the world's cheapest energy source: saving energy.

Your Place in the Sun

If you've been thinking about solar energy, you've most likely been thinking about it in terms of how to apply it to your own house. To complete the energy tour you can make a survey of all your home's solar surfaces, all the potential sites for solar improvements. Once again, take a few notes on what you see; don't hesitate to record your initial impressions of what you might want to do with some of those surfaces. Is the sun shining?

That south roof pitch, if you have one (two?), is potentially a valuable solar surface. For the record, how steep is it? (Roof pitch angles can be figured by a simple exercise described in Section II in "Project: A Flat Plate Collector.") Take a couple of measurements to find the area of the roof pitch. When you come to estimating your home's potential for solar heating (see "Finding Your Heat Gain" in this section), you'll be needing this information, so hang onto it and to all the measurements of your solar surfaces. Even if your roof is flat, don't discount it as a possible site for solar improvement. For example, solar domestic water heating systems can use roofs at any pitch, and there are a couple of effective space heating retrofit options that can be built up from flat roofs.

Unless you've got an A-frame with a roof that runs to the ground, the next stop on your solar survey is vertical walls. Get out your camp compass, put it right up to the wall surface, and record the indicated bearing. Repeat the operation a couple more times at different points to make sure the compass hasn't been influenced by ferrous metals that might be hidden in the wall. This bearing is your *solar aspect,* which may or may not be close to the optimum solar aspect, known as *true south.* True south marks the location of the sun when it is exactly halfway between sunrise and sunset, which occurs at *solar noon.* Usually true south isn't the same as the 180-degree bearing indicated by your compass, but a few degrees east or west of that. The map shown in figure I-6 shows how much the compass indication of magnetic south can deviate from true south in different parts of the country. For example, if you live in eastern Pennsylvania, true south is really 189 degrees on the compass, 9 degrees west of magnetic south. If the compass tells you your solar aspect is 170 degrees, then you are 19 degrees east of true south. It's easy to figure.

To be a useful solar surface, it is generally recommended that aspect deviations be no more than 30 degrees east or west of true south. Of course, the ultimate usefulness of the wall depends on other factors, especially whether or not it "sees" direct sunshine or is shaded by trees or other buildings. The classic dilemma is what to do with a house that only has a south-facing corner. As figure I-7 shows, even surfaces that are 45 degrees east or west of south can still receive an appreciable amount of the energy available in a day's worth of sunlight. The decision about which wall to utilize (or whether to utilize both walls) will again depend on other factors that will have to be sorted out, such as morning and afternoon weather patterns and the proximity of shade trees and other buildings.

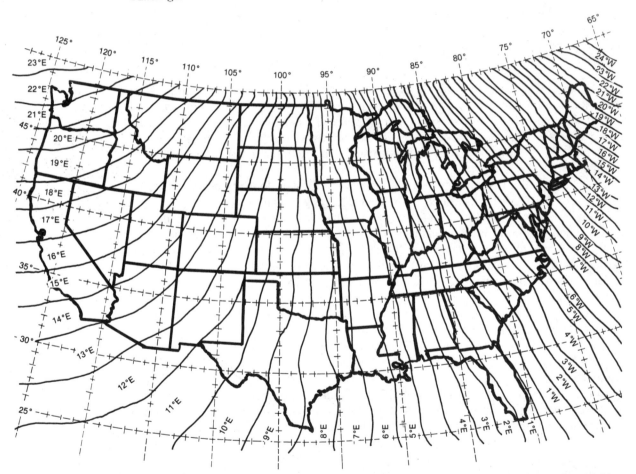

Figure I-6: This map shows the approximate compass deviations from true south for your area. For example, Washington, D.C., is shown as having a compass deviation of 8°W, which means that true south is actually eight degrees west of what your compass indicates to be south.
SOURCE: *Redrawn from the Isogonic Chart of the United States, U.S. Department of Commerce, Coast and Geodetic Survey, 1965.*

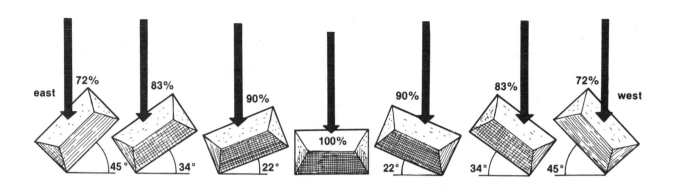

Figure I-7: When a surface is turned more than 45 degrees away from south, its ability to collect solar energy in winter is greatly reduced. Theoretically the energy input from the sun is the same east and west of true south, but other factors may conspire to favor one aspect over another. For one thing, by the time the sun reaches the western sky, air temperature is higher than it was in the morning. This can help to increase energy collection on a wall that faces west of south. On the other hand, local weather patterns might show that cloudiness is primarily an afternoon phenomenon, which could make a wall facing east of south a more viable solar surface.

Now find the area of all your vertical solar surfaces, which is most easily done by measuring on the outside. While you're at it, you might want to take some quick measurements of the ground areas that could be possible sites for a solar addition such as a greenhouse or a sunspace. Sometimes when a house is heavily shaded, a useful solar surface has to be created by building out from the east or west sides. In other cases, the long axis of a house may run in a north-south direction, which puts the smallest wall on the south side. Here again, a south-facing extension to the east or west may be desirable.

The total area of your solar surfaces may add up to several hundred square feet, which can mean a high potential for solar space heating. If the solar surface area exceeds 20 to 25 percent of the heated floor area of your house, you may be able to use a solar system to supply half or more of its space heating load (which you're going to be estimating in the essay, ''Finding Your Heat Loss'' in this section).

Solar Heat: What Does It Do?

So far you've read a little about how different modes of heat transfer (conduction, convection, radiation) are involved with your home's heating system and its thermal envelope. In solar heating systems the very same modes of heat transfer play equally important, if not more important, roles. Only the fuel has changed; the laws of thermodynamics remain the same. As you'll see in later projects and essays, controlling heat in solar systems (i.e., controlling or designing for the right heat transfer results) is just as important as collecting it. A good place to start is by simply getting a feel for what the sun does in your house. It may well turn out that what you end up doing for, say, a space heating retrofit is essentially a magnification of what's going on right now.

From that trip up to the attic, when you may have felt a warm roof, you were feeling results of a couple of interacting modes of heat transfer. In order for you to feel the warmth that was conducted to the inside, the outside surface had to absorb the sun's radiation. That's where it all begins: radiation and absorption. The energy beaming away from a blazing woodstove doesn't become heat until it strikes and is absorbed by an object, and radiant solar energy works the same way. When it's absorbed by "collector" surfaces, it becomes heat.

Everything under the sun is a collector: people getting suntans, leaves converting sunlight for their food and our oxygen, broiling asphalt streets on a summer day. These collectors absorb different amounts of energy with different results, and although there are no perfect collectors in the natural or man-made world, we can get pretty close. *Absorptance* is the ratio of the energy absorbed by a surface to the energy that is made available to it. If, for example, the sun were beaming ten units of energy onto a square foot of collector, and the collector absorbed eight units, its absorptance would be 80 percent. In solar retrofitting, and in any solar application, the primary goal is, of course, to optimize the quality of collector surfaces so they absorb as much energy as possible. Absorption is affected (and controlled) primarily by the color and texture of the surface and whether the surface is shiny or dull. Light colors reflect, and dark colors absorb. Dull surfaces absorb better than shiny ones, and thus you'll be seeing that dull black or dark-colored surfaces are basic to most of the collector surfaces that are used in the projects that come later.

Thus, a dark tarred roof gets somewhat hotter than a silvery metal one, but if the dark roof were a perfect absorber, it would just keep collecting and collecting, getting hotter and hotter and ultimately burn up or melt or do whatever roofs do. Not to worry though, solar roof fires are about as common as solar eclipses. The roof's less-than-perfect absorptance is limited in one way by its *reflectance.* A percentage of the solar light is literally bounced back to the atmosphere. Solar collector surfaces aim to minimize reflective losses in order to maintain higher degrees of absorptance, on the order of 85 to 95 percent absorption of the available solar energy. That's how close we can get to perfect absorption.

Reflection, however, can also become a positive element in the control of solar light. Because a highly reflective roof won't be as hot in summer as a dark one, the attic stays 20 to 30 degrees cooler. Reflectors can also be used to bounce more solar light onto a collector, increasing its energy gain by 10 to 30 percent. Snow is a reflector; white gravel is a reflector; Reynolds Wrap is a reflector.

Another reason why the roof doesn't vaporize relates to its ability to get rid of excess heat by conduction and *reradiation.* You met conduction up there in the attic.

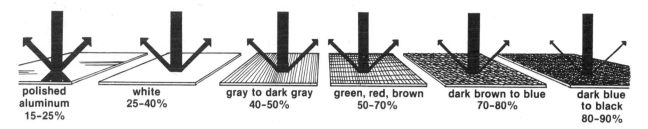

| polished aluminum 15-25% | white 25-40% | gray to dark gray 40-50% | green, red, brown 50-70% | dark brown to blue 70-80% | dark blue to black 80-90% |

(% = percentage of available solar energy absorbed)

Figure I-8: Color alone can have a great effect on the amount of solar energy a surface absorbs, and as the polished metal surface shows, surface shininess also increases reflection.
SOURCE: *Adapted from Edward Mazria,* The Passive Solar Energy Book, *1979, with permission of Rodale Press, Emmaus, Pa.*

Heat was conducted through the shingles, the decking and whatever else was in contact with these materials, effectively distributing the heat and to some degree, limiting excessive heat buildup. *Conductance* is a measurement of how fast the heat passes through a solid material, and it can vary widely depending on the material. This means that materials' selection is a primary means of controlling conduction. Insulation is a horrible conductor, which is fine as far as your home's thermal envelope is concerned. Metals are at the opposite end of the scale. They conduct heat much faster than wood, plastic or masonry materials, and among common metals silver is a better conductor than copper, which is a better conductor than aluminum, which is a better conductor than steel. In collectors that are used to heat water, a sheet of copper is the preferred material for it can rapidly conduct absorbed heat to the water flowing through copper tubes that are attached to the sheet. In collectors that directly heat air, sheets of aluminum provide adequate conduction at somewhat less cost than copper.

But speed isn't everything. The fact that masonry and water are relatively poor conductors compared to metals is actually a positive control factor in some solar applications. A thick masonry wall or a wall composed of water-filled containers can be used as a collector that absorbs and slowly conducts heat from the outside surface to the inside. These massive materials are used to store solar heat so that it will be available for space heating after sundown. By the end of the day, heat is reradiated from the heat storage wall. Because they have so much mass, masonry and water walls can store a lot more heat than a thin roof section can. This is a function of the *specific heat* value of these different materials. Specific heat compares equal weights of different materials and measures their heat storage capacity. A pound of copper can store about one-half the amount of heat that a pound of brick can, but a pound of water can store about five times as much heat as the brick, which makes it a very desirable solar heat storage component. For different reasons masonry is also commonly used for solar storage, usually because it can perform the dual functions of a heat storage and a structural element for a building. As you'll see in later projects, both have their uses in retrofit applications.

Back up on the roof, the roof section, with its low mass, rapidly fills up with heat to the point where its cup runneth over, and it starts to reradiate back to the outside and the inside. If you held your figurative hand a couple of inches away from our figurative south roof pitch, you'd feel radiant heat again. Reradiation is really radiation, but it helps to differentiate the two conditions. Radiation emanates from an energy source: the sun, a woodstove, an electric radiant heating panel. In this book we will use reradiation to describe the flow of energy from a storage source: an uninsulated tank filled with solar-heated water, a hot roof shingle, the hot interior of a "solarized" car that is wonderfully warm in winter or agonizingly sweltering in summer. In most new solar homes, heat storage and reradiation are important space heating design elements because they can give the building the ability to be self-heating after sundown. In retrofitting for space heating, the same elements can be built in to carry the sun's warmth into the night (see "Moved-In Mass" in Section III-D).

Reradiation has its "bad" side too. Like any mode of heat transfer, radiant energy takes the easiest routes first, and in many solar applications reradiation of heat back to the outside is a well-worn path. It stands to reason: When solar energy is collected on an exterior surface, it concentrates there. If heat isn't pulled into the house, the temperature of these surfaces builds up to the point where they reradiate back to the atmosphere. Since that can greatly reduce the efficiency of the collector, there is a need for a way to block some of that energy loss. Let's step back into our solarized car, with all the windows rolled down. If it's sunny and hot or sunny and cold outside, it's the same in the car. Now roll up all the windows. If it's sunny and hot outside, you'll be starting to sweat inside. If it's cold, the car will become comfortably warm. The windows! It's the windows (call it *glazing*) that are the primary agents for heat retention in the vehicular collector and in most other solar collectors.

Glazing is the greatest. Its transparency, of course, allows it to transmit radiant solar energy, which is mostly a high-energy, short wavelength part of the spectrum. (Light consists of a wide range of wavelengths which comprise the total spectrum.) When that energy is transformed into heat, the wavelength shifts to a longer-wave, lower-energy part of the spectrum. The good news is that glazing (glass and various plastics) does not readily transmit this longer wavelength, and the result is that heat is trapped, which leads to a temperature buildup in the collector.

An equally important function of glazing is that as a solid material it creates a dead airspace between itself and the absorber. This greatly reduces the rate of convective heat loss from the collector. A hot collector surface heats the layer of air just above it and puts it into convective motion. With a glazing-absorber air gap of anywhere between ½ to 2 inches, the air is kept still, and that layer of air gains some insulating value. Double-pane insulating windows use the same dead airspace to reduce heat loss.

Now we have the essential solar collector: An absorber with a glazing layer sealed over it. The remaining element that's needed is a heat transfer medium. Without heat transfer all you have is a hot collector and no delivery of heat to interior space or to storage for later use. Water and air are two primary heat transfer media. In this book water-heating collectors are used to heat domestic water (see Section II). Air heating collectors (see Section III-E) are used mostly to heat space, but they can be used to heat

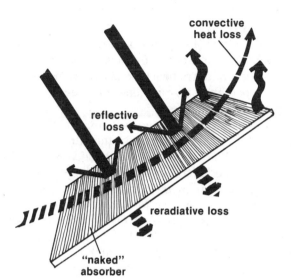

convective
heat loss

reflective
loss

reradiative loss

"naked"
absorber

Figure I-9: When a heat-absorbing surface is unprotected, reradiative and convective losses deliver most of the heat back to the atmosphere. What's needed, as shown by the improved version, is glazing on the outside to trap heat, insulation on the back side to trap heat and a heat transfer medium to deliver heat directly to use or to storage. In some collector systems you'll see that the back side insulation is not included so that heat can radiate directly into a room from the back of the collector, which also eliminates the need for a formal heat transfer and storage component.

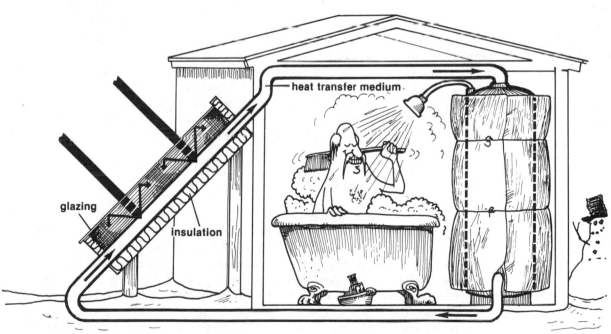

heat transfer medium

glazing

insulation

water. In both types of collectors, heat building up in the absorber is conducted or reradiated to flowing air or water and thence to direct use or to storage. Either fluid can be circulated by forced convection with a blower or pump or they can be circulated by convection, the way the old furnaces and boilers used to work.

Active and Passive Solar

Forced-circulation systems are called *active systems* because they use electrical controls to move heat around. In some cases it is quite advantageous to make solar systems work this way because relative to the amount of heat (energy) that is moved from the collector to the "point of good use" perhaps only one-tenth or one-twentieth as much electric energy is consumed by a fan or pump.

Systems that run by convection are called *passive systems,* for they work without electric controllers. Simplicity and lower cost are part of the attractiveness of passive systems, but as you'll see there are other factors that will guide your selection of the system type. For example, windows placed in a south-facing wall become a part of a passive system in that they simply transmit solar energy directly into a room, where it strikes objects and heats the place up—no place nor need for a fan. Perhaps there are some of these passive "mini-systems" in your house right now. Other passive collector systems make use of existing masonry walls (with an added layer of glazing) as a combination heat transfer-heat storage medium. You can't pump heat through brick! Still other systems use a little active and a little passive control to get the job done. Systems that employ this marriage of controls are appropriately called *hybrid systems.*

These systems can be built to almost any scale, from that of the smallest "window box" air heating collector to the scale of a whole house. In new construction solar homes are commonly referred to as being primarily "active," "passive," or "hybrid." And the same is true with regard to retrofitted houses. In Section VI, "The Whole House," you can see stellar examples of houses that have received several active, passive and hybrid solar improvements. The general tendency for space heating is to develop passive and hybrid systems, for they are generally lower in cost, better suited to the do-it-yourself crowd and sometimes more efficient.

Time to Get Started

Right now, your place in the sun is probably in the minor leagues as far as its use of solar heat goes. But hopefully you've found that the old homestead isn't as far from being solarized as you might have thought it was. Now that you've cataloged your solar surfaces and learned about the basics of solar heating, well, the die is cast. Frankly, we hope you'll become hopelessly addicted to the idea of solarizing your present home, or at least ready to take action. If you need some more input to firmly root your interest, skim through the following chapters and glance at the photographs and illustrations. They're certain to incite your imagination. Beyond that, it's time to get started, and the articles that follow in this section will help you start out right.

Joe Carter

References

Books about Solar Design

AIA Research Corp. 1976. *Solar dwelling design concepts.* Washington, D.C.: U.S. Government Printing Office. no. 023-000-00334-1.

Anderson, Bruce. 1977. *Solar energy: fundamentals in building design.* New York: McGraw-Hill Book Co.

Anderson, Bruce, and Riordan, Michael. 1976. *The solar home book: heating, cooling and designing with the sun.* Harrisville, N.H.: Cheshire Books.

Anderson, Bruce, and Wells, Malcolm. 1981. *Passive solar energy.* Andover, Mass.: Brick House Publishing Co.

Argue, Robert. 1980. *The well-tempered house: energy-efficient building for cold climates.* Toronto: Renewable Energy in Canada.

Argue, Robert; Emanuel, Barbara; and Graham, Stephen. 1978. *The sun builders: a people's guide to solar, wind and wood energy in Canada.* Toronto: Renewable Energy in Canada.

Baer, Steve. 1977. *Sunspots.* Seattle: Cloudburst Press.

Barnaby, C. S., et al. 1977. *Solar for your present home.* Sacramento: California Energy Commission.

Buckley, Shawn. 1979. *Sun up to sun down.* New York: McGraw-Hill Book Co.

Butti, Ken, and Perlin, John. 1980. *A golden thread: 2500 years of solar architecture and technology.* Palo Alto, Calif.: Cheshire Books.

Daniels, Farrington. 1974. *Direct use of the sun's energy.* New York: Ballantine Books.

Davis, Norah Deakin, and Lindsey, Linda. 1979. *At home in the sun.* Charlotte, Vt.: Garden Way Publishing Co.

Dawson, Joe. 1976. *Buying solar.* Washington, D.C.: Federal Energy Administration. no. PB 262 134. Available from National Technical Information Service, Springfield, Va.

Eccli, Eugene, ed. 1976. *Low-cost energy efficient shelter for the owner and builder.* Emmaus, Pa.: Rodale Press.

The Energy Task Force. 1977. *No heat, no rent: an urban solar and energy conservation manual.* New York.

Farallones Institute. 1979. *The integral urban house.* San Francisco: Sierra Club Books.

Halacy, D.S., Jr. 1975. *The coming age of solar energy.* New York: Avon Books.

Heschong, Lisa. 1979. *Thermal delight in architecture.* Cambridge, Mass.: MIT Press.

Leckie, Jim, et al., eds. 1975. *Other homes and garbage: designs for self-sufficient living.* San Francisco: Sierra Club Books.

Lyons, Steve, ed. 1978. *Sun: a handbook for the solar decade.* San Francisco: Friends of the Earth.

Mazria, Edward. 1979. *The passive solar energy book.* Emmaus, Pa.: Rodale Press.

Merril, Richard, et al., eds. 1975. *Energy primer: solar, water, wind and biofuels.* Menlo Park, Calif.: Portola Institute.

Michels, Tim. 1979. *Solar energy utilization.* New York: Van Nostrand Reinhold Co.

Montgomery, Richard H. 1978. *The solar decision book: a guide for heating your home with solar energy.* New York: John Wiley & Sons.

Rahders, Richard R. 1979. *Your house can do it: the passive approach to free heating and cooling.* Santa Cruz, Calif.: Thacher & Thompson.

Reif, Dan. 1980. *Solar retrofit: adding solar to your home.* Andover, Mass.: Brick House Publishing Co.

Reynolds, J. 1975. *Energy conservation and solar retrofitting for existing buildings in Oregon.* Eugene, Oreg.: University of Oregon, School of Architecture and Allied Arts.

Scully, Dan, et al. 1978. *The fuel savers: a kit for solar ideas.* Harrisville, N.H.: Total Environmental Action.

Shurcliff, William A. 1978. *Solar heated buildings of North America: 120 outstanding examples.* Andover, Mass.: Brick House Publishing Co.

————. 1979. *New inventions in low cost solar heating.* Andover, Mass.: Brick House Publishing Co.

Spetgang, Irwin, and Wells, Malcolm. 1976. *Your home's solar potential: the do-it-yourself solar survey.* Barrington, N.J.: Edmund Scientific Co.

Sunset Books and *Sunset Magazine,* eds. 1979. *Sunset homeowner's guide to solar heating: space heating and cooling; hot water heaters; pools, spas and tubs.* Menlo Park, Calif.: Lane Publishing Co.

Van Dresser, Peter. 1979. *Homegrown sundwellings.* Sante Fe: Lightning Tree.

Wade, Alex. 1977. *30 energy-efficient houses you can build.* Emmaus, Pa.: Rodale Press.

Wells, Malcolm, and Spetgang, Irwin. 1978. *How to buy solar heating . . . without getting burnt!* Emmaus, Pa.: Rodale Press.

Williams, J. Richard, ed. 1977. *Solar energy technology and applications.* rev. ed. Ann Arbor, Mich.: Ann Arbor Science Publishers.

Wilson, Tom, ed. 1981. *Home remedies.* Philadelphia: Mid-Atlantic Solar Energy Association. This is a collection of articles about conservation and solar retrofitting by experts in the field, and it is available from MASEA, 2233 Gray's Ferry Avenue, Philadelphia, PA 19146.

Wing, Charles. 1979. *From the walls in.* Boston: Little, Brown and Co.

Wright, David. 1978. *Natural solar architecture: a passive approach.* New York: Van Nostrand Reinhold Co.

Books about Energy Issues

Archer, Victor, and Rom, William. 1980. *Health implications of new energy technologies.* Ann Arbor, Mich.: Ann Arbor Science Publishers.

Argue, Robert; Emanuel, Barbara; and Graham, Stephen. 1978. *The sunbuilders: a people's guide to solar, wind and wood energy in Canada.* Toronto: Renewable Energy in Canada.

Brand, Stewart. 1974. *Whole earth epilog.* Baltimore: Penguin Books.

Branley, Franklyn. 1975. *Energy for the twenty first century.* New York: Thomas Y. Crowell.

Clark, Wilson. 1975. *Energy for survival: the alternative to extinction.* Garden City, N.Y.: Anchor Press.

Commoner, Barry. 1970. *Science and survival.* New York: Penguin Books.

————. 1971. *The closing circle: nature, man and technology.* New York: Alfred A. Knopf.

————. 1977. *The poverty of power.* New York: Bantam Books.

————. 1979. *The politics of energy.* New York: Alfred A. Knopf.

Commoner, Barry, et al., eds. 1975. *Energy and human welfare: a critical analysis.* 3 vols. New York: Macmillan Publishing Co.

Congdon, R. J., ed. 1977. *Introduction to appropriate technology: toward a simpler life-style.* Emmaus, Pa.: Rodale Press.

Cook, Earl. 1976. *Man, energy, society.* San Francisco: W. H. Freeman & Co.

Courrier, Kathleen, ed. 1980. *Life after '80: environmental choices we can live with.* Andover, Mass.: Brick House Publishing Co.

Goodman, Paul, and Goodman, Percival. 1960. *Communitas: means of livelihood and ways of life.* rev. ed. New York: Random House.

Halacy, Daniel S., Jr. 1973. *The coming age of solar energy.* rev. ed. New York: Harper and Row Publishers.

Hammond, Allen L.; Metz, William D.; and Maugh, Thomas H. II. 1973. *Energy and the future.* Washington, D.C.: American Association for the Advancement of Science.

Hanley, Wayne, and Mitchell, John. 1979. *The energy book: a look at the death throes of one energy era and the birth pangs of another.* Brattleboro, Vt.: Stephen Greene Press.

Hayes, Denis. 1976. *Energy: the case for conservation.* Washington, D.C.: Worldwatch Institute.

————. 1977. *Energy: the solar prospect.* Washington, D.C.: Worldwatch Institute.

————. 1977. *Rays of hope: the transition to a post-petroleum world.* New York: W. W. Norton & Co.

————. 1978. *The solar energy timetable.* Washington, D.C.: Worldwatch Institute.

Lovins, Amory. 1972. *The Stockholm Conference: only one earth.* San Francisco: Friends of the Earth.

———. 1975. *World energy strategies: facts, issues and options.* Cambridge, Mass.: Ballinger Publishing Co.

———. 1979. *Soft energy paths: toward a durable peace.* New York: Harper and Row Publishers.

Lovins, Amory, and Price, John. 1975. *Non-nuclear futures: the case for an ethical energy strategy.* Cambridge, Mass.: Ballinger Publishing Co.

McCallum, Bruce. 1977. *Environmentally appropriate technology: renewable energy and other developing technologies for a conserver society in Canada.* Ottawa: Fisheries and Environment Canada.

Meadows, Donella, and Meadows, D.L. 1974. *The limits to growth: a report for the Club of Rome's project on the predicament of mankind.* 2d ed. New York: Universe Books.

Metz, William, and Hammond, Allen. 1978. *Solar energy in America.* Washington, D.C.: American Association for the Advancement of Science.

Newman, Dorothy K., and Day, Dawn. 1974. *A time to choose: America's energy future.* Energy Policy Project of the Ford Foundation. Cambridge, Mass.: Ballinger Publishing Co.

Odum, Howard T., and Odum, Elizabeth C. 1976. *Energy basis for man and nature.* New York: McGraw-Hill Book Co.

Okagaki, Alan, and Benson, Jim. 1979. *County energy plan guidebook.* Fairfax, Va.: Institute for Ecological Policies.

RAIN, eds. 1977. *Rainbook: resources for appropriate technology.* New York: Schocken Books.

Schumacher, E.F. 1976. *Small is beautiful: economics as if people mattered.* New York: Harper & Row Publishers.

Skurka, Norma, and Naar, Jon. 1976. *Design for a limited planet.* New York: Ballantine Books.

Wolfe, Ralph, and Clegg, Peter. 1979. *Home energy for the eighties.* Charlotte, Vt.: Garden Way Publishing Co.

Good Solar Magazines

Alternative Sources of
Energy
107 S. Central Ave.
Milaca, MN 56353

Solar Engineering Magazine
Solar Engineering Publishers, Inc.
8435 N. Stemmons Freeway
Suite 880
Dallas, TX 75247

Popular Science
Times Mirror Magazines, Inc.
380 Madison Ave.
New York, NY 10017

Solar Heating and Cooling
Gordon Publications, Inc.
20 Community Place
Morristown, NJ 07960

RAIN Journal of Appropriate
Technology
Rain Umbrella, Inc.
2270 N.W. Irving
Portland, OR 97210

Solar Living and Solar Greenhouse
Digest
P. O. Box 2626
Flagstaff, AZ 86003

Rodale's New Shelter
Rodale Press, Inc.
33 E. Minor St.
Emmaus, PA 18049

Solar Times
3 Old Post Rd.
Madison, CT 06443

Solar Age
Solar Vision, Inc.
Church Hill
Harrisville, NH 03450

American Section, International Solar Energy Society

Consider joining the American Section of the International Solar Energy Society, and then contact your local chapter of AS of ISES to put you in touch with people in your area who have experience in solar retrofitting.

American Section of the
International Solar Energy Society
P. O. Box 1416
Killeen, TX 76541
(817) 526-1300

Local Chapters of AS of ISES:

Alabama Solar Energy Association
c/o University of Alabama
Center for Environmental Energy Studies
UAH/JEEC RI Annex D
P. O. Box 1247
Huntsville, AL 35807
(205) 895-6257

Arizona Solar Energy Association
Arizona Solar Energy Commission
1700 West Washington St., Room 502
Phoenix, AZ 85007
(602) 255-3682

Colorado Solar Energy Association
P. O. Box 1284
Alamosa, CO 81101
(303) 589-5184

Eastern New York Solar Energy
Association
P. O. Box 5181
Albany, NY 12205
(518) 270-6301

Florida Solar Energy Association
P. O. Box 248271
University Station
Miami, FL 33124
(305) 284-3438

Georgia Solar Energy Association
P. O. Box 32748
Atlanta, GA 30332
(404) 894-3448

Hoosier Solar Energy Association, Inc.
Gordon Clark Associates, Inc.
6523 Carrolton Ave.
Indianapolis, IN 46204
(317) 259-7711

Illinois Solar Energy Association
P. O. Box 1592
Aurora, IL 60507

Iowa Solar Energy Association
1433 Wildwood Drive, N.E.
Cedar Rapids, IA 52402
(319) 365-6103

Kansas Solar Energy Association
P. O. Box 8516
Wichita, KS 67208

Metropolitan New York Solar Energy
Association
P. O. Box 2147
Grand Central Station
New York, NY 10163

Michigan Solar Energy Association
201 E. Liberty St.
Suite 15
Ann Arbor, MI 48104
(313) 663-7799

Mid-Atlantic Solar Energy Association
2233 Gray's Ferry Ave.
Philadelphia, PA 19146
(215) 545-2150

Minnesota Solar Energy Association, Inc.
P. O. Box 762
Minneapolis, MN 55440
(612) 535-0305

Mississippi Solar Energy Association
225 W. Lampkin Rd.
Starkville, MS 39759
(601) 323-7246

Nebraska Solar Energy Association
University of Nebraska
Department of Electrical Technology
60th and Dodge Sts.
Omaha, NE 68182
(402) 554-2769

Nevada Solar Advocates
P. O. Box 8179
University Station
Reno, NV 89507
(702) 323-1267

New England Solar Energy Association
P. O. Box 541
Brattleboro, VT 05301
(802) 254-2386

New Mexico Solar Energy Association
P. O. Box 2004
Santa Fe, NM 87501
(505) 983-5338, 983-2861 or 983-2887

North Carolina Solar Energy Association
7001 Buckhead Dr.
Raleigh, NC 27609
(919) 566-3111

Northern California Solar Energy
Association
P. O. Box 1056
Mountain View, CA 94042
(415) 526-4594

Ohio Solar Energy Association
Environmental Studies Program
Secretariat
Wright State University
Dayton, OH 45435
(513) 767-7324 ext. 78 or 873-2169

Oklahoma Solar Energy Association
Solar Energy Laboratory
University of Tulsa
Tulsa, OK 74104
(918) 592-6000

Pacific Northwest Solar Energy
Association
Puget Sound Power & Light
Puget Power Building
Bellevue, WA 98008
(206) 454-6363 ext. 2315

Solar Energy Resource Association of
Wisconsin
1121 University Ave.
Madison, WI 53715
(608) 251-4447

South Dakota Renewable Energy
Association
P. O. Box 782
Pierre, SD 57501
(605) 224-1115

Tennessee Solar Energy Association
P. O. Box 448
Jefferson City, TN 37760
(615) 397-2594

Texas Solar Energy Society
1007 South Congress, Suite 359
Austin, TX 78704
(512) 443-2528

Virginia Solar Energy Association
c/o Piedmont Technical Associates
300 Lansing Ave.
Lynchburg, VA 24503
(804) 846-0429

Btu's, Thousands of Them

Introducing a basic unit of measurement in the energy numbers game

Can you imagine people who don't speak the same language trying to develop a coherent policy? That appears to be one of our basic problems with energy. Different energy forms have traditionally been paired with different units of measurement. We buy our natural gas in cubic feet or therms (100 cubic feet). Our electricity comes to us in kilowatt-hours. Crude oil is counted by the barrel (42 gallons per), while fuel oil, gasoline and propane, are usually metered a gallon at a time. The government speaks of quads, or kilojoules, or "barrels of oil equivalent." How can we the people make sense of this energy equivalent of the Tower of Babel?

We could learn a lesson from weight watchers. Whether you consume meat, vegetables, cereal, or dessert, all food energy is measured in calories. Calorie charts are vital to anyone who wants to establish a personal food policy. We could adopt a similar approach to energy measurement. Onward to a Btu standard!

Btu: British thermal unit. Sounds civilized. What is it?

Burn a wooden kitchen match to the end. You've just released 1 British thermal unit (Btu) of energy. Officially speaking, 1 Btu is the amount of energy required to raise the temperature of 1 pound of water 1°F. There are slightly more than 8 pounds of water in a gallon, 8.35 pounds to be exact. (Remember the saying "a pint's a pound the world around"?) Thus it would take 8.35 Btu to raise a gallon of water 1°F.

The federal government tells us that as a nation we consumed 77 quadrillion Btu (quads) of energy in 1979. That's 77 with 15 zeros after it and more energy by far than any other nation used. It's hard to get your arms around a number that big, but the Btu can also be a useful measurement, with much smaller numbers, in our own personal energy planning. We can begin by examining our daily shower. How many Btu's do we need to take a hot shower; or more directly, how many cubic feet of gas or gallons of oil, or square feet of solar collector do we need for our daily ablutions?

We need some information. What is the temperature of cold tap water and what is the temperature of the water we want for our shower? How many gallons of water will we consume during our shower? With this basic information we can answer the Btu-per-shower question.

The temperature of tap water, which comes from groundwater, varies between about 45 and 55°F (7–13°C), depending on the season and the location. The temperature of the hot water should be between 110 and 120°F (43–49°C). (Yours may vary. Take a look at the thermostat at the bottom of your gas hot water tank or inside the service plate of your electric water heater. Many people keep the thermostat at 140 or even 160°F, 60–71°C. That wastes thousands of Btu. You're making very hot water, much too hot for the skin, and then mixing it down with cold water to get the proper temperature in the shower. It's like driving down the street with one foot on the accelerator and one foot on the brake.)

How much hot water do we use for our shower? We tend to use about 10 gallons per shower. (This can vary widely. The Japanese, for example, use only one-fourth of what Americans average, yet oddly enough, the Japanese are renowned for their personal cleanliness.) Ten gallons or about 83 pounds of water must rise in temperature from 50 to 120°F, a 70-degree rise (temperature difference or delta T, ΔT). Since 1 pound of water consumes 1 Btu to increase its temperature 1°F, we need 83 (pounds) $\times$ 70 (ΔT) = 5810 Btu.

A cubic foot of natural gas contains about 1000 Btu, so we would need slightly less than 6 cubic feet of natural gas to take our shower, if the water heating system were 100 percent efficient. But it's not, and since a certain portion of the heat generated by the gas will be lost to the environment, the actual amount we would need to burn to get 5810 Btu into our shower water would be higher. For example, if the water heating system is 65 percent efficient, then our shower actually consumes 5810 ÷ 0.65 or about 8900 Btu (energy needed ÷ system efficiency = total energy consumption). That's about 9 cubic feet of gas. A gallon of fuel oil contains 139,000 Btu and is usually used at about 60 percent efficiency for domestic water heating. Thus our shower consumes 5810 ÷ 0.60 or 9680 Btu, which is 9680 ÷ 139,000 or about seven hundredths of a gallon. There are 3413 Btu in one kilowatt-hour (KWH) of electricity, which is consumed at about 90 percent efficiency in an electric water heater. How many KWH does our 5810 Btu shower need? A little consolidation speeds up the math: 5810 ÷ (3413 $\times$ 0.90) = 1.90 KWH.

We can also estimate how much roof area must be covered by a solar collector to capture sufficient heat for a shower. In June in Washington, D.C., a square foot of properly oriented collector intercepts about 2000 Btu per square foot per sunny day. With a 50 percent solar system efficiency, about 6 square feet of solar collector would be needed to power the daily shower: 5810 ÷ (2000 $\times$ 0.5) = 5.8.

Energy arithmetic can be fun. There are, for example, 4 Btu in 1 food calorie. (One food calorie actually equals 1000 "energy" calories.) The average person consumes 2500 calories or 10,000 Btu per day of food energy, and about 80 percent of it is used to keep the body at 98.6°F. Thus the body needs 8000 Btu (or 2000 food calories) of energy per day, or slightly more than used to take our hot shower. The body consumes energy at approximately the rate of a 100-watt light bulb (341 Btu per hour).

Although we use energy for myriad different applications, the major single function of energy is for human comfort, to keep our buildings warm enough or cool enough. Because of the unprecedented increase in the price of energy in the 1970s, architects now design homes and office buildings to take into account the waste heat given off by people and lights. (Only 10 percent of the energy used by an incandescent light bulb is given off as light; the rest is heat. The standard fluorescent light bulb doubles that efficiency: only 80 percent of the energy is lost as heat.) We burn off calories as we work and play. A roomful of people can make the building warm because the body gives off heat at the average rate of about 400 Btu an hour.

The Btu is also used to measure "cooling." This can be confusing—using a heat measurement to define cooling—but what cooling really means is the withdrawal of heat. Thus a 9000 Btu window air conditioner is rated to take 9000 Btu of heat energy out of a room each hour. New air conditioners must carry energy efficiency ratings (EER's), which are determined by the following formula: EER = energy removed in Btu per hour ÷ energy consumed in watts. Air conditioners have EER's ranging from about 6 to 13, and the higher the number, the more cooling effect you are getting per KWH of electricity you put in. For example, if the air conditioner provides 10,000 Btu of cooling capacity and uses 1000 watts (3413 Btu), its EER is 10, an excellent rating. You might find that another, possibly cheaper brand gives the same amount of cooling power, but draws 2000 watts (6826 Btu) to do it, giving it an energy efficiency rating of only 5. During the lifetime of the air conditioner, the more efficient appliance could save the owner much more than was paid for the unit in the first place.

Make friends with the British thermal unit. Let it through a south-facing window by pulling aside the drapes on a sunny winter's day. Depending on where you live, you could pick up 400 to 800 Btu a day for space heating for every square foot of window glass. Keep it in when the sun goes down by closing the drapes. Keep warm by clothing your body in wool, which captures the body's heat in its numerous air pockets, and you can turn down your thermostat and save thousands of Btu. Cool yourself by opening a window to catch the breeze, which blows against the skin, evaporating perspiration, which takes heat from the body at the rate of 62.5 Btu per ounce of sweat.

The next time you fill up at the gas station, ask for 2 million Btu (slightly more than 15 gallons at 130,000 Btu per gallon). If you get 20 miles per gallon, your energy efficiency rating is 130,000 ÷ 20 = 6500 Btu per mile. Another measure of transportation efficiency is Btu per passenger mile. Thus the car that holds five people has an efficiency of 1300 Btu per passenger mile instead of 6500 Btu with one driver. For comparison, what is our energy efficiency if we ignore the family car and walk to work? I might need 200 calories to walk a mile. The efficiency is thus 800 Btu per mile plus the benefits of exercise. If I ride my bicycle, the efficiency is even greater. I can get the equivalent of 1100 miles per gallon on my bike, about 120 Btu per mile. (I consume only slightly more calories biking than walking, but travel much farther.)

Unfortunately, most of the energy we use to get our Btu's isn't reuseable. We have but one opportunity to utilize that cubic foot, gallon or KWH, and then it's gone. This is the law of entropy. The universe is running down. Even the richest renewable source, the sun, is destined to poop out, though we have a few hundred million years to figure that one out. Our goal is to make the most efficient use of the Btu before it dissipates into so many spread-out heat molecules. For instance, an electric power plant uses 3 Btu to provide us only 1 Btu of electricity. Thus 65 percent of the energy is lost, mostly as waste heat. Some of the waste could be recycled back into power production by a technique called cogeneration. It could also be used to heat and cool nearby buildings. Or we might finally decide that it makes little sense to burn fossil fuels at thousands of degrees Fahrenheit simply to generate electricity to keep our homes at a mere 65°, or our hot water at 120°. We may decide that electricity, because of its inherent inefficiencies, should be limited to uses such as the operation of lights or motors which aren't used for heating.

As we deplete our fossil fuels and they become more expensive, we will have strong incentives to use them more efficiently and to replace them with more and more use of direct solar energy and other renewable resources to heat and power our society. Fossil fuels are concentrated forms of energy, while solar energy is diffuse. (One square foot of United States turf receives 10,000 to 25,000 Btu annually.) One gallon of gasoline contains the equivalent energy potential of the daily output of 150 square feet of a typical solar collector. Because of the compactness of fossil fuels and their ease of distribution and storage, society has used them first, but as we near the end of the fossil fuel era, we must become more inventive, using a more diffuse energy source more efficiently to capture the Btu's we need to survive and live comfortably.

In the United States, we have tremendous opportunities to increase the efficiency of our energy use. For example, producing 1 Btu of vegetable food energy nowadays can require 8 Btu of petroleum products in its growing, fertilizing, spraying with pesticides and herbicides, harvesting and marketing; but the backyard, rooftop or windowsill organic garden uses nowhere near that much energy. The food's better, too. It takes 75 million Btu to make an aluminum can, but only 3 million to make one from recycled aluminum, and we as consumers can see to it that much more does get recycled. As was mentioned above, a window on the south side of a house collects more Btu's than it loses, but with an insulating panel put up at night the gain is increased by at least 30 percent, a big improvement over the uninsulated window.

For those of us who like to take hot baths in winter, it's best if the hot water is left in the bathtub for a few minutes, so that the heat goes into the house, rather than into the sewer. The pilot lights on a gas stove consume more than one-third the total cooking energy, but they can be turned off and the burners can be lit with a sparker to save thousands of Btu. Well-built and well-insulated houses need far fewer Btu's, millions fewer for cooling and heating. Within the first five years of use, a solar collector can generate sufficient energy to match the energy required to build it in the first place.

Even though every energy use can be boiled down to a matter of Btu's, it's probably too much to hope that we, as a nation or a planet, will ever go on a Btu standard. But in the realm of our own homes and communities, we must develop a familiarity with energy in order to create personal energy plans, and the Btu is a most convenient vehicle for accomplishing this.

The Btu is the standard of energy measurement used throughout this book to help you find out things like how much energy your house is using, how much less it could use if it were well rigged with conservation treatments, and how much income energy you can expect to gain from solar retrofitting.

David Morris

The Turned-Off House

The R$_X$ for energy upgrading starts with conservation

There is no denying that solar energy has a great potential for energy savings and security in a time of rapidly rising heating costs. But don't let your vision of the sun eclipse another energy "source" that can produce a greater return on your energy investment as well as make your home more comfortable. This resource isn't as glamorous and it isn't likely to make your neighbors stand up and take notice, but the smart homeowner won't ignore it: energy conservation. It is not, of course, delivered in a pipeline or over the wires; it is rather a term that stands for a wide range of energy improvements that have in recent years been shown to offer the cheapest routes to home energy savings.

As a people, Americans like to think of themselves as an extremely efficient bunch. This image of efficiency is justified in some areas, but it certainly doesn't apply to energy efficiency, where we have failed miserably. A decade ago, just a few experts believed energy efficiency was necessary; today it is dubbed the "key energy source" by the prestigious Energy Project that was conducted by members of the Harvard Business School.[1] As the price controls that made our energy prices so cheap are being lifted, we are discovering via our energy bills just how inefficient our buildings, our cars and some of our appliances really are, and some of us are wondering how we could have been so deceived. One answer lies in the marketplace, where until recent times energy of all forms became consistently cheaper over time. Little attention was given to efficiency because cheap energy was a minor element in residential, commercial and industrial energy budgets.

That situation and the attitudes it fostered has now reversed itself to the point where some homeowners are paying more for energy than for their entire monthly mortgage payment, and they're finding that investments in conservation are often better than putting their money into a high-yield bond or savings account. Even in locales where energy costs have not increased much in recent years, high levels of investment in conservation have proven worthwhile in terms of return on investment, competing favorably with increasing the supply of energy from any other source, whether it be the sun, oil or electricity.

For most homeowners the future of energy price and supply is bleak. A third of this country's residences were constructed before 1940, a time when insulation was available but little used. In the 1940s, electric space heating came into being, and with it came the beginning of the insulation industry. Electricity was so much more costly than other heat sources that utilities often made insulation mandatory in all-electric homes. This helped to lessen the shock customers would get from astronomical electric bills resulting from poorly insulated homes. Homes built in the 1950s and 1960s usually have some insulation, but the efficiency of most of the dwellings constructed during this period is highly questionable. Even our newest houses could use more insulation and tightening up, especially in northern climates. Building efficiency is such a new and underfunded field of study that even many builders and designers are often unaware of what levels of conservation are the most cost-effective optimums for today and the future. It is because energy prices have risen so rapidly that most of this country's housing can be considered substandard from the standpoint of efficiency.

Energy Prices: Up, Up and More Up

Predicting future energy costs has been an exercise of constant revision for anyone who has attempted it in recent years. The year 1979, in particular, caused many bureaucratic faces to redden when, in six months, oil prices jumped to the level previously predicted for the 1990s! We got our first taste of higher oil prices in 1973–1974 with the Arab oil embargo, a price hike that followed nearly six years of stable energy costs. In 1967, the federal Department of Labor's Consumer Price Index (CPI) was set at 100 for all items, in order to track the inflation of the future. Energy prices began their runaway course in 1973. From that year to 1979, all items taken together on the CPI increased in price by 163 percent, or at an annual rate of 8.5 percent. At the same time, however, natural gas and electricity (lumped together in a single index category) increased by 204 percent or 12.6 percent per year, and fuel oil, bottled gas and coal (another combined category) zoomed up by 296 percent, an annual rate of 19.8 percent.

A closer look at the rates of increase between December 1978 and December 1979 indicates that prices for fuel oil and natural gas are increasing faster than electricity prices, and the year as a whole saw the greatest price increase in decades, much worse than during the 1973 embargo. In 1978, fuel oil jumped by 62 percent, natural gas by 23.5 percent and electricity by 12.4 percent, while all items in the CPI went up by 13.3 percent.

Note, however, that these numbers are national averages. A wide disparity in energy prices usually exists between different regions, with some prices actually decreasing or staying the same in the last decade, while others have soared by several hundred percent. Utilities differ greatly in their pricing and supply situations, but for the nation as a whole, the 1970s was a tough decade, and the future looks grim.

There are no "safe" predictions of future energy prices, given the potentially explosive situation with Middle Eastern oil-producing nations, but one fact is clear: Oil decontrol is underway in the United States, and most energy prices are directly or indirectly related to the price of oil. In 1985, oil price decontrol in the United States will be completed, and we will still be at the mercy of foreign oil producers. A growing number of nations are entering the world oil market, and the greater demand will push prices to ever-higher levels. But no matter how outrageous energy prices become in the future, you can feel utterly safe with an investment in conservation. It's a good buy even at today's prices. Energy conservation measures are simple to understand and execute, and dramatic savings can be realized.

The key is to put your dollars where they will do the most good: Stop the excessive heat loss from your building envelope before adding a new heat source in a solar retrofit. If you do so, you'll also get an extra bonus, because recent studies have shown that if you stop heat losses, you'll save more energy than you might expect. Researchers at Princeton University's Center for Energy and Environmental Studies have shown that by cutting heat losses in a series of modern townhouses by 50 percent, the furnace energy required to heat the homes was reduced by 67 percent.[2] This represents a 33 percent "extra bonus" over what could be expected by cutting heat losses. Similar results were reported by the National Bureau of Standards, which found that the bonus was 25 percent more than the heat loss reduction.[3] Even more dramatic is the work of Canadian researchers at the University of Saskatchewan, who built an experimental house in 1977 that saved 50 percent more furnace energy than expected by cutting heat losses.[4]

What's the trick? Simply this: When you cut heat loss to a low level, heat sources that were previously unimportant—your body, appliances, lights—provide a greater percentage of the heat load. As you sit reading this book, your body is giving off about 500 Btu per hour. Your hound, sleeping in the corner, is providing about 100 Btu an hour, your 100-watt light bulb is radiating another 341 Btu an hour and your appliances as much as 2000 Btu. In a house that is neither well-sealed nor well-insulated, these gains are insignificant, lost in a matter of minutes through cracks in the walls and thin ceilings. But, in a super-insulated, sealed dwelling, these sources have been shown to provide more than 40 percent of the heat needed in a severely cold climate.

In the Saskatchewan Conservation House, Canada's highly successful experimental model home in Regina, walls are insulated to R-40, attic to R-60, and infiltration losses are cut to a fraction of those found in our newest houses. The result is that internal gains and sunshine through a few south-facing windows provide nearly 85 percent of all the heat needed for this large, 2000-square-foot home. While some builders and architects in this country are still scoffing at foot-thick walls and other conservation features of the Saskatchewan Conservation House, Canadian builders got the message. Within a year's span, more than 60 similar low-energy houses were built. Lucky home-buyers are finding that their heating costs have dropped by as much as 90 percent under what they were paying in their old houses.

Most people don't live in a climate as cold as Saskatchewan's, which has an average of 11,000 degree-days a year, but spin-off studies from the Saskatchewan Conservation House have shown that these levels of insulation are appropriate in many northern climates, including such cities as Minneapolis and Bismarck, North Dakota. By contrast, pending federal building regulations call for insulation levels of only R-27 in the walls and R-38 in the ceiling for these areas, far below what has been shown to be an optimum level.

Of course, each individual situation will vary, but to begin with, money put into efficiency improvements will yield greater returns than the same amount spent on a solar retrofit. At a certain level, however, that solar retrofit will become more cost-effective than further conservation improvements. Later in this chapter, we will present a simplified approach to help you find the optimal R-value for your insulation jobs. More complex studies done recently show that in most cases the optimal mix between conservation and solar calls for an investment of $3000 to $4000 in conservation before solar looks economically attractive.

Where the Heat Goes and How to Control It

Although efficiency improvements are simple to understand and fairly easy to accomplish, deciding how much to spend on which improvements isn't always so clear. Researchers have found that houses differ so much in design and construction that it isn't possible to devise a catchall plan to achieve reduction in heat losses.

All major utilities have been mandated to perform expert energy audits on request of the customer, and you should take advantage of this service if it is available in your area. It is always wise to study your dwelling's heating and cooling requirements before beginning a solar project or spending funds on any energy-related improvements. Another essay in this section, "Finding Your Heat Loss," outlines methods for calculating heat losses from your house and will prove a helpful guide through this process. Our discussion will focus on the improvements that will give you the most benefit for your bucks so that when you do calculate your heat loss, it will have been minimized with conservation.

Stopping Infiltration

Conservation improvements are aimed at stopping the two types of heat loss: infiltration and conduction. Infiltration serves two purposes in your home, one helpful and one disastrous. Cracks around doors, windows and in the structure itself provide fresh air to the house, but they also allow precious heat to be sucked away at an incredible rate, particularly on windy days. Also, when indoor temperature is greater than that outdoors, heated air escapes from upper-level leaks and is replaced by cold air that is sucked in through lower-level cracks. Conduction, the other heat loser, is the process by which heat passes through solid materials such as doors, windows, ceilings and walls.

How these two types of heat loss affect your heating bill will depend on the condition of your house. If your dwelling isn't well-insulated, caulked or weatherstripped, you're probably losing more heat by conduction than by infiltration, and an investment in insulation will be extremely cost-effective. However, if yours is a newer house, fairly well-insulated and apparently well-sealed, infiltration could represent as much as 50 percent of your heat loss. The newest, best-insulated houses built today lose half of their total enclosed air volume to infiltration every hour. This rate of exchange between inside and outside air is expressed in terms of "air changes per hour." Thus, the above rate of infiltration is expressed as 0.5 air changes per hour. That means that

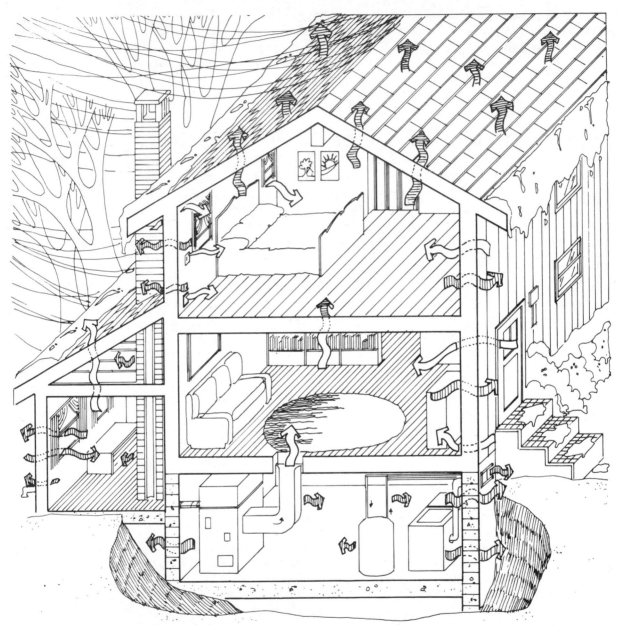

Figure I-10: This house represents a "zero" house with no insulation or other conservation improvements, and within its walls are myriad money-wasting energy leaks. Most of the heat is being conducted through the uninsulated roof pitches, which does a great job of melting off snow that would itself have had some insulating value. Leaky windows, doors and even unsealed wall switches and outlets are all major conspirators in heat loss, and so are uninsulated walls, windows and foundations. If this house were real, and there are many like it, it would be a drafty, chilly and expensive place to live. But it's not a hopeless case; see figure I-11 in this essay to see how common improvements make it all better.

every two hours all the heated air in the house has been completely replaced by fresh, cold air, which of course has to be heated to maintain comfort. Because infiltration losses are by far the cheapest to correct, they should always be attended to first, no matter what the condition of the house.

Economic studies of various options in stopping infiltration have shown that replacing broken glass, installing a tight threshold on your outer door and sealing structural cracks are the most cost-effective of all options, no matter if you live in a warm climate or a freezing one. In climates of 4000 or more annual degree-days, extensive infiltration work is recommended. (What's a "degree-day"? See the box on page 42.) This includes caulking windows, doors, adding weather stripping to windows, doors and any openings to the attic, in addition to the first-mentioned jobs. Some people prefer to skip some of the window sealing work by using high-quality, clear plastic over the entire interior window frame to stop both leaks around windows and reduce conductive losses.

It is fairly obvious that the main sources of infiltration are windows, doors and openings to the attic, but infiltration can be as devious as it sounds, and your home may have some hidden heat losers that will foil the usual anti-infiltration program. The Princeton University research team, mentioned previously, discovered that sneaky infiltration problems can wreak havoc on heating bills, a determination that came from an extensive study of infiltration losses in reasonably well-built town houses in Twin Rivers, New Jersey. These wood-frame structures were separated from their neighbors by masonry walls, which were found to be major heat losers. This was because wallboard attached to the interiors left an open airspace from the basement to the attic, bypassing the insulation. Once this hidden shaft was sealed, heat requirements dropped considerably. In addition, the builder had left considerable space around the flue that ran from the basement furnace to the attic and out through the roof, creating another unwanted and expensive source of heat loss.

Other hidden heat losers that were discovered in the Twin Rivers experiment are common in many homes. They include gaps between the wall and ceiling joists, leaks all around the ceiling to the attic, including spacing around pipes, wires and light fixtures, and leaks between the wood and masonry at the top of the house foundation. Watch for heat losses around baseboard trim installed on concrete slab floors. Often, nothing is added to stop infiltration directly into the house under the baseboards.

Once you have tackled obvious infiltration problems, the hidden or unusual ones will become more apparent. A general rule to remember is that special attention should be paid to any area where two different building materials meet, such as concrete and wood, or door framing next to metal siding.

It is highly unlikely that your infiltration work will ever make your home too tight, cutting off needed fresh air ventilation. It is possible, though, that the house's builder did do a good job sealing the house, and you may get to the point where infiltration losses could be cut to a level where your comfort could actually decrease because of an increase in humidity and "stale" air. It used to be that house design stressed maximizing ventilation for health reasons, in the days when diseases like tuberculosis were much more prevalent. But there never was any direct correlation between the two, and it is now known that to a point, very low rates of infiltration, or air change,

Degree-Days_____

Since no two climate zones are exactly alike, we need to have ways of knowing just how cold or how hot a certain region is so that our energy conservation and solar retrofit improvements are neither under- nor oversize. The *degree-day* (DD) is one measurement used to "rate" climate. Like the British thermal unit (Btu), the degree-day is another one of those little units of measurement that in general use usually adds up into the thousands. But while Btu's easily run up counts into the millions and quadrillions, degree-days seldom exceed the 10,000 mark, and while Btu's represent a unit of energy, the degree-day is a unit used to record the "temperature history" of a given climate.

In exact terms, heating degree-days represent the difference between the average outside temperature for a given day and the base reference indoor temperature of 65°F (18°C). If, for example, the average daily outdoor temperature is 30°F, 35 degree-days have accumulated for each day of that average temperature. The assumption is that when the outdoor temperature drops below 65°F, a space heating load exists. Thus the total accumulation of degree-days over a heating season can be used to calculate the size of a house's space heating load, based on the floor area of the heated living space and the area and insulation levels of exterior walls, windows, and doors, of roof pitches and foundation walls and of any other building element that is exposed to the indoor-outdoor temperature difference. If we know the space heating load, which is usually expressed in terms of Btu per degree-day per square foot of heated floor (Btu/DD/ft²), then we can see the extent to which a building's thermal performance needs upgrading with conservation or can benefit from solar retrofitting. If identical houses were built in Bismarck, North Dakota (9044 DD), and Daytona Beach, Florida (897 DD), the Btu/DD/ft² value for the former would be about ten times that of the latter, and the house in Bismarck would probably need a lot more energy upgrading than would its twin basking in Florida sunshine.

In later essays in this section, you'll be using degree-day values to make simple calculations about how much heat your home presently uses and about what kind of energy benefits you can expect from whatever solar retrofit option you're planning to employ. To help you with the conservation recommendations discussed in this essay, a list of degree-day totals for several locations in North America is provided in table I-8 of the essay, "Finding Your Heat Gain," which is found later in this section.

do not pose a health problem. So you needn't let ideas of the past stop you from going the extra distance in sealing your home, especially if you live in a cold climate (over 5500 degree-days a year). An extensive anti-infiltration campaign will drop the infiltration rate down to 0.3 to 1 total air change per hour. Below the 0.3 level, it may be necessary to actually induce fresh air ventilation to maintain adequate air quality, and that can be done without increasing the heat loss you've worked so hard to minimize.

Canadian researchers have invented a novel central ventilation device for use on homes that have infiltration rates of less than 0.1 air changes an hour. The "air-to-air heat exchanger" is a simple-to-build unit that works by exhausting stale house air to the outside while pulling cold fresh air in. The heated stale air preheats the incoming air, although the two never actually mingle. In fact about 80 percent of the heat in the stale air is recovered with this device, which is available commercially or which can be built with plywood and plastic sheeting for under $150 in materials. The developers of the do-it-yourself design stress that it is truly cost-effective only when used in an extremely well-sealed house (0.2 to 0.3 air changes per hour) in a colder climate. Measuring the air change rate in a house is done with some pretty sophisticated instrumentation and is by no means a generally available service. But if you feel you've gone over your home's air leaks with a fine-tooth plug, and if it seems to be excessively stuffy or humid inside, the air-to-air heat exchanger could be a useful item.

Blanket Your House with Insulation

With infiltration problems solved, your next step is adding insulation to cut conduction losses. Attic insulation is usually the most cost effective improvement after cutting infiltration losses, followed closely by window treatments. How much attic insulation to apply depends on the cost of your insulation, the climate and fuel costs. Fiberglass batts, mineral wool and cellulose are among materials of choice for this job, considering cost and performance. Loose-fill insulation, whether it is cellulose, fiberglass or mineral wool, sometimes tends to settle and must be checked every few years.

Window treatments include storm windows and insulating curtains and shutters. The heat resistance of the single-pane window is about R-1. Thus, you can see that if the walls surrounding the window have a higher R-value, your windows are the big heat losers. In colder climates, double or triple glazing and insulating curtains or shutters for nighttime use are recommended.

As mentioned earlier, some people prefer installing clear plastic film over the entire window frame. This measure has the advantage of accomplishing both infiltration and conduction solutions in a single step. Disadvantages, however, include not being able to open the window and having to replace the plastic every few years.

With these jobs accomplished, improving the insulation in your walls is usually the next most cost-effective step. In most existing houses, the wall cavity will allow only about 3 inches of insulation, which makes for a resistance rating of R-11 if fiberglass is used, plus the insulating value of the existing building materials. If you live in a cold climate, more wall insulation would be helpful, but because it will usually demand a major improvement, such as adding an extra outside wall, it is generally not cost-effective to go beyond filling the existing wall cavity. Again, fiberglass, mineral wool or cellulose are the favored materials for this job, which most often is accomplished by drilling holes and blowing the insulation into the wall. Also available are foams that are sprayed into wall cavities where they expand and solidify into high-quality insulation. But all foams aren't the same, and one variety has had its share of problems. Urea-formaldehyde foam has been at the center of some controversy in recent years, with the substance being completely banned in some states. Other states are also considering a similar ban because of shrinkage. Another problem reported with the foam is adverse health effects from gases emitted by the foam, a situation usually arising from improper mixing and installation. The safest bet for getting reliable insulation service is to thoroughly question several insulation contractors or dealers before making a purchase.

Another consideration in adding wall or attic insulation is whether to include a vapor barrier to avoid moisture damage. Recent studies in the moist climate of Oregon have shown that moisture damage hasn't been problematic in walls that are retrofitted with blown-in insulation without vapor barriers. If you think moisture will be a problem, paint interior walls with a coat of vapor barrier paint. For attic insulation jobs, consult your contractor about adding a vapor barrier before the job is done. In cold climates, a vapor barrier is crucial if high levels of insulation are added. Warm air escaping from your home carries moisture with it, and in cold climates this moisture can accumulate in thick insulation and ultimately ruin it. The decision to add a vapor barrier depends on climate and the amount of insulation to be installed. It also depends on what kind of siding or flooring is outboard or above the insulation, because insulation has to be able to pass any accumulated moisture through whatever covers it.

In most climates, insulating the basement walls above grade is another cost-effective task. This can be accomplished with sheets of polystyrene or polyurethane foam glued or tacked onto the walls, either inside or outside. If interior basement walls are insulated in occupied areas, the foam should be covered with a material that has a one-hour fire rating, because when they burn, foams give off toxic gases. Applying foam insulation to the exterior of a wall is another option. When this is done, the foam must be protected from the elements with siding or stucco.

Further basement insulation has been shown to be cost-effective in extremely cold climates with 9000-plus degree-days annually. Experts generally agree that insulating below floors isn't advisable in very cold climates because most water pipes are located in the basement where they could freeze if the air is allowed to cool too much. In many homes, furnace plants are located in the basement, and if the basement ceiling is sealed and insulated, waste heat generated from the furnace won't rise into the house as it usually does, though if the floor were insulated, it could be used to keep the basement above freezing temperatures. You can insulate all the pipes and heating ducts in the basement, but this may not be practical if the pipes aren't all easily accessible. If ducts are insulated alone, frozen pipes could again be a result. As you see, there are a few catch-22s in basement treatments. Therefore, insulating basement walls, whether the basement is used for living space or not, is the best course of action, and insulating under the floor may be feasible if the plumbing is protected as well. Of course, heating ducts should always be checked for leaks around joints to ensure that a majority of the heat produced by the furnace reaches the living area upstairs.

One final little insulation job should be mentioned as one of the most cost-effective you can do. Your water heater could be losing an amazing amount of energy, even though the tank is insulated with 1 or 2 inches of fiberglass. Wrapping the water tank with foil-faced, 6-inch fiberglass batts is typically such a good investment that it can pay for itself in a few months.

Better Heating Systems

Once your home's "thermal envelope" has been thoroughly attended to, it's time to examine the heating plant itself. Annual maintenance is called for with any fossil fuel furnace or boiler, including the changing of filters and a check of all moving parts. Researchers have shown that heat loss reductions usually save more energy than major modifications to the heating system, but if you think your furnace has big problems, ask your heating contractor to perform a "steady state furnace efficiency test." This test measures how much heat is leaving the furnace compared to how much energy went in to produce that heat. An efficiency rating of 75 percent or more is common on most newer furnaces. If your heating plant is significantly below this level after it has been tuned up, you should consider replacing it.

Although a furnace efficiency test is a helpful thing, such a test doesn't measure how well your furnace performs in actual practice. The steady state test shows how well your furnace operates when it is running at its maximum capacity, but it can't predict how much heat is actually delivered to your living areas because too many other factors intervene. These factors include where the furnace is located in the house, how well it performs at less than full capacity, how well the heating ducts are sealed and where they are located. A furnace placed in a ventilated crawl space can lose a large amount of heat through the foundation vents, which should always be closed in the winter, despite what you may have heard before. Your furnace can pull the extra air it needs for combustion from some other source. If you pay attention to these little factors and remedy any apparent problems, you'll significantly raise the overall or seasonal efficiency of your heating system.

Getting the Most for Your Investment in Conservation

The preceding guidelines for conservation upgrading should help you in making a survey of your house to see what needs to be done. Once you've thoroughly examined your house for heat losses and have an idea of what to do, then you're ready for more specific suggestions on how to best invest your conservation dollar.

Your savings from conservation improvements in the home depend on several factors. Your climate, local fuel or electricity costs and the costs for materials and labor all figure into the formula that will determine how much you should invest in conservation and solar devices to get the greatest benefit.

If you feel uncomfortable with numbers, relax. A comprehensive economic comparison of the benefits and costs of various conservation options in a variety of climates was published in 1980 by the National Bureau of Standards.[5] Fifteen cities were studied, ranging from Miami to Fargo, North Dakota, and an "optimum weatherization package" was described for each city. These conservation packages are a safe investment bet because the Bureau used actual prices for fuels, conservation materials and labor in determining the highest net savings that could be achieved in each city.

Economists describe the "optimal level" of investment as that amount which provides the greatest monetary benefits versus costs over a period of time. For its study period the Bureau chose a 20-year life span, with the further restriction that all weatherization options must pay for themselves in 11 years or less to be included in a final package of recommendations. The goal of the study was to determine the government's best investment level in conservation for low-income homes, and the results are highly useful for the homeowner who wants a good outline of the specific conservation improvements that will yield the greatest long-term savings.

The Bureau noted that a two-to-one difference within a city could be found between quotes for contractor-installed weatherization improvements; even greater differences were found from region to region. The Bureau chose the lowest reasonable estimate for their calculations. Even larger differences were found for energy prices, which led to the examination of more than 60 different options in each city to determine those that were the most economically attractive. The Bureau also calculated how future price increases would affect investment in the present. This was done by comparing the rates of fuel price increase and the value of money, or the discount rate. The value of money is determined by prevailing interest rates and, in the Bureau's vision of our energy future, the price of energy was going to increase 1.9 percent faster than the value of money.

An important number for the Bureau's calculation was the total cost figure for heat over a 20-year period, with the assumption that heat can be purchased either from the utility or oil company or through permanent conservation savings. Every item listed in a final weatherization package was found to be cost-effective, have little life-cycle maintenance and pay for itself in 11 years or less. These optimal packages were shown to be remarkably stable, despite differing initial fuel and conservation costs. The law of diminishing returns comes into play, with the bulk of each conservation package staying the same no matter which fuel is used for heat.

Let's look at an optimal weatherization package for a wood-frame house in Chicago, a northern-climate city with about 6100 heating degree-days. If the homeowner uses fuel oil for the home's prime heating source, the optimal package would include the following work, contractor-installed:

1. Replace broken glass.
2. Reset window glazing where necessary.
3. Install new thresholds under exterior doors.
4. Seal structural cracks.
5. Weather-strip windows, doors and attic hatch.
6. Caulk windows and doors.
7. Add storm windows and insulate interior window shutter (R-7).
8. Add a wooden storm door, with up to 60 percent glass in it.
9. Increase attic insulation to R-30 total.
10. Install R-11 insulation in the walls.
11. Insulate above-grade basement walls to R-7.

These options, the Bureau found, would yield higher benefits than any other conservation options studied, such as more layers of glass on the windows or heavier insulation.

Now, assume our Chicago homeowner uses natural gas for heat. Gas is a source that has traditionally been cheaper than fuel oil or electricity in most areas. The Bureau's analysis shows that the homeowner should still complete all the above-mentioned infiltration measures and insulate the walls, attic and above-grade basement walls. But with a cheaper fuel, two items changed for the new optimal package. A wooden storm door would no longer be cost-effective, nor would insulating window shutters, and both were dropped from the package.

By comparing optimal weatherization packages for all 15 cities, it is clear that all the infiltration options mentioned should be done in climates with 2000 degree-days or more a year, no matter which fuel is used. Another feature of nearly all the packages, for climates with 2000 degree-days or more, was the option of insulating basement walls above grade, work that is highly cost-effective. The major differences in the optimal packages come with attic insulation and window and door treatments. For the most part, walls should be insulated to at least R-11 in all climates above 2000 degree-days. Other guidelines are broken down by climate zones.

In climates with 2000 to 4000 degree-days a year: Attic insulation optimum levels vary from R-11 to R-19. Window treatments range from no improvement (over single glazing) to triple glazing or double glazing plus window insulation, depending on fuel costs. Door treatments for these climates varied only between no improvement and adding a storm door, depending on fuel costs.

In climates with 4000 to 6000 degree-days a year: Minimum attic insulation is R-30 and may be as high as R-38, if fuel prices are high. Storm windows are the minimum window treatment, and an interior plastic film is also recommended in areas with high fuel prices in the lower 4000-degree-day ranges. Triple glazing in the form of a double storm or a storm and a glazing layer inboard of the existing one is recommended in some climates near the 6000-degree-day mark. An alternative to the triple glazing (R-2.7) is double glazing with some kind of insulating panel, curtain or roll-up shade (R-5 to R-10). It must be emphasized, however, that as the R-value of window insulation increases, care must be taken to prevent as completely as possible any transfer of indoor air to the space between the insulation and the window. Without a virtually airtight seal around the perimeter of the insulation, severe condensation problems can develop, especially in climates with more than 6000 degree-days. Storm doors are recommended, with wooden storm doors (which conduct less heat than aluminum) advisable for colder climates with high fuel prices.

In climates with 6000 to 8000 degree-days a year: Our Chicago example fits in this category. Variations for climate and fuel costs include triple glazing, adding a second wooden storm door with no more than 30 percent glass, and R-38 insulation in the attic.

In climates with 8000 degree-days or more a year: In the attic, R-38 insulation is the rule, and storm windows are the minimum window treatment. A second wooden door or a new insulating door will be cost-effective in many climates, and in the extremely cold areas, triple-glazed windows with night insulation are in order. If you're in a place with more than 9000 degree-days a year, it would be cost-effective also to insulate your entire basement along with the above-grade insulation described earlier.

At today's energy prices, natural gas ranks as the cheapest fuel, on the average,

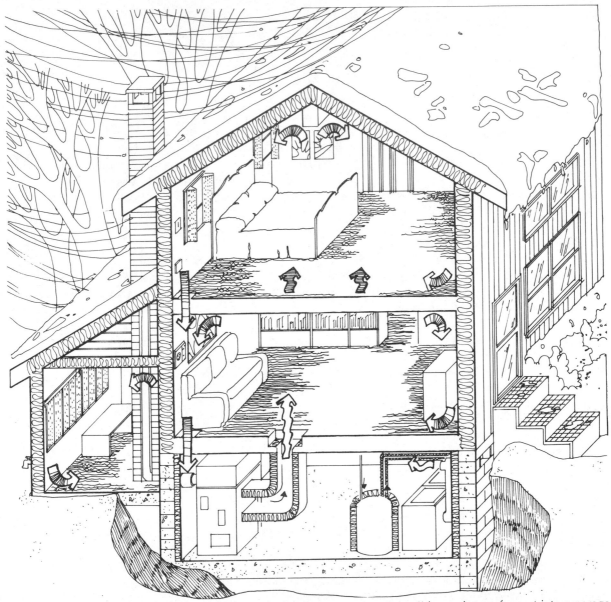

Figure I-11: By making a wide array of conservation improvements, our "zero" house is transformed into a paragon of energy efficiency. Insulation goes everywhere: into the roof pitches, the exterior walls, over windows at night, over basement walls, around hot air ducts, hot water pipes, the water heater. Air leaks are stopped with caulking and weather stripping; storm windows and doors are installed; floor vents are added to improve heat distribution; clothes dryer exhaust air is diverted to heat the basement; a stack damper is installed in the furnace flue; and a dual setback thermostat replaces the standard unit. There's even been an increase in south glass area for a little more solar heat gain. The net effect of all these improvements is to give the occupants more control over the heating energy they're buying.

followed by fuel oil and electricity. Propane and kerosene are the most expensive fuels for space heating, on the average. In New England, fuel oil is often the only fuel available other than wood, and conservation programs should follow the maximum course for the climate in these areas.

The basic guidelines described by the Bureau's work will suffice for homeowners who don't want to bother with additional calculations. However, it isn't too difficult to determine your own optimum weatherization package with the help of methods detailed in the next section of this essay. A little extra calculation can help you find the maximum level of investment for your conservation improvements, which is where the solar retrofits you have in mind begin to be cost-effective.

Finding the Optimum Level of Conservation

The economics of adding insulation or solar devices to your home will depend on the balance between the initial cost and the future fuel savings that can be expected with each option. In this section, we will present a simplified method for determining the optimum R-value of various insulation options. This method will allow you to do your own life-cycle cost analysis, and in the process you can play "energy futurist" by picking the level of fuel price increase which you think will most closely approximate circumstances in your local area. To accomplish this analysis, you'll need your pocket calculator with a square root function. And you'll also need to gather a few recent fuel bills and find out from table I-8 in "Finding Your Heat Gain," found later in this section, what the heating degree-days are for your location.

To best illustrate how the analysis is done, let's turn to the saga of a Mr. Baker, an imaginary man in his early 40s, who is concerned about his fast-growing heating bills and what his future holds.

Mr. Baker lives in a two-story, wood-frame farmhouse in a climate with 6000 heating degree-days a year. His home measures 25 feet by 40 feet, has approximately 1000 square feet of floor area on each floor and in the attic, and about 2200 square feet of exterior wall area that includes 16 windows which total 200 square feet of glass. He has a full basement, where an oil furnace is located. Approximately 130 square feet of the basement wall is located above grade.

Built in 1920, the Baker farmhouse has little insulation, and it is getting extremely expensive to heat. Baker does some investigation and discovers that his ceiling has an R-value of about 5, his walls about 4, his single-pane windows about 1, and his above-grade basement wall about 1.6.

To sum the situation, these are the essential factors of the Baker house:
attic area: 1000 square feet, rated R-5
wall area: 2000 square feet (without windows), rated R-4
window area: 200 square feet, single pane, rated R-1
concrete basement above grade: 130 square feet, rated R-1.6.

Like Baker, you should tally your area that is eligible for insulation work. Our simplified method of figuring the cost-benefit for conservation items doesn't apply to infiltration measures, but suffice it to say that these items are almost always cost-effective if you live in a climate with 2000 degree-days or more a year. Baker knows this, and his first job on his house is to weather-strip and caulk his windows and doors, sealing structural cracks and replacing the thresholds on his doors. He spends $280 on these items.

Finding The Magic Number

Now Baker wants to get some exact figures on the savings and costs he can expect from tackling the insulation jobs. He begins by looking at his fuel bills since 1972. Fuel oil prices nationally have gone up by an average of about 20 percent a year since 1972, and in 1979 they went up by more than 60 percent, so Baker knows that he had better consider possible fuel price increases in his analysis. Baker has faith that the price situation will improve somewhat, and he decides that he can expect price hikes of 10 percent a year for the period he wishes to analyze. He knows he will live in the house for the next 15 years for sure, so he picks 15 years as the period to study.

To do this calculation, Baker has to determine his "time value of money" and compare it to what he thinks future fuel prices will be. The value of the money he has available to spend on conservation or on a solar system is sitting in a bank in the form of a long-term savings certificate, which nominally earns 13 percent interest a year, but only 10 percent in real earnings after taxes. The comparison between the rate of fuel price escalation and the growth rate (in the case of money on hand) or the discount rate (in the case of interest on a loan) of money is made with this formula:

$$\frac{1 + e}{1 + d}, \text{ where } e = \text{fuel price escalation and } d = \text{the discount or growth rate.}$$

Thus in Baker's case the ratio is:

$$\frac{1 + 0.10}{1 + 0.10} = 1.00.$$

This number is taken to the table of "Present Worth Factors" (table I-1) and looked at in relation to the time period being studied, which in Baker's case is 15 years. The table shows that Baker's *present worth factor* is 15, which is a multiplier that will be used to help Baker find out how much insulation to install and what the dollar benefits will be over time.

Continuing the calculation, Baker looks at the price he is currently paying for fuel and sees that #2 fuel oil costs $0.88 a gallon. Next he has to translate that cost into a price per million Btu (MBtu), which he can do by using the conversion table shown in table I-2. The multipliers in this table take into account the fact that no heating system operates at 100 percent efficiency, which in effect makes the energy cost more. For example, there are 139,000 Btu of potential heat in a gallon of fuel oil, but when it's burned, about 90,000 Btu of useful heat are actually delivered to the heated space because the heating system is operating at around 65 percent overall efficiency. Thus, if the $0.88 oil were operating in a perfect system, the cost per MBtu would be $6.29, but in reality the cost is $0.88 $\times$ 11.07 = $9.74. The assumed efficiency for natural gas systems in the table is 70 percent, and for baseboard electric heat it is 100 percent. Actual efficiencies in home heating systems can vary widely depending on the age and condition of the furnace or boiler and the heat distribution system. If your central heater is an old rattletrap, the whole system might be only 50 percent efficient or lower. The best place to start finding out is to get a tune-up from a qualified service, which will have the tools to make an efficiency check.

Once he knows his energy cost, Baker next has to work his climate into the calculation. There are 6000 heating degree-days in Baker's little town, and for the purpose of this calculation that figure is translated into degree-hours, which is done by multiplying degree-days times 24 hours per day, or in Baker's case:

$$6000 \times 24 = 144,000 \text{ degree-hours.}$$

Now Baker has all the factors he needs to complete the problem, which will yield "The Magic Number," a multi-purpose figure that Baker can use in determining things like how much insulation to install in his attic, in his walls, in his basement and in his window insulation. The equation runs like this:

$$\frac{\text{present worth factor} \times \text{fuel cost per MBtu} \times (\text{degree-days} \times 24)}{1,000,000} = \text{The Magic Number.}$$

For Baker:

$$\frac{15 \times \$9.74 \times (6000 \times 24)}{1,000,000} = 21.03.$$

How Much Insulation?

Now Baker is ready to determine what his optimum insulation values are. He'll be able to calculate how much he will save over 15 years by making these improvements, and he'll also be able to come up with a cost-benefit ratio for each particular improvement, which can be compared with cost-benefit ratios of the solar improvements he'd like to make (calculated in "Finding Your Heat Gain" found later in this section).

Baker's Attic

Baker calls local contractors and discovers that one will install 3½-inch fiberglass batts (rated at R-11) for $0.24 a square foot. How many layers of fiberglass batts

TABLE I-1   PRESENT WORTH FACTOR

Our friend Mr. Baker decided that his ratio of fuel price inflation to the cost or the interest value of money was 1.00 and that the time period of his investment was 15 years, thus giving him a present worth factor of 15. What if fuel prices were predicted to increase at 20 percent per year, while the cost of money or interest rates were 10 percent? The ratio would then be 1.09, and for a 15-year period the present worth factor would be about 33.

Period Ratio	10 Yr	15 Yr	20 Yr	25 Yr	30 Yr
1.10	17.53	34.95	63.00	108.18	180.94
1.08	15.65	29.32	42.42	78.95	122.35
1.06	13.97	24.67	38.99	58.16	83.80
1.04	12.48	20.82	30.97	43.31	58.33
1.02	11.17	17.64	24.78	32.67	41.38
1.00	10.00	15.00	20.00	25.00	30.00
0.98	8.96	12.81	16.29	19.34	22.27
0.96	8.04	10.99	13.39	15.35	16.94
0.94	7.23	9.47	11.12	12.33	13.22
0.92	6.50	8.21	9.33	10.07	10.56
0.90	5.86	7.15	7.91	8.35	8.62

should he order, and what's the optimum R-value for the attic? First (and with all other examples), Baker calculates the cost of an R per square foot by dividing $0.24 by R-11, which results in a cost of $0.0218 per square foot per R. The simple formula for calculating his optimum R-value is:

$$\sqrt{\frac{\text{The Magic Number}}{\text{cost per square foot per R}}},$$

which in Baker's case works out to be

$$\sqrt{\frac{21.03}{0.0218}} = \sqrt{965}; \sqrt{965} = 31.$$

Thus, Baker's optimal R-value for the attic for the 15-year period is about R-31. Since he already has R-5 in the attic because of the insulating value of the building materials, he would need 2.3 layers of R-11 batts to reach R-31. He decides that two layers, bringing the attic up to R-27, will suffice. Three layers would raise the R-value to 38, which is beyond the optimum. This means the top 2 inches or so of the insulation

TABLE I-2 THE TRUE COST OF ENERGY

The first step in finding your optimum insulation level is to find out how much energy is costing you. Refer to a past energy bill and see how many energy units you used and what the cost was in dollars (not cents) per unit. Multiply the cost per unit times the indicated numbers corresponding to oil, electricity, gas, propane, wood or coal. For example, if electricity costs $0.05 per KWH, then the cost per million Btu (MBtu) is $14.65. If natural gas costs $0.50 per therm, it costs $7.15 per MBtu.

Fuel	Cost ($) per Unit (Btu per unit; assumed system efficiency, %) ×	Multiplier =	Cost per MBtu
Fuel oil	$/gallon (139,600 Btu/gallon; 65%)	11.03	_____
Electricity	$/KWH (3413 Btu/KWH; 100%)	293	_____
Natural gas	$/therm (100,000 Btu/therm; 70%)	14.29	_____
Propane	$/gallon (94,000 Btu/gallon; 70%)	15.20	_____
Wood	$/cord (average 20,000,000 Btu/cord seasoned softwood-hardwood mix; 50%)	0.10	_____
Coal	$/ton 30,000,000 Btu/ton; 50%)	0.07	_____

would not pay for itself in 15 years, so Baker decides to go with two layers at $0.48 per square foot of R-11 batts. The attic job totals $480 for 1000 square feet.

How much will Baker save by making this investment, and what is the ratio of benefits to cost? Baker does the following calculations to find the answer. He calculates the attic heat loss *before* the insulation and compares this to the heat loss *after* the insulation is installed:

Before: $\dfrac{\text{area} \times \text{degree-days} \times 24}{\text{original R-value}}$, which translates to

$\dfrac{1000\ (\text{ft}^2) \times 6000 \times 24}{\text{R-5}} = 28.8$ MBtu lost a year.

After: $\dfrac{\text{area} \times \text{degree-days} \times 24}{\text{new R-value}}$, which translates to

$\dfrac{1000\ (\text{ft}^2) \times 6000 \times 24}{\text{R-27}} = 5.33$ MBtu lost a year, which results in a total savings of 23.47 MBtu.

Thus, he will save 23.47 MBtu a year if he insulates his attic to R-27. He then multiplies this figure by current cost per MBtu ($9.74) and by 15 (years) to determine the *life-cycle benefit* for the investment. Thus:

$$23.47 \text{ MBtu} \times \$9.74 \times 15 = \$3430.$$

His life-cycle benefit is $3430, a number that is used to determine the ratio between total life-cycle benefit and total life-cycle cost. To do this, he simply divides the benefits by the costs. Thus:

$$\frac{\text{(benefits)} \; \$3430}{\text{(costs)} \; \$ \; 480} = 7.15 \text{ benefit-cost ratio,}$$

a favorable ratio indeed since any ratio above 1 means a positive benefit and thus a feasible investment.

His *net benefit* over the period is $3430 — $480 or $2950. With a little "streamlined" economic analysis, Baker sees that he's saving an average of $2950 ÷ 15 or about $197 a year for a one-time investment of $480, which is far above what Baker could make from any interest-bearing investment. Seeing that this is a very lucrative investment, Baker decides to go on to calculate what he can save by insulating his walls, basement and windows.

Baker's Walls

The best estimate for insulating Baker's walls is $0.90 a square foot for 3½ inches of blown-in insulation, which is the maximum he can have without building onto the wall. Is this move cost-effective?

Baker uses the same formulas to find out:

R-11 divided into $0.90 gives a cost of $0.0818 per R per square foot. This is divided into The Magic Number, 21.03, which yields 257, the square root of which is about 16. Thus R-16 is the optimum level of insulation.

Baker calculates that he already has R-4 in his walls because of the insulating value of the building materials. He will lose an R-1 (the value of the empty airspace in the wall) when he insulates because the wall cavity will be filled. Thus, he considers his wall to be R-3. By adding R-11, his total will be R-14, close to the optimum of R-16.

The contractor on the job follows the usual habit of charging Baker to insulate 2200 square feet of wall even though that includes 200 square feet of windows. (Using the gross footage is common because of the extra work required around windows and doors.) Therefore, the total cost of the job is $1980. How does this investment compare with insulating the attic? Again Baker grabs his calculator to figure the heat loss through 2000 square feet of wall area before and after the job:

$$\frac{2000 \times 6000 \times 24}{\text{R-4}} = 72 \text{ MBtu lost a year } \textit{before.}$$

$$\frac{2000 \times 6000 \times 24}{\text{R-14}} = 20.6 \text{ MBtu lost a year } \textit{after,} \text{ which means}$$
51.4 MBtu saved annually.

The annual heat savings make for a total savings over 15 years of $7460. The job costs $1980, so Baker's net saving is $5480 and the ratio of benefits to costs is 3.77. While this investment is attractive, it has a lower cost-benefit ratio than the attic, which was 7.09. This means the attic job is more cost-effective than the walls and should be done first, if Baker's improvements are being done on a pay-as-you-go basis. If the money is at hand to do all the improvements, then a benefit-cost ratio of 3.77 is good enough reason to go ahead with the work right away.

Baker's Windows

Next Baker turns to window treatments. Single-pane glass has an R-value of 1. He currently has single-pane windows. Would storm windows be cost-effective? His estimate per square foot of storm window is $2.30. Since the storm windows will add about R-1, no further calculation is necessary to figure the cost per R. Thus:

$$\frac{21.03}{2.30} = 9.083; \ \sqrt{9.083} = 3.$$

R-3 is the optimum window R-value at this price.

Baker realizes that he can get the optimum level of R-value for his windows by installing two storms, thus triple glazing his windows at a cost of $4.60 per square foot. With 200 square feet of glass, his investment would be $920 for R-3 on all the windows. Before he shells out the money, however, he checks into other window treatments. He looks into curtains that could offer twice the R-value of two storm windows for the same price. However, he notes that he can trust storm windows to last 15 years, and he's not so sure about the claims of the curtain manufacturers. He decides to go ahead with the investment in triple glazing.

He calculates that triple glazing will save him 19.2 MBtu a year, with his life-cycle savings for window treatments at $2785, compared to a cost of $920. Net savings are $1865 and the benefit-to-cost ratio is 3.03, still good though not as high as those of the attic or the walls.

Baker's Basement

Baker gets a surprise when he calculates savings for insulating the 130 square feet of above-grade foundation wall. Concrete is such a poor insulator that his 8-inch-thick basement wall has an R-value of only 1.6.

Again, he pulls out The Magic Number and goes to town with his calculations. For 1½ inches of polystyrene foam (R-7) at a cost of $0.70 a square foot:

$$\frac{\$0.70}{R\text{-}7} = \$0.10 \text{ per R}; \quad \frac{21.03}{0.10} = 210.3; \quad \sqrt{210.3} = 14.5, \text{ the optimum R-value.}$$

Thus, adding the R-7 is definitely cost-effective. Going from R-1.6 to R-8.6 will save 9.52 Btu a year. The total benefit would be $1381 compared to a total cost of the job of only $91. Net benefits are just $1290 over the period, but the cost-benefit ratio is a whopping 15.2, far outstripping any other options.

In the final analysis, all the insulation jobs would cost Baker $3741, and his dollar benefit over 15 years would be $15,429, with a net benefit of $11,688.

Baker's Bonus

But, Baker remembers that hidden "extra bonus" from the increased importance of heat gains from appliances, his body and other interior heat sources that he will receive because of his insulation and infiltration work. That bonus is figured conservatively at 35 percent for our example, which means Baker will actually get 1.35 times the benefit he has figured this far if he takes into account the "heat savings" from increased internal gains.

It was easy for Baker to reap a fat 35 percent bonus from his improvements because his house basically started with no insulation or weather stripping. But even newer houses that are relatively tight and well-insulated stand to gain some bonus from internal gains if they are further weatherized. In these cases gains of around 10 to 20 percent can be expected, which you can work into your equation if you're going through it for your house.

Thus, his total benefits jump from $15,429 to $20,829, and his net benefits from $11,688 to $17,088. Internal gains give him a bonus of about $5400 for the 15-year period.

Now, Baker wonders how an attached solar greenhouse would compare with these insulation jobs. He chooses a design that is 8 feet deep and 20 feet long, for a total of 160 square feet of floor area. He reads in a solar book that such greenhouses can produce 4 to 10 MBtu a year in heat savings, and he is very interested in growing vegetables to supplement his family's food purchases. He also knows that the greenhouse will be providing another benefit: extra space. But, he wants to make his decision on the basis of heat savings, not the other benefits, so he does a bit more calculating. First, he looks at how much heat he would save if the greenhouse produced only 4 MBtu a year. Then he compares this with the high figure of 10 MBtu a year:

		present cost of				present value of
MBtu	×	oil per MBtu	×	years	=	energy savings
4	×	9.74	×	15	=	$580
10	×	9.74	×	15	=	$1450.

The greenhouse will cost $1200. If the greenhouse produces only 4 MBtu in savings a year, Baker will take a loss of $620 over 15 years, considering heat benefits only. If the greenhouse performs well, he'll have a net benefit of $250 over 15 years, for a positive cost-benefit ratio of 1.21. Baker sees that this ratio doesn't come close to those produced by the insulating jobs. But, since he decides that food produced in the greenhouse would make up for any possible monetary losses if the greenhouse performs on the low side, he goes ahead with the investment anyway.

Some solar options may not make such easy comparisons with conservation as in our greenhouse example. But the above analysis may prove helpful if you can get a rough calculation of how many MBtu you will save on an annual basis from your intended solar retrofit. The essay "Finding Your Heat Gain," which is found later in this section, continues the discussion of solar economics with more simple do-it-yourself calculation methods.

You may be luckier than Baker in the overall cost category. If your home is already well-insulated and sealed, you may be closer to readiness for a solar retrofit than our friend. But before beginning any solar project, you should check to see if conservation improvements wouldn't be a better buy. In the final analysis, you will spend the amount that corresponds to how much energy and money you wish to save in the future, whether it be savings from the solar energy or its less exciting but more lucrative partner, conservation.

Larry Palmiter and Barbara Miller

Notes

1. Robert Stobaugh and Daniel Yergin, eds., *Energy Future: Managing and Mismanaging the Transition* (New York: Random House, 1979).

2. Robert Socolow, ed., *Saving Energy in the Home: Princeton's Experiments at Twin Rivers* (Cambridge, Mass.: Ballinger Publishing Co., 1978).

3. D.M. Burch and C.M. Hunt, *Retrofitting an Existing Wood-frame Residence for Energy Conservation: an Experimental Study,* no. 003-003-01885-6 (Washington, D.C.: National Bureau of Standards, 1978).

4. Robert Dumont, Robert Besant and Greg Schoenau, "Saskatchewan House: 100 Percent Solar in a Severe Climate," *Solar Age* (May 1979), pp. 18–24.

5. Robert Chapman et al., *Optimizing Weatherization Investment in Low-income Housing: Economic Guidelines and Forecasts* (Washington, D.C.: National Bureau of Standards, 1980).

References

Albright, Roger. 1978. *547 easy ways to save energy in your home.* Charlotte, Vt.: Garden Way
   Publishing Co.

Burt Hill Kosar Rittleman Associates. 1977. *Planning and building the minimum energy dwelling.*
   Solana Beach, Calif.: Craftsman Book Co.

Consumer Guide, eds. 1977. *The energy saver's catalog.* New York: G.B. Putnam's Sons.

Gay, Larry. 1980. *The complete book of insulating.* Brattleboro, Vt.: Stephen Greene Press.

Hand, A.J. 1977. *Home energy how to.* New York: Popular Science/Harper & Row Publishers.

Higson, James D. 1977. *Building and remodeling for energy savings.* Solana Beach, Calif.: Craftsman Book Co.

Peter, Hotton. 1979. *So you want to fix up an old house.* Boston: Little, Brown & Co.

Vandervort, Donald, ed. 1978. *Do-it-yourself insulation and weatherstripping.* Menlo Park,
   Calif.: Lane Publishing Co.

Wing, Charles. 1979. *From the walls in.* Boston: Little, Brown & Co.

Hardware Focus

Air-to-Air Heat Exchangers

Enercon Industries
2073 Cornwall St.
Regina, Saskatchewan
Canada S4P 2K6
(306) 585-0025

Enercon
Efficiency claimed to be as high as 85 percent; can be integrated with existing duct system or installed with independent ducting.

Melco Environmental Products
3030 E. Victoria St.
Compton, CA 90221
(213) 537-7132

Lossnay
Distributor for heat exchanger manufactured by Mitsubishi of Japan; efficiency claimed to be as high as 83 percent.

U-Learn
University of Saskatchewan Extension
Division
Saskatoon, Saskatchewan
Canada S7N OWO
(306) 343-5974

Do-it-yourself plans available for $1.00.

Dick Van Ee
R.R. #3
Saskatoon, Saskatchewan
Canada S7K 3J6

Prefabricated version of the University of Saskatchewan plans.

The Sun in Your Place

Understanding climate conditions and sun paths at your site

Your house probably means many things to you; so much of people's lives revolves around "these four walls." When you're considering making energy improvements, however, it's important to look at your home's primary function—that of a buffer between you and the outside environment, particularly the climate—that multifaceted interaction between the atmosphere and the surface of the earth, all of it powered by the sun.

There are four basic climatic elements that most affect human comfort: sun, air temperature, humidity and wind. Most houses are designed to provide comfort by separating people as much as possible from the undesirable effects of climate. By resisting great changes in the weather, the building envelope helps to maintain a comfortable interior, and when the weather change is extreme, energy imported through a wire or pipe is used to make up the difference between what the climate has to offer and what it takes to keep you comfortable.

But simply resisting the effects of climate is really only going halfway, for there are potentially significant benefits that can be derived by utilizing *desirable* climatic effects, which is what this book is all about. This essay discusses how the basic climatic elements affect your site and your building, and ultimately your comfort. You can use this information as a general planning aid in order to make better use of the climate's resources (particularly the sun) and thereby reduce your dependence on nonrenewable energy sources.

General climate conditions result from long-term patterns of interaction between the atmosphere and the varied character of the earth's surface. Weather is climate's little brother, the daily result of these interactions. Climate and weather are, of course, closely related since the climate over a long period of time is essentially the sum of all the daily weather. You don't have to be a weather expert or topography whiz to realize your solar plans, but for however far your site extends—be it to the backyard or the "back 40"—it's important to be aware of what's going on around it.

Many traditional house designs evolved in direct response to climate, but much of this country's mass housing output hasn't maintained those traditions. Thus in the broadest sense solar design, or redesign in the case of retrofitting, aims at resensitizing our buildings to climate and to the shape of the land. So just as you toured the inside of your house to discover its energy assets, you can continue the tour with a survey of outside resources that can be used to maintain a comfortable living space. And just as you probably found some energy liabilities on the house tour, you're certain to get a better handle on the undesirable elements of the local climate and of the site around your home.

Climate

Climate is always related to a given locale, although the size of the reference area can vary enormously. We can talk about the climate of a large region such as the Gulf Coast or the Rocky Mountains as the *macroclimate.* We can talk about *local climate* as being that of a city or county, and we can talk about the small-scale *microclimate* at your house site, or even surrounding your body. The basic climate elements have intricate interactions on all these levels, but for the purposes of this essay we can look at them as having distinct individual effects on building thermal performance and on comfort. And to some extent we can quantify these effects both for their positive and negative impacts. The solar resource, our primary concern, has been extensively quan-

Macroclimate   **Local Climate**   **Microclimate**

Figure I-12: If you get your weather news on TV, you see the satellite transmissions of the daily weather patterns of the macroclimate. *In scale, macroclimate encompasses huge land areas—New England, the Southwest, the Pacific Northwest—and huge weather patterns—fronts, pressure zones, jet stream wiggles. But the national weather report doesn't give an accurate account of the* local climate, *which involves events like quick-and-over thunderstorms, local cloud cover and winds, all of which can be affected by regional hills and mountains, lakes and urban development. The weatherman will never file a report on your* microclimate, *which you've got to study yourself to figure out the climatic qualities of your homesite—how open it is to sunlight, or how shaded, how it receives or blocks wind, how the design and condition of your house promotes or detracts from your comfort. All these climatic scales will warrant some examination as you find out more about the sun in your place.*

tified, and in the essay, "Finding Your Heat Gain," you'll be able to make a fairly accurate estimate of just how much sun there is in your place.

The sun is the primary climatic element; it provides all the energy that makes climate and weather happen. The earth's tilt in relation to the sun causes the regular seasonal changes in climate and, as we stand on the earth, the location of the sun in the sky changes predictably throughout the day and throughout the year. It is important to understand these daily and seasonal patterns in order to optimize the collection of the sun's energy for use in our buildings. There are some simple methods for visualizing the sun's location throughout the year, and they are presented later in this essay. The predictability of the sun's path has also been charted for different latitudes, and you can use these charts as a first step toward determining the sun's comings and goings.

Air temperature is a climatic factor that affects the heat gain and heat loss of our buildings and thus our comfort. It is directly linked to the amount of solar energy striking the earth's surface, which varies greatly with latitude. It is also affected by geographic elements such as water bodies, mountains and large cities.

Humidity is important because it both affects your perception of air temperature and directly affects your comfort, as we know all too well from the feeling of a hot, muggy summer day. The amount of humidity in the air is directly affected by the air temperature and the general weather patterns. The location of large bodies of water and the levels of precipitation in your region have the most effect on humidity at your site, although these factors do combine with large general wind and weather patterns that spread humidity out over huge regions.

Wind greatly affects the rate of heat loss from your building in the winter, and it can be an important source of cooling in the summer. Wind is the most site-dependent of all the weather elements in that the lay of the land and the natural or built objects around the house can greatly affect wind speed. Landforms can both accelerate and break up wind streams. For example, wind will accelerate up a slope, but just over the crest it is diminished into more erratic turbulences. The height of surrounding objects affects wind on your site, and proper use of landscaping and tall trees can help you control wind reaching your house. Tree and hedgerow windbreaks are the norm for protecting many a Midwest farm compound.

The Sun and the Earth

The sun is earth's primary power plant. Its daily delivery of energy drives the general circulation patterns of the atmosphere and is the primary influence of all climatic conditions. The sun's radiant energy helps plants recycle carbon dioxide and soil nutrients to make oxygen and become the first link of the food chain. Our fossil-fuel supplies are nothing more than solar energy stored by plants millions of years ago. If the sun stopped shining, it would take only a few days for all life on the planet to cease.

We can look at the sun's impact on the earth on both the macro and the micro scale. Solar energy reaching the earth warms everything that sees it. It also evaporates water in huge quantities, and the resulting water vapor enters the vast system of atmospheric circulation that we experience as weather. The warming of the surface causes air to rise until it cools and falls, creating large air circulation patterns that flow from the equator to the poles and back. These flow patterns have certain irregularities because of the earth's rotation. They are further complicated by the irregular arrangement of oceans and continents, which have different rates of heating and cooling and, of course, a different moisture content.

What happens when the sun's rays strike a surface? The angle between the surface and the rays is called the *angle of incidence,* and it governs how much energy is actually striking a given surface area. Energy from the sun is most concentrated when it strikes directly perpendicular to the surface (0-degree angle of incidence). As we tilt the surface, the same amount of energy gets spread over a larger area and thus, the more the receiving surface is tilted away from the sun's rays, the less energy it receives per unit of surface area.

This simple relationship has large impacts on the earth's climate systems since the curvature of the planet causes varying amounts of solar energy to be received at different latitudes. Thus it is that the lower latitudes around the equator are more perpendicular to the sun's rays and receive the highest concentration of energy, while the higher latitudes are more tilted relative to the sun's rays and receive less energy. The same condition occurs with the change in seasons. As summer becomes fall, then winter, the declining angle of the sun spreads the same amount of energy over a larger surface area.

The difference in climate between the tropical equatorial belt and the progressively colder higher latitudes is further accentuated by seasonal change and day-length change that is due to the tilt of the earth's axis. An imaginary line drawn from pole to pole through the earth is called the *axis of rotation.* An imaginary disk passing through the equator of the sun is called the *plane of the ecliptic,* and it is along this plane that the earth travels in its path around the sun. The earth's axis is tilted 23½ degrees away

Figure I-13: The angle of incidence is an important design factor in retrofitting and new construction. As the angle approaches 0 degrees (when the sun's rays are perpendicular to the surface), the collection of energy increases (see table I-4). Thus in winter steeply angled or perpendicular surfaces are preferred over shallow angles. If summer heat collection were the primary design goal, as it would be for a solar pool heater, the shallow roof pitch would be the ideal collector location.

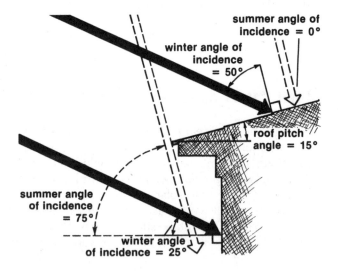

from the plane of the ecliptic. The sun shining on the spherical earth illuminates only part of the earth's surface and this part is called the *circle of illumination.* As the earth travels around the sun, the axis tilt causes the center of the circle of illumination to shift from the Northern Hemisphere to the Southern Hemisphere, and this change causes the seasonal variation.

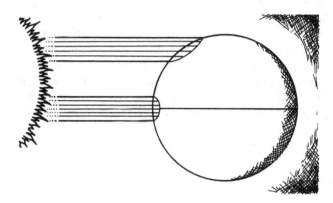

Figure I-14: Now you can see why the equator is tropical, simply because it collects more solar energy. If the earth were a basketball and the sun were a flashlight, you would see how the beam of light remained concentrated around the basketball's "equator" and how the same beam, the same amount of energy, was increasingly dispersed at higher latitudes until it got to the coldest spot, the North (and South) Pole.

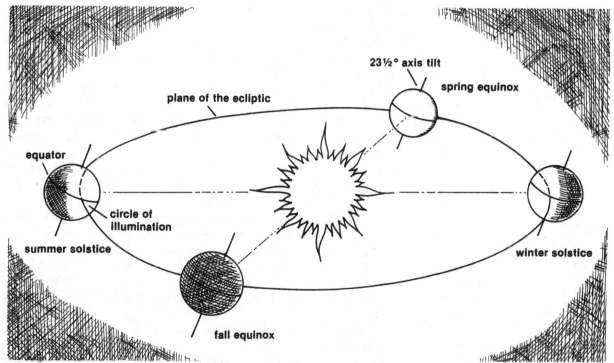

Figure I-15: Why is it summer at 40 degrees south latitude when it's winter at 40 degrees north latitude? The shift in the circle of illumination is the primary agent for seasonal change and, as is shown, the Southern Hemisphere receives the most concentrated rays from the sun at the winter solstice, while the Northern Hemisphere gets the least share. At the summer solstice, conditions are reversed.

The most accurate method for noting this change is by noting the latitude where the sun is directly overhead at *solar noon.* (Solar noon is the time when the sun reaches its highest point in the sky on a given day; it doesn't necessarily coincide with noon by the clock.) The *spring equinox,* which occurs on or very close to March 21, marks the time when the sun is directly over the equator at noon, when day and night are of equal length everywhere on earth. As the earth progresses in its orbit around the sun, the sun's position moves northward until the *summer solstice,* when the sun is directly overhead at 23½ degrees north latitude. The solstice generally occurs on June 21, and it marks the longest day of the year in the Northern Hemisphere. From the solstice onward, the sun's noontime position moves south until it crosses the equator on the *autumn equinox,* usually around September 22. Once again, day and night are equal. The sun's noontime overhead position continues south until it reaches 23½ degrees south latitude—the *winter solstice.* This is the shortest day of the year in the Northern Hemisphere and the longest day in the Southern Hemisphere.

Sun Angles and Your Site

Getting an idea of the Big Picture can help to make sense of the earth-sun relationship in your own particular place. How then do seasonal and daily changes in the sun's position affect your site and your ability to use the sun? This section will show you how to locate and predict the sun's position at various times in the year—vital information for planning your solar designs.

Two terms are used to locate the sun's position in the sky relative to a fixed surface. The *altitude angle* describes the sun's height above the horizon. If you stood facing the sun with one hand pointing at the horizon and the other hand pointing at the sun, the angle formed by your arms would be the altitude. The sun's altitude is always highest at solar noon.

If you stand with one hand pointing south, and the other hand pointing to the place on the horizon where the sun rises, the angle between your hands is the *azimuth* angle of the sun at sunrise. Thus, we say that the sun rises 90 degrees east of south and sets 90 degrees west of south, two azimuth measurements. At solar noon the azimuth angle of the sun is 0 degrees, since the sun is in a *true south* (nonmagnetic) position from where you are standing. To determine the compass orientation of *true* south in your area, refer to the map in figure I-6. In most parts of the United States and Canada, your compass orientation gets you within 5 to 15 degrees of true south, but the deviations noted on the map must be added to or subtracted from 180 degrees to get a compass correlation to true south.

A few more facts to help you get "oriented" to sun angles: The sun's path across the sky is always symmetrical about true south, which means that for any day, the azimuth angles at sunrise and sunset are always equal. There's more symmetry: If, for example, the sun's arc is 50 degrees (altitude) above the horizon, and 10 degrees (azimuth) east of south one hour before solar noon, it will be 50 degrees above the horizon and 10 degrees west of south one hour after solar noon.

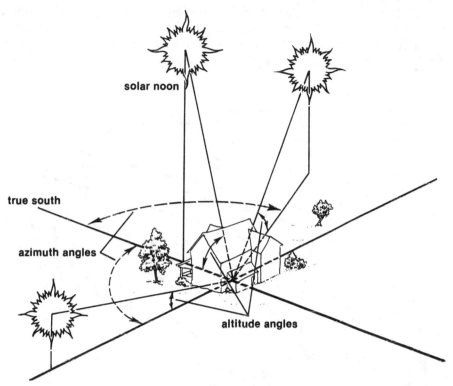

solar noon

true south

azimuth angles

altitude angles

Figure I-16: Solar angles can be "frozen" at any time of the day in order to study how they relate to a given site. What is the bearing of the south side of a house relative to solar noon, when the sun's azimuth is 0 degrees? How do low solar altitude angles in morning and afternoon affect a house in summer, and where should shade trees be located? Knowing seasonal altitude and azimuth angles (as shown in the sun path charts, figures I-17 through I-24) is of primary importance in site and building evaluation.

The range of the sun's altitude and azimuth also changes dramatically with the seasons. During summer the sun traces a wide and high arc across the sky, rising in the northeastern sky and setting in the northwestern sky. The altitude angle of the sun at solar noon is much higher in the summer than in the winter. For example, on New Year's Day at 40 degrees latitude the highest altitude is about 27 degrees. On the Fourth of July the highest angle has increased to 73 degrees. The maximum altitude and azimuth angles are reached at the summer solstice, the longest day of the year. During the winter, the sun traces a much shallower and narrower arc across the sky, rising in the southeast and setting in the southwest. The altitude angles at solar noon are quite low, and the lowest sun angles are reached on the winter solstice.

Sun angles vary greatly with latitude due to the earth-sun relationship explained earlier. The closer you are to the equator, the higher the altitude angles of the sun will be, both in winter and in summer. For example, the altitude of the sun at noon on

December 21 (winter solstice) is 22 degrees in Minneapolis–St. Paul (45-degree latitude), but down in Miami (25-degree north latitude) at the same time the altitude is 40 degrees. Above the arctic circle the sun is so low in the sky on this day that it literally never reaches above the horizon, which is why it is dark or at least very dim for several months of the year.

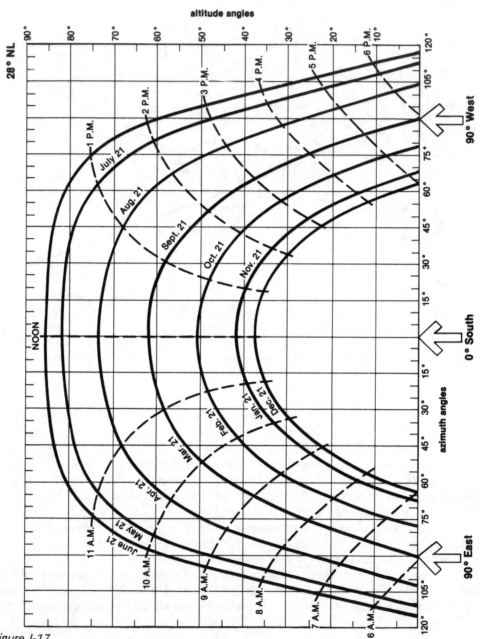

Figure I-17

Figures I-17 through I-24: These sun path charts can answer many questions about the sun in your place. They cover all latitudes from 28 to 56 degrees north latitude (NL) at 4-degree intervals. The horizontal axis of the graph indicates azimuth angles and the vertical axis shows altitude. Note that the designation of south means true south. You can see how dramatically the range of both the azimuth and altitude angles changes through the year. These charts can be your basic tool for studying the movement of the sun around your site.
Source: Adapted from Edward Mazria, The Passive Solar Energy Book, *1979, with permission of Rodale Press, Emmaus, Pa.*

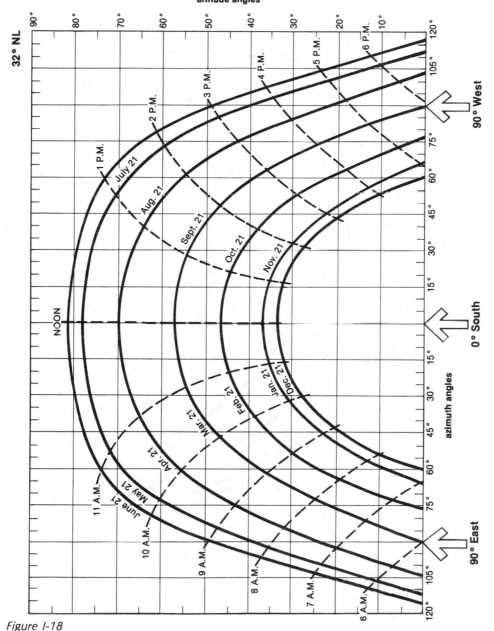

Figure I-18

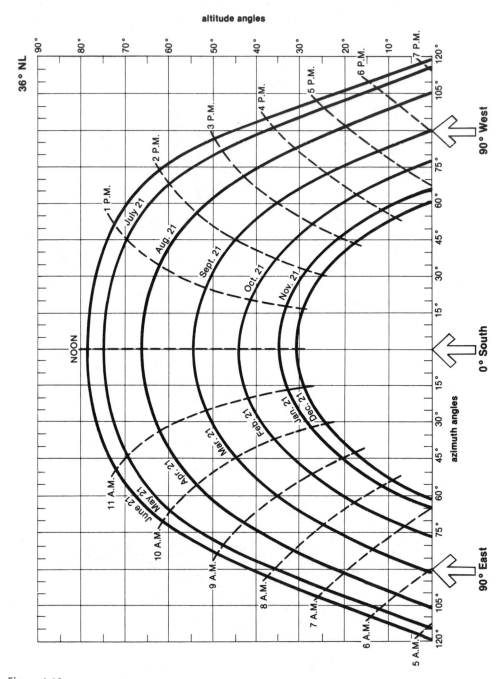

Figure I-19

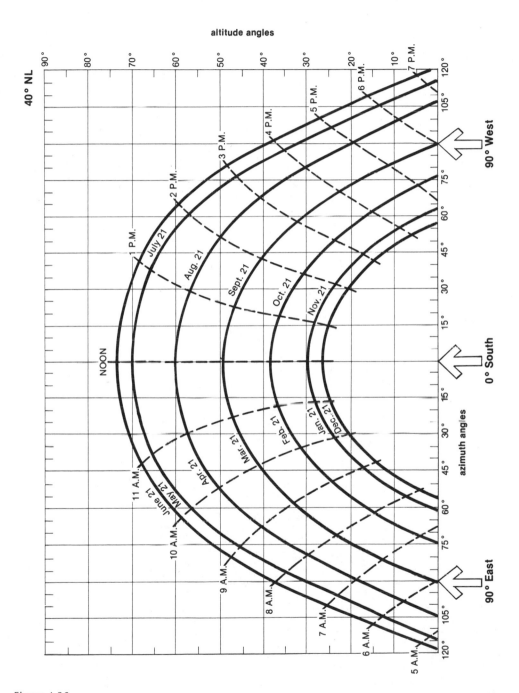

Figure I-20

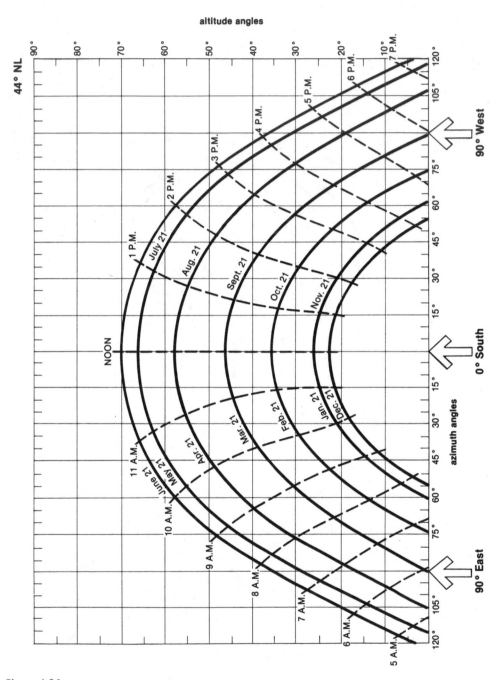

Figure I-21

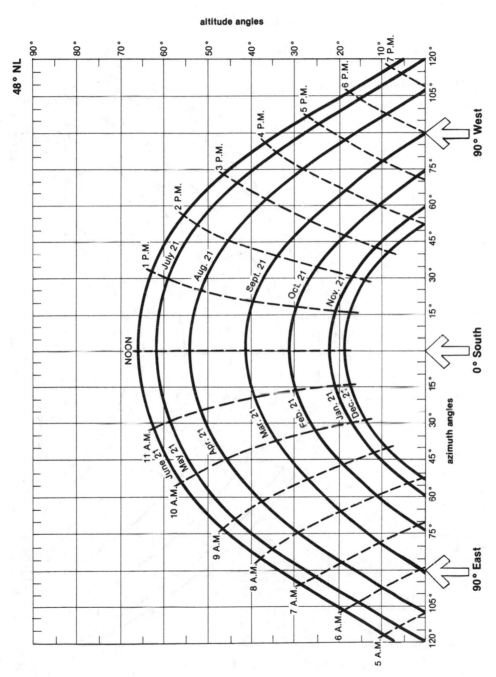

Figure I-22

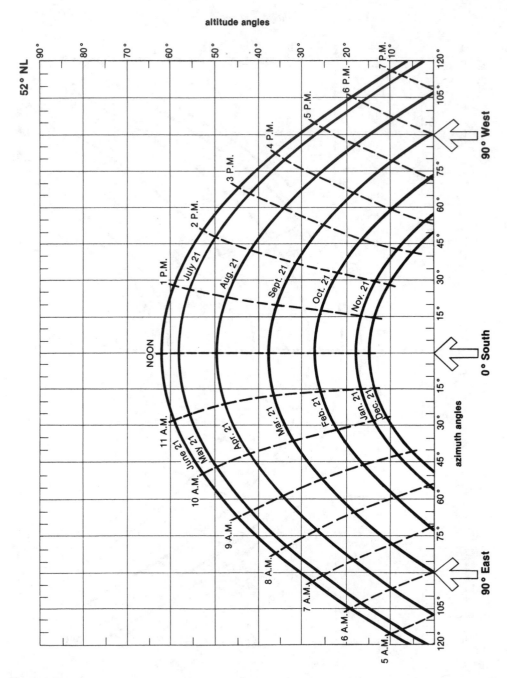

Figure I-23

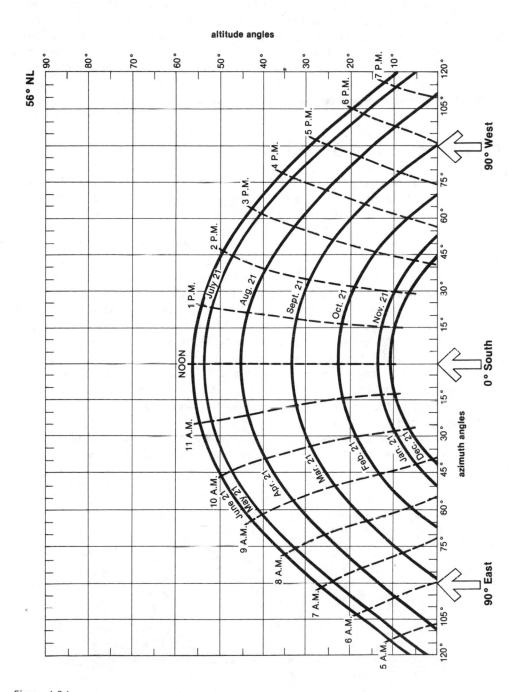

Figure I-24

Azimuth angles, and consequently the length of day, also vary greatly with latitude. Generally, the closer you are to the equator, the less difference there will be between the length of day and the length of night. The further north you go, the more seasonal variation there will be in the length of day. The previous sun path charts showing altitude and azimuth for various latitudes and seasons are primary design tools. With these you can begin to "see" a variety of seasonal sun paths at your site and start to zero in on which surfaces of your house have the best collecting potential.

Solar Angles and Shadows

Your ability to use the sun on your site depends on having a properly oriented, unshaded collector surface. This section discusses the nature of shading and how you can use the sun angle data to determine whether your potential collector surfaces receive sun during the right time of the year.

Your building site may be surrounded by buildings, trees and fences that can shade potential collector areas, which for space heating need to be unshaded in winter. Collector surfaces used for domestic water heating need to be exposed throughout the year. Take a look at objects near your house that may block the sun from reaching your collector surface. There are three factors that affect the length of a shadow: (1) the angle of the sun, (2) the height of the shading object and (3) the distance from the collector surface to the object. Just as sun angles are constantly in flux, so too does the shadow cast by an object shift throughout the day. If we were to plot these patterns as they fall on the ground, they would resemble a boomerang. While it is not necessary for you to plot formal shadow patterns for your site, it is important to be aware that shadow patterns change throughout the day; therefore, you might be deceived by just noting shadow lengths at any one time of the day.

Locating collector surfaces away from shade can be more difficult at higher latitudes because the lower winter sun angles cause nearby objects to cast very long

Figure I-25: Want to do something crazy? If you have a tree and some spare time, plot the tip of its shadow, with lime or sticks or stones, as it shifts through the day. It will be a piece of solar art!

shadows. Steep north-facing slopes, for example, can prevent little if any solar energy from reaching a south-facing surface. The hill itself is a shading object, or it can cause an effective increase in the height of trees and other objects that are above the building. In contrast, south-facing slopes can be very advantageous sites, since they can act to minimize the effective height of any objects located south of the building. East-facing slopes receive most of their solar energy in the morning, and collector surfaces may be shaded by the hill or the objects to the south by the afternoon. A west-facing slope may be shaded in the morning and will receive solar energy mostly in the afternoon. (In the summer, east and west slopes can be vulnerable to intense but unwanted gains from the morning and afternoon sun, which can present a cooling problem. The same is true for the east- and west-facing walls of a house.)

The height of the collector surface above the ground and adjacent objects can affect the amount of shading it receives with higher locations, generally meaning that there will be fewer shading problems. That naturally often makes roofs the least subject to shading by other buildings or trees. Sometimes the placement of a tree or a stand of trees can put you in the middle of a tough decision, namely: to cut or not to cut, or perhaps to trim. Evergreen trees, of course, cast shadows year-round, and they should be studied in relation to the full range of altitude angles for your latitude. Deciduous trees, on the other hand, don't necessarily guarantee a shadow-free winter exposure for your wall or roof. Depending on the species, the naked branches of a tree can block as much as 25 percent of the available sunlight from reaching your collector surface. The solutions to the problem of branch shading are simply cutting trees down or trimming and thinning branches to make them less dense. Perhaps you'll be able to replant a small tree out of the way of your southern exposure. If cutting is the only solution, hopefully you can plant some saplings elsewhere on your site to replace the loss.

Visualizing Sun Angles

After eyeballing your site, you'll want to find out more precisely if the annual or seasonal path of the sun will cause excessive shadowing of the collector locations you're planning. Naturally, you don't want to spend a year painstakingly plotting shadow lines, so you can use the sun path charts (see figures I-17 through I-24) to correlate altitude and azimuth angles with the real world of your site.

The basic method is to look at the specific portion of the southern sky that corresponds to the daily period of availability of truly useful solar energy. The time period is generally considered to be the six hours between 9:00 A.M. and 3:00 P.M., or 10:00 A.M. and 4:00 P.M. during daylight saving time. Any object that is within this *skyspace* may be a potential shadow on your solar plans. As the sun path charts show, the area of the skyspace changes with different latitudes, and the skyspace will be different if you are looking at it relative to a year-round collector location (domestic water heating) or to a heating season collector location. If, for example, you live near the 40th parallel and you want to study the skyspace for a rooftop solar water heater and for a space heating collector on a vertical wall, you see two very different skyspaces from the sun path chart.

When looking at the charts, it is important to visualize the relationship of azimuth to altitude. The skyspace is not flat like lines on a page, but curves around your house from east to west.

Once you've identified the skyspace you want to study, you can make a simple viewer to help you scan your site by copying the solar access viewer diagram shown in figure I-27, strengthening it with a thin cardboard backing and assembling it according to the instructions. To use the viewer:

1. Place it on a level surface. The top step of a ladder or a tabletop works well, or tape it to a camera tripod with a panning head. If you're evaluating the skyspace relative to your south windows, you can place the viewer on your windowsill. Whatever the case, try to locate yourself as close to and at the same height as the collector areas you're planning to develop.

2. Line the south arrow of the viewer up with true south. With the viewer at eye level and your eye centered at the south arrow line, sight along the top of the viewer for any objects that may appear in the skyspace.

Bear in mind that this viewer is not a precision tool and is meant to help you make a rough survey of your skyspace. A little more precision can be had using a builder's transit or an inclinometer and a compass. Use the compass to locate the solar azimuth at 9:00 A.M. and the other tool to sight the altitude. Note any objects protruding above your line of sight, since they will shade your collector in the morning. Then repeat the process, rotating through south to the west, correlating azimuth and altitude as many

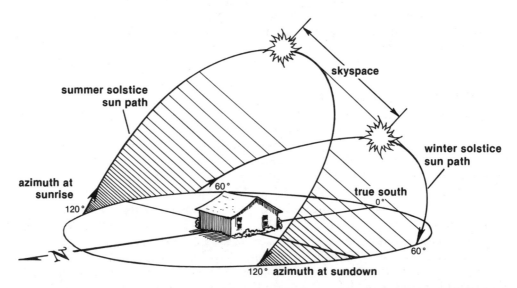

Figure I-26: This is a representation of the skyspace for a site at 36 degrees north latitude according to the solstice azimuth angles. If you're studying your site primarily for space heating, you're probably interested in finding shadow makers in the lower half of the skyspace band. For year-round domestic water heating, you should look for objects throughout the band, particularly those nearest to true south, where at solar noon insolation peaks.

times as you need to see where the potential shadows will be. If you can see that you've just got a few possible shadow makers, just "shoot" them individually to see if they inhabit your skyspace.

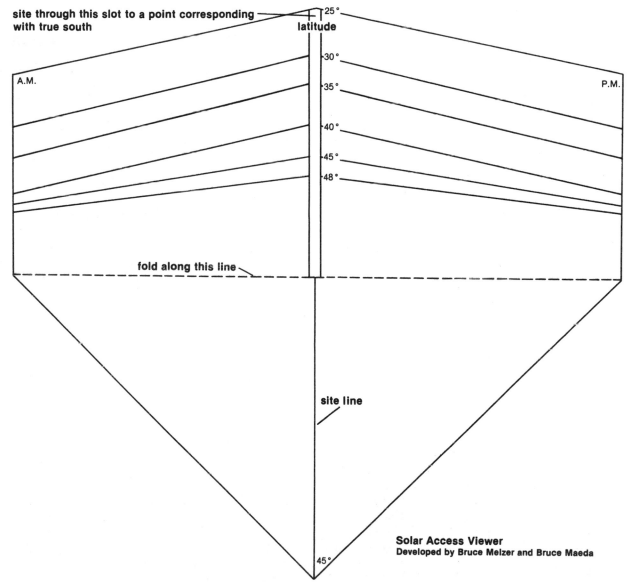

Figure I-27: After you have photocopied this page, reinforce the copy with a thin sheet of cardboard and fold it up along the indicated bend line. Cut the top of the viewer along the latitude line closest to the one corresponding to your site. As indicated on the viewer, you can now study your site within 45 degrees east or west of true south.

Since neither a transit nor an inclinometer (a useful $10 tool) may be available, you can make a simple angle-finder with a protractor and a piece of straight, stiff wire or thin rod. Bend one end of the wire or rod to form a small U and hook it into a hole at the midpoint of the protractor's straight edge. To use this, point yourself at various solar

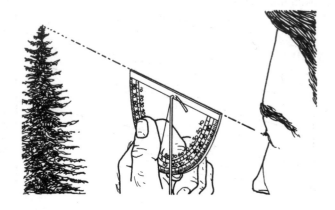

Figure I-28: To find the altitude angle of nearby objects, you can use this simple device—a protractor with a piece of straight wire hooked at the center point—by putting yourself as close as possible to the solar surfaces you're studying. Once you know the altitude, you can find the approximate height of the object by multiplying the tangent of the angle times the number of feet between you and it. (The tangents for several angles are listed in figure I-29.) To that result add the distance (in feet) between the protractor and the ground to get "object height." You can use the object height in the calculation in figure I-29 to calculate shadow lengths at different times of the year.

$$\text{shadow length} = \frac{\text{object height}}{\text{tan solar altitude angle}}$$

$$\text{shadow height on building} = (\text{shadow length} - \text{distance between objects}) \times \frac{\text{object height}}{\text{shadow length}}$$

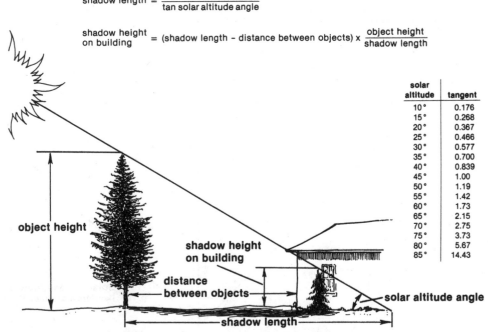

solar altitude	tangent
10°	0.176
15°	0.268
20°	0.367
25°	0.466
30°	0.577
35°	0.700
40°	0.839
45°	1.00
50°	1.19
55°	1.42
60°	1.73
65°	2.15
70°	2.75
75°	3.73
80°	5.67
85°	14.43

Figure I-29: This calculation is all you need to determine the height of a shadow on a vertical wall. But what about the length of a shadow cast upon a roof pitch? Some of the information from this calculation can be plugged into the calculation in figure I-30 to find out.

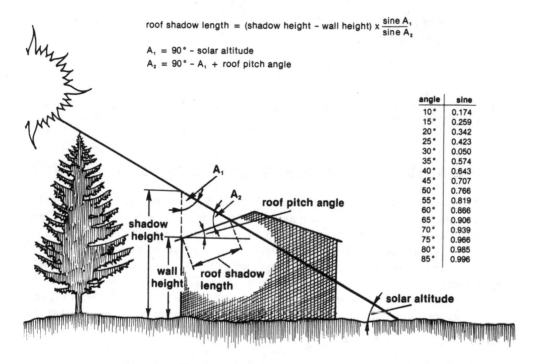

roof shadow length = (shadow height − wall height) x $\frac{\text{sine } A_1}{\text{sine } A_2}$

A_1 = 90° − solar altitude
A_2 = 90° − A_1 + roof pitch angle

angle	sine
10°	0.174
15°	0.259
20°	0.342
25°	0.423
30°	0.050
35°	0.574
40°	0.643
45°	0.707
50°	0.766
55°	0.819
60°	0.866
65°	0.906
70°	0.939
75°	0.966
80°	0.985
85°	0.996

Figure I-30: Use the above equation to calculate the length of the roof shadow cast by an object. For example, if the shadow height calculated in figure I-29 is 15 feet, the vertical wall height of the building is 10 feet, the roof pitch angle is 20° and the solar altitude is 30°, then:

$A_1 = 90° − 30° = 60°$; sine 60° = 0.866
$A_2 = 90° − 60° + 20° = 50°$; sine 50° = 0.766

$$\text{roof shadow length} = (15 \text{ feet} − 10 \text{ feet}) \times \frac{0.866}{0.766} = 5.65 \text{ feet.}$$

This information can help you with the placement of rooftop collectors.

azimuths as before and sight along the straight edge of the protractor, raising it until the rod crosses the proper altitude angle. You may need a friend to help give you angle readings more conveniently. There are also several sun angle tools available in a wide range of prices, some of which are listed in the "Hardware Focus" section which is located at the end of this essay.

If you're still wondering about what some tree or building is going to do to your collector surface and you're willing to do a little math, use the sun path charts (see figures I-17 through I-24) with some simple trigonometry to determine the length of a shadow at any time of day or year. As figure I-29 shows, you also need to know the height of the object, which is divided by the tangent of the solar altitude angle you're studying. Simple, but what if you calculate a shadow length to be 50 feet for an object that's just 35 feet away from your house? How far up the building will the shadow run? This is a simple matter of proportions. If a 25-foot-high tree casts a 50-foot shadow, then at a distance of 35 feet the shadow height is:

$$\frac{(50-35)}{X} = \frac{50}{25} : \text{thus, } X = 7.5 \text{ feet.}$$

Using this method will give you total accuracy in plotting shadow lengths and heights on and around your house.

The Solar Resource—How Much Energy at Your Site?

The total amount of solar energy striking the top of the earth's atmosphere has been measured at 429.2 Btu per square foot per hour, which is called the *Solar Constant*. From that point in space the atmosphere acts as a giant filter that somewhat reduces the intensity of the solar radiation finally striking the earth's surface, to anywhere from about 5 to 25 percent of the Solar Constant, depending on the season and location. Some of the radiation is reflected back to space by clouds and is absorbed by particles and water vapor in the atmosphere. Some of this absorbed radiation is reradiated back to space, and it is also reradiated down to the earth's surface. Solar radiation is also scattered by

Figure I-31: Even though the Solar Constant is reduced by anywhere from 75 to 95 percent by the time solar radiation reaches the earth, there's still enough energy remaining to get a tan, heat houses, boil water, melt steel, etc., etc., etc.

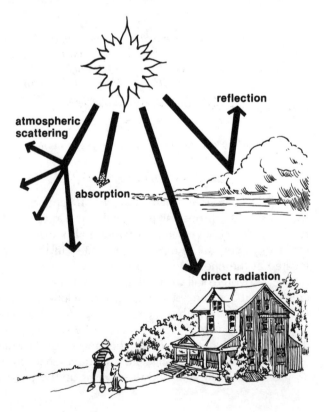

the atmosphere, which diffuses the sunlight in the way that a frosted light bulb diffuses incandescent light. Thus from the sky we receive primarily direct and diffuse radiation.

Direct radiation is a beam of light that has specific direction and is capable of casting a shadow if you place an object in its path. *Diffuse radiation* is scattered evenly about the sky, and it does not appear to originate from a specific source. It's the type of sunlight you experience on a cloudy day. There is useful energy even in diffuse sunlight, since on a bright cloudy day a surface will receive about one-third of the energy it receives on a clear day.

Daily and Seasonal Variation

The intensity of the solar resource changes throughout the day and the seasons. During the course of a day *insolation (in*coming *sol*ar radi*ation)* on a south-facing surface increases through the morning, reaches a maximum at noon, and falls off in the afternoon. Any variations in this general pattern are primarily due to changes in the local weather.

Figure I-32: This map gives the total amount of solar energy available, in millions of Btu's per square foot of horizontal surface area, through the entire heating season. The length of the heating season varies throughout the country and this was taken into account in the map. Florida has more sunshine than Minnesota, but a much shorter heating season, which explains why they have equivalent potentials for solar space heating.

Seasonal variations in the solar resource are due to changes in the day length and weather patterns and to changes in sun angles. Winter days are shorter, so there is less time to receive the solar radiation. This is especially noticeable in high latitudes where winter day length may be less than five hours, and the number of hours available for effective energy collection are even fewer. In the Northern Hemisphere, winters are cloudier and stormier than summers, and this will of course affect the amount of available solar energy.

Variations in sun angles greatly affect our ability to use the sun. For low-winter sun angles radiation is greatest on vertical or nearly vertical tilted surfaces, which means for example that south-facing windows make good collectors. For high-summer sun angles there is more energy available on horizontal surfaces, or tilted surfaces that are close to horizontal, which makes shallow roof pitches good collector locations.

The map in figure I-32 shows solar radiation available on a vertical surface during the heating season. The numbers are useful for giving you a sense of how much solar heat is available during the heating season, which you can compare with the Btu's your house presently needs to stay warm. These exercises are carried out in detail in the essays "Finding Your Heat Loss" and "Finding Your Heat Gain" which follow in this section.

Temperature

Air temperature is the main component of both our comfort and the rate of heat loss from buildings. The daily patterns for temperature change are closely linked to the changing levels of insolation. Unless local weather changes considerably, the coolest part of the day is usually right at sunrise, until which time all the exterior surfaces have been radiating heat to the sky. At sunrise, the earth begins to heat up again as the land, water bodies, buildings and the atmosphere start to absorb solar radiation.

Air temperature increases throughout the day and peaks one to three hours after solar noon. As the intensity of solar energy striking the earth decreases in the later afternoon, air temperatures slowly start to drop, and the cycle repeats itself. Daily variations in local air temperature are also greatly influenced by the weather patterns. Storms and fronts move across the earth's surface, sometimes bringing in warmer air, sometimes cooler air. Cloudiness affects air temperatures, since the water vapor in the clouds absorbs the radiant heat emitted by objects during the night. The clouds then reradiate part of this heat back to earth. Consequently, temperatures on cloudy nights are warmer than temperatures on clear nights.

Seasonal variations in temperature are also heavily influenced by the seasonal solar patterns. Shorter days and the more oblique solar altitude angles mean less solar energy strikes the surface in the winter than in the summer, and the decrease in available solar energy causes a decrease in average air temperatures. There are also several geographic factors that will affect air temperatures at your site. These include the effects of slopes and mountains, bodies of water, large cities and paved areas, and plants and landscaping.

Air temperatures can vary greatly in hilly and mountainous areas, due to a wide variety of factors including altitude, slope orientation and cold air drainage patterns. Altitude is directly related to air temperature since the higher you go, the colder it gets. In fact, as far as temperature is concerned, each 1000-foot increase in elevation is equivalent to traveling 1 degree north in latitude.

Slope direction is extremely important. As you would expect, south slopes tend to be much warmer than slopes of any other orientation, and often they are warmer than adjacent flatlands or valleys. North slopes are the coldest. East slopes are warmer in the morning, while west slopes are warmer in the afternoon.

Cold air drainage is an important microclimatic influence on air temperature. Since cold air is heavier than warm air, it tends to sink to the lowest point in the landscape. During the evening, cold air moves down slopes and mountains as a breeze or light wind and accumulates in the valleys.

Bodies of water can have a strong moderating effect on air temperatures in surrounding areas. This is because the water stores a large amount of heat during the day and releases it at night, keeping the environs warmer. Also, the combination of heat storage and evaporation that occurs from bodies of water helps to keep peak daytime air temperatures down. The heat storage effect works on a more protracted seasonal basis, too, since all water bodies, including oceans, heat up gradually during the summer and release heat in the winter. The temperature-moderating effects of water operate on all scales, from oceans to large lakes to a water thermal mass inside a solar house. Generally, the larger the body of water, the wider is its area of thermal influence.

In recent years studies have confirmed that air temperatures in cities are higher than temperatures in surrounding areas. This is due in part to the large masses and surface areas of concrete and asphalt that readily soak up heat during the day and store it through the night. Wide and unshaded asphalt streets and parking areas can raise our temperature in summer and actually increase air conditioning costs significantly. It is also due to the relative lack of vegetation that cools air by the evaporation of water through leaves. Thus in summer, areas that have an abundant flora will be cooler than nearby "bare spots." Cities are also a concentrated center for consumption of fossil fuel to run cars and factories and to heat and cool buildings. The large amounts of waste heat generated from these processes increase average air temperatures.

Humidity

Like temperature, humidity affects our comfort in both extremes: There can be too much and too little. The effects of high humidity are most noticeable when it is hot, although too-low humidity can also negatively affect your comfort in cold weather.

Relative humidity levels change with changes in air temperature. At any given temperature, there is a maximum amount of water vapor that can be held in the air. The warmer the temperature, the more water vapor the air can contain. Air that contains all the water vapor that it possibly can is called saturated. If saturated air is cooled, the amount of water vapor the air can hold is reduced, and some of the water vapor in the air condenses to become liquid. The temperature at which water vapor in the air will condense is called the *dew point.* On a hot, muggy day, the water vapor on the outside of your glass of ice water condenses because the air next to the surface of the glass has been cooled to the dew point.

   Most people associate humidity with summertime comfort, when they wish the humidity would drop, but humidity affects human comfort in more than one temperature range. High humidity in hot weather reduces the ability of your skin to evaporate perspiration, and since the effectiveness of this important cooling mechanism is reduced, you feel uncomfortable. High humidity when it is moderately cold—between 40 and 58°F (4 and 14°C)—will make you feel clammy, yet overly dry air can make you uncomfortable by drying out mucous membranes in your throat and nose. At indoor temperatures around 65°F (18°C) in winter, increasing humidity will increase your sense of comfort.

   Temperature and humidity can be combined into a discomfort index known as

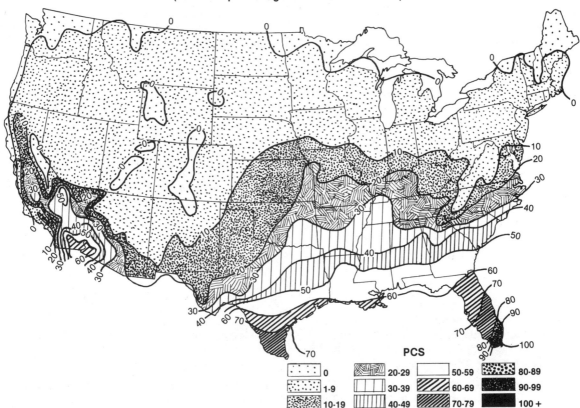

PROPORTIONAL CUMULATIVE STRESS (PCS)
(based on percentage of annual heat stress)

PCS

0	20-29	50-59	80-89
1-9	30-39	60-69	90-99
10-19	40-49	70-79	100 +

Figure I-33: This map expresses discomfort using an index that combines the effects of temperature and humidity. The higher the values on the map, the more uncomfortable a place is during the summer.
Source: Adapted from the Geographical Review, *vol. 57, 1967, with permission of the American Geographical Society, New York.*

the proportional cumulative stress index. This index (see figure I-33) generally reflects the fact that you can be more comfortable at high temperatures with low humidity as, for example, in a place like southwest Texas where the climate is hot, but where the index shows a relatively comfortable zone. The humidity in southern Illinois, however, could be more uncomfortable than it is in southwest Texas.

There are two main influences on humidity related to geography. Proximity to bodies of water has the most influence on humidity levels because they are a source of evaporation and thus increased moisture in the air. In Ojai, California, for example, a dam created a large lake, and now the area's residents complain that the once bearable hot and dry summers have become hot and muggy. Orographic influences refer to the effects of mountain ranges on precipitation and humidity. The east side of a range of mountains or hills is usually drier than the west side.

Another quasi-geographic influence is vegetation. An abundance of flora affects humidity and air temperature due to the continuous transpiration of moisture from ground to air through leaves. Perhaps more importantly, the sprinkler and irrigation systems that are used with residential landscaping and in agriculture may cause a noticeable increase in the humidity of the surrounding areas because, as with the Ojai dam, they introduce large surface areas of evaporating water.

Wind

Wind can dramatically affect both your comfort and your home's energy use. No matter what the season, wind is an agent for cooling. In winter, wind can greatly increase the rate of heat loss from buildings and from people. Summer breezes, on the other hand, can provide beneficial natural ventilation and cooling. Wind is the most site-dependent of all the climatic variables. While there are large-scale atmospheric patterns that affect the weather and local wind speed, local geography and other physical factors (e.g., flora and the built environment) have a major controlling influence on the presence of wind.

In the winter, wind increases building heat loss in two ways: by increasing the infiltration of cold air into the living space and by increasing the flow of cold air over your walls, which literally strips away heat from the house. Wind blowing around your house creates an air pressure difference between your house and the outside. This pressure difference causes the air in your house to flow out through the cracks around windows and doors. Because the warm air is leaving, it sucks cold air in, creating cold drafts that make you feel uncomfortable and increase your heating bill.

Infiltration is also linked to the simple fact that just heating a building causes a pressure difference with higher pressure on the inside than that on the outside. Warm pressurized air wants to leave the building, again to be replaced by cold outside air. In a poorly sealed house, wind is a coconspirator, aiding and abetting infiltration heat loss. Wind also cools your building to a small extent by breaking the film of still air that sits next to the walls. This still-air film has some insulating value and, when it is broken by strong winds, there is a small increase in a building's conductive heat loss.

In the summer wind flowing through buildings and around your body can make you feel more comfortable. For one thing, wind passing over your skin speeds the rate at which your body evaporates water, and this can increase your comfort even on a humid day if the temperature is below 85°F (29°C). Also, with all your windows open wind infiltration is maximized and helps to push collected hot air out of the building.

Daily and Seasonal Variations

The two main components of wind are its speed and direction, both of which can change somewhat through a single day and with the seasons. Most locations, however, do have fairly regular seasonal wind patterns, so if in one location the prevailing winter winds come from the northwest, the summer winds can shift around and come in from the southwest. Changes in average wind speed are also often related to the seasons, with regions such as the East Coast, Midwest, Mountain States and the Southwest experiencing the highest average speeds in winter, while the West Coast generally has stronger winds in summer. You can find out from local weather data if the wind's direction and force will be friend or foe in your plans for energy upgrading.

Land-Sea and Mountain-Valley Winds

The land-sea breeze is a very distinct wind pattern that is driven by the daily temperature differences that develop between the land and the ocean or large lakes. During the day, the land heats up faster than the sea and, as a result, the air over the land is heated and rises. Cool air from the sea is sucked inland to replace the warm rising air. The warm air is pulled toward the ocean where it cools and sinks. This cycle repeats itself throughout the day; as a result, we feel the cool breezes coming in from the ocean.

During the evening, this heating and cooling cycle reverses itself, and the land becomes cooler than the water. The warm air rises off the ocean, which pulls cool air from the land, and the cycle repeats itself. What a nice heat exchange system this all is: Water discharges its "coolth" to the land by day to help temper a hot afternoon, and by night it radiates heat collected during the day back to space.

Mountains and valleys also share distinct patterns that vary throughout the day. During the day there is an upslope wind caused by the heating of sun-exposed slopes and the valley floor. Upslope winds begin at sunrise, reach their peak at midday, and diminish in the afternoon. While they are strongest on south-facing slopes, they also occur to a lesser extent on slopes with other orientations due to valley heating and some solar gain on the slope itself. The direction of the air flow reverses after sunset due to heavier cold air literally draining down the slope and into the valley.

Your location on a slope can make a great deal of difference in the wind patterns at your site. Wind speeds tend to be highest just below or atop the crest of a hill. A pass or low point between two hills can often get higher wind speeds than even the hill crest. The orientation of the slope relative to the wind direction is also important.

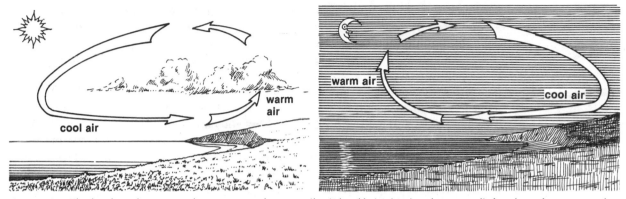

Figure I-34: The land-sea breeze cycles can extend many miles inland bringing in a breezy relief on hot afternoons when the land has reached the peak of its daily heating.

Figure I 35. If you have ever stood at the base of a hill on a warm summer's evening, you might have felt cooler, denser and therefore heavier air literally falling down the hillside. This type of wind is called katabatic. *By day when valley air heats up, you can feel convective upflow of air, called an* anabatic *wind.*

For example, wind can accelerate as it travels up the windward side of a hill, yet on the leeward side the windstream breaks up, which can create a calmer zone. You can sometimes spot these calm areas by looking at the vegetation. Grasses and plants may appear bent over in windy sections, while they are more upright in calm sections.

Backyard, Front Yard, All around the House

For most people "site" really means one or a couple hundred feet this way or that way, which delineates your basic city, town or suburban building lot. There are things all around and within its bounds: trees and other buildings primarily, and hedgerows, fence runs, big and little streets. As has been discussed, these elements may be a boon or a bust to your solar retrofitting plans, and the same is true for the effects they have on the presence of wind at your site.

Once you gain some familiarity with local wind speed and direction patterns, either by studying recorded data and/or by implementing your own on-site "wet-finger" data collection routine, it should be easy to see how nearby flora and the built environment are limiting winds you don't want in winter or are allowing passage for cooling summer winds. You'll also be able to see how you can modify your site to better control winds and breezes. Landscaping and fence placement are respectively long- and short-term options, and your energy upgrading or remodeling or expansion plans for your house should take wind as well as sun into consideration for things like window and door placement or removal. Building height is an important planning consideration, since the effective wind speed can increase from ground level to the top of a two- or three-story building, depending on the height and proximity of surrounding objects. Fences of course wouldn't work way up there, and trees would be a long time in coming (but go ahead and plant them anyway!), so in the upper levels of a house, the locations of openings in the wall should be carefully considered. Fixed glass, for example, might suffice on walls that face strong prevailing winter winds if operable windows or vents can be used elsewhere to allow for adequate ventilation. If prevailing summer winds are shifted from the prevailing winter direction, you can think and plan in terms of taking advantage of their cooling and ventilating potential, again with window and tree placement (for shading). At ground level a strategically placed solid fence run or dense hedgerow could be used to direct winds and breezes, or actually funnel them to the desired location at your building.

Comfort—the Bottom Line

The main reason for understanding a little bit about your climate is to see the strong link that it can have with your comfort. You can manipulate your building to optimize its interaction with beneficial climatic elements to create a more pleasant living environment. And by the same token you can manipulate your four walls to resist more effectively seasonal and climatic conditions that would decrease your comfort.

In one sense your comfort is as individual as the clothes you wear and the foods you enjoy: Comfort is a perceived and physical sense of well-being, a quality that is different for different people. Some people find that a 60°F (16°C) indoor temperature in winter is quite adequate with a sweater and warm slippers, while others literally need to see 70°F (21°C) on their thermostats to feel secure in their comfort. If comfort is compared to a habit, then it can be said that our individual notions of well-being are the sum of many parts or influences: Climate, other people, psychological factors, economic factors—a veritable gamut of inputs has fashioned our personal comfort scales, and as these influences change, so can our attitudes toward comfort. The almighty dollar, for one thing, can work (or wreak) astonishing changes on our habits; witness the effects of

OPEC on our driving and home heating habits. We can also learn a little more about what our thermal comfort really consists of, knowledge that can positively influence change in our comfort habits and also be a help in planning home energy improvements.

Winter comfort is, of course, warmth, and summer comfort is being cool. Both are perceptions of relative temperature difference between the indoors and outdoors, but more precisely between any surrounding air temperature and the way your body feels. Your sense of temperature is governed by three factors, two of which you are already familiar with: air temperature and humidity. The third factor is no less significant than air temperature and is called *mean radiant temperature*. As was discussed in a previous essay, radiation is a form of heat transfer. An object radiates heat if its surface temperature is higher than that of other surfaces and the surrounding air temperature. Since your body is usually warmer than the indoor air temperature, it is radiating heat omnidirectionally throughout the room, and whatever objects are warmer than air temperature are also radiating heat at you. Since your body is receiving heat from a mix of radiators that have varying radiant temperatures, you are experiencing the effects of the room's average or mean radiant temperature.

Your perception of temperature is governed equally by air and mean radiant temperature, with humidity having a lesser influence. The first two factors each account for about 40 percent of your comfort perception, while humidity affects about 20 percent. Wind, by the way, is not generally included as a direct factor in your perception of comfort, although it can certainly play both an actual and a psychological role.

Mean radiant temperature is important to our understanding of comfort because relatively warm radiant environments allow you to tolerate lower air temperatures. For each 1°F (0.56°C) increase in the mean radiant temperature, the air temperature can be reduced 1.4°F (0.79°C), and you will feel equally comfortable. This ties in directly with our use of solar heating systems, since many of them involve heat storage methods that increase the radiant temperatures inside the building. As you read through Section III, "Space Heating," you'll see that much of the emphasis is on using common household elements such as walls, floors and furniture to store and ultimately reradiate heat. Another emphasis is in adding a heat storage component such as a masonry wall or large containers of water directly to the living space to further promote a higher mean radiant temperature.

You can use the climate to keep yourself comfortable simply by opening your building to beneficial climatic elements and excluding those harsh climatic elements. It sounds amazingly simple, and in concept it is. You're working at its simplest level when you open windows when it gets too warm, or close all the windows and doors and put up the storms in cold weather. But to go further, we have to carefully redesign our buildings to take maximum advantage of our climatic resources.

You have started here, by learning about your home's current energy status with regard to its own structure and to the local climate, by learning about the sun in your place. And from here we begin examining the many options for increasing the efficiency of your home's comfort systems.

Bruce Melzer

References

Aronin, Jeffrey. 1953. *Climate and architecture.* Reprint. New York: AMS Press. 1977.

Baldwin, John. 1973. *Climates of the United States.* Washington, D.C.: U.S. Government Printing Office. no. 003-017-00211-0.

Berdahl, Paul; Grether, Donald; Martin, Marlo; and Wahlig, Michael. 1978. *California solar data manual.* Berkeley, Calif.: Lawrence Berkeley Laboratory, University of California.

Fisher, Stephen S. 1954. *Climatic atlas of the United States.* Cambridge, Mass.: Harvard University Press.

Geiger, Rudolf. 1965. *The climate near the ground.* 4th ed. Cambridge, Mass.: Harvard University Press.

Givoni, Baruch. 1976. *Man, climate and architecture.* 2d ed. Philadelphia: International Ideas.

Keyes, John H. 1979. *Consumer handbook of solar energy for the United States and Canada.* Dobbs Ferry, N.Y.: Morgan & Morgan.

Mather, John R. 1974. *Climatology: fundamentals and applications.* New York: McGraw-Hill Book Co.

National Climatic Center
Federal Building
Asheville, NC 28011
This center prepares state-by-state records on temperature, wind, snow, cloud cover, rain and other weather information that are available for a minimal cost.

Officials of the National Oceanic and Atmospheric Administration, U.S. Department of Commerce. 1974. *Climates of the states.* Port Washington, N.Y.: Water Information Center.

Olgyay, Victor. 1963. *Design with climate.* Princeton: Princeton University Press.

United Nations. 1971. *Climate and house design.* vol. 1. New York: Department of Economic and Social Affairs.

U.S. Department of Commerce, Environmental Data Service. 1978. *Climatic atlas of the United States.* Asheville, N.C.: National Climatic Center.

Hardware Focus

Sun Tools

Robert Bennett
6 Snowden Rd.
Bala Cynwyd, PA 19004
(215) 667-7365

Bennett Sun Angle Chart
17 × 22″ charts for designated latitudes that show position of sun for any time of day and any time of year. A good aid for the do-it-yourselfer.

Crystal Productions
107 Pacific Ave.
Aspen, CO 81611
(303) 925-8160

Cushing Instruments
7911 Herschel Ave.
Suite 214
La Jolla, CA 92037
(714) 459-3433

Dodge Products, Inc.
P. O. Box 19781
Houston, TX 77024
(713) 467-6262

Kalwall Corp.
Solar Components Division
P. O. Box 237
Manchester, NH 03105
(603) 668-8186

Lewis & Associates
105 Rockwood Dr.
Grass Valley, CA 95945
(916) 272-2077

Passive Solar Institute
P. O. Box 722
Davis, CA 95616

Solar Energy Corp. (SOLEC)
Dept. DM
Box 3065
Princeton, NJ 08540
(609) 924-1879

Solar Pathways, Inc.
Valley Commercial Plaza
3710 Highway 82
Glenwood Springs, CO 81601
(303) 945-6503

Sun Site
Solar transit designed to determine sun's position, altitude and azimuth relative to site. Laboratory manual available. Educational device.

Solar Radiation Meter
Measures direct and diffuse solar radiation. Records watts/m² and Btu/ft². Indicates incident angles at 10, 20 and 30 degrees. Educational device.

Solar Energy Meter, Model SM-12A
Readout in "percent of full sun." Self-powered by silicon solar cell. Professional instrument.

Portable Solar Meter Model 776
Measures direct plus diffuse solar radiation. Measures langleys/hr, Btuh/ft², mw/cm². Professional instrument.

Sun Scope
Magnetic compass with solar altitude indicator. Determines sun's location within a few degrees for any season. Inexpensive aid for the do-it-yourselfer.

Solar Site Selector
Calculates solar day/hours from sun paths and hour segments; silk-screened on curved transparent grid. Determines shading patterns. Professional instrument. Oriented by compensating compass.

Solar Simulator
For shading, daylighting and solar penetration studies. In kit, ready-made and deluxe models.

Inclinometer
For finding levels and angles. Sensitive to ½ of 1 degree. Place on surface to be checked. Attaches magnetically to square.

Solar Pathfinder: The Energy Evaluator
Provides site analysis including sunrise and sunset times and directions, shading patterns, energy loss consequence of shadows and each month's average daily radiation. Professional instrument. Expensive.

Teamworks Manufacturing Co.
331 Harvard St.
Cambridge, MA 02139
(617) 661-2081

The Sun Machine, Model No. 50-A
Traces path of sun across sky for any latitude, month or hour of day. Measures azimuth and altitude bearings of potential solar obstructions. Professional instrument. Expensive.

Zomeworks Corp.
P. O. Box 712
Albuquerque, NM 87103
(505) 242-5354

Sun-Calculator
Provides site analysis. Indicates sun's path across the sky for 20th of each month, the altitude and azimuth readings for any time of day or year and potential obstructions. Professional instrument.

Energy Mail-Order Catalogs

If local suppliers don't have some of the special solar- or energy-related hardware items you need, the following energy mail-order catalogs can probably deliver the goods. Following the mail-order catalogs is a list of directories of current solar hardware manufacturers that will lead you directly to the source.

Brother Sun
Rte. 6, Box 10A
Santa Fe, NM 87501
(505) 471-1535

Solar Building Materials and Products
Glazing products, glazing accessories, movable window insulation products, SolaRoll components, water conservation devices, greenhouse vent openers, thermal storage, selective surfaces.

Energy House
P. O. Box 5288
Salem, OR 97304
(503) 364-7718

Energy House Catalog
Solar components, stoves, fireplaces, wind-powered hardware, air handling devices, pumps, valves, controls, energy and water conservation devices.

Energy Market Place
Conservation Consumer
19143 15th Ave., N.W.
Seattle, WA 98177

Energy Marketplace: Energy Saving Products for People Who Plan to Stay in the City
Solar appliances, novelties, water conservation devices, solar hot water system, shades, films, woodburning appliances, books.

Energy Shack
2211 S. Dort Highway
P. O. Box 7305
Flint, MI 48507
(313) 235-3553

Energy Shack
Pipe insulation, water-conserving devices, films, shades, fuel saver thermostats, water heater insulation, water heater time controls, devices for redistributing heat.

W. W. Grainger, Inc. (Main Office)
5959 W. Howard St.
Chicago, IL 60648

Grainger's Wholesale Net Price Motorbook
Ventilating fans, blowers, pumps, thermostatic controls. This is a national distributor of a wide variety of equipment useful in solar systems. Check your phone book for an outlet in your area.

Kalwall Corp.
Solar Components Division
P. O. Box 237
Manchester, NH 03105
(603) 668-8186

R. D. Associates
P. O. Box 99
New Rochelle, NY 10804

Solar Usage Now, Inc.
P. O. Box 306
420 E. Tiffin St.
Bascom, OH 44809
(419) 937-2226 or toll-free number:
(800) 537-0985 (except Ohio)

The Warming Trend
Box Q
Dorset, VT 05257
(802) 867-5754

Zomeworks Corp.
P. O. Box 712
Albuquerque, NM 87103
(505) 242-5354

Kalwall Solar Components
Glazing products, sealants, fasteners, absorber plates, coatings, insulation, differential controllers, circulation devices, storage containers, miscellaneous hardware, books.

Energy Savers
Light, water and heat conservation devices, books.

People's Solar Sourcebook
Collectors, controls, tanks, pumps, pipe, insulation, glazing products, foils, frames, fluids, pool equipment, window films, skylights, greenhouse equipment, tools.

*The Warming Trend: Items and Ideas for
   Energy Efficient Comfort*
Three different catalogs per year of over 200 items including solar greenhouse accessories, solid fuel burning accessories, water conserving devices, pipe insulation, air circulation devices, fuel conserving devices.

Nightwall clips, Beadwall panels, Skylid louvers, domestic hot water heaters, sun angle calculators, Exolite glazing; plans for domestic hot water heater, breadbox and pro heater plans, drumwall, Beadwall panels.

Directories of Solar Equipment Manufacturers

Camden House Publishing, Ltd.
Queen Victoria Rd.
Camden East
Ontario, Canada K0K 1J0
(613) 378-6661

Gordon Publications
P. O. Box 2106R
20 Community Place
Morristown, NJ 07960
(201) 267-6040

The Harrowsmith Sourcebook
Lists manufacturers, distributors and agencies in Canada that provide information or products on wood, wind, solar and alternative technologies.

Solar Buyers Guide
Each March issue of *Solar Heating and Cooling* is a comprehensive guide of manufacturers and distributors of complete systems and components. Subscribers receive this issue; can be purchased separately.

Passive Solar Institute
P. O. Box 722
Davis, CA 95616

Solar Age
Church Hill
Harrisville, NH 03450
(603) 827-3347

Solar Energy Publishers, Inc.
2636 Walnut Hill Lane
Suite 257
Dallas, TX 75229
(214) 350-1370

Solar Energy Institute of America
1110 6th St., N.W.
Washington, D.C. 20001
(202) 667-6611

The Second Passive Solar Catalog
Includes manufacturers of and descriptions for over 400 passive solar components and energy conserving products.

The Solar Age Resource Book
Contains hundreds of product listings, a manufacturers' directory and articles on alternative energy.

Solar Products Directory
Included in the April, 1980, issue of *Solar Age*. Includes product listing by category, description of product, manufacturer's name, address and phone number.

Third Annual Deskbook Directory of Solar Product Manufacturers
Included in the December, 1979, issue of *Solar Engineering Magazine*. Includes collectors and collector components, solar systems and solar system components, passive solar systems and components, power generation, solar-related products. Lists name, address and phone number of manufacturers.

The Solar Energy Source Book
Lists manufacturers and their products, architects, engineers, installers, other services, reading lists, educational materials, schools.

Finding Your Heat Loss

A simple calculation method for determining how much energy your home uses for space heating

Sometimes the hardest part of retrofitting a home with a solar heating system is deciding just how much solar heating capability it really needs. One important aid in making that decision is the *energy audit*. An energy audit answers the questions, "How much energy for space heating does my house need every winter, and what is the space heating load?" Assuming that you have lightened the load as much as possible with conservation improvements (see "The Turned-Off House" found earlier in this section), an audit will give you a fix on your home's new energy status and help you set realistic goals for solar improvements. It may also be that you haven't yet gone far enough with conservation, and the audit will tell you that as well.

The most common approach to performing home energy audits is to calculate heat losses through the various surfaces of the home, a method that is used nowadays by numerous public agencies that are performing energy audits. Unfortunately, the method is not always accurate. What is often forgotten is that the methodology for calculating heat losses was developed to fill the need of engineers in the *design* phase of a building. It helps them determine the size of heating, ventilating and air conditioning systems for new buildings. When a building is just lines on a piece of paper, about the only way to get an idea of the building's energy needs is through a mathematical calculation process. Using surface area measurements and R-values associated with those surfaces, plus assumptions about the average temperatures inside and outside, an engineer can estimate heat flows accurately enough to decide whether to include 10 or 20 tons of air conditioner capacity. But there are entirely too many fixed assumptions in that calculation process to make an accurate decision about whether to use, say, a 10-ton or an 11-ton unit. To play it safe, systems for heating and cooling often tend to be oversized for the loads they have to satisfy. In calculating loads for solar improvements, we want to be as close to the mark as possible because nobody wants to go through the hassle of building an undersized system or suffer the needless expense of one that's oversized.

Computerized Energy Audits: Not Always Smart Enough

Another approach to studying heating loads is with computerized "blackbox" audits which, unfortunately, have become a very popular means of determining whether or not a home needs additional conservation effort. These audits are based on some computer model of a "typical" home and a "typical" family. The auditor fills out a form and feeds the information into a computer. The result is a number that is supposed to represent the energy use of the home. *If* all the homes audited had similar construction and *if* all the families living in them were "typical," then the audit system could provide reasonably accurate profiles of energy use. But most homes do not fit the prototype model, which makes standardized audits only academically interesting. They really don't work very well for our undoubtedly atypical homes occupied by atypical people (us) with our mixed bags of life-style and energy use patterns.

One solution is to greatly increase the detail of the information included in the heat loss calculation model. Some more detailed versions do exist. They have enough questions about the home and its occupants to make a census taker blush. It's certainly a challenge—and, for some computer fans, fun—to try to devise calculation models that reflect real life and really do predict performance, but for homeowners not in possession of computer expertise, there is a better way.

Audit Thyself: Let Your Fuel Bills Do the Talking

Why try to predict when you already have real-life historical data about your home's energy use? Associated with living in that home is a long, possibly painful trail of bills, bills, and more bills—not a few of which relate to energy consumption. By properly analyzing those bills, you can accurately determine what your specific home uses in energy. "Wow!" you say, "a great idea. Sounds simple. It'll be so easy to do!" Unfortunately (there is always an "unfortunately" lurking behind every great idea), there are some complications.

The basic problem is that most fuels are used for more than heating. For example, natural gas is also used for cooking and water heating. Electricity is used for everything. Since our analysis is aimed at determining space heating loads, we somehow have to deduct the nonheating usage from our billed totals. Bulk delivery of fuel causes trouble, too. Oil, propane, coal and wood are generally delivered in bulk, and purchases in the summer are common since prices are often reduced. Also, at the end of a year's heating season, the leftovers remain to confuse us next year.

Another problem with determining actual fuel deliveries is that with metered utilities (gas and electricity), bills are often estimated. Meter readers don't like to wade around in snow up to their posteriors any more than you or I. So, winter gas or electric bills may either be duplicates of the previous month or estimated by some arcane formula probably known only to some green-visored gnome in the basement of your local utility company.

Keeping track of all these different variables can be pretty confusing. To help, here is a simplified fill-in-the-blanks-and-do-just-what-it-says form that will lead you to the conclusion—the Home Heating Index. The Home Heating Index is a number in Btu per degree-day per square foot of heated floor area $(Btu/DD/ft^2)$. That is the heating efficiency index for your home.

Before you start, one more thing: You will have to have a record of all the fuel used for at least one year; the more years you can study, the better will be the accuracy of the result. Naturally, your bills highlight the dollars and cents column, but for our purposes we also need a record of how many energy units were bought in gallons, therms, cubic feet, kilowatt-hours (KWH), cords or tons. If you are like me, reinsulating the whole house would probably be easier than finding bills that are over a month old. In that case a call, letter or trip to your fuel supplier should get you the numbers you need. They might even give you heating degree-day data for the years you're analyzing. If not, the nearest United States Weather Bureau office will have those numbers. It might be interesting to compare with the degree-day averages listed in table I-8 of the following essay, "Finding Your Heat Gain."

There's a fair amount of work involved here, so give yourself the space of a couple hours to get the job done. In case you're wondering what lies ahead, you're not getting into anything more complicated than addition, subtraction, multiplication and division. So, pencils sharpened? Relaxed? Calculator batteries at full charge? Let's go!

Select a 12-month period (hereinafter referred to as a year!) starting with June. To repeat, the more years you can go back, postconservation, the more accurate will be the determination of your average use because you can repeat the calculation for different years,. Then, obtain fuel use and weather data for that period. The weather data needed for all fuels are heating degree-days (HDD) and for electricity, if it is used for both heating and cooling, you need both HDD and cooling degree-days (CDD). Additionally, if you heat with either bulk-delivered propane or fuel oil, you will need to know fuel usage and weather data from six months before to six months after the year you're studying. The following worksheets (I-1 through I-5) are devoted to the Btu per HDD computation for natural gas, fuel oil, propane, electricity and wood and coal. Fuel oil worksheet I-2 is a little trickier than some of the others, so we've included a sample calculation there to show you how they're all done.

Worksheet I-1: Natural Gas _____

Step 1: Compile energy use and weather data.

Fill in the following table with your data. (See the box in worksheet I-2 for guidance on what to do.) ***Confusion Factor*** Your bills may show the units you buy in therms (100,000 Btu), MBtu (1000 Btu), Mcf (1000 cubic feet), ccf (100 cubic feet), scf (standard cubic feet) or simply "units." Unless the units billed are clearly identified, a call to the gas company for clarification is in order.

Month	Year	Units Used	Heating Degree-Days (HDD)
June	19____	_____	_____
July		_____	_____
August		_____	_____

Month	Year	Units Used	Heating Degree Days (HDD)
September		_____	_____
October		_____	_____
November		_____	_____
December		_____	_____
January	19___	_____	_____
February		_____	_____
March		_____	_____
April		_____	_____
May		_____	_____
		_____ **(1)** total units used	_____ **(2)** total HDD

Step 2: Calculate the deduction for nonheating use of natural gas.

If you use natural gas *only* for space heating, proceed to *Step 3*. If you use natural gas for any purpose other than space heating, complete this part.

First, add "Units Used" for June, July and August:

$$\frac{\quad}{June} + \frac{\quad}{July} + \frac{\quad}{August} = \frac{\quad}{3 \text{ months of nonheating load}} \textbf{(3)}.$$

Multiply the 3-month figure times 4 to get the 12-month estimated nonheating load:

$$\frac{\quad}{3 \text{ months of nonheating load}} \textbf{(3)} \times 4 = \frac{\quad}{\text{annual estimated nonheating use}} \textbf{(4)}.$$

Step 3: Compute the portion of natural gas usage directed only to heating.

Subtract the annual estimated nonheating use**(4)** from the total units used**(1)** to get the total annual heating units**(5)**:

$$\frac{\quad}{\text{total units used}} \textbf{(1)} - \frac{\quad}{\text{annual estimated nonheating use}} \textbf{(4)} = \frac{\quad}{\text{total annual heating units}} \textbf{(5)}.$$

Step 4: Convert units to Btu.
Fill in the line in the following table which corresponds to the units used for your bills:

total annual heating units**(5)** $\times$ Btu/unit = Btu consumed for heating **(6)**

$$\underline{\text{therms}} \textbf{ (1) or (5)} \times \frac{100,000}{\text{Btu/therm}} = \frac{\quad}{\text{total heating Btu}} \textbf{(6)}$$

$$\underline{\text{MBtu}} \ \textbf{(1) or (5)} \times \ \frac{100,000}{\text{Btu/MBtu}} \ = \underline{\hspace{2cm}} \ \textbf{(6)}$$
$$\text{total heating Btu}$$

$$\underline{\text{Mcf}} \ \textbf{(1) or (5)} \times 1,035,000 = \underline{\hspace{2cm}} \ \textbf{(6)}$$
$$\text{Btu/Mcf} \quad \text{total heating Btu}$$

$$\underline{\text{ccf}} \ \textbf{(1) or (5)} \times \ \frac{103,500}{\text{Btu/ccf}} \ = \underline{\hspace{2cm}} \ \textbf{(6)}$$
$$\text{total heating Btu}$$

$$\underline{\text{scf}} \ \textbf{(1) or (5)} \times \ \frac{1,035}{\text{Btu/scf}} \ = \underline{\hspace{2cm}} \ \textbf{(6)}.$$
$$\text{total heating Btu}$$

Step 5: Compensate for system efficiency.
If your heating system is in poor shape (get it tuned up!), multiply your total heating Btu **(6)** times 0.5, your system efficiency factor:

$$\underline{\hspace{2cm}} \ \textbf{(6)} \times 0.5 = \underline{\hspace{3cm}} \ \textbf{(7)}.$$
$$\text{total heating Btu} \qquad\qquad \text{Btu delivered for heating}$$

If your heating system is about average, multiply your total heating Btu**(6)** times 0.7, your system efficiency factor:

$$\underline{\hspace{2cm}} \ \textbf{(6)} \times 0.7 = \underline{\hspace{3cm}} \ \textbf{(7)}.$$
$$\text{total heating Btu} \qquad\qquad \text{Btu delivered for heating}$$

If you have a new, high-efficiency heating system with an automatic pilot and a stack damper, multiply your total heating Btu**(6)** times 0.8, your system efficiency factor:

$$\underline{\hspace{2cm}} \ \textbf{(6)} \times 0.8 = \underline{\hspace{3cm}} \ \textbf{(7)}.$$
$$\text{total heating Btu} \qquad\qquad \text{Btu delivered for heating}$$

Step 6: Calculate Btu/HDD.

Divide the Btu delivered for heating**(7)** by the total HDD**(2)**:

$$\underline{\hspace{3cm}} \ \textbf{(7)} \div \underline{\hspace{1.5cm}} \ \textbf{(2)} = \underline{\hspace{1.5cm}} \ \textbf{(8)}.$$
$$\text{Btu delivered for heating} \qquad \text{total HDD} \qquad \text{Btu/HDD}$$

Step 7: Complete your audit.
    If gas is your only heating fuel, go on to "The Last Step: Finding Your Heat Loss," found later in this essay. If you have another heating fuel, such as electricity or wood, turn to the corresponding worksheet and complete those calculations before proceeding to the last step.

Worksheet I-2: Fuel Oil _____

Step 1: Compile energy use and weather data.
    Fill in the following table. (Use the box on page 103 as a guide.)

Because fuel oil is delivered in bulk, you have to start the usage calculation with a full tank and end with a full tank (just like figuring miles-per-gallon for a car). Prior to the June starting month of your sample 12-month period, when was the tank *last* filled? On the table fill in the date of delivery and the HDD for the entire month. After May, the last month of the sample period, when was the tank next filled? On the table fill in the date, the number of gallons and the HDD for the entire month. Fill in the information for all deliveries in between those dates (whether the tank was completely filled or not), noting dates, gallons and HDD for every month.

Month	Year	Date of Delivery	Gallons	Heating Degree-Days (HDD)
List month of last fill-up before June.				
_____	19___	_____	X	_____
List all months and HDD between last fill-up and June.				
_____		X	X	_____
_____		X	X	_____
_____		X	X	_____
June		_____	_____	_____
July		_____	_____	_____
August		_____	_____	_____
September		_____	_____	_____
October		_____	_____	_____
November		_____	_____	_____
December		_____	_____	_____
January	19___	_____	_____	_____
February		_____	_____	_____
March		_____	_____	_____
April		_____	_____	_____
May		_____	_____	_____
List month of first fill-up after May.				
_____		_____	_____	_____
List all months and HDD between May and first fill-up.				
_____		X	X	_____
_____		X	X	_____
			_____ (1) total gallons	_____ (2) total HDD

Total the gallons entered for this period. In this case the period includes both the whole and fractional months between the last fill-up before June and the first fill-up after May.

Total the HDD's for the whole months of the period; then add the HDD amounts for the fractional months. The fractional months, if any, will occur on the first and last months of the period. For example, if the last fill-up before June was made on May 20, then one-third of the HDD for that month would be added in since only one-third of the month is in that period. If the first fill-up after May was on June 20, then two-thirds of the HDD for that month would be added since two-thirds of the month is in the period.

Step 2: Calculate the deduction for nonheating use of oil.

If you use oil *only* for space heating, proceed to *Step 3*. If you use oil for heating your domestic water (summer-winter hookup), complete this part.

First, figure the number of days between the first and the last deliveries of the period:

$$\frac{\qquad\qquad}{\text{number of days}}\ \textbf{(3)}.$$

Multiply:

$$\frac{\qquad\qquad}{\text{number of people in household}}\times 10{,}000\ \text{Btu/person/day} = \frac{\qquad\qquad}{\text{Btu/family/day}}\ .$$

Then add 5000 Btu per day to account for system inefficiency:

$$+\ 5000\ \text{Btu/day} = \frac{\qquad\qquad}{\text{Btu/day for water heating load}}\ \textbf{(4)}.$$

Using the total number of days from the first to the last delivery(**3**), compute the total water heating Btu load for the period by multiplying:

$$\frac{\qquad}{\text{number of days}}\ \textbf{(3)}\times\frac{\qquad\qquad}{\text{Btu/day for water heating load}}\ \textbf{(4)} = \frac{\qquad\qquad}{\text{total water heating Btu}}\ \textbf{(5)}.$$

If you use a dishwasher, take long showers, or generally use lots of hot water, multiply:

$$\frac{\qquad\qquad}{\text{total water heating Btu}}\ \textbf{(5)}\times 1.25 = \frac{\qquad\qquad}{\text{nonheating use Btu}}\ \textbf{(6)}.$$

Or, if you consider your household pretty much average in its hot water usage, use **(5)** for **(6)**.

Or, if you are very careful about hot water use, have restrictor shower heads, use cold water in the laundry, etc., multiply:

$$\frac{\qquad\qquad}{\text{total water heating Btu}}\ \textbf{(5)}\times 0.8 = \frac{\qquad\qquad}{\text{nonheating Btu.}}\ \textbf{(6)}.$$

Step 3: Compute the Btu delivered only for heating.

Multiply:

$$\underline{\hspace{2cm}} \text{ (1)} \times \frac{138{,}700}{\text{Btu/gal}} = \underline{\hspace{2cm}} \text{ (7)}.$$
$$\small\text{total gallons}\hspace{5.5cm}\text{total Btu}$$

If water heating is part of oil use, subtract nonheating Btu(**6**) from total Btu(**7**):

$$\underline{\hspace{2cm}} \text{ (7)} - \underline{\hspace{2cm}} \text{ (6)} = \underline{\hspace{2.5cm}} \text{ (8)}.$$
$$\small\text{total Btu}\hspace{1.8cm}\text{nonheating Btu}\hspace{1.8cm}\text{total heating Btu}$$

Step 4: Compensate for system efficiency.
If your oil burning system is old and not up to snuff (snuff it up!), multiply your total heating Btu (**8**) times 0.5, your system efficiency factor:

$$\underline{\hspace{2.5cm}} \text{ (8)} \times 0.5 = \underline{\hspace{3cm}} \text{ (9)}.$$
$$\small\text{total heating Btu}\hspace{3cm}\text{Btu delivered for heating}$$

If your oil burning system is not new, but is in good shape, multiply your total heating Btu(**8**) times 0.6, your system efficiency factor:

$$\underline{\hspace{2.5cm}} \text{ (8)} \times 0.6 = \underline{\hspace{3cm}} \text{ (9)}.$$
$$\small\text{total heating Btu}\hspace{3cm}\text{Btu delivered for heating}$$

If your system is new and properly installed and has a stack damper, multiply your total heating Btu(**8**) times 0.7, your system efficiency factor:

$$\underline{\hspace{2.5cm}} \text{ (8)} \times 0.7 = \underline{\hspace{3cm}} \text{ (9)}.$$
$$\small\text{total heating Btu}\hspace{3cm}\text{Btu delivered for heating}$$

Step 5: Calculate Btu/HDD.

Divide the Btu delivered for heating(**9**) by the total HDD(**2**):

$$\underline{\hspace{3cm}} \text{ (9)} \div \underline{\hspace{2cm}} \text{ (2)} = \underline{\hspace{2cm}} \text{ (10)}.$$
$$\small\text{Btu delivered for heating}\hspace{1.8cm}\text{total HDD}\hspace{1.8cm}\text{Btu/HDD}$$

Step 6: Complete your audit.
 If oil is your only heating fuel, go on to "The Last Step: Finding Your Heat Loss," found later in this essay. If you have another heating fuel, such as electricity or wood, turn to the corresponding worksheet and complete those calculations before proceeding to the last step.

How It's Done: A Sample Worksheet_____

For worksheet I-2 we've filled in all the blanks and done all the calculations to show you how it's done with the fuel oil calculation. When you are ready to do your own, no matter which fuel you use, it will probably be a good idea to make photocopies of the blank worksheets.

Step 1: Compile energy use and weather data.

Month	Year	Date of Delivery	Gallons	Heating Degree-Days (HDD)
List month of last fill-up before June.				
March	1980	3/15	X	406
List all months and HDD between last fill-up and June.				
April		X	X	205
May		X	X	164
June		X	X	25
July		7/20	150	0
August		X	X	9
September		X	X	89
October		10/20	100	379
November		11/20	100	740
December		12/20	200	1076
January	1981	1/20	200	1132
February		2/5	200	1120
March		3/20	200	406
April		X	X	205
May		X	X	164
List month of first fill-up after May.				
August		8/31	150	9
List all months and HDD between May and first fill-up.				
June		X	X	25
July		X	X	0

(continued on next page)

Total the gallons entered for this period. In this case the period includes both the whole and fractional months between the last fill-up before June and the first fill-up after May.

Total the HDD's for the whole months of the period; then add the HDD amounts for the fractional months. In this case one-half of March's HDD ($406 \times 0.5 = 203$) and all of August's HDD are added:

$$\underbrace{1300}_{\text{total gallons}} \text{ (1)} \quad \underbrace{5951}_{\text{total HDD}} \text{ (2).}$$

Step 2: Calculate the deduction for nonheating use of oil.
For our example, oil *is* used for domestic water heating.

First, figure the number of days between the first and the last deliveries of the period:

$$\underbrace{384}_{\text{number of days}} \text{ (3).}$$

The number of people in this household is five; therefore multiply:

$$\underbrace{5}_{\text{number of people in household}} \times 10,000 \text{ Btu/person/day} = \underbrace{50,000}_{\text{Btu/family/day}} .$$

Add 5000 Btu/day to account for system inefficiency:

$$5000 + 50,000 = \underbrace{55,000}_{\text{Btu/day for water heating load}} \text{ (4).}$$

Using the total number of days from the first to the last delivery(**3**), compute the total water heating Btu load for the period by multiplying:

$$\underbrace{384}_{\text{number of days}} \text{ (3) } \times \underbrace{55,000}_{\text{Btu/day for water heating load}} \text{ (4) } = \underbrace{21,120,000}_{\text{total water heating Btu}} \text{ (5).}$$

This family is very careful about their hot water use, have restrictor shower heads, use cold water in the laundry, etc.; thus:

$$\underbrace{21,120,000}_{\text{total water heating Btu}} \text{ (5) } \times 0.8 = \underbrace{16,896,000}_{\text{nonheating use Btu}} \text{ (6).}$$

Step 3: Compute the Btu delivered only for heating.

Multiply:

$$\underbrace{1300}_{\text{total gallons}} \text{ (1) } \times 138,700 \text{ Btu/gal} = \underbrace{180,310,000}_{\text{total btu}} \text{ (7).}$$

Since water heating is part of oil use, subtract nonheating Btu(**6**) from total Btu (**7**):

$$\frac{180,310,000}{\text{total Btu}}\ (\textbf{7})\ -\ \frac{16,896,000}{\text{nonheating Btu}}\ (\textbf{6})\ =\ \frac{163,414,000}{\text{total heating Btu}}\ (\textbf{8}).$$

Step 4: Compensate for system efficiency.
This oil burning system is not new, but is in good shape; therefore, the system efficiency factor is 0.6.

Multiply:

$$\frac{163,414,000}{\text{total heating Btu}}\ (\textbf{8})\ \times\ 0.6\ =\ \frac{98,048,400}{\text{Btu delivered for heating}}\ (\textbf{9}).$$

Step 5: Calculate Btu/HDD.

Divide the Btu delivered for heating(**9**) by the total HDD(**2**):

$$\frac{98,048,400}{\text{Btu delivered for heating}}\ (\textbf{9})\ \div\ \frac{5951}{\text{total HDD}}\ (\textbf{2})\ =\ \frac{16,476}{\text{Btu/HDD}}\ (\textbf{10}).$$

If oil was your only heating fuel, you would follow this calculation to its end by finding the Home Heating Index (HHI).

$$16,476\ (\textbf{10})\ -\ \frac{1600}{\text{ft}^2\ \text{of heated floor area}}\ =\ \frac{10.3}{\text{HHI}}$$

This home probably has some insulation in the attic and perhaps some in the vertical walls, but it probably isn't as tight as it could be and could benefit from more insulation and some kind of solar space heating retrofit.

Worksheet I-3: Propane

Situation A: Your propane is metered.

Step 1: Compile energy use and weather data.
Fill in the following table with your data. (See the box in worksheet I-2 for guidance on what to do.)***Confusion Factor***As with natural gas your bills may show usage units in therms (100,000 Btu), MBtu (1000 Btu), Mcf (1000 cubic feet), ccf (100 cubic feet), scf (standard cubic feet) or simply "units." Unless the units billed are clearly identified, a call to the gas company for clarification is in order.

Month	Year	Units Used	Heating Degree Days (HDD)
June	19____	_____	_____
July		_____	_____
August		_____	_____
September		_____	_____
October		_____	_____
November		_____	_____
December		_____	_____
January	19____	_____	_____
February		_____	_____
March		_____	_____
April		_____	_____
May		_____	_____
		_____ (1) total units used	_____ (2) total HDD

Step 2: Calculate the deduction for nonheating use of metered propane.

If you use metered propane *only* for space heating, proceed to *Step 3*. If you use metered propane for any purpose other than space heating, complete this part.

First, add "Units Used" for June, July, and August:

$$\frac{\quad}{\text{June}} + \frac{\quad}{\text{July}} + \frac{\quad}{\text{August}} = \frac{\quad}{\text{3 months of nonheating load}} \text{ (3).}$$

Multiply the 3-month figure times 4 to get the 12-month estimated nonheating load:

$$\frac{\quad}{\text{3 months of nonheating load}} \text{ (3)} \times 4 = \frac{\quad}{\text{annual estimated nonheating use}} \text{ (4).}$$

Step 3: Compute the portion of metered propane usage directed only to heating.

Subtract the annual estimated nonheating use(**4**) from the total units used(**1**) to get the total annual heating units(**5**):

$$\frac{\quad}{\text{total units used}} \text{ (1)} - \frac{\quad}{\text{annual estimated nonheating use}} \text{ (4)} = \frac{\quad}{\text{total annual heating units}} \text{ (5).}$$

Step 4: Convert units to Btu.
Fill in the line in the following table which corresponds to the units used for your bills:

total annual heating units (**1**) or (**5**) $\times$ Btu/unit = Btu consumed for heating (**6**)

$$\text{therms (1) or (5)} \times \underset{\text{Btu/therm}}{100,000} = \underset{\text{total heating Btu}}{\underline{\qquad\qquad}} \text{ (6)}$$

$$\text{MBtu (1) or (5)} \times \underset{\text{Btu/MBtu}}{1000} = \underset{\text{total heating Btu}}{\underline{\hspace{3cm}}} \textbf{(6)}$$

$$\text{Mcf (1) or (5)} \times \underset{\text{Btu/Mcf}}{2,500,000} = \underset{\text{total heating Btu}}{\underline{\hspace{3cm}}} \textbf{(6)}$$

$$\text{ccf (1) or (5)} \times \underset{\text{Btu/ccf}}{250,000} = \underset{\text{total heating Btu}}{\underline{\hspace{3cm}}} \textbf{(6)}$$

$$\text{scf (1) or (5)} \times \underset{\text{Btu/scf}}{2500} = \underset{\text{total heating Btu}}{\underline{\hspace{3cm}}} \textbf{(6)}.$$

Step 5: Compensate for system efficiency.
If your heating system is in poor shape, multiply your total heating Btu**(6)** times 0.6, your system efficiency factor:

$$\underset{\text{total heating Btu}}{\underline{\hspace{3cm}}} \textbf{(6)} \times 0.6 = \underset{\text{Btu delivered for heating}}{\underline{\hspace{3cm}}} \textbf{(7)}.$$

If your heating system is about average, multiply your total heating Btu**(6)** times 0.7, your system efficiency factor:

$$\underset{\text{total heating Btu}}{\underline{\hspace{3cm}}} \textbf{(6)} \times 0.7 = \underset{\text{Btu delivered for heating}}{\underline{\hspace{3cm}}} \textbf{(7)}.$$

If your heating system is a new, high-efficiency unit with an automatic pilot and a stack damper, multiply your total heating Btu**(6)** times 0.8, your system efficiency factor:

$$\underset{\text{total heating Btu}}{\underline{\hspace{3cm}}} \textbf{(6)} \times 0.8 = \underset{\text{Btu delivered for heating.}}{\underline{\hspace{3cm}}} \textbf{(7)}.$$

Step 6: Calculate Btu/HDD.

Divide the Btu delivered for heating**(7)** by the total HDD**(2)**:

$$\underset{\text{Btu delivered for heating}}{\underline{\hspace{3cm}}} \textbf{(7)} \div \underset{\text{total HDD}}{\underline{\hspace{2cm}}} \textbf{(2)} = \underset{\text{Btu/HDD for metered propane}}{\underline{\hspace{3cm}}} \textbf{(8)}.$$

Step 7: Complete your audit.
        If metered propane is your only heating fuel, go on to "The Last Step: Finding Your Heat Loss," found later in this essay. If you have another heating fuel, such as electricity or wood, turn to the corresponding worksheet and complete those calculations before proceeding to the last step.

Situation B: Your propane is bulk delivered by the pound or by the gallon.

Step 1: Compile energy use and weather data for bulk-delivered propane.
        Fill in the following table with your data. (See the box in worksheet I-2 for guidance on what to do.) Because of bulk delivery, you have to start the usage calculation with a full tank and end with a full tank (just like figuring miles-per-gallon for a car). Prior to the June starting month of your sample 12-month period, when was the tank *last* filled? On the table fill in the date of

delivery and the HDD for the entire month. After May, the last month of the sample period, when was the tank next filled? On the table fill in the date, the number of units (gallons or pounds) and the HDD for the entire month. Fill in the information for all deliveries in between those dates (whether the tank was completely filled or not), noting dates, units and HDD for every month.

Month	Year	Date of Delivery	Units Used (gal or lbs)	Heating Degree Days (HDD)
List month of last fill-up before June.				
_____ 19__		_____	_____	_____
List all months and HDD between last fill-up and June.				
_____		___ X ___	___ X ___	_____
_____		___ X ___	___ X ___	_____
_____		___ X ___	___ X ___	_____
June		_____	_____	_____
July		_____	_____	_____
August		_____	_____	_____
September		_____	_____	_____
October		_____	_____	_____
November		_____	_____	_____
December		_____	_____	_____
January	19__	_____	_____	_____
February		_____	_____	_____
March		_____	_____	_____
April		_____	_____	_____
May		_____	_____	_____
List month of first fill-up after May.				
_____		_____	_____	_____
List all months and HDD between May and first fill-up.				
_____		___ X ___	___ X ___	_____
_____		___ X ___	___ X ___	_____
			_____ (1) total units used (gal or lbs)	_____ (2) total HDD

Total the gallons or pounds listed over the period. In this case the period includes both the whole and fractional months between the last fill-up before June and the first fill-up after May.

Total the HDD's for the whole months of the period; then add the fractional months' HDD amounts. The fractional months, if any, will occur on the first and last months of the period. For example, if the last fill-up before June was made on May 20, then one-third of the HDD for that month would be added in since only one-third of the month is in that period. If the last fill-up after May was on June 20, then two-thirds of the HDD for that month would be added since two-thirds of the month is in the period.

Step 2: Calculate the deduction for nonheating use of bulk-delivered propane.

If you use bulk-delivered propane *only* for space heating, proceed to *Step 3.* If you use bulk-delivered propane for any purpose other than space heating, complete this part.

First, figure the number of days between the first and the last deliveries of the period:

$$\frac{\qquad\qquad}{\text{number of days}} \text{ (3)}.$$

Multiply:

$$\frac{\qquad\qquad}{\text{number of people in household}} \times 10{,}000 \text{ Btu/person/day} = \frac{\qquad\qquad}{\text{Btu/family/day}}.$$

Then add 5000 Btu/day to account for system inefficiency:

$$+ \ 5000 \text{ Btu/day} = \frac{\qquad\qquad}{\text{Btu/day for water heating load}} \text{ (4)}.$$

If you use a dishwasher, take long showers, or generally use lots of hot water, multiply:

$$\frac{\qquad\qquad}{\text{Btu/day for water heating load}} \text{ (4)} \times 1.25 = \frac{\qquad\qquad}{\text{nonheating use Btu}} \text{ (5)}.$$

If you consider your family pretty much average in its hot water use, use **(4)** for **(5)**.

If you are very careful about hot water use, have restrictor shower heads, use cold water in the laundry, etc., multiply:

$$\frac{\qquad\qquad}{\text{total water heating Btu}} \text{ (4)} \times 0.8 = \frac{\qquad\qquad}{\text{nonheating Btu}} \text{ (5)}.$$

If you use propane for cooking or for drying clothes, calculate daily total for nonheating use:

$$\frac{\qquad\qquad}{\text{nonheating Btu/day}} \text{ (5)} + \underset{\text{(if used for cooking)}}{19{,}000/\text{Btu/day}} + \underset{\text{(if used for drying clothes)}}{10{,}500 \text{ Btu/day}}$$

$$= \frac{\qquad\qquad}{\text{total nonheating Btu/day}} \text{ (6)}.$$

Using the total number of days from the first to the last delivery(**3**), compute the total nonheating use of bulk-delivered propane for the period by multiplying:

$$\underline{\hspace{3cm}} \text{ (3)} \times \underline{\hspace{3cm}} \text{ (6)} = \underline{\hspace{3cm}} \text{ (7)}.$$

number of days   Btu/day nonheating load   total nonheating Btu

Step 3: Compute the portion of bulk-delivered propane usage directed only to heating.

Multiply:

$$\underline{\hspace{2.5cm}} \text{ (1)} \times \underline{\hspace{4cm}} = \underline{\hspace{1.5cm}} \text{ (8)}.$$

total units used   Btu/unit (gal = 91,500 Btu;   total Btu
                          pound = 21,630 Btu)

If bulk-delivered propane is also used for nonheating purposes, subtract nonheating Btu(**7**) from total Btu(**8**):

$$\underline{\hspace{1.5cm}} \text{ (8)} - \underline{\hspace{2.5cm}} \text{ (7)} = \underline{\hspace{3cm}} \text{ (8)}.$$

total Btu   nonheating Btu   total heating Btu

Step 4: Compensate for system efficiency.
If your propane system is old and not up to snuff, multiply your total heating Btu(**8**) times 0.5, your system efficiency factor:

$$\underline{\hspace{2.5cm}} \text{ (8)} \times 0.5 = \underline{\hspace{3.5cm}} \text{ (9)}.$$

total heating Btu   Btu delivered for heating

If your propane system is not new but is in good shape, multiply your total heating Btu(**8**) times 0.7, your system efficiency factor:

$$\underline{\hspace{2.5cm}} \text{ (8)} \times 0.7 = \underline{\hspace{3.5cm}} \text{ (9)}.$$

total heating Btu   Btu delivered for heating

If you have a new, high-quality system with an automatic pilot and a stack damper, multiply your total heating Btu(**8**) times 0.8, your system efficiency factor:

$$\underline{\hspace{2.5cm}} \text{ (8)} \times 0.8 = \underline{\hspace{3.5cm}} \text{ (9)}.$$

total heating Btu   Btu delivered for heating

Step 5: Calculate Btu/HDD.

Divide the Btu delivered for heating by the total HDD(**2**):

$$\underline{\hspace{3.5cm}} \text{ (9)} \div \underline{\hspace{2cm}} \text{ (2)} = \underline{\hspace{5cm}} \text{ (10)}.$$

Btu delivered for heating   total HDD   Btu/HDD for bulk-delivered propane

Step 6: Complete your audit.
       If bulk-delivered propane is your only heating fuel, go on to "The Last Step: Finding Your Heat Loss," found later in this essay. If you have another heating fuel, such as electricity or wood, turn to the corresponding worksheet and complete those calculations before proceeding to the last step.

Worksheet I-4: Electricity _____

Step 1: Compile energy use and weather data.

Fill in the following table with your data. (See the box in worksheet I-2 for guidance on what to do.) If you use electricity for cooling—that includes attic fans or window fans that run over 12 hours per day as well as air conditioning, you will also need cooling degree-days (CDD).

Month	Year	KWH Used	Heating Degree-Days (HDD)	Cooling Degree-Days (CDD)
June	19__	_____	_____	_____
July		_____	_____	_____
August		_____	_____	_____
September		_____	_____	_____
October		_____	_____	_____
November		_____	_____	_____
December		_____	_____	_____
January	19__	_____	_____	_____
February		_____	_____	_____
March		_____	_____	_____
April		_____	_____	_____
May		_____	_____	_____
		_____ (1) total KWH used	_____ (2) total HDD	_____ (3) total CDD

Step 2: Calculate the deduction for nonheating KWH usage.

If you do *not* use electricity for cooling, add the "KWH Used" for June, July and August:

$$\frac{}{\text{June}} + \frac{}{\text{July}} + \frac{}{\text{August}} = \frac{}{\text{3 months of nonheating load}} \textbf{(4)}.$$

Multiply the 3-month figure times 4 to get the 12-month estimated nonheating load:

$$\frac{}{\text{3-months of nonheating load}} \textbf{(4)} \times 4 = \frac{}{\text{annual estimated nonheating use}} \textbf{(5)}.$$

To get the total annual use of electricity for space heating when electrical cooling is not used, subtract your nonheating use**(5)** from your total KWH use**(1)**:

$$\frac{}{\text{total KWH use}} \textbf{(1)} - \frac{}{\text{annual estimated nonheating use}} \textbf{(5)} = \frac{}{\text{total annual heating KWH}} \textbf{(6)}.$$

Skip to *Step 4* and enter **(6)** in the blank marked **(13)**.

If you *do* use electricity for cooling as well as heating, examine your energy use table and find those months which have small values for HDD and CDD. Usually they'll be found in the late spring and early fall. "Small" HDD and CDD values mean numbers that are less than 10 percent of the DD value for the largest month. For example, if January is your major heating month with 1500 HDD, then a month with just 150 HDD or less is "small." These "model" months will be different depending on your climate. If you live in a 2000–4000 HDD climate, look at April, May, September and October as possibly meeting the above criterion. If your climate has 4000–6000 HDD, look at April, May, August and September; and if you're up north in 6000–9000 HDD, May, June, July and August may be your model months. After you pick two or three months that fit the bill (small values for both HDD and CDD), total the KWH for those months and divide by the number of months you used. That gives the average nonheating and noncooling use per month. For example, if the loads for April, May and September were 353, 294 and 331 KWH respectively, then the average use would be:

$$353 + 294 + 331 \div 3 = 326 \text{ KWH.}$$

Months chosen are _____ .
The total KWH usage for those months is _____ **(7)**.
The number of months totaled is _____ **(8)**.

Compute the average usage per month**(9)**:

$$\frac{\rule{3cm}{0.4pt}}{\text{total KWH usage}}\text{ (7)} \div \frac{\rule{3cm}{0.4pt}}{\text{number of months totaled}}\text{ (8)}$$

$$= \frac{\rule{5cm}{0.4pt}}{\text{average nonheating or cooling usage per month}}\text{ (9).}$$

Step 3: Compute the portion of KWH usage directed only to heating.
Total the KWH usage for those heating season months which each have HDD that are over 10 percent of the month with the highest HDD. From the previous example using January (1500 HDD), any month with over 150 HDD is used for finding **(10)**.

The total KWH for those months is $\dfrac{\rule{4cm}{0.4pt}}{\text{KWH used during heating months}}$ **(10)**.

The number of months totaled to get **(10)** is $\dfrac{\rule{4cm}{0.4pt}}{\text{number of months with over 150 HDD}}$ **(11)**.

The total in **(10)** represents the electricity used for heating and noncooling purposes for that year. To compute the part of that total that went for nonheating use, multiply:

$$\frac{\rule{4cm}{0.4pt}}{\text{number of months with over 150 HDD}}\text{ (11)} \times \frac{\rule{4cm}{0.4pt}}{\text{KWH used during heating months}}\text{ (10)}$$

$$= \frac{\rule{4cm}{0.4pt}}{\text{nonheating usage during heating season}}\text{ (12).}$$

The number for **(12)** must be subtracted from the total electricity usage during the heating season to get the figure for electricity use for heating only:

$$\frac{\rule{4cm}{0.4pt}}{\text{KWH used during heating months}}\text{ (10)} - \frac{\rule{4cm}{0.4pt}}{\text{nonheating usage during heating season}}\text{ (12)}$$

$$= \frac{\rule{3cm}{0.4pt}}{\text{total annual heating KWH}}\text{ (13).}$$

Step 4: Convert KWH to Btu.

If you do *not* heat with an electric heat pump, to convert to Btu, multiply:

$$\underset{\text{total annual heating KWH}}{\underline{\qquad\qquad}}\textbf{(13)} \times \underset{\text{Btu/KWH}}{3414} = \underset{\text{Btu delivered for heating}}{\underline{\qquad\qquad}}\textbf{(14)}.$$

If you *do* use an electric heat pump for heating, you have to allow for the increased efficiency of electrical usage, which is affected by outside temperatures.

Situation A: You live in a winter climate where temperatures commonly fall below zero.

To convert KWH to Btu, multiply:

$$\underset{\text{total annual heating KWH}}{\underline{\qquad\qquad}}\textbf{(13)} \times \underset{\text{Btu/KWH}}{3414} \times \frac{2.0}{\text{seasonal coefficient of performance (COP)}}$$

$$= \underset{\text{Btu delivered for heating}}{\underline{\qquad\qquad}}\textbf{(14)}.$$

Situation B: You live in a climate where most of the winter temperatures do not fall below zero.

To convert KWH to Btu, use a COP of 2.5 in the above equation.

Step 5: Calculate Btu/HDD.

Divide the Btu delivered for heating(**14**) by the total HDD(**2**):

$$\underset{\text{Btu delivered for heating}}{\underline{\qquad\qquad}}\textbf{(14)} \div \underset{\text{total HDD}}{\underline{\qquad\qquad}}\textbf{(2)} = \underset{\text{Btu/HDD}}{\underline{\qquad\qquad}}\textbf{(15)}.$$

Step 6: Complete your audit.

If electricity is your only heating fuel, go on to "The Last Step: Finding Your Heat Loss," found later in this essay. If you have another heating fuel, such as oil or wood, turn to the corresponding worksheet and complete those calculations before proceeding to the last step.

Worksheet I-5: Wood and Coal _____

Determining the number of Btu's delivered to a home from burning wood or coal is very complex and the resulting accuracy is not as good as for the other fuels. One problem is measuring the amount of fuel used. (A cord is not a cord is not a cord, or how much is left over from the last ton of coal you bought?) Even more difficult is determining the Btu content of the fuel. Also, accurately estimating the operating efficiency of a wood- or coal-burning appliance would require an exercise in ESP. Still, if you have a reasonable idea of your usage and the type of wood or coal burned, it should be useful to make the computations.

Be sure you make the appropriate computations for each solid fuel appliance used. That is, if you have two woodstoves and one coal stove all in use over the sample year, you will need to estimate the cords and tons going to each unit.

Step 1: Compile weather data.
Fill in the following table with your data. (See the box in worksheet I-2 for guidance on what to do.)

Month	Year	Heating Degree-Days (HDD)
June	19___	_____
July		_____
August		_____
September		_____
October		_____
November		_____
December		_____
January	19___	_____
February		_____
March		_____
April		_____
May		_____
		_____ (1)
		total HDD

Step 2: Calculate energy use.
Calculate Btu consumed by estimating cords or tons used only during the heating season:

cords of seasoned hardwood (oak, maple, etc.) burned: $\dfrac{\quad}{\text{cords}} \times \dfrac{24{,}000{,}000}{\text{Btu/cord}} = \dfrac{\quad}{\text{Btu}}$

cords of nonseasoned hardwood burned: $+ \dfrac{\quad}{\text{cords}} \times \dfrac{16{,}000{,}000}{\text{Btu/cord}} = \dfrac{\quad}{\text{Btu}}$

cords of seasoned softwood (pine, fir, etc.) burned: $+ \dfrac{\quad}{\text{cords}} \times \dfrac{16{,}000{,}000}{\text{Btu/cord}} = \dfrac{\quad}{\text{Btu}}$

cords of nonseasoned softwood burned: $+ \dfrac{\quad}{\text{cords}} \times \dfrac{11{,}000{,}000}{\text{Btu/cord}} = \dfrac{\quad}{\text{Btu}}$

tons of anthracite coal: $+ \dfrac{\quad}{\text{tons}} \times \dfrac{\quad}{\text{Btu/ton*}} = \dfrac{\quad}{\text{Btu}}$

tons of bituminous coal: $+ \dfrac{\quad}{\text{tons}} \times \dfrac{\quad}{\text{Btu/ton*}} = \dfrac{\quad}{\text{Btu}}$

total Btu from   total Btu consumed
wood and coal   for heating(2)

*Ask your supplier what type of coal you're buying and what the Btu content is.

Step 3: Calculate the Btu delivered for heating.
Determine your wood- or coal-stove's efficiency factor by matching it with one of the categories listed below:

Appliance Description	Efficiency Factor(3)
Open fireplace, masonry, no doors or heat exchanger	0.05
Open fireplace with circulator heat exchanger	0.15
Fireplace with doors and draft control, no heat exchanger	0.15
Fireplace with doors, draft control and heat exchanger	0.20
Fireplace with sealed firebox insert unit	0.25
Nonairtight stove, circulator type	0.30
Nonairtight stove, radiant type	0.35
Airtight stove, circulator type	0.50
Airtight stove, radiant type	0.55
Airtight furnace	0.60

Enter your efficiency factor in blank(**3**) below.

To determine Btu delivered by the stove, multiply:

$$\underline{\qquad\qquad} \textbf{(2)} \times \underline{\qquad\qquad} \textbf{(3)} = \underline{\qquad\qquad} \textbf{(4)}.$$
Btu consumed for heating   your efficiency factor   Btu delivered for heating

If you have more than one wood- or coal-burning appliance, go back to *Step 2* and run through the calculations again; otherwise continue.

Step 4: Calculate total Btu delivered from all wood- or coal-burning appliances.

Appliance 1: Enter(**4**)   _____
                              Btu delivered for heating

Appliance 2: Enter(**4**) + _____
                              Btu delivered for heating

Appliance 3: Enter(**4**) + _____
                              Btu delivered for heating

Appliance 4: Enter(**4**) + _____
                              Btu delivered for heating

Appliance 5: Enter(**4**) + _____
                              Btu delivered for heating

_____ (**5**)
total Btu delivered for heating from wood or coal

Step 5: Calculate Btu/HDD.
Divide the Btu delivered for heating from wood and/or coal(**5**) by the total HDD(**1**):

$$\underline{\qquad\qquad} \textbf{(5)} \div \underline{\qquad} \textbf{(1)} = \underline{\qquad} \textbf{(6)}.$$
total Btu delivered for heating from wood or coal   total HDD   Btu/HDD

Step 6: Complete your audit.

    If solid fuel is your only heating fuel, go on to "The Last Step: Finding Your Heat Loss," which follows below. If you have another heating fuel, such as electricity or oil, turn to the corresponding worksheet and complete those calculations before proceeding to the last step.

The Last Step: Finding Your Heat Loss

    You've already found out a lot. The sum of the Btu's delivered for heating, for all heating fuels used, is your reference number to compare with what you can gain every heating season with a solar retrofit. (See "Finding Your Heat Gain" found later in this section.)

    The next step is to take this home heat loss audit to completion. First, add up all the Btu/HDD you've calculated (if you're heating with more than one fuel) to get the total Btu/HDD(**1**); then divide the total Btu/HDD by the square footage of your heated floor area(**2**):

$$\frac{}{\text{total Btu/HHD}}\ (\mathbf{1}) \div \frac{}{\text{ft}^2 \text{ of heated floor area}}\ (\mathbf{2}) = \frac{}{\substack{\text{Btu/DD/ft}^2 \\ \text{home heating index (HHI)}}}\ (\mathbf{3}).$$

    You can use the HHI to see what kind of energy-shape your home is in:
    1. If the HHI is less than 5, you have a very energy-efficient home.
    2. If the HHI is between 5 and 10, it is pretty good, but with careful selection of conservation and solar options, you can cut it in half.
    3. If the HHI is between 10 and 15, you are in the average range, and a lot can be done to make it better.
    4. If the HHI is between 15 and 20, immediately reread the essay "The Turned-Off House," found earlier in this section!
    5. If the HHI is greater than 20, you should move out of that tent you call home.
    If your calculations came out somewhere in the above range, you probably did everything right. If your HHI is over 30 or less than 3, you may have goofed or, in the over-30 case, your home is an absolute, utter energy disaster; and, in the under-3 case, your home is so ultra efficient you'll have to be careful not to overheat it with a solar retrofit or even by baking cookies. It is not possible for the HHI to be negative, but if you had a 100 percent solar-heated home, then the HHI would be 0.

Post Figurum Contemplatensis

    This calculation method is intended primarily for determining the relative energy efficiency for space heating. Thus, it doesn't really reflect problems in cooling, but you can assume that a high HHI can be a sign of problems in cooling as well. If you use an air conditioner and your electricity bills shoot up after May and before October, then your home is probably a candidate for some cooling energy conservation. (See "Air Conditioning without Air Conditioners" in Section V.)

Since this computational method can be repeated for any year, it is particularly revealing to perform this "internal audit" for several past years and see what effects changes in your family, their life-style and your home have had on energy use.

Herb Wade

Do you own a personal computer? If you would like to run this computation through it, Herb Wade has a program for you. If you have a TRS-80, a PET, an Apple or any unit that uses Microsoft BASIC, you can plug in this program. Send Herb a self-addressed, stamped envelope to P. O. Box 104, Ashland, Missouri 65010.

References

American Society of Heating, Refrigerating and Airconditioning Engineers. 1972. *ASHRAE handbook of fundamentals.* New York.

———. 1979. *Cooling and heating load calculation manual.* New York.

Finding Your Heat Gain

A simple calculation method for predicting how much useful heat your solar retrofit will deliver

Perhaps one of the most exciting parts of planning your retrofit is finding out how much heat it will put into your house and how much that heat is worth in energy-dollar savings. You have a vision of how the retrofit will look, how it will be fitted onto your house and how it will operate, and now you can complete the scene by predicting its annual energy benefits. You can then compare the annual savings against the initial cost of the system to see when the accrued value of the energy savings equals the cost of the retrofit, the time when your solar heat gain truly becomes free.

"Calculations, solar heat gain, cloudy days, partly cloudy days, cold weather, cool weather, latitude, solar aspect"—your mind may begin to imagine a maze of variables that enter into a calculation like this, and there are in fact dozens of them. But fear not; not only is finding your heat gain exciting, it's also an easy calculation, far easier, say, than the convoluted confusion of the dreaded IRS 1040 and all its ancillary subforms. We've developed a simple worksheet method for doing this calculation in which most of the variables have been boiled down to just a few basic factors. This is a fill-in-the-blanks calculation where the only variables you have to come up with are those that have to do with the particular retrofit you have in mind: how big it will be, what its solar aspect is, whether the glazing is vertical, horizontal or somewhere in between. Everything else you need is here in the following tables, and to show you how it's all done, a sample calculation runs through each worksheet step by step.

To make an economic analysis of space heating retrofit, you'll also need the results of your calculation in "Finding Your Heat Loss" (see the previous essay in this section), particularly the number you came up with for the total space heating energy consumption. It's assumed that you've already reduced this load by installing (or by planning to install) conservation measures to the greatest possible extent. As a rule of thumb, solar retrofitting is truly economical only after your Home Heating Index (HHI) is down around 8 (Btu per degree-day per square foot) or less. For domestic water heating, a solar system really makes sense after your water heating load is down to about three million Btu (MBtu) per person per year after conservation improvements have been made.

It can of course be argued that putting solar heat into any system, whether or not it's been "conservatized," reduces the use of other fuels and that's true, as far as it goes. But as has been emphasized several times before, if you're interested in putting your money where it will be the most effective for saving energy, conservation investments should be your first investment.

In this essay you'll be finding out how to do the heat gain calculations for both solar space heating and domestic water heating retrofit systems. The calculations are separate and slightly different mainly because the water heating load exists all year long in all parts of the country and remains fairly constant from month to month. Space heating loads of course begin and end in the fall and spring; they peak in January and February and the length of the heating season varies greatly in different climate zones. In both worksheets, though, the basic elements of the heat gain calculation are the same. You start with the average amount of solar energy that falls on a square foot of horizontal surface area in your specific location during the heating season (space heating calculation) or through the whole year (domestic water heating). This amount of energy is known as the *incident solar energy.* You then multiply this by certain adjustment factors to account for the fact that the *tilt angle* and *orientation* (solar aspect) of the collecting surface (as determined by the plane of the glazing) have an important effect on the amount of energy it receives. You then multiply by a *system efficiency factor,* which expresses the fraction of the incident solar energy that is actually converted into useful heat and delivered to where you want it. You then multiply by the number of square feet you have available for collection (glazed) area to determine how ''big'' your energy gain will be for the system you want to build. The end result is the *total annual energy savings,* in MBtu's per year, which can be expressed as a percentage of your after-conservation space or water heating load to obtain the *solar fraction* for your system. You can then find your annual dollar savings by simply multiplying your annual energy savings times the cost (in dollars per MBtu) of purchased energy that solar energy is displacing.

I, for one, am always excited by doing this calculation because it brings the system to life. That pile of lumber, hardware and blueprints, inert until now, suddenly starts producing energy in amounts that you can get a handle on: a specific number of Btu's of solar energy for your home and your morning shower, before you even lift a hammer. There probably aren't too many other kinds of investments you can make for which you can calculate, in advance, what the return will be, year after year. But this is the beauty of solar energy: So much is now known about its availability and its performance in different kinds of heating systems that performance predictions can be made with great accuracy. What you're doing here will give you a ''good ballpark'' with about a plus-or-minus 20 percent degree of accuracy. The results may make you all the more ready to forge ahead with your solar plans, or they may make you reevaluate and modify what you've had in mind. You shouldn't expect the solar source to supply 100 percent of your heating needs right off the bat; only a few *new* solar homes have achieved that level of performance. If your calculations show a 20 to 50 percent solar space heating fraction and a 50 to 90 percent solar domestic water heating fraction, you've probably done them correctly and you're within realistic performance parameters for solar retrofits. But don't think you're necessarily wrong if you come up with a 60 to 80 percent space heating fraction. In small- to medium-size homes that are tight and very well insulated, it's altogether possible that a sizeable solar improvement could reduce a fossil-fueled space heating load by that much.

The sample calculations that run with the worksheets are described in the following instructions. Space is provided on both space heating and water heating worksheets for two sets of calculations for the retrofit possibilities you want to study. You may find it convenient to photocopy the blank worksheets now and then do your figuring on the copies. All you need from here is a calculator (or a patient, mathematical brain) and a pencil. We hope you find the solar fraction you're looking for.

How to Figure Your Domestic Hot Water (DHW) Load

It would be so convenient, but most unlikely, if your water heater were metered separately for its energy use. You could easily convert the energy units you are paying for (kilowatt-hours, therms, gallons) into Btu's, and there you'd have it, your DHW load. But because a water heater is normally part of the house's total energy system, we have to do a little more figuring in order to "separate" its energy consumption from other household loads. This involves working with realistic averages to create a model of DHW consumption that allows us to determine a load without taking actual measurements. The result is valid to about the same degree of accuracy as the results of the calculations that follow in the space heating calculation, and they'll give you a basis from which to determine, among other things, the necessary collector area for a solar domestic water heating system. You'll also be using the result in the calculation of solar heat gain for the DHW system you may be planning to retrofit.

Worksheet I-6 has several slots for numbers that will lead you to the final number, the total energy required for DHW, but the basic relationship is simple:

total annual DHW energy use = energy delivered at the tap as hot water + standby losses.

"Standby losses" refer to all the points of heat loss in DHW systems, such as heat loss from the tank and from pipes. The term also embraces combustion inefficiencies in gas- and oil-fired water heaters, i.e., whatever doesn't get to you as hot water but goes in as energy. If you were to divide the energy delivered by the total energy used, you would come up with some factor—0.50, 0.60, 0.70—that would represent the overall efficiency of the system—50, 60, 70 percent. The goal, of course, is to maximize that efficiency to save energy dollars and also to reduce the first cost of a solar water heater. In the latter case it's simple math: If you need less energy for domestic water heating, you can get what you need from the sun with a smaller, less-expensive collector system. In the next section there is a short piece, "Lightening the DHW Load," that ticks off several highly cost-effective conservation measures. Also, the worksheet for this calculation (I-6) presents two sample calculations: one that is based on "average" use and one based on reasonably conservative use. In the latter case, energy consumption is about one-third that of the nonconservation load.

The Calculation

You can't really plug in the factors in the above equation because they are as yet too general and amorphous, but they can be broken down to a workable format:

$$\begin{array}{c}\text{total annual}\\\text{DHW energy use}\end{array} = \begin{array}{c}\text{gallons of hot}\\\text{water used}\end{array} \times \begin{array}{c}\text{temperature}\\\text{rise}\end{array} \times \begin{array}{c}\text{numerical}\\\text{constants}\end{array} + \text{standby losses.}$$

This breakdown introduces some of the opportunities for conservation: using fewer gallons of hot water, reducing the temperature difference between cold incoming water and the hot water stored in the tank (thermostat setback) and reducing standby losses (adding insulation to the water heater and hot water pipes). We refer again to the essay "Lightening the DHW Load" in the next section.

Worksheet I-6 begins with estimates of your gallonage requirements. Use table I-3, "Typical Hot Water Consumption," to get the numbers for lines **1** through **6**. For line **7** you'll have to make a guesstimate of the gallonage of other household uses, based on the averages in table I-3. Add up your gallonage estimates on line **8** and divide the result by the number of people in the household to get per person use levels (line **9**). You may be surprised at just how much hot water everybody uses in just one day. The "average" American uses 17 to 22 gallons of hot water per day, but conservation measures can reduce that down to 7 to 10 gallons a day without hardship.

"Temperature rise" is essentially the temperature difference between water entering the water heater (line **10**) and hot water leaving it (line **11**) for various points of use. The average cold water inlet temperature (line **10**) in your locale is about equal to the annual average air temperature that is listed in column A of table I-8, "Sunshine and Weather Conditions," which is located at the end of this article. You can obtain the hot water temperature (line **11**) by actually measuring it at the tap that's nearest to the water heater. Pick a time of day when there is little or no hot water use, and let the water run for a couple of minutes into a cup. Then submerge a thermometer into the flowing water and take a reading after another couple of minutes. The temperature rise (line **12**) is simply the difference between cold and hot. Note how the temperature rise is greatly reduced in the conservation example.

TABLE I-3   TYPICAL HOT WATER CONSUMPTION

Automatic washing machine, hot cycle	21 gal/load
Automatic washing machine, warm cycle	11 gal/load
Automatic washing machine, cold cycle	0 gal/load
Automatic dishwasher	15 gal/load
Dish washing by hand	4 gal/load
Food preparation (4 people)	3 gal/day
Household cleaning (4 people)	2 gal/day
Hand and face washing	2 gal/day/person
Wet shaving	2 gal/day/person
Tub bath	15 gal/bath
Showering, regular shower head	25 gal/shower
Showering, low-flow shower head	12 gal/shower

The annual energy use from hot water consumption (line **13**) is found by multiplying the factors shown. The 8.33 factor is the weight, in pounds, of a gallon of water, and the number 365 is days per year. It's clear now that the conservation example is a tortoise of energy consumption compared to the rate of consumption in the average case. Standby losses (line **14**) add from 5 to 20 percent to the amount calculated on line **13**. If no insulation has been added to the water heater or to hot water pipes, and if the water heater thermostat hasn't been set back to 110 to 115°F (43–46°C), use a factor of 0.20. If the thermostat has been set back, the water heater covered with a 6-inch layer of fiberglass (R-19), and the pipes insulated to R-4, use a factor of 0.05. Multiply the loss factor times the result obtained on line **13**, and then add lines **13** and **14** to obtain the total annual DHW energy use (line **15**). If you're interested, you can again divide that by the number of people in the household to see how much one person uses per year (line **16**) and per day (line **17**). (Of course, people-reduction isn't usually a viable conservation method. . . .)

The final calculation is of what the DHW load costs every year (line **18**). Refer to table I-2 "The True Cost of Energy" (see the essay "The Turned-Off House" found earlier in this section), to find the cost per MBtu of the energy you're using. That number times line **15** gives you the annual DHW energy cost. This is where the benefits of conservation are most apparent—in the old pocketbook.

Solar Heat Gain for Domestic Water Heating Retrofits

For the sample calculation in worksheet I-7 we'll assume this situation: A solar water and space heating retrofit is planned for a home in Boston, Massachusetts. The water heater is a passive "breadbox" or batch-type system, and the space heating retrofit is a solar greenhouse. The batch-type water heater includes a 40-gallon tank set into a box that carries 34 square feet of single glazing at a 45-degree tilt. Two insulating shutter-reflector doors are mounted on the box to increase heat gain (with the reflectors) and to reduce nighttime heat loss (with the shutters closed). The greenhouse has about 240 square feet of double glazing that is tilted at a 60-degree slope. As the illustration shows, the water heater is installed inside the greenhouse where it is totally protected from freezing, assuming that the greenhouse space will not be allowed to drop below 32°F (0°C). This placement of the water heater is a good strategy, one of several freeze protection schemes discussed in the next section. The orientation of both systems is 5 degrees west of true south, a nearly optimum solar aspect.

The Calculation

Step 1: You are now working with worksheet I-7. Find the annual total energy received on the collector, which in this case is the water heater itself plus its reflectors.

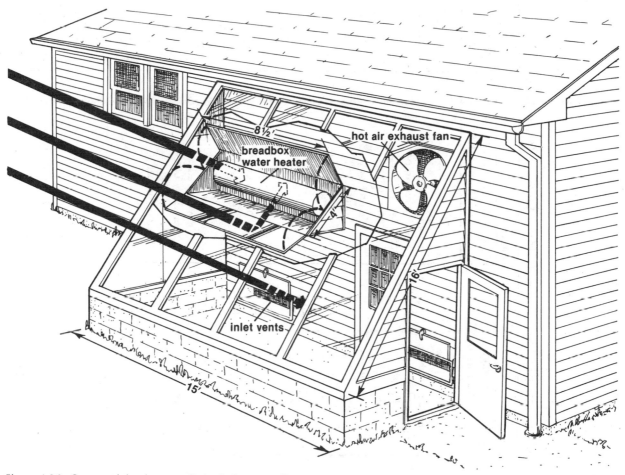

Figure I-36: Our model solar retrofit includes a medium-size greenhouse with a glazing pitch of 60 degrees, and a breadbox or batch-type domestic water heater installed inside the greenhouse for positive freeze protection. In this model the short vertical wall at the bottom of the greenhouse glazing is opaque, but some greenhouse designs call for this wall, the kneewall, to be glazed. If it were glazed, you could figure the added solar gain by going through the space heating calculation again and adding the result to the figure obtained for the tilted glazing. Similarly, if the west or east wall were glazed, you could run the calculation based on those specifications.

Start by consulting table I-8, "Sunshine and Weather Conditions in the United States and Canada," which is located at the end of this article. Column D in this table gives you the numbers to plug in for water heating systems, and column E is used for space heating systems. For Boston, column D reads 423,000 Btu per square foot per year (423 KBtu/ ft²/yr). Enter 423 on line **1**. The next factor adjusts that energy gain for variations in glazing tilt and orientation, and for this refer to table I-4, "Adjustment Factors for Collector Tilt and Orientation." The tilt angle column in this table denotes six conditions. The 45-degree glazing tilt in the example is very close to the 42-degree north latitude

(continued on page 126)

Worksheet I-6 Estimating Your Domestic Hot Water (DHW) Energy Requirements

Calculation	Line No.	How Obtained	Examples		Your Household
			Average Use (family of 4)	Conservative Use (family of 4)	
GALLONAGE FOR: Clothes washing	1	see table I-3	automatic washing machine, hot cycle, 1 load every 2 days — 11 gal/day	automatic washing machine, warm or cold cycle, 1 load every 2 days — 4 gal/day	
Dish washing	2	see table I-3	automatic dishwasher, 1 load per day — 15 gal/day	dish washing by hand — 4 gal/day	
Food preparation	3	see table I-3	3 gal/day	2 gal/day	
Household cleaning	4	see table I-3	2 gal/day	2 gal/day	
Hand and face washing and shaving	5	see table I-3	10 gal/day	with flow controls in faucets — 8 gal/day	
Baths and showers	6	see table I-3	regular shower head, shower or bath every other day — 40 gal/day	with low-flow shower head shower every other day, quicker showers; no baths — 20 gal/day	
Other uses in your family	7	see table I-3	0 gal/day	0 gal/day	
DAILY TOTAL GALLONAGE	8	add 1 through 7	81 gal/day	40 gal/day	
DAILY TOTAL GALLONAGE/ PERSON	9	$\dfrac{8}{\text{no. of people in household}}$	$\dfrac{81}{4} = 20.25$ gal/person/day	$\dfrac{40}{4} = 10$ gal/person/day	

#				
10	COLD WATER INLET TEMPERATURE (°F)	see text	50°F	50°F
11	HOT WATER DELIVERY TEMPERATURE (°F)	see text	140°F	110°F
12	TEMPERATURE RISE	subtract **10** from **11**	140°−50°=90°F	110°−50°=60°F
13	ENERGY USE FROM HOT WATER COMSUMPTION	$\dfrac{8 \times 12 \times 8.33 \times 365}{1,000,000}$	$\dfrac{81 \times 90 \times 8.33 \times 365}{1,000,000} = 22.17$ MBtu/yr	$\dfrac{40 \times 60 \times 8.33 \times 365}{1,000,000} = 7.30$ MBtu/yr
14	STANDBY LOSSES	see text	$0.20 \times 22.17 = 4.43$ MBtu/yr	$0.05 \times 7.30 = 0.37$ MBtu/yr
15	TOTAL ANNUAL DHW ENERGY USE	**13** + **14**	$22.17 + 4.43 = 26.6$ MBtu/yr	$7.30 + 0.37 = 7.67$ MBtu/yr
16	TOTAL ANNUAL DHW ENERGY USE/PERSON	$\dfrac{\mathbf{15}}{\text{no. of people in household}}$	$\dfrac{26.6}{4} = 6.65$ MBtu/person/day	$\dfrac{7.67}{4} = 1.92$ MBtu/person/day
17	TOTAL DHW ENERGY USE/PERSON/DAY	$\dfrac{\mathbf{16}}{365}$	$\dfrac{6.65}{365} = 0.01821$ MBtu $= 18,210$ Btu/person/day	$\dfrac{1.92}{365} = 0.052$ MBtu $= 5250$ Btu/person/day
18	ANNUAL DHW ENERGY COST	**15** × energy cost from table I-2	$26.6 \times \$14.65/\text{MBtu} = \$390/\text{yr}$ (electricity @ 5¢/KWH)	$7.67 \times \$14.65/\text{MBtu} = \$112/\text{yr}$ (electricity @ 5¢/KWH)

location of Boston. Thus we go down to the line marked "Latitude." With an orientation that is 5 degrees west of true south we use the "True South" column to arrive at an adjustment factor of 1.10. Enter this on line **2**. The collector's glazed area (aperture) totals about 34 square feet, and this number is entered on line **3**. Line **4** is a multiplication of lines **1**, **2**, and **3**, with the result divided by 1000 to convert the energy units from KBtu (1000 Btu) to MBtu (1,000,000 Btu). Line **4** thus represents the total annual amount of solar energy available at the collector before any deductions have been made for energy losses, in the conversion of available energy to useful hot water. As we'll see in the next step, the energy value of this "raw" sunshine is reduced by several unavoidable factors.

Step 2: There are several factors that affect overall system efficiency. The number that is finally obtained on line **15** expresses the fraction of available solar energy that the retrofit system will actually deliver as hot water heat. To find this efficiency factor you must have some specific ideas about the type of system you will use (flat plate collector or batch-type system), whether it's active (pumped circulation) or passive (convective circulation or batch-type), the kind of glazing material used and the number of glazing layers, whether or not some kind of insulation is used at night to cover the collector, and so forth. If you want to make a wild guess at your system efficiency, instead of going through this phase of the calculation, use a 0.3 factor for systems used in cold climates (more than 4000 degree-days) and a 0.6 factor for systems used in warm climates. Then proceed directly to line **15**. Of course, wild-guessing wipes out all the sensitivity to design details that we've built into this calculation method, so we would urge you to specify your plans for a system to get more accuracy out of the result. You

TABLE I-4   ADJUSTMENT FACTORS FOR COLLECTOR TILT AND ORIENTATION
Ratio of yearly total solar energy received on surfaces of various tilt and orientation angles to that received on a horizontal surface

Tilt Angle of Collector Surface	Orientation Angle of Collector Surface				
	0° True South	15° East or West of True South	30° East or West of True South	45° Southwest or Southeast	Winter Correction Factor for Space Heating Collectors (use for worksheet I-8, line 3)
Vertical (90°)	0.72	0.69	0.65	0.52	2.00
Latitude +25°	0.98	0.95	0.90	0.73	1.45
Latitude +15°	1.03	1.01	0.97	0.78	1.37
Latitude	1.10	1.09	1.06	0.86	1.27
Latitude −15°	1.10	1.10	1.08	0.88	1.09
Horizontal (0°)	1.00	1.00	1.00	1.00	1.00

NOTE: Factors from these columns are used for worksheet I-7, line **2** and worksheet I-8, line **2**.

could use the rough estimate efficiency factors as a starting point in sizing your system, and then go back through the calculation the "long" way.

To summarize, system efficiency equals a heat gain factor (line **8**) minus a heat loss factor (line **14**). Sounds quite plausible, doesn't it? In this step, the heat gain factor is determined entirely by the glazing materials that are used. Glazing transmittance factors for different glazing materials are found in table I-5, and there is a line for factoring the outer glazing (line **5**) and one for the inner glazing (line **6**). If your system is single glazed, enter 1.0 on line **5** and then enter the transmittance factor for the glazing on line **6**. In the example the fiberglass reinforced plastic glazing for the water heater is considered to be an inner glazing to the outer glazing of the greenhouse, so a value of 0.85 is entered on line **5**. The outer glazing is a double layer of glass, which has a transmittance of 0.71 (0.84 times 0.84). This is entered on line **6**. The *glazing averaging factor* accounts for whether the collector is single or double glazed. The glazing averaging factor for double glazing is 0.90, entered on line **7**; in this case the system is, in effect, double glazed. Line **8** is the product of all these factors, which in this case is 0.54. To review, what you have found here is the net amount of heat gained by the collector. If you're interested, you can take this factor and multiply it times line **4** to get an energy value for this factor. In the example it would be 8.54 MBtu.

Step 3: The heat loss factor accounts for how well the overall system design "holds on" to the heat it can gain and prevents it from being lost back to the atmosphere. In the case of a domestic water heating system, we're really talking about the system's ability to put useful heat into the hot water system. An important factor is the temperature of the cold water that enters the system with every hot water draw. Naturally the colder the incoming water is, the more energy is needed to make it hot. This groundwater temperature can vary greatly around the country, but a convenient and reasonably accurate rule of thumb is to equate groundwater temperature with the average annual air temperature, listed in column A in table I-8 (located at the end of this article), and entered on line **9**. The average temperature for Boston is 51.4°F (11°C). From there, several heat loss factors are worked in on lines **10** through **13**. The *collector loss factor* (line **10**) is found in table I-6, and in the example we are studying—a batch-type water heater located in the building envelope with insulating doors—the factor is 0.50. The *glazing loss factor* (line **11**) is related to heat loss through the glazing. For the example, 0.80 is entered for a double-glazed collector (greenhouse glazing plus collector glazing).

The surface treatment that the absorber receives can increase overall efficiency. Regular, flat black paint, the coating used in the example, rates a factor of 1.0, but there are absorber coatings that increase heat absorption and thus reduce the *absorber loss factor* to 0.9. The factor 1.0 is entered on line **12**. If a collector system uses a heat exchanger, heat loss is increased. With no heat exchanger the loss factor (line **13**) is 1.0 in the example. Some domestic water heating systems do use a heat exchanger, which would mean plugging in a factor of 1.1.

Line **14** puts all the heat loss factors together. The calculation starts with the number 80, which is the "balance temperature" for domestic water heating systems. This is the temperature at which the collector begins to lose heat back to the atmosphere. Of course the actual temperature of the system varies considerably, but the balance temper-

(continued on page 131)

Table I-5   Glazing Factors for Various Solar Glazing Materials

Glazing Material (generic)	Common Description or Brand Name	Transmittance of a Single Layer (use for worksheet I-7, lines 5 and 6 and worksheet I-8, lines 6 and 7)
Window glass	double strength	0.84
Low-iron glass	ASG Sunadex	0.91
Fiberglass	Sunlite Premium II Filon Filoplated Lascolite Glasteel	0.85
Acrylic	Du Pont Lucite Rohm & Haas Plexiglas	0.89
Acrylic (double wall)	Cy-Ro Exolite SDP	0.83
Polycarbonate	Lexan Merlon	0.87
Polycarbonate (double wall)	Tuffak-Twinwall Qualex	0.79
Polyester	LLumar (UV resistant)	0.88
	Flexigard (laminate)	0.89
Fluorocarbons	Du Pont Teflon FEP	0.96
	Du Pont Tedlar PVF	0.90
Polyethylene	6-mil VisQueen	0.85
	Monsanto 602	0.87

Glazing Averaging Factor   = 0.93 for single-glazed systems
(for worksheet I-7, line **7**   = 0.90 for double-glazed systems or double-wall glazings
and worksheet I-8, line **8**)

Table I-6   Heat Loss Factors for Various Solar System Designs

Description of System		Collector Loss Factor (use for worksheet I-7, line 10 and worksheet I-8, line 11)
Window	with night insulation (R-5 or better)	0.20
Skylight	with heavy curtains or drapes	0.35
Clerestory	without night insulation or curtains	0.75
Attached greenhouse	thermosiphoning air flow	0.40
Sunspace	fan-forced air flow	0.25
Enclosed porch or breezeway (R-3 window insulation assumed for all cases)		
Glazed masonry-mass wall	stagnating (unvented)	0.70
	thermosiphoning air flow	0.50
	fan-forced air flow	0.48
Glazed water-mass wall	with night insulation (R-5 or better)	0.25
	without night insulation	0.70
Batch-type water heater	freestanding outside with insulating doors	0.50
	in building envelope with insulating doors	0.50
	in building envelope without insulating doors	0.54
Liquid-heating flat plate collectors	thermosiphoning air flow	0.52
	pumped flow	0.50
Air-heating flat plate collectors	thermosiphoning air flow	0.55
	fan-forced air flow	0.45

Glazing Loss Factor   = 1.0 for single-glazed collectors
(for worksheet I-7, line **11**   = 0.8 for double-glazed collectors or double-wall glazings
and worksheet I-8, line **12**)

Absorber Loss Factor   = 1.0 for nonselective absorbers, windows, etc.
(for worksheet I-7, line **12**   = 0.9 for selective absorbers (black chrome, etc.)
and worksheet I-8, line **13**)

Heat Exchange Loss Factor = 1.0 for all systems without heat exchangers
(for worksheet I-7, line **13**   = 1.1 for systems with heat exchangers
and worksheet I-8, line **14**)

TABLE I-7   DIMINISHING RETURNS FACTOR

A	B	C
If your gross solar fraction is: from worksheet I-7, line **17** or worksheet I-8, line **18**	then for DHW systems, your net solar fraction is: enter on worksheet I-7, line **18**	then for space heating systems, your net solar fraction is: enter on worksheet I-8, line **19**
0.05	0.05	0.05
0.10	.10	.10
0.15	.15	.15
0.20	.20	.20
0.25	.25	.25
0.30	.30	.30
0.35	.35	.35
0.40	.40	.39
0.45	.45	.42
0.50	.49	.46
0.55	.53	.49
0.60	.56	.52
0.65	.59	.54
0.70	.61	.57
0.75	.64	.59
0.80	.66	.61
0.85	.68	.63
0.90	.69	.65
0.95	.71	.67
1.00	.72	.68
1.10	.74	.70
1.20	.76	.72
1.30	.78	.74
1.40	.80	.76
1.50	.82	.78
1.60	.84	.80
1.70	.86	.82
1.80	.88	.84
1.90	.90	.86
2.00	0.92	0.88

ature establishes an appropriate average for DHW systems. From there the calculation proceeds as shown. The final factor, 0.027, is simply a fixed conversion factor that changes temperatures and heat loss factors to energy units. In line **15** you subtract the system heat loss factor (line **14**) from the system heat gain factor (line **8**) to get the net system efficiency, in this case 0.23. In cold climates certain systems might produce a negative efficiency factor, which would mean that the system was not suitable for that climate. For example, a batch-type water heater installed outside a house with no insulating shutters could yield a negative efficiency or one that was lower than 0.20, making it unsuitable. If you come up with a low efficiency factor for a system that seems right for your climate, go back through the calculation. If the result remains the same, plug in a different system design. The primary variable for water heaters will be the area of the glazing. Of course, if you increase the glazing area, you should be planning to increase the absorber area as well.

   Step 4: Find the solar fraction. You can relax now. At this point you've specified all the essential design parameters for the system, and there are no more of these decisions to make, just the final number-crunching for your calculator. We're simply going to compare the energy output from the solar system against the requirements of the domestic water heating load, which you calculated in worksheet I-6 (line **15**). Enter this result on line **16**. The gross or unadjusted solar fraction is found on line **17** by multiplying the annual total solar energy (line **4**) times the net system efficiency and dividing by the water heating load. In this case the result is 0.48. Now the *gross solar fraction* must be converted to the net solar fraction by using table I-7 to account for the *diminishing returns factor.* We know about diminishing returns from both real-world and model calculations that have shown that solar energy gain does not increase in direct proportion to increases in glazing area. If, for example, 1 square foot of glazing area provides a 1000-Btu gain, it doesn't follow that 2 square feet will provide a 2000-Btu gain. Several factors conspire to reduce the real energy gain from solar systems, but one primary reason is that heat collection occurs during a limited time period that doesn't necessarily coincide with peak loads such as night or cold, cloudy days. Thus we experience diminishing returns with larger collector areas, and as table I-7 shows, with higher gross solar fractions the effect of diminishing returns becomes more pronounced.

   In the example, starting with a gross solar fraction of 0.48, we find that the *net solar fraction* for domestic water systems (see table I-7, column B) is 0.47, entered on line **18**. This means that the solar system in question will supply 47 percent of the water heating load. Finally, we come to the Big Reward: How much energy and money does the system save every year? Line **19**, annual energy savings, is found by multiplying the net solar fraction (line **18**) by the water heating load (line **16**). To find the annual dollar savings (line **20**), you multiply the annual domestic water heating energy cost (worksheet I-6, line **18**) by the net solar fraction.

   In the example, the batch-type water heating system will save a calculated $52.64 per year. In the "real world" you can expect that figure to vary by 20 percent in either direction, making the range of annual savings $42 to $62. For a simple determination of the "payback period"—when the accrued value of energy savings equals the cost of the system—you could divide the cost by the annual savings. If the system had

Worksheet I-7 Solar Domestic Hot Water (DHW) System Performance Calculation

Calculation	Line No.	How Obtained	Example	Your Home Case A	Case B
Annual total solar energy received on horizontal surface	1	see table I-8, column D	423 KBtu/ft²/yr	_____	_____
Adjustment factor for collector tilt and orientation	2	see table I-4	1.10	_____	_____
Collector area (clear aperture)	3	your own measurements	34.0 ft²	_____	_____
Annual total solar energy received on collector	4	$\dfrac{1 \times 2 \times 3}{1000}$	$423 \times 1.10 \times 34.0 = 15.82$ MBtu/yr	_____	_____
Inner glazing transmittance	5	see table I-5	0.85	_____	_____
Outer glazing transmittance	6	see table I-5	0.71	_____	_____
Glazing averaging factor	7	see table I-5	0.90	_____	_____
System heat gain factor	8	$5 \times 6 \times 7$	$0.85 \times 0.71 \times 0.90 = 0.54$	_____	_____
Average annual temperature	9	see table I-8, column A	51.4°F	_____	_____
Collector loss factor	10	see table I-6	0.50	_____	_____
Glazing loss factor	11	see table I-6	0.80	_____	_____
Absorber loss factor	12	see table I-6	1.0	_____	_____
Heat exchange loss factor	13	see table I-6	1.0	_____	_____
System heat loss factor	14	$(80-9) \times 10 \times 11 \times 12 \times 13 \times 0.027$	$(80-51.4)(0.5)(0.8) \times (1)(1)(0.027) = 0.31$	_____	_____
System efficiency	15	$8 - 14$	$0.54 - 0.31 = 0.23$	_____	_____
DHW load (including standby losses)	16	see worksheet I-6, line 15	7.67 MBtu/yr	_____	_____
Gross solar fraction	17	$\dfrac{4 \times 15}{16}$	$\dfrac{15.82 \times 0.23}{7.67} = 0.47$	_____	_____
Net solar fraction	18	see table I-7, column B	0.47	_____	_____
Annual energy savings	19	18×16	$0.47 \times 7.67 = 3.60$ MBtu/yr	_____	_____
Annual dollar savings	20	$18 \times$ worksheet I-6, line 18	$0.47 \times \$112 = \$52.64$/yr	_____	_____

a materials' cost of $350, solar energy would have paid for it in six to eight years. If you want to look a little deeper into the cost-benefit ratio of a particular system, you can go back to the calculation in "The Turned-Off House," found earlier in this section. All you have to do is plug in the annual energy savings, and away you go.

Solar Heat Gain for Space Heating Retrofits

In this calculation we will get the same kind of result as in the DHW calculation; thus, the space heating calculation is essentially the same with a few minor differences. To restate the conditions, the glazing area is 240 square feet, facing 5 degrees west of true south and tilted at 60 degrees from horizontal. The greenhouse is equipped with window insulation that is put up at sundown and put away at sunrise.

The Calculation

Step 1: You are now working with worksheet I-8. The winter total solar energy received on a horizontal collector surface is found from table I-8, column E. Space heating systems, of course, do not benefit from gains during the warm months, so to develop these numbers we totaled up the gains only for the heating season months for the specific region. For Boston the amount is 222 KBtu/ft²/yr entered on line **1**. From table I-4 the adjustment factor for tilt and orientation is 1.03. At a 60-degree tilt we use the "latitude + 15°" line, and with an orientation that is 5 degrees west of true south, we again use the "0° true south" column and enter the factor on line **2**. For studying space heating systems we also need to include the winter correction factor (line **3**) because vertical and near-vertical surfaces receive more solar energy in winter than in summer. A 60-degree tilt is certainly closer to a vertical than to a horizontal plane. In this step, factors that are greater than 1.0 increase the amount of available solar energy.

The collector area is again taken to be the net glazing area. In the large glazing areas used in space heating systems, a percentage of the glazing is blocked and reflected away by framing members and trim pieces. Our rule of thumb is that the actual *clear aperture area* is 85 percent of the gross area (length times width). In the example the "window wall" of the greenhouse is 16 feet long and 15 feet wide, giving it a gross area of 240 square feet. Multiplied by 0.85, the clear aperture area becomes 204 square feet, entered on line **4**. (In the DHW calculation we didn't need to use this clear aperture factor because there were no obstructing framing or trim pieces.) The winter total solar energy received on the collector is again the product of all the above factors, divided by 1000. In this case 63.27 MBtu per year is entered on line **5**.

Step 2: The system efficiency calculation begins with glazing factors. Since the double glazing involves two layers of glass, both the inner and outer glazing transmittance are the same; 0.84 (from table I-5) is entered on lines **6** and **7**. With double glazing the glazing averaging factor is 0.90 (table I-5), entered on line **8**. The product of all these factors is entered on line **9** as the system heat gain factor.

Worksheet I-8 Solar Space Heating System Performance Calculation

Calculation	Line No.	How Obtained	Example	Your Home Case A	Case B
Winter total solar energy received on horizontal surface	1	see table I-8, column E	222 KBTU/ft²/yr	_____	_____
Adjustment factor for collector tilt and orientation	2	see table I-4	1.03	_____	_____
Winter correction factor	3	see table I-4	1.37	_____	_____
Collector area (clear aperture)	4	gross collector area × 0.85	204 ft²	_____	_____
Winter total solar energy received on collector	5	$\dfrac{1 \times 2 \times 3 \times 4}{1000}$	$\dfrac{222 \times 1.03 \times 1.37 \times 204}{1000}$ $= 63.91$ MBtu/yr	_____	_____
Inner glazing transmittance	6	see table I-5	0.84	_____	_____
Outer glazing transmittance	7	see table I-5	0.84	_____	_____
Glazing averaging factor	8	see table I-5	0.90	_____	_____
System heat gain factor	9	$6 \times 7 \times 8$	$0.84 \times 0.84 \times 0.90 = 0.64$	_____	_____
Average winter temperature	10	see table I-8, column B	40.0°F	_____	_____
Collector loss factor	11	see table I-6	0.25	_____	_____
Glazing loss factor	12	see table I-6	0.80	_____	_____
Absorber loss factor	13	see table I-6	1.0	_____	_____
Heat exchange loss factor	14	see table I-6	1.0	_____	_____
System heat loss factor	15	$(60 - 10) \times 11 \times 12 \times 13 \times 14 \times 0.04$	$(60-40)(0.25)(0.80) \times$ $(1)(1)(0.04) = 0.16$	_____	_____
System efficiency	16	$9 - 15$	$0.64 - 0.16 = 0.48$	_____	_____
Space heating load	17	see text	63.10 MBtu/yr	_____	_____
Gross solar fraction	18	$\dfrac{5 \times 16}{17}$	$\dfrac{63.91 \times 0.48}{63.1} = 0.49$	_____	_____
Net solar fraction	19	see table I-7, column C	0.45	_____	_____
Annual energy savings	20	17×19	$63.10 \times 0.45 = 28.40$ MBtu/yr	_____	_____
Annual dollar savings	21	$19 \times$ annual heating cost	$0.45 \times \$740 = \$333/yr$	_____	_____

Step 3: The system heat loss factor is calculated on lines **10** through **15**. For space heating systems, we're concerned about the average air temperature during the heating season, which is found in table I-8, column B, and entered on line **10**. In the example the winter average is 40°F (4°C). Table I-6 shows that for a greenhouse the collector loss factor (line 11) is 0.25 if warm air is moved into the main house with a fan. (With a natural convection air flow the loss factor increases to 0.40, which means that more heat is lost.) The following factors for lines **12** through **14** are taken from Table I-6. The glazing loss factor (line **12**) for double-glazed systems is 0.80, and the absorber loss factor (line **13**) is 1.0. Since there is no formal heat exchanger in the greenhouse, the heat exchange loss factor is also 1.0, entered on line **14**.

The system heat loss calculation starts with the number 60. As in the DHW calculation this is the balance temperature we use for space heating systems. Subtract line **10** from 60 and multiply the result by the factors on lines **11**, **12**, **13** and **14**, and by the conversion factor, 0.04. In the example the result entered on line **15** is 0.16. By subtracting line **15** from line **9** we get the net system efficiency factor, entered on line **16**. The example shows a net efficiency factor of 0.48 or 48 percent.

Step 4: Find the solar fraction. Assume that the space heating load for the home in question was calculated in the preceding essay, "Finding Your Heat Loss," to be 63.10 MBtu per year (line **17**) which represents a Home Heating Index of 8 (8 Btu per square feet per degree-day) for a 1400-square-foot home in a 5634 degree-day climate. The gross solar fraction is thus the product of lines **5** and **16**, divided by the heating load (line **17**). The gross solar fraction for the example becomes 0.48, entered on line **18**. With a fraction like this, the effect of diminishing returns is more noticeable than with lower solar fractions. Thus the net solar fraction (line **19**) becomes 0.45 as shown by table I-7, column C.

The annual energy savings (line **20**) is again the product of the net solar fraction times the space heating load and, in the example, we see that the greenhouse will provide (save) 28.40 MBtu per year which represents a dollar savings of $333 (net solar fraction times annual heating cost).

Now for the sake of true accuracy, it should be noted that because the solar DHW is located inside the greenhouse, the greenhouse's energy gain is reduced by the amount of energy collected by the batch-type heater. Subtracting 3.60 MBtu from 28.40 MBtu leaves a net greenhouse gain of 24.80 MBtu and many many hot showers.

Once again you can plug the annual savings into the calculation described in "The Turned-Off House," found earlier in this section, to determine the cost-benefit ratio. From that essay we saw that our friend Baker came out a little low on the cost-benefit ratio because the greenhouse was a relatively expensive heat collector. It was also, however, a food grower and provided some additional living space, so that with all things considered the investment was worthwhile.

Finding your heat gain is an important planning tool. Using it may help you choose from among several retrofit possibilities, or it may affect the final size of the collector you build. It will certainly shed some light on the bottom line in energy improvements: the amount of energy saved and the value, compared to the cost of saving it.

Bob Flower

TABLE I-8   SUNSHINE AND WEATHER CONDITIONS IN THE UNITED STATES AND CANADA

State or Province	Location	Ambient Temperature (°F)		Annual Total Heating Degree-Days	Average Solar Energy Received on Horizontal Surfaces*	
		Annual Average	Winter Heating Season Average		Annual Total (KBtu/ ft²/yr)	Winter Heating Season Total (KBtu/ ft²/season)
		A	B	C	D	E
Alabama	Birmingham	64.1	54.2	2551	534	145
	Montgomery	65.0	55.4	2291	543	153
Alaska	Fairbanks	40.0	6.7	14279	321	172
Arizona	Flagstaff	45.6	35.6	7152	691	326
	Phoenix	69.0	58.5	1765	711	335
	Tucson	67.7	58.1	1800	697	339
	Yuma	72.5	64.2	974	682	325
Arkansas	Fort Smith	61.8	50.3	3292	519	140
	Little Rock	61.7	50.5	3219	526	141
Northern California	Eureka	52.3	49.9	4643	422	177
	Red Bluff	63.5	53.8	2515	552	234
	Sacramento	60.4	53.9	2502	587	233
	San Francisco	56.9	54.4	2419	533	240
Southern California	Bakersfield	65.1	55.4	2122	784	300
	Fresno	63.0	53.3	2611	615	215
	Los Angeles Airport	no data	57.4	2061	602	230
	Los Angeles City	61.9	60.3	1349	614	240
	San Diego	63.2	59.5	1458	548	228
	Santa Maria	57.0	54.3	2967	657	251
Colorado	Denver	49.5	40.8	5524	601	269
	Grand Junction	52.5	39.3	5641	621	269
	Pueblo	52.7	40.4	5462	625	285
Connecticut	Hartford	49.8	37.3	6235	457	192
	New Haven	50.2	39.0	5897	474	199
Florida	Jacksonville	69.5	61.9	1239	544	173
	Miami	76.2	71.1	214	606	214
	Pensacola	68.4	60.4	1463	560	166
	Tallahassee	68.0	60.1	1485	604	189
	Tampa	72.2	66.4	683	618	170
Georgia	Atlanta	61.4	51.7	2961	555	155
	Macon	65.6	56.2	2136	556	161
	Rome	60.4	49.9	3326	530	153
	Savannah	66.4	57.8	1819	546	162

*Solar energy data (columns D and E) refer to average conditions, with clear and cloudy days combined.

Table I-8—Continued

State or Province	Location	Ambient Temperature (°F)		Annual Total Heating Degree-Days	Average Solar Energy Received on Horizontal Surfaces*	
		Annual Average	Winter Heating Season Average		Annual Total (KBtu/ ft²/yr)	Winter Heating Season Total (KBtu/ ft²/season)
		A	B	C	D	E
Idaho	Boise	51.0	39.7	5809	538	272
	Pocatello	47.0	34.8	7033	559	290
Illinois	Cairo	59.6	47.9	3821	529	226
	Chicago	50.8	38.9	5882	468	184
	Granite City	55.3	43.1	4900	500	210
	Moline	50.1	36.4	6408	474	191
	Springfield	53.6	40.6	5429	496	201
Indiana	Evansville	57.0	45.0	4435	515	212
	Fort Wayne	50.3	37.3	6205	480	190
	Indianapolis	52.1	39.6	5699	471	190
Iowa	Des Moines	49.2	35.5	6588	487	200
	Sioux City	49.1	34.0	6951	512	215
Kansas	Concordia	54.1	40.4	5479	533	232
	Dodge City	55.4	42.5	4986	600	271
	Topeka	54.9	41.7	5182	500	214
	Wichita	57.1	44.2	4620	546	241
Kentucky	Louisville	55.7	44.0	4660	484	200
Louisiana	Lake Charles	68.6	60.5	1459	568	168
	New Orleans -	68.6	61.0	1385	537	159
	Shreveport	66.1	56.2	2184	554	154
Maine	Caribou	38.4	24.4	9767	432	231
	Portland	45.0	33.0	7511	477	254
Maryland	Baltimore	55.2	46.2	4111	477	204
Massachusetts	Boston	51.4	40.0	5634	423	222
	Worcester	46.8	34.7	6969	434	231
Michigan	Alpena	42.1	29.7	8506	438	220
	Detroit	50.1	37.2	6232	452	230
	Escanaba	41.9	29.6	8481	435	226
	Grand Rapids	47.6	34.9	6894	447	224
	Lansing	47.6	34.8	6909	447	227
	Marquette	42.6	30.2	8393	433	216
	Sault Ste. Marie	40.6	27.7	9048	454	236
Minnesota	Duluth	37.9	23.4	9250	440	226
	Minneapolis	43.7	28.3	8382	445	231

Table I-8—Continued

State or Province	Location	Ambient Temperature (°F)		Annual Total Heating Degree-Days	Average Solar Energy Received on Horizontal Surfaces*	
		Annual Average	Winter Heating Season Average		Annual Total (KBtu/ ft²/yr)	Winter Heating Season Total (KBtu/ ft²/season)
		A	B	C	D	E
Mississippi	Jackson	65.5	55.7	2239	534	148
Missouri	Columbia	55.0	42.3	5046	520	218
	Kansas City	56.8	43.9	4711	511	215
	St. Louis	55.3	43.1	4900	500	210
	Springfield	56.5	44.5	4900	525	228
Montana	Billings	47.5	34.5	7049	510	307
	Glasgow	41.4	26.4	8996	518	316
	Great Falls	44.7	32.8	7750	591	300
	Havre	42.1	29.8	8182	560	271
	Helena	43.4	31.1	8129	511	306
	Missoula	43.2	31.5	8125	462	267
Nebraska	Lincoln	52.8	38.8	5864	503	275
	North Platte	49.2	35.5	6684	529	227
	Omaha	51.5	35.6	6612	494	205
	Valentine	46.9	32.6	7425	540	231
Nevada	Ely	44.3	33.1	7733	638	355
	Las Vegas	65.7	53.5	2709	695	399
	Reno	48.4	39.3	6332	645	353
	Winnemucca	47.4	36.7	6761	583	306
New Hampshire	Concord	45.6	33.0	7383	435	177
New Jersey	Atlantic City	54.1	43.2	4812	497	216
	Trenton	53.9	42.4	4980	479	205
New Mexico	Albuquerque	56.6	45.0	4348	707	330
New York	Albany	47.6	34.6	6875	447	181
	Binghamton	45.8	36.6	6451	445	173
	Buffalo	46.7	34.5	7062	447	164
	New York	54.5	42.8	4871	432	179
	Schenectady	no data	35.4	6650	386	157
North Carolina	Asheville	54.7	46.7	4042	542	247
	Charlotte	60.8	50.4	3191	548	250
	Greenville	no data	no data	no data	597	271
	Raleigh	59.5	49.4	3393	515	236
	Wilmington	63.8	54.6	2347	531	247
North Dakota	Bismarck	42.2	26.6	8851	502	263

Table I-8—Continued

State or Province	Location	Ambient Temperature (°F)		Annual Total Heating Degree-Days	Average Solar Energy Received on Horizontal Surfaces*	
		Annual Average	Winter Heating Season Average		Annual Total (KBtu/ ft²/yr)	Winter Heating Season Total (KBtu/ ft²/season)
		A	B	C	D	E
North Dakota (continued)	Devils Lake	40.0	22.4	9901	457	235
	Fargo	41.1	24.8	9226	445	231
	Williston	40.9	25.2	9243	467	237
Ohio	Cincinnati	53.6	45.1	4410	479	195
	Columbus	52.0	41.5	5211	438	172
	Toledo	49.0	36.4	6494	468	182
Oklahoma	Oklahoma City	60.3	48.3	3725	595	228
	Tulsa	59.7	47.7	3860	525	193
Oregon	Eugene	52.5	45.6	4726	443	210
	Medford	52.6	43.2	5008	531	258
	Portland	52.9	47.4	4109	391	187
Pennsylvania	Allentown	51.1	38.9	5810	461	194
	Harrisburg	53.3	41.2	5251	441	176
	Philadelphia	53.5	44.5	4486	479	205
	Pittsburgh	53.0	42.2	5053	471	192
Rhode Island	Providence	50.1	38.8	5954	462	249
South Carolina	Charleston	65.2	57.9	1794	544	220
	Columbia	64.0	54.0	2484	548	216
	Spartanburg	61.2	51.6	2980	540	208
South Dakota	Huron	44.7	28.8	8223	498	258
	Rapid City	46.8	33.4	7345	537	292
Tennessee	Chattanooga	61.2	50.3	3254	515	189
	Memphis	61.5	51.6	3015	535	192
	Nashville	60.0	48.9	3578	497	174
	Oak Ridge	58.2	47.7	3817	482	170
Northern Texas	Abilene	64.3	53.9	2624	608	240
	Amarillo	58.7	47.0	3985	635	251
	Dallas	65.8	55.3	2363	533	210
	El Paso	63.3	52.9	2700	720	296
	Fort Worth	65.8	55.1	2405	609	232
Southern Texas	Austin	68.3	59.1	1711	554	160
	Brownsville	73.7	67.7	600	603	181
	Corpus Christi	71.8	64.6	914	588	172
	Houston	69.2	61.0	1396	561	161

Table I-8—Continued

State or Province	Location	Ambient Temperature (°F)		Annual Total Heating Degree-Days	Average Solar Energy Received on Horizontal Surfaces*	
		Annual Average	Winter Heating Season Average		Annual Total (KBtu/ ft²/yr)	Winter Heating Season Total (KBtu/ ft²/season)
		A	B	C	D	E
Southern Texas (continued)	San Antonio	68.7	60.1	1546	602	181
Utah	Salt Lake City	50.9	38.4	6052	558	291
Vermont	Burlington	43.2	29.4	8269	427	218
Virginia	Lynchburg	56.7	46.0	4166	509	225
	Norfolk	59.7	49.2	3421	515	229
	Richmond	58.1	47.3	3865	501	218
Washington	Seattle-Tacoma	51.1	44.2	5145	394	192
	Walla Walla	54.2	43.8	4805	476	228
West Virginia	Parkersburg	55.0	43.5	4754	454	181
Wisconsin	Green Bay	44.3	30.3	8029	440	172
	Madison	45.0	30.9	7863	460	274
	Milwaukee	45.1	32.6	7635	464	182
Wyoming	Cheyenne	45.9	34.2	7381	581	319
	Lander	44.4	31.4	7870	586	320
	Sheridan	45.2	32.5	7680	522	272
Washington, D.C.		57.0	45.7	4224	477	206
Alberta	Calgary	38.2	22.9	9520	433	324
	Edmonton	34.5	19.6	10320	381	285
	Medicine Hat	41.2	24.9	8650	456	341
British Columbia	Prince Rupert	45.8	34.9	6910	271	156
	Vancouver	49.7	42.0	5520	387	197
Manitoba	Brandon	35.1	19.6	10930	441	330
	Winnipeg	36.2	20.4	10658	435	240
New Brunswick	Fredericton	41.8	28.6	8830	397	226
	Moncton	41.6	29.2	8830	401	300
Newfoundland	Gander	39.7	29.5	9440	332	191
	St. Johns	40.8	32.4	8940	340	254
Nova Scotia	Halifax	42.8	33.4	7585	393	223
	Sydney	42.8	32.9	8220	387	290
	Yarmouth	44.6	35.7	7520	391	292
Ontario	London	45.5	34.1	7380	431	323
	Ottawa	42.5	28.6	8740	423	240
	Sudbury	38.5	24.4	9870	411	307
	Toronto	45.5	33.9	7008	434	247
	Timmons	34.6	18.7	11480	417	311

Table I-8—Continued

State or Province	Location	Ambient Temperature (°F)		Annual Total Heating Degree-Days	Average Solar Energy Received on Horizontal Surfaces*	
		Annual Average	Winter Heating Season Average		Annual Total (KBtu/ ft²/yr)	Winter Heating Season Total (KBtu/ ft²/season)
		A	B	C	D	E
Prince Edward Island	Charlottetown	42.0	30.5	8710	402	226
	Summerside	42.5	30.9	8486	410	307
Quebec	Quebec	39.9	26.4	9070	394	295
	Sherbrooke	39.2	24.1	8610	399	299
Saskatchewan	Prince Albert	32.3	15.5	11430	400	232
	Regina	35.7	19.5	10770	439	328
	Saskatoon	34.9	18.4	10960	431	322

SOURCES:

1. "Monthly Normals of Temperatures, Precipitation and Heating Degree Days," U.S. Weather Bureau publication reprinted in *American Society of Heating, Refrigerating and Airconditioning Engineers Handbook: 1976 Systems* (New York: ASHRAE, 1976), pp. 43.2–43.7.

2. *HUD Intermediate Minimum Proper by Standards: Solar Heating and Domestic Hot Water Systems,* no. 4930–2 (Washington, D.C.: U.S. Government Printing Office, 1977).

3. T. Kusuda and Ku Ishi, *Hourly Solar Radiation Data for Vertical and Horizontal Surfaces on Average Days in the United States and Canada,* no. 003-003-01698-5 (Washington, D.C.: U.S. Government Printing Office, 1979).

4. *Copper—Brass—Bronze Design Book: Solar Energy Systems* (New York: Copper Development Association, 1978).

5. Data for Canadian cities obtained from Canadian Climate Center, Climatological Applications Branch, Downsview, Ontario, Canada.

Working

Some thoughts on doing it right

This roof I'm standing on is downright dangerous. The winter sun has been up for a couple of hours, but it hasn't melted off a devilish coating of frost that wants to send me back to the sidewalk the fast way. What's worse, the old asphalt shingles are completely embrittled, and every step is like a spin of the chamber in Russian roulette. Will this shingle carry my boot or will it disintegrate and conspire with the frost to send me flying? Why were these collectors laid over this mess? What the heck am I doing here clinging to this crumbling, slippery accident-waiting-to-happen?

The repair was finally made, but on a scale of 10, job safety for that little outing rates about 2 (a 1 would actually imply taking the fast lane to the sidewalk, a 0—heaven forbid). Job planning gets maybe a 4: Flaky as it was, the roof wasn't a total failure until a few maintenance excursions finally did it in. Finally it was replaced, the job done right.

Solar retrofitting, like window washing, can put you in some pretty high places, and the higher the place the more important job planning and safety become. But at any altitude a well-planned job saves time, increases your enjoyment and probably improves the result. As for job safety, well, your body is your business, but every ounce of prevention you can work into the job plan will return peace of mind in abundance, along with all the other above-mentioned benefits. (Besides, the price of a solar retrofit plus doctor bills can really reduce cost-effectiveness. . . .)

Planning

Job planning simply means using forethought to minimize the number of questions a job presents so you can start from a state of readiness instead of uncertainty. The projects in this book list the basic steps for doing the retrofit work, but they can't, of course, speak directly to the vagaries of your job at your site. What will it take, for example, just to get to project Step One? If you are more or less new to do-it-yourself home improvement, try to visualize and write down what it's really going to take to get from A to B and finally to Z, based on what the project tells you and on what your house tells you. Naturally, your experience in home improvement should dictate which project you select, and with the simpler projects you'll be able to see from A to Z without too much difficulty. With more experience you can literally take a more complex project apart in your mind's eye and see most of what's going to be happening. But it will always be helpful to work up a little list of major and minor steps, reminders on little details that might drop from the view of your mind's eye. Always give yourself a break now and again too, not just for rest, which is important, but also so you can sit back and survey the scene, modify your plans, and check details. Energy retrofitting is very much a matter of details. Insuring against air, water and thermal leaks is one of the best ways to optimize a heating system and get the most Btu's for the bucks you're putting in. A 10 or 15 percent difference in thermal performance can depend solely on workmanship.

Scheduling

Time is, of course, not likely to be in great supply, and some degree of scheduling may help you to better utilize that not-so-plentiful resource. Experience will also improve your scheduling, but for starters figure that a job can easily take one-and-a-half to two times longer than you think. No kidding. It's easy to be tricked by the apparent simplicity of a project. But the simplicity is more likely to be in the appearance and operation of the finished product, while the product itself requires close attention to several time-consuming details. This is exciting work, building your own energy self-reliance, but don't let your eagerness boil down to impatience and frustration when things seem to take too long.

In larger jobs it's important to schedule your progress so that by the end of the day you can leave work knowing that all is stable and, where necessary, protected from the elements (wind and rain being the most formidable). Larger jobs are also likely to involve more than one worker, which puts even more emphasis on the importance of planning and scheduling the sequence of events.

Safety

Odds are that most do-it-yourselfers scoff at or don't even think about the most basic job safety measures. I've been a flagrant offender myself, and I've suffered enough to yield to what is probably the most important safety requirement: eye protection. Six or seven dollars gets you a pair of ventilated goggles and a pair of clear plastic safety eyeglasses. When you are doing work over your head, cutting into ceilings or whatever, the dust and debris can fly anywhere, and glasses alone won't keep it out of your eyes. Goggles do, and they give you much more freedom to watch what you're doing.

Power tools or any kind of impact work such as hammering or chiseling can propel little particles along all sorts of random trajectories, and the only safe bet is safety glasses, which can be either low cost and nonprescription or tailored to your eyesight. Both types will have extra large plastic lenses. Once you have them, try not to forget them. Keep them in your tool pouch, and they'll follow you everywhere. Put them on when you start working, and then forget about them.

Ears can use a little protection from the high-pitched screams of saws, routers and other noises. Unlike eye damage, hearing impairment is something you won't notice because it happens so slowly, over a period of years. But when you turn around one day and say, "What?"—it's too late. "Ear goggles" are another five to ten dollar investment that protect your ears and also help your workmanship. The din of power tools can be quite distracting; it can make you hurry when you shouldn't. Ear protection muffles the racket without eliminating sounds you should hear, like other people's voices. It makes work more enjoyable.

The list of personal job safety items also includes hard hats, respirators and plain old gloves. Protecting your skull naturally becomes more of an issue in larger projects when you're working below other people. Mistakes do occur, but with good job planning you should be able to keep people out of each other's way and out of the path of things that might be falling. Respirators, gloves and eye protection are a must when handling fiberglass insulation, which routinely sends tiny glasslike fibers into any part of you that isn't covered. The dusty environments of old attics and basements are another possible source of lung irritation, as are closed indoor spaces where volatile fluids such as paint are being handled. As with ears, lung damage takes time, so if you're frequently involved with these conditions, you will do well to filter out potential troublemakers.

Materials

The materials checklists included in the projects appear in front of the construction steps for one simple reason: Get *all* your materials together before you start. I'll never forget the time when a simple plumbing alteration took about ten six-mile round trips between home and the hardware store, all for picayune but absolutely necessary parts. It was a classic case of planning by "revelation." If the job had been completely considered, it might have been one trip for most of the stuff, with one "cleanup" excursion for the unforeseeable items.

Buying materials should be one of the most enjoyable parts of the job, not one of the most frustrating. It's part of getting psyched up: There you are, list in hand, compiling all this potential! And all the while your mind's eye is connecting that odd assortment of lumber, glazing and hardware with the finished masterpiece.

When you're gathering materials, especially lumber, be picky. Some dealers will cull any twisted or bowed pieces out of your order, and some won't bother. You're better off making your own selections. Lumber quality is going down if it's going anywhere these days, and there are few things more bothersome than having to work a mutant board into an otherwise precision structure. Green lumber can go from marginal to unusable after drying, usually because of knots. Remember, too, that wood in a solar retrofit is likely to experience high temperatures and/or all the worst that nature has to throw at it. Good quality wood helps to ensure longevity as well as fewer problems in construction. Don't compromise: Avoid knots; look at 20 pieces to select 10 you really like; for exposed pieces use rot-resistant woods like cedar, redwood, cyprus or pressure-treated pine.

The finished appearance of your retrofit design can be as much a factor in its success as its energy output. This will be especially true when the home enhancement and the energy improvement are the same project, as with a greenhouse or a building addition. Glazing, interior and exterior trim, paint and stain color, even the selection of nails and other exposed fasteners—these are all material decisions you can control, and if spending a little more means gaining a much better appearance, it could really be worth it. Along with choosing materials for durability, you're also aiming for a good integration of your solar improvement and your house. The angularity and the expanses of glazing in solar design are striking enough, and they should be deemphasized rather than exaggerated. The sections in this book on solar additions and whole-house retrofits show how materials' selection and use, along with design, can help the project blend with as well as enhance the rest of the house.

Tools, Tools, Tools

If tools could talk, they would recount tales of abuse and abandonment, fair treatment, tender loving care and even of being worshiped—all at the hands of their owners. For better or worse their fate is in your hands. Tools don't "do"; they are "done with," an important distinction when you're ready to punish one that's "misbehaved." Be nice to your tools, and they'll perform faithfully. Tools don't make mistakes, people do; so be nice to yourself on the job—plan, schedule, rest—and your tools won't become instruments of ill-preparedness or fatigue.

There's no direct relationship between the quality of the finished project and the quality of tool handling, but you can make a job somewhat more or less difficult by the way you manage your tools. Clutter is progress enemy number one. Spread all over a workshop or a job site, tools lose their accessibility, and you lose time. It may be totally impossible for you to keep your tools together (it's my own worst tool offense), but try to acquire the habit. Putting a tool back in the rack or in your pouch may seem time-consuming when you're in the fray of a project, but it's a true time-saver. (I've even lost some favorite tools forever, by putting them "nowhere.")

Another time eater and prime frustrator are tools that are out of shape. Consider a dull circular saw blade: It's harder to cut straight lines; you have to exert yourself more to push the saw through the stock; the motor has to exert itself more to make up for the dull blade; you both burn out sooner. Depending on the tool, it will want to be sharpened, cleaned, oiled and adjusted sooner or later; it's preventive medicine that forestalls early retirement and promotes job happiness.

If you're just beginning to build your tool collection, start with the basics and buy more specialized tools as you need them. I've never been impressed by tools that purport to have ten different functions, especially measuring and layout tools where accuracy is often sacrificed in favor of gimmickry. The following tool list breaks down the parts of large and small projects into plumbing, electrical and woodworking categories, and it may be useful if you're beginning or enlarging your collection.

A final word on tools: You've probably noticed that all tools are *not* created equal. Quality can vary tremendously, and it's safe enough to say that price generally varies with quality, when you're buying new. With power tools the major manufacturers often rate their various lines by a duty grade; homeowner's, medium, heavy, super and continuous-duty are some of the labels. With hand tools you've got to give a close look at any number of details when comparing for quality: the heaviness of cast parts, the way parts fit together, the way they're finished and so forth.

For my own money I will tend to spend more to buy higher quality tools, both hand and power, because time and time again I've seen cheap tools that self-destruct

Tooling Up

If you're preparing for any of these kinds of jobs, use these checklists to make sure you're tool-ready. It's a good idea to have a separate box or compartment in your toolbox or storage space for each class of tools; that way you may only have to rummage through 20 tools instead of 100 to find the one you want.

Plumbing

pipe handling
two pipe wrenches, one at least 14 inches long
locking pliers (e.g., Vise Grips)
adjustable wrenches, one at least 12 inches long

adjustable plier-wrench (e.g., Channel-Lok)
tubing bender (for soft copper tubing)

pipe cutting
tubing cutter (capacity up to 1-inch diameter copper)
hacksaw

pipe reamer (for up to 1-inch ID pipe)

pipe joining
teflon thread seal tape or pipe thread compound for sweating (soldering) copper pipe:
gas torch (propane, butane, Mapp gas) and sparker
solder, referred to as 60–40 (60 percent lead, 40 percent tin)
coarse steel wool (for preparing outside pipe surfaces) or emery cloth

cylindrical wire brush (½- or ¾-inch diameter for cleaning the inside of pipe fittings)
paste flux
PVC solvent for joining plastic pipe

Electrical

wirecutter-stripper
simple 120 VAC circuit tester (test light), or volt-ohm-milliammeter (measures voltage drop, continuity, resistance)

cable sheathing cutter
needle-nose pliers
screwdriver (for small slotted screw heads)
"fish tape" (a tool for running wire through stud walls)

Woodworking

Hand tools or power tools? If you're thinking that way, stop. It's really more appropriate to talk about *which* hand tools and *which* power tools because, depending on the situation, both classes can be a boon to job progress.

or at least become marginally useful when the going gets rough. The better tool performs better, is more accurate, lasts longer, and thus makes itself a better investment. Drop a cheap circular saw and you can throw it out of whack, even out of commission. A better model will have ball and roller bearings instead of bushings, heavier cast or machined parts, and more motor power from a better quality motor. With hand tools the 99¢ bin in your hardware store is not the place to shop. Find out where people who work in the

Generally the basic power tools, such as a circular saw, a saber saw and a drill, will become indispensable for saving time and muscle energy. Remodeling often means cutting into tight corners or other impossible zones where the only (neat) way in is via a certain power tool. Many remodelers swear by the convenience of a reciprocating saw for making light work of ever-present tight spots. A great deal of convenience is also afforded by stationary power tools, namely a table saw or radial arm saw. They can help speed up and improve the quality of finish work and are very handy when there's a need for nonstandard lumber dimensions. As you peruse this list of woodworking tools, look at not only your present but also your future plans for do-it-yourself projects. Maybe it's a good time to buy some of the more "advanced" tools now.

Tools for Simple Add-On Projects
(e.g., small air or water heating collector systems)

20-foot tape measure
combination square
24-inch level
bevel gauge
chalk line
16-ounce straight claw hammer
10-point crosscut saw
¾-inch wood chisel
nail puller ("catspaw")
screwdrivers
hacksaw

sheet metal cutters
keyhole saw
caulking gun
adjustable wrench
⅜-inch variable speed-reversible drill
1/16- to ¼-inch high-speed steel
   drill bit set
¼- to 1-inch paddle-type wood
   boring bits
countersink bit, screwdriver bits

Tools for Light Remodeling
(e.g., increasing south glass area or installing a skylight)

carpenter's framing square
plumb bob
flat bar-prybar
cold chisel
5-pound hand sledge (the
   quintessential remodeler's tool)

block plane
smoothing plane
circular saw (at least 7-inch
   diameter blade)
saber saw

Tools for Extensive Remodeling or Building Additions

reciprocating saw (nice but
   not absolutely essential)
table saw or radial arm saw

belt sander
orbital sander

trades buy their gear. That's where you'll find quality and possibly even discounts. Of course, a lot of people will insist that the old tools are better than their descendants, and that's very often true. You can get your buy-list together and very successfully hit the junk stores, tag sales and auctions and come away with nice stuff at righteous prices. It may take more time, but then again treasure hunting can be a lot of fun.

Joe Carter

II DOMESTIC WATER HEATING

Photo II-1: Solar DHW, circa 1911, Pomona, California. Only the clothes are the giveaway; otherwise this could be now: a nicely installed flat plate collector system for a sizable family. There's more: a little sunroom, under the collector array, and what appears to be a little ventilation cupola at the roof ridge for some summer cooling. Add the garden and self-reliance looks like business-as-usual in 1911. (Photograph from A Golden Thread, by Ken Butti and John Perlin [Palo Alto, Calif.: Cheshire Books, 1980].)

DHW

DHW is domestic hot water. It's a separate little subsystem of your home's total energy system, and in terms of its monthly fuel cost it often doesn't appear to be much of an energy liability. Maybe people only notice the cost of energy when the big winter bills start rolling in. The DHW bill is buried in there somewhere, and only when spring comes to the rescue does energy start to look "cheap" again. But "cheap" is in the eye of the beholder. If we take a straight look at DHW economics, maybe it isn't so cheap, especially in the light of what conservation and solar improvements can do to dramatically reduce DHW energy cost. At average consumption levels, a household of four people can run up an annual DHW load of 25 to 30 million Btu, which can cost anywhere from $200 to $450, depending on the fuel used. If that doesn't seem so cheap, here's the pitch: One-half to two-thirds of that cost can be saved every year with a one-time conservation investment of around $125. Reliable solar DHW systems can pick up a high percentage of the remaining load, and although they cost more than conservation, the systems described in this section can still pay for themselves in less than ten years (less than six years when the 40 percent federal solar tax credit is applied). In short, the opportunities for belittling your DHW load are many and ripe for the picking. Solar DHW systems are among the most cost effective of residential solar applications. Doing one yourself requires a basic knowledge of plumbing, electricity and carpentry, a little native canniness, and a lot less investment than for franchised, store-bought systems. But don't kid yourself: You've got to do these systems all right, or they're likely to be all wrong.

The Big Picture is always fun to look at. The residential sector uses about 3 percent of this country's total energy use for DHW. If that load could be cut in half (easy, conservation alone could take care of that), a lot of energy could be saved: 350 billion kilowatt-hours of electricity, 12 billion therms of natural gas or 8.6 billion gallons of fuel oil. Taken 40 or 50 gallons at a time, DHW is small; take it across the country and it can play a significant role in our efficiency growth.

Photo II-2: There are at least two water-saving techniques shown here: taking a shower instead of a bath and doing it with a low-flow shower head. There aren't any data on the water-saving value of showering with a friend, but the fun value ought to rank high. (Photograph courtesy of Resources Conservation Inc., Greenwich, Conn.)

Lightening the DHW Load
Some simple techniques that save plenty

Like most home energy systems, DHW systems don't include high efficiency as a built-in feature. That of course isn't surprising. Conventional DHW systems share the same roots as their space heating and cooling brethren: cheap energy. "And it isn't cheap anymore . . ." etc., etc., etc., the same old song. What's nice about DHW systems, though, is that their inefficiencies can be easily corrected with a minimal investment, while bringing space conditioning systems up to snuff takes more time and money. Maybe lightening the DHW load is a good place to begin your home energy upgrading. You will get immediate results and gratification in your very first postconservation energy bill.

The DHW load can be cut by 25, 50, even 75 percent with mostly standard conservation measures that will pay for themselves in two to three years. And the benefits go further: If you only need three-quarters or one-half as much energy to have hot water, a solar water heater will provide a proportionately higher fraction of the remaining load, further reducing your need for conventional DHW fuels. Better still, a greatly reduced DHW load needs a smaller and therefore less expensive solar component to get the job done. Instead of building four collectors for about $250 apiece, you might need just two collectors, which saves you $500 right off the bat.

Fewer Gallons Get You Just As Wet

Saving water heating energy involves two basic categories of conservation opportunities: reducing the gallons of hot water used and reducing heat loss from various points in the system. Reducing your consumption of hot water doesn't mean going without; it means getting the same service with less hot water, restricting the flow. The place to start is right where the cold water main enters your house. This is where you may be able to reduce the line water pressure. In municipal water systems pressure is often too high, which is any pressure over 50 psi (pounds per square inch). I've played around with pressure reduction and found that domestic water flows just fine with as little as 30 psi behind it. To find out what your line pressure is, call the people in your area who pump the water. If they pump it up too high, your system is a candidate for a pressure-reducing valve, a $20 to $35 item that is installed just downstream of the main shutoff valve on the cold water line. There are types that can be soldered onto copper pipe and others that can be threaded onto steel pipe. These valves are adjustable for the desired pressure, and the best way to keep track of your throttling-down is with a pressure gauge ($5 to $10), which is installed just downstream of the pressure-reducing valve. Try 40 psi; try 30; try 25! Lower pressure means a decreased flow rate and thus fewer gallons running out the tap. Cutting the pressure in half gives about a 25 to 35 percent reduction in flow rate. You save hot *and* cold water; a water heater operating under reduced pressure lasts longer; water hammer is quelled. If you run your own water system, try to leave one or two outside spigots isolated from pressure reduction. In case of fire you want to be able to shoot water long and high.

The next point for gallonage reduction is at the other end of the system, the outlets. Flow-restricting devices for taps and showers can greatly reduce flow rates. A low-flow shower head passes water at 1½ to 2½ gallons per minute (gpm) instead of the usual 4 to 6 gpm. These units simply replace the existing shower head; you need a crescent wrench, some Teflon plumber's tape for the pipe threads and a little elbow grease. Low-flow faucet aerators, screwed in place by hand, also cut flow to 1½ to 2½ gpm, beating out standard aerators that run at 3 to 3½ gpm. The greatest gallonage savings will be realized in the shower, where people often tend to stand around just for the enjoyment of it all. Low-flow shower heads don't eliminate the luxury, just the waste involved. Instead of putting 10 to 20 hot gallons per shower down the tubes, you can enjoy yourself with 4 to 6 gallons (five-minute shower). With flow reduction improvements you can save on both your water and your sewer bills if the latter is determined by how much water you use.

Another major hot water load is the washing machine. A standard-size washer uses about 25 gallons of hot water per full load on a hot-wash, warm-rinse cycle. All hot water use could be eliminated with the flick of a switch to the cold-cold cycle. Cold water detergents work, so you're not compromising your whitest whites if you make the switch.

Smaller gallonage reductions can be had by paying attention to the way you use water. Technically it could be called habit modification; actually it just means not running the water when you don't need to. If you catch the conservation bug, things like that will become mindlessly habitual. Average American rates of hot water consumption are in the range of 17 to 22 gallons per person per day (gpd). By instituting these conservation measures, that average can be cut in half, with an equivalent cut in your energy use.

Less Energy Keeps It Just As Hot

Obviously reducing your hot water gallonage makes your water heater work less, yet there remain several more opportunities for giving the old tank a break. The quickest and easiest of them is thermostat setback. As they come from the factory, electric water heater thermostats are usually set at 140°F (60°C), much, much too hot. Even if you enjoy maximum scald in the shower, you're experiencing at the most 120 to 125°F (49–52°C) water; most people do it at 100 to 115°F (38–46°C). You can experiment with the thermostat setting to get it as low as you like. First, turn off the power to the water heater. (There should be a separate circuit breaker for it.) Then open the two side cover plates, one at the top and one at the bottom, corresponding to the location

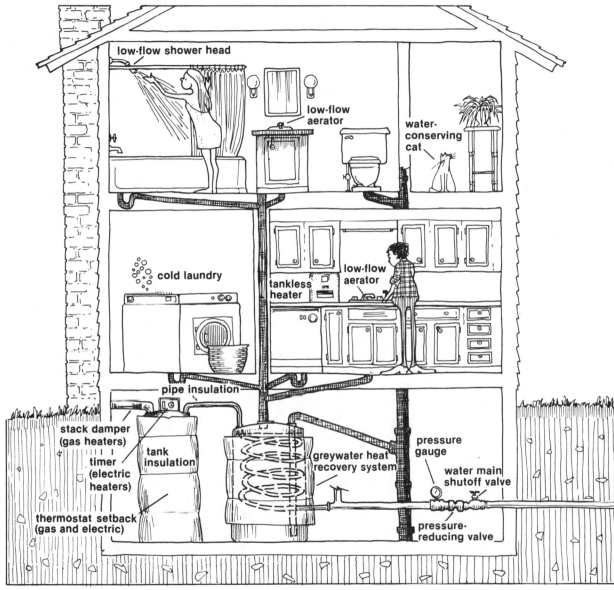

Figure II-1: Converting domestic water heating systems into energy misers doesn't take a whole lot of work. The quickest retrofits are the low-flow devices. The stack damper for a gas heater will take 10 minutes to install. Tank insulation takes another 15 minutes, and a setback thermostat takes 0 (gas heaters) to 15 (electric) minutes. The pressure-reducing valve and pressure gauge might take an hour or two, depending on conditions, and a timer for an electric heater might take another one or two hours to install. It's a small investment of time that can save $10 to $30 every month.

of the heating elements. Inside you'll see the thermostats. They need only a thin screw-driver or a sturdy fingernail to turn them back. Always set the top thermostat 5 degrees higher than the bottom one. On gas- and oil-fired water heaters go just below the "warm" setting and see how it suits you. But, you say, there are five people here who literally languish in the shower; I need super-hot water or there'll be none left for me at the end of the line. Well, ask your people to languish a little less, but remember this too: By cutting flow rates by 50 to 75 percent, your group doesn't use the same volume anymore, and it can probably get by with a "cooler" tank. The total cost for this improvement: $0.

Thermostat setback reduces the rate of heat loss from the water heater tank and the hot water pipes. Insulation does a good job of that too. Add 6 inches of fiberglass insulation around the tank and over the top. You can completely cover electric water heaters. With oil- and gas-fired units you have to leave a little space around the exhaust stack and also make sure that insulation doesn't block any combustion air intake passages. This improvement will cost about $12, the price of a roll of insulation. The combined energy savings of thermostat setback and added insulation will cut about 25 percent from the remaining DHW load after flow reduction.

Depending on the hot water use patterns in your household, pipe insulation may be a significant energy-saver. If hot water draws are frequent throughout the day, hot water will always be present in the pipelines, losing heat. By insulating pipes to at least R-4, you can cut another 10 percent from your water heating load. If your hot water use is concentrated in the mornings and evenings, pipe insulation will still be a benefit, but less so, reducing the remaining load (after low flow, setback and tank insulation) by about 3 percent. The cost for pipe insulation is around 25¢ to 50¢ per foot.

Another way to reduce heat loss from gas-fired water heaters is with a small stack damper. Just as boiler and furnace efficiency is increased with a stack damper in place, you can slow down convective heat loss up the water heater's center tube with an automatic damper that opens when the burner is on and closes when the burner shuts down. To date there is just one product available that is appropriate for water heaters. The damping action in this unit is done with four bimetallic leaves attached inside a stainless steel stove pipe section. Heat (burner on) causes them to spread open and vice versa. There's no wiring or fancy hookup involved, which makes installation very simple. The "AmeriTherm" is made by American Metal Products Co., 6100 Bandini Boulevard, Los Angeles, California 90040.

Automatic timers can be used to control the on-times of electric water heaters. Again, if your hot water use periods are fairly concentrated, a timer can be used to turn on the heater during those periods and to keep it off during limited-use periods. This improvement will cost about $25 in materials and can save 5 to 10 percent of a reduced DHW load. Timers are often used when a solar water heater is part of the DHW system. Most of the sun's energy input occurs during a limited daytime period, usually 9:00 A.M. to 3:00 P.M. (standard time), which means that a substantial amount of hot water is put into storage for evening and morning use. The timer is used to keep the heater off during the day, when solar energy is replenishing the depleted hot water supply. If there is no solar gain on a given day, you suffer only a slight delay in getting electrically heated water after the timer turns the heater back on.

Boiler systems that use fuel oil are often equipped with a domestic water heating coil in the boiler water containment. When the boiler is puffing away in winter, the boiler water is always hot, and the heat exchanger DHW system operates with adequate efficiency. But when there is no space heading load, the boiler is kept running anyway just to feed the DHW load. The system efficiency plummets from about 70 percent in winter to between 15 and 25 percent during nonspace heating periods, and the cost of DHW energy shoots up. For example, if oil at $1.00 a gallon is used at 70 percent efficiency, the cost per million Btu (MBtu) is $10.23. But when it's used at 20 percent efficiency, the cost for the energy delivered as useful hot water is $35.82 per MBtu, which is not cheap. You can't turn off the hot water, but you can turn off the boiler in summer if you install a separate water heater (gas or electric). When you don't need space heat, you can turn on the water heater, bypass the boiler and save energy dollars. For example, at 5¢ per kilowatt-hour the cost of heating water electrically is $16.28 per MBtu (90 percent efficiency). The cost of adding the water heater will be paid back with energy savings in about three years. Another plus: If the boiler isn't running in summer, it isn't heating up your basement or your house. Boilers are for winter.

There's a very different sort of water heater now available, called a tankless heater or demand-type heater, which can further reduce system heat loss. Tankless means that the unit doesn't hold any hot water, but rather heats it only on demand, as water passes through it. These gas or electric units can produce temperature rises of up to 60°F from incoming water temperature at lower flow rates of around 1 gpm. This makes them possibly an ideal companion to solar DHW systems, which can always be counted on to maintain at least 70 to 80°F (21–27°C) water in the storage tank. With or without solar, they can also be used to boost water temperature to dishwashers, and they can allow further thermostat setback of the standard water heater, which further reduces tank heat loss.

There is one more opportunity for DHW energy conservation that could lighten the load by 25 to as much as 50 percent: *greywater heat recovery.* Greywater is water that goes down the drain after you've used it. When you shower or do laundry, a lot of energy in the form of waste hot water is simply drained away. A system for heat recovery would get some of that energy back into the DHW system before it was lost to the sewer. If that sounds a bit odd or complicated or just plain unsanitary, well, it may be odd because nobody in your town is doing it, but it's essentially a simple, completely hygienic energy-saving possibility. Divert certain greywater sources (the hottest ones: shower; laundry, if it's still hot; dishwasher; bathroom sinks) to a small 50- to 100-gallon holding tank. As figure II-2 shows, this tank is basically just a bulge in your sewer line, a place for hot and warm water to collect before it runs out to the town sewer line or your septic tank. It's a heat storage tank, so it should be insulated. To get the heat out of the tank and into your water heater, you need a heat exchanger, and that means diverting the cold line that supplies the water heater. Run it through the tank in the form of a coil of 60 to 120 feet of ½-inch copper tubing, and then connect it back to the water

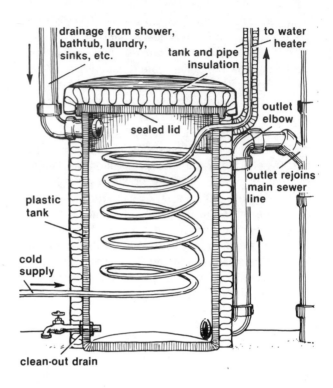

Figure II-2: First of all, there is nothing but idea in this illustration; it has never been tried, but it looks like it could work. The tank could be some kind of heavy plastic, such as a garbage can, and the lid could be gasketed to make a tight seal to prevent odors from escaping. The inlet line is placed above the elbow in the outlet line, and it's that elbow that determines the water level in the tank. The outlet comes from the bottom of the tank because that's where the coldest water is. If the outlet were at the top, the hottest water would run out first. One standard coil of copper tubing is 60 feet long, a good length to experiment with. Most of the coils should be near the top of the water line, though none should be above it (above the outlet elbow). Some sort of drain would probably be needed at the bottom for flushing out sediment that may collect. A coat of insulation completes the scene.

heater. The water entering the heater tank will thus be preheated as it extracts heat from the greywater, without ever touching the greywater itself.

That 25 to 50 percent heat recovery estimate isn't just wishful thinking. In Sweden the system has been successfully commercialized. Recent research in this country has shown that this level of heat recovery is quite possible. Unfortunately, you can't go to the store and buy the Acme Greywater Goddess, not yet at least. So if you're interested, as only a fanatic could be, it'll be up to you.

Of course, you don't have to do the greywater thing in order to make a big dent in your DHW load. The standard measures described here will cause a minimum 25 percent reduction. When I worked in all these improvements, my load was unloaded by about 60 percent, at a cost of about $120, which will be recovered in about three years. I no longer need 60 square feet of solar collector to get a reasonable solar DHW fraction. With 30 to 40 square feet I can all but eliminate natural gas from my DHW system. If I did the greywater thing, I could probably close the gas valve. We can't sing the praises of conservation any more loudly than that.

Joe Carter

References

Milne, Murray. 1976. *Residential water conservation.* California Water Resources Center Report no. 35. Davis, Calif.: University of California at Davis.

———. 1979. *Residential water re-use.* California Water Resources Center. Davis, Calif.: University of California at Davis.

Palla, Robert L. 1979. *The potential for energy savings with water conservation devices.* Washington, D.C.: National Bureau of Standards.

Hardware Focus

Instantaneous Water Heaters

Chronomite Laboratories, Inc.
21011 S. Figueroa St.
Carson, CA 90745
(213) 320-9452 or 533-0409

Instant-Flow Water Heater (electric)
Use as booster unit or total hot water system. Units available from 2300 to 9000 watts and operate at either 110, 208 or 220 VAC. UL approved.

Kiley Co.
67 River Rd.
Cos Cob, CT 06807
(203) 869-4750

Thorn Instant Heater (gas)
Heater comes in three models ranging from 30,000 to 114,000 Btu/hr input.
Elf (electric)
Single outlet, shower or sink heater; 6 kw (220 volts).

Paloma Industries, Inc.
241 James St.
Bensenville, IL 60106
(312) 595-8778

Paloma Constant Flo (gas)
Japanese-made heater comes in six models ranging from 30,000 to 178,500 Btu/hr input. Brochure available.

John Condon Co., Inc. (distributor)
1103 N. 36th St.
Seattle, WA 98103
(206) 632-5600

Little Giant Manufacturing Co., Inc.
P. O. Box 518
907 Seventh St.
Orange, TX 77630
(713) 883-4246

Little Giant (electric)
Comes in five models ranging from 1.5 to 18 kw (110 volts). UL approved.
Hotomatic (gas)
Comes in four models ranging from 40,000 up to 240,000 Btu/hr input.

Pressure Cleaning Systems, Inc.
    (distributor)
612 N. 16th Ave.
Yakima, WA 98902
(509) 452-6607

Junkers (gas)
German-made heater comes in four models ranging from 37,200 to 126,000 Btu/hr input.

PROJECT
A Flat Plate Collector

There are several options for solarizing your DHW system, and most of them start with the basic, classic solar collector.

Today there must be millions of them lying on the rooftops of the world, but they're by no means newcomers to the built environment. The flat plate collector has been here before. In this country, beginning around the turn of the century, flat plate DHW systems became a popular necessity, especially in southern California and Florida. Solar businesses flourished until cheaper fossil fuels became available, against which solar fuel couldn't compete. The hardware wasn't at fault, just the economics; systems that were installed in the 1920s and 1930s can be found in operation today. But now that the economics are favorable, the hardware has come back like gangbusters.

The systems that were in place decades ago were the simplest of the flat plate collector systems. *Thermosiphon systems* ran by convection; water heated in the collector would rise into an insulated storage tank, and colder water from the tank would enter the collector in a continuous cycle. Nowadays thermosiphon systems are joined by half-a-dozen other generic system types that have evolved in recent years. As you'll see in "A Catalog of Flat Plate Collector Systems," found later in this section, they're all good and useful for different reasons, and they all use that same basic solar component.

The designs for flat plate collectors are many and varied in their details, but in their essential parts they're all the same. A flat plate collector is a slim, rectangular box containing some insulation and a metal heat-absorbing plate with water conduits and glazing. It's all assembled into a weather-tight unit

that you can put on your roof, on one of your walls, out in the yard or wherever the sun is in your place.

From that point you basically want to forget about the thing. A properly made collector should stand up to the weather for years with just minimal maintenance, and the type of system you connect it to should be equally reliable and trouble-free. In all of the flat plate collector systems that will be described in "Catalog," the collector is just another repeatable item, and like a pump or a valve or a pipe fitting, you can go out and buy what you need from a local solar business or even from a mail-order source. But as with anything that's built, you pay for materials, labor, overhead, profit and whatever else works into the deal. Store-bought collectors can be expensive, but you can get into the deal for less by making your own. If you can build a bookcase, a do-it-yourself collector is in your league. Its parts are few and simple—some wood, some metal, some glazing—yet the sum of these parts is an energy producer that will deliver anywhere from 140,000 to 360,000 Btu per square foot per year. The materials' cost for a good quality, home-built collector will be in the area of $10 to $12 per square foot, well below commercial price ranges. In terms of performance, a home-built unit will deliver as much or nearly as much energy as many ready-mades.

We can start by describing all the parts in this particular design, and then talk about how they all go together. This collector design consists of a wood frame with a plywood back and rigid foam insulation, an all-copper absorber plate and single or double glass or plastic glazing. A few screws and nails, some silicone caulk and metal angle round out the collection.

The Box

Collectors that you can buy usually have metal frames, but a wood frame can be just as substantial. Wood is also a better insulator than metal, and it helps to reduce heat loss out the sides of the collector. In this design a standard 1 × 4 stock is used. It must be a high grade wood, preferably clear, kiln-dried boards that are unbent and unbowed. The best woods to use are those that are naturally rot-resistant, such as redwood, cedar or cypress. It won't be cheap wood, but it's very much worth it to have maximum insurance against warping or rotting. If these varieties aren't available, stick with the clear, kiln-dried rating for whatever you can get and treat it with wood preservative or paint if you want to match the box with your roof or wall color. With proper fastening, sealing, and maintenance, a wooden collector is going to last a long time.

The number of feet you'll need depends on the size of the absorber plate, which will be 2, 3, even 4 feet wide and 6, 8, or 10 feet long, give or take a few inches. After the 1 × 4 is cut to the right lengths, each piece gets a ⅜ × ⅜-inch rabbet along the bottom inside edge. This groove receives the plywood backing. The plywood should be a standard ⅜-inch exterior grade (CDX). The only other wood milling is on the ends of the two short frame pieces, which get ⅜ × ¾-inch rabbets to make stronger corner joints.

After the wood pieces are assembled, the insulation is laid into the box. Insulation is very important for reducing heat loss through the back of the collector. In normal operation the absorber can reach temperatures of upward of 160°F (71°C), and if it's 50°F (10°C) outside, that's a big temperature difference. A good selection for collector insulation is a rigid foam type, generically known as isocyanu-

rate, with trade names such as Thermax, R Max, High R, all of which have reflective foil bonded to both sides of a 4 × 8 sheet. This foam is rated at about R-7 per inch and has an adequate service temperature rating of 200°F (93°C). Styrofoam, on the other hand, is not to be used in this application. It has large shrinkage problems at high temperatures. Another option is fiberglass, rated at about R-3 per inch, which has an even higher service temperature rating. If you decide to use 2 or 3 inches of fiberglass, you should use 1 × 6 boards for the collector box. The advantage of the foam is that with its higher R-value the depth of the box is minimized.

Finally a few small pieces of wood measuring around ½ × ¾ inch are needed to separate the absorber plate from the insulation to create an airspace. These absorber support sticks are simply tacked onto the foam insulation, which can easily bear the weight of the absorber. If you use fiberglass insulation, the support sticks will have to be 2 or 3 inches wide so they can sit on the plywood back and thus maintain the loft of the insulation after the absorber is put in the box. The last piece of wood is the glazing support stick, which is installed widthwise across the middle of the collector. A 2 by 10-foot collector, for example, should be glazed with two 2 by 5-foot pieces of glass, not one awkward (and expensive) 2 by 10-foot piece. Plastic glazings will also need support to minimize buckling from expansion and contraction.

The Absorber

Here of course is the heart of the collector, the point where radiant solar energy is converted into hot water. A water-heating absorber plate essentially consists of a sheet of metal to which water pipes have been bonded. Solar energy striking the metal becomes heat, and the heat is conducted to water flowing through the pipes. The heated water returns to the middle or top of a storage tank and is replaced by cooler water from the bottom of the tank. Just

SolaRoll

Every solar consumer wants systems that are versatile, durable and easy to install. Bio-Energy Systems, Inc. (BESI) of Ellenville, New York, developed SolaRoll, a unique and patented solar water heating system that comes close to meeting these ideals. In addition to domestic water heating, SolaRoll is suitable for other applications including space, pool and radiant floor heating systems.

The primary component of the DHW system is a heat exchanger/absorber mat, which is made of an extruded synthetic rubber called ethylene-propylene-diene-monomer (EPDM). In many respects, this substance is remarkable for what it won't do as well as for what it will. EPDM can withstand temperatures from −80 to 325°F (−60 to 160°C), and it is inert to ozone, ultraviolet light and weathering. This tough material will not rot, crack, rust, corrode or become brittle. So

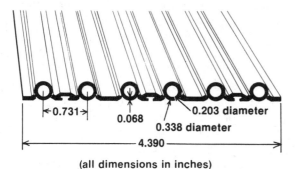

(all dimensions in inches)

Figure II-3: The top illustration shows a cross section of the SolaRoll absorber mat. The basic module is about 4 3/8 inches wide, and you can buy virtually unlimited lengths of the stuff, up to 600 per roll. The illustration to the right shows how SolaRoll is handled as absorber. The webbing is cut away from between the water tubes to allow the mat to be doubled back. This puts both headers on the same side of the collector. Thus, for example, a 20-foot length of absorber mat would make about a 9-foot-long by 9-inch-wide absorber. Several mats can be assembled side by side to make a wider collector.

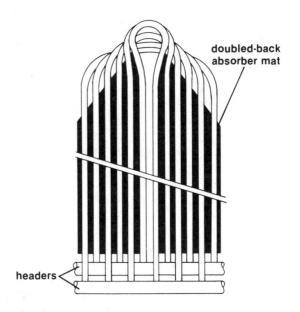

about every common metal has been used in absorber fabrication: steel, stainless steel, aluminum, copper; even plastics and a special synthetic rubber are used as absorber materials. (See the box, "SolaRoll.") Absorbers can be home-built, but purely on the basis of efficient use of your time, we don't recommend it. Instead we suggest that you buy all-copper absorbers from the most convenient source. (The "Hardware Focus" section at the end of this article lists several manufacturers of all-copper units.) Absorbers are made with special machines, and home builders just can't hope to duplicate the quality that goes into factory fabrication. The all-copper feature is specified because, of all the common collector materials, copper is the best heat conductor, and it is more corrosion-resistant than anything else except stainless steel (a poor conductor) and synthetic materials.

BESI confidently guarantees its SolaRoll collector material under a 15-year limited warranty.

Because it is rubber and not metal, EPDM is not as good a thermal conductor as other absorber materials such as copper or aluminum, but BESI claims that this is not a major consideration. Tests conducted by the University of Connecticut's Solar Energy Evaluation Center indicate that SolaRoll's system could provide the same or better efficiency than many metal absorber collectors on the market. The SolaRoll collector compensates for the low conductivity of EPDM because the absorber tubes are just ¾ inch apart compared to 4 or 5 inches for other absorbers. With more closely spaced tubes, heat is more readily transferred to the water. The SolaRoll heat exchanger mat comes on rolls up to 600 feet in length. BESI sells it by the linear foot, and it can easily be cut to suit any length or area requirement. Each absorber mat is 4.4 inches wide and consists of six circulation tubes (3/16-inch ID).

The SolaRoll collector system is designed with the do-it-yourselfer in mind. For completing the absorber, BESI supplies specially drilled copper (domestic water heating) or plastic (pool, space and radiant-floor heating) header pipes and all the hardware needed to connect them to the EPDM absorber tubes. They also make available some auxiliary components such as valves, controllers and temperature sensors. The absorber can then be used with most standard water heating systems as a glazed or unglazed collector.

One of the nice things about SolaRoll's absorber mat is that it stands up so well against freezing. Thus there is no need to drain the absorber tubes because they are flexible and freezing will not burst them. There is no need for antifreeze, corrosion inhibitors or any special fluids. It is important, however, that the copper or plastic header manifolds be protected from freezing such as by insulating them heavily to isolate them from the absorber. The maximum rated operating pressure for the SolaRoll collector is 45 psi, which does not make it suitable for use in most open-loop DHW systems, but fine for closed-loop systems.

Indoors, the SolaRoll absorber mat can be used in concrete floors, walls or ceilings to provide radiant heating. In this application BESI claims that its product is compatible with heat pumps, woodstoves and hot water heaters as well as solar heat sources.

The durable qualities of EPDM make SolaRoll components worthy of consideration by the do-it-yourselfer. BESI sells its components through a national network of distributors and dealers. Additional information is available from Bio-Energy Systems, Inc., Box 87, Ellenville, New York 12428. Telephone: (914) 647-6700.

Margaret J. Balitas

You'll probably have to beat the bush a little bit to find what you want. Stalking the naked absorber plate may prove to be time-consuming; you may end up at your local solar business, or you may have to put a check in the mail, or you may have to drive a fair piece to get the thing. But be assured they're out there, and they're for sale. Once you know what you're getting, you can start estimating the sizes and costs for all the other materials. Everything depends on the size of the absorber.

If fate smiles upon you, you'll get an absorber that's already been painted the requisite flat black at the factory. If it comes without paint, you must do a very careful job of preparing and painting the copper, or you may have trouble later in the form of peeling paint. The first step in surface preparation is to rub both sides of the absorber down with a solvent to clean off oils or other foreign substances.

Then wash the plate with detergent. Finally the copper must be etched with vinegar or muriatic acid to ensure a better bond with the paint. There are several paints that are suited for use on collectors. Basically you'll need a high-temperature, flat black paint. Don't use "satin" or "semigloss" finishes, since they will have a higher reflectance than a truly flat finish. Several absorber paints are listed in the "Hardware Focus" section located at the end of this article.

The Glazing

As you'll be seeing throughout this book, there is more than one way to glaze a collector, more by at least a dozen, if not two dozen. And that's not counting ways that you can think of. The first question to answer is whether or not to have single or double glazing. Generally speaking, the colder the climate, the more double glazing becomes necessary. The trade-off is simple: The glazing is there to reduce reradiative and convective heat loss from the absorber. One layer of glazing rates about an R-1 in insulating value; two layers is roughly R-2, which means conductive heat loss through the glazing is cut in half. *But,* a second glazing layer also reduces the transmission of solar energy to the absorber; thus reduced heat loss also means reduced heat gain. It's a small dilemma with a simple solution. As mentioned, the main criterion for deciding on the number of glazing layers is climate. In climates with more than 6000 heating degree-days, double glazing produces a net heat gain and generally outperforms single glazing in terms of overall energy collection in a liquid flat plate collector. Because of the big temperature swings inside a double-glazed collector, the inside layer of a double-glass glazing should be tempered glass or a plastic such as a thin film. There are also some fiberglass-reinforced plastics that are suitable for single or double glazing but you must make sure that they're designed for use in flat plate collectors. In milder climates (fewer than 4000 degree-days) where the rate of heat loss from the collector isn't as great, a second glazing layer doesn't produce

a significant energy advantage. In between 4000 and 6000 degree-days it's a toss-up; you might as well go with single glazing.

The next decision is what kind of glazing material to use. There is of course glass, and there are several types of rigid and semirigid plastics and thin plastic films that can be used in single or double

Photo II-3: Before the glass goes on, the collector should be all but finished, clean and vacuumed free of dust. The inside surface of the glass should be polished clean.

glazing configurations. (See "All about Glazing" in Section III-A for a bundle of information about the advantages and disadvantages of all types of glazing.) For many kinds of solar collectors, not just water heaters, glass is the preferred glazing because of its visual clarity and permanence. Some plastics are easier to work with; they don't break, and they can be drilled for fastening. But some plastics don't have the visual clarity of glass (although they transmit as much or more solar energy than regular glass), and plastics can degrade after years of solar exposure. In the construction steps below, we'll show ways of handling both kinds of material.

The most important steps in glazing are making an airtight seal to the collector frame and protecting that seal with some kind of perimeter trim; we'll call it the glazing cap. A commonly used material is some type of metal angle or corner flashing. Standard galvanized roof edging can be used successfully. Extruded aluminum angle makes for a very neat look and effective weather sealing. Even wood corner trim can be used. Silicone is the preferred sealing agent. If you're beginning to feel that there's a lot of room for option in collector construction, you're feeling in the right direction. The following construction steps will specify a very basic design which can be modified to suit your own needs and inspirations.

Construction Steps

Materials Checklist

1 × 4 clear, kiln-dried redwood, cedar, etc.	metal or wood corner trim
#8 × 1½ inch brass or stainless steel flathead wood screws	metal or wood flat trim (for glazing center break)
one sheet ⅜-inch CDX plywood	#8 × ¾-inch panhead sheet metal screws
1¼-inch ring-shank nails	(aluminum screws for aluminum corner trim)
1-inch isocyanurate rigid foam insulation	flat black paint
all-copper absorber plate	wood sealer/preservative or paint
glass or plastic glazing	silicone caulk

For Double-Glazed Collectors

¾ × 1-inch wood sticks for glazing frame	#6 × ⅝-inch flathead wood screws
glazing film	#6 × 1½-inch flathead wood screws
packing or duct tape	closed-cell foam weather stripping
aliphatic resin glue (yellow carpenter's glue)	1¼-inch staples

1. Cut the side and end pieces for the collector frame. These lengths will of course depend on the size of the absorber plate; make the cuts leaving at least a ½- to 1-inch space between the edge of the absorber and the inside face of the frame pieces.

2. Take the two short end pieces and cut ¾ × ⅜-inch rabbets across the inside face at both ends. Then take all four of the frame pieces and cut ⅜ × ⅜-inch rabbets along the lower inside edges to receive the plywood backing. Note: Sight along the edges of the frame pieces to see if there is any slight bowing in the wood. The up side of the bow, the "crown," should become the top edge, which carries the glazing.

3. Use the #8 wood screws to join the frame pieces at the corners. Use a little dab of silicone caulk to seal the corner joint. Smear more silicone across all exposed end grain, and fill all the countersunk screw holes with silicone. Note: If the absorber inlet and outlet are located at opposite ends or sides of the

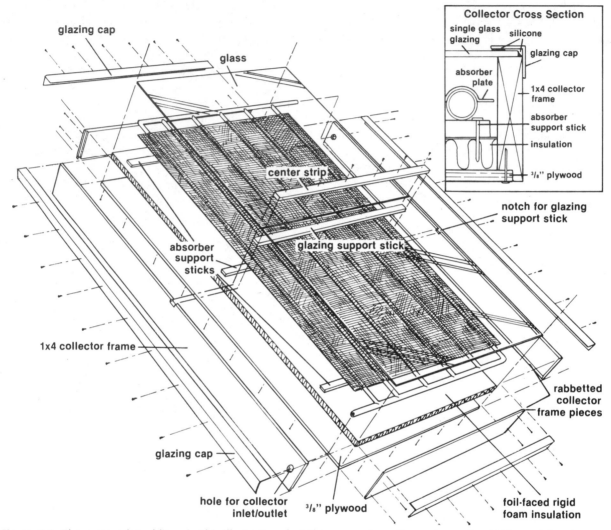

glazing cap

glass

Collector Cross Section

single glass glazing

silicone

glazing cap

absorber plate

1x4 collector frame

absorber support stick

insulation

³/₈" plywood

center strip

notch for glazing support stick

absorber support sticks

glazing support stick

1x4 collector frame

rabbetted collector frame pieces

glazing cap

hole for collector inlet/outlet

³/₈" plywood

foil-faced rigid foam insulation

Figure II-4: There are a lot of lines in this illustration, but if you count up the major parts of this module, there are only 18 pieces in all, not including the fasteners. In the layout, cutting and assembly of these pieces, precision is important for having a weather-tight, long-lived collector.

plate, one of the end or side pieces must be left unattached until the plate is laid into the box.

4. Cut the plywood backing to size and nail it into the rabbetted opening with the ring-shank nails. Run a thin bead of silicone along where the plywood meets the frame, and smooth the bead into the joint. Flip the box over and lay in the foam insulation,

which has been cut to make a very tight fit inside the frame. Cut the absorber support pieces to length and lay them across the insulation. For an 8- to 10-foot-long collector four support sticks will be adequate. To secure the support sticks, drive a ring-shank nail through both ends and then press the stick down onto the foam so that the nails penetrate it.

5. Lay the absorber plate into the box to determine the points where the absorber inlet and outlet will pass through the frame. Mark the spots and drill holes of the appropriate diameter.

6. Now you can lay in the absorber plate, slipping the inlet and outlet through the holes just drilled. If the absorber pipe nipples were at opposite ends or sides, you must now join the fourth collector frame piece (which has been drilled to allow passage of the nipple). Don't forget to nail the plywood to it underneath and run a slim bead of silicone between the frame piece and the plywood.

7. If you want to really minimize heat loss through the wood, cut down some of the foil-faced foam to a ½-inch thickness and tack it inside the perimeter of the collector box, between the absorber and the top edge of the box.

8. Rip the glazing support stick to the appropriate dimension and cut its length equal to the exact width of the frame. The thickness of the stick must be such that the lower face rests on the absorber and the upper face is dead flush with the top edge of the collector frame. (Because the stick is screwed to the collector frame, it can also serve as a hold-down for the absorber.) Find the exact midpoint of the collector length and cut notches in both sides of the collector frame. The glazing support stick should fit tightly into the notches, and it should lie flat on the absorber. Screw, don't nail, the support stick into the collector frame, making sure that the screw heads are positively countersunk below the plane of the wood.

9. All that remains now is the glazing, but before doing that the collector has to be "baked." Tack on a piece of polyethylene plastic sheeting so that it completely covers the collector and put the unit out in the full sun for a couple of days. The heat buildup inside the collector will completely dry out the inside of any moisture, residual paint vapors or other vapors that may be lurking. It's important to bake out vapors now since they could leave a film on the inside surface of the glazing, which could reduce light transmission.

10. The first glazing procedure we'll discuss is single glazing with glass. For collectors that are up to

2 feet wide, double strength (⅛ inch thick) glass can be used, and with wider spans triple strength (3/16 inch thick) should be used. The glass width should be exactly ¼ inch *less* than the total width of the collector. With a two-piece glazing, the length of each piece should be ¼ inch less than *one-half* the total length. These dimensions will allow for a ¼-inch gap between the two panes and a ⅛-inch inset of the glass around the entire perimeter of the collector.

11. Thoroughly clean the inside surface of both pieces. Try to end up with no streaks or blotches that could mar the finished appearance of the collector. Then turn the glass panes over onto the collector (after thoroughly vacuuming the inside of the collector), and position them to get the ⅛-inch perimeter inset and the ¼-inch gap where they lie on the glazing support stick. Run a thin bead of silicone along all the glass-wood seams. There's no need to caulk between the glass and the wood.

12. Cut the glazing cap material to the proper lengths. Run a thin bead of silicone along the face that will actually be in contact with the glass. Have a helper press the cap down firmly onto the glass while you drill into the side and secure the cap with the panhead screws. It's important to maintain the downward pressure on the glass to get a mild compression effect once the cap material is secured.

13. Cut the center strip to fit the width between the inside edges of the glazing caps. Predrill the strip along a precise center line. Run two beads of silicone along the "down" side of the strip and lay it over the glazing support stick, perfectly centered over the gap between the two panes. The screws in the center strip must not have any contact with the glass as it can crack just from a little point of pressure on the edge.

14. Before it cures, clean off any excess silicone that has been squished out from under the glazing cap and generally clean the glass. Brush on the wood sealer/preservative over all wood surfaces. It's also a good idea to run silicone around where the inlet and outlet nipples protrude through the collector frame.

15. If a single layer of plastic glazing is used, the

procedure doesn't change. Since plastic can be drilled for fasteners, you can screw down the glazing cap from the top and the side. But because plastics expand and contract a lot more than does glass, it's important to drill holes in plastic glazing slightly larger than the shaft diameter of the screw. This way expansion and contraction can occur without undue stresses at the fastening point or on the glazing, which can lead to fracturing.

16. To develop a double glazing, you'll need to make a glazing frame assembly out of ¾ × 1-inch sticks with a center divider that is in line with the glazing support stick. A good choice for the inner glazing is one of the lightweight films that are now available which have excellent light transmittance and resistance to ultraviolet radiation. (See ''All about Glazing'' in Section III.) These films can be stretched

tight across the glazing frame to produce a strong, wrinkle-free second glazing.

First, make the frame to the exact outside dimension of the collector frame. Then lay the film over the frame and cut it roughly to size, leaving some slack for the final trimming. As figure II-5 shows, the edges of the film that receive the fastening staples are strengthened with either fiber-reinforced packing tape or duct tape. A strip of tape is also needed for the surface that meets the collector frame. The stapling is then done through the tape and film, into the frame. Pulling the film tight over the edge of the glazing frame (which has been sanded free of sharp edges) creates a uniform tension across the whole sheet, which helps to prevent wrinkling. Fasten one long side first and then start in the middle of the second long side and work toward both ends.

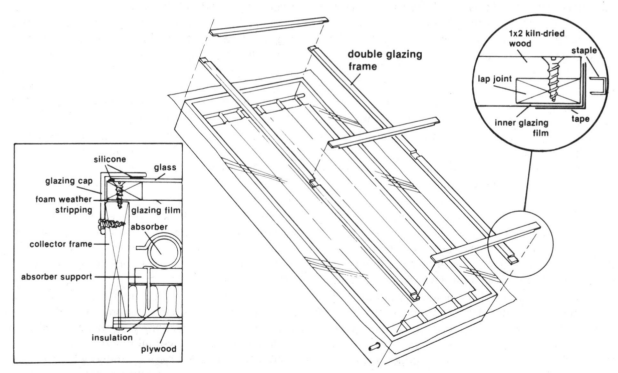

Figure II-5: A good glazing frame begins with dry, straight pieces of wood. The lap joints must be finished totally flat to ensure that the top glazing will lie flat, and to the same end all screws must be countersunk below the surface of the frame. The glazing film can easily show wrinkles, so try to stretch it as evenly as possible.

Then fasten the two short sides. To receive the glazing frame the top edge of the collector frame needs a strip of closed-cell foam weather stripping to ensure a proper seal. Use a type that is self-sticking. Lay the frame onto the collector and fasten it with #6 × 1½-inch flathead wood screws that are positively countersunk below the surface of the wood. The outer glazing, be it glass or plastic, is then installed as it was in steps 11 and 15, and the glazing cap and center strip are also fastened in the same way. The width of the side face of the glazing cap material will probably have to be increased to completely cover the edge of the glazing frame.

17. In the operation of the collector itself, the only problem that may arise is moisture condensation on the inside of the single glazing, and possibly between two glazing layers. If this happens and doesn't go away after a week or two, check all the seams of the collector box to make sure they're tight and sealed. If the condensation persists, drill a couple of ¼-inch holes at the low end of the collector or the glazing frame to create an outlet for water vapor. Better still, along with the low-end holes, drill holes on both sides of the collector near the top in such a way that they won't capture rainwater.

How Many Collectors?

Worksheet I-7, "Solar DHW System Performance Calculation" (see Section I), is where the answer lies. If you are planning to install a flat plate collector system, you can do that calculation to get an idea of how much energy X square feet of collector will deliver in a year. (When you work in the collector tilt factor, the best year-round angle is your latitude.) At that point you can do a couple of calculations using different collector areas to achieve a higher or lower (preferably higher!) solar fraction. The decision of whether or not you have the space or the budget to get the solar fraction you'd like is, of course, up to you. (Don't forget the 40 percent federal tax credit!!!)

What if your calculations tell you that you need 50 square feet of collector area to obtain a 75 percent solar fraction, and the only absorber you can find comes in 20-square-foot modules? You can't very well build half a collector, so naturally you ought to build three collectors. The "extra" area isn't really extra; it's more solar energy for your DHW system. If you want to be a little more conservative now, see if it's possible to leave space in your installation zone for more collector area to be added later.

If you find that you need three or four or more collector modules, and you're ready to build them all right away, you might want to consider enclosing two, three or four absorber plates in a single big box. The basic collector design presented here just has to be stretched, and you can actually save a little on materials. When putting the collector together, the absorber plates are usually joined inside the box, which means they have to be pressure-tested before the glazing goes on. Along with materials' saving there's a little efficiency boost in a big collector because the perimeter has been shortened relative to the total perimeter of three or four separate modules. Less perimeter means less heat loss. In the end you have a single, large collector, and if you've got a place to put it, maybe it's right for your application.

Joe Carter

References

Bryenton, Roger; Cooper, Ken; and Mattock, Chris. 1980. *The solar water heater book: building and installing your own system.* Toronto: Renewable Energy in Canada.

Campbell, Stu, and Taff, Douglas. 1978. *Build your own solar water heater.* Charlotte, Vt.: Garden Way Publishing Co.

Franklin Research Center. 1979. *Installation guidelines for solar dhw systems in one- and two-family dwellings.* Washington, D.C.: U.S. Government Printing Office. no. 023-00-00520-4.

Goldberg, Alan. 1980. *Design manual for solar water heaters.* North Hollywood, Calif.: Horizon Institute.

Lucas, Ted. 1978. *How to build a solar heater.* New York: New American Library.

Partington, William M., Jr., and Root, Douglass E. 1976. *Build your own solar water heater.* Winter Park, Fla.: Environmental Information Center.

Valdez, A., and Kawanabe, A. 1977. *Building a solar water heater: a do it yourself manual with blueprints.* Available from Valdez and Kawanabe, P. O. Box 1014, Alamosa, CO 81101.

Hardware Focus

All-Copper Absorber Plates

Columbia Chase Corp.
Solar Energy Division
55 High St.
Holbrook, MA 02343
(617) 767-0513

Sunplate
32½″ × 74″; 32½″ × 94″; 33″ × 75¼″; 33″ × 95¼″. Custom sizes available. Low pressure model to 40 psi. High pressure model to 150 psi. Manifold: ¾″ type M. Risers: ½″ × ⅛″ flat oval tubes.

Phelps Dodge Brass Co.
Solar Division
2665 Woodland Dr.
Anaheim, CA 92801
(714) 761-4260

Phelps Dodge Solar Absorber Plate
34″ × 76″ or 34″ × 96″. Custom sizes available. 150 psi pressure rating. Manifold: 1″ type M. Risers: ½″ OD.

Solar Development, Inc.
Garden Industrial Park
3630 Reese Ave.
Riviera Beach, FL 33404
(305) 842-8935

Solar Absorber Plate
4′ × 10′ sinusoidal or 4′ × 10′ parallel with ½″, ¾″ or 1″ headers. Other sizes on special order. Risers: ½″ ID. Operating pressure 150 psi.

Solpower Industries, Inc.
10211 C Bubb Rd.
Cupertino, CA 95014
(408) 996-3222

Solpanel Solar Absorber
46″ × 93½″; 46″ × 120″; 48″ × 93½″. Test pressure 160 psi. Manifold: 1½″ ID. Risers: ⅜″ ID.

Specialty Manufacturing, Inc.
7926 Convoy Court
San Diego, CA 92111
(714) 292-1857

Pool Solar Collector Model 24-125U
34″ × 94″. Maximum operating pressure 150 psi. Manifold: 1¼″ type L. Risers: ⅜″ tubing on 2″ centers.

Sunburst Solar Energy, Inc.
P. O. Box 490
Elk Grove, CA 95624
(916) 739-8485

Terra-Light, Inc.
30 Manning Rd.
P. O. Box 493
Billerica, MA 01821
(617) 663-2075

Thermatool Corp.
280 Fairfield Ave.
Stamford, CT 06902
(203) 357-1555

Western Solar Development
1236 Callen St.
Vacaville, CA 95688
(707) 446-4411

Sunburst Solar Absorber Panels
45.6″ × 94″; 45.6″ × 117″. Manifold: 1¼″ or 1½″ ID. Risers: ½″ OD, type M.

Terra-Light Absorber Plate
22″ × 94″; 34″ × 74″; 34″ × 81½″; 34″ × 94″; 44″ × 94″; 44″ × 91″. Maximum operating pressure to 150 psi. Manifold: ¾″, 1″ or 1¼″ ID. Risers: ⅜″ OD.

Thermafin
34″ × 74″; 34″ × 94″; 47″ × 96″; 47″ × 120″. Maximum operating pressure to 150 psi. Manifold: any size or configuration. Risers: ⅜″ or ½″ OD.

WSD-1 Copper Absorber Plate
WSD-2 Copper Absorber Plate
4′ × 8′, 4′ × 10′ or custom sizes. 125 psig pressure rating. Manifolds: ½″, ¾″, 1″ or 1½″. Risers: ½″ type M tubes soldered to fins 4.5″ or 4.6″ on centers.

Insulation

Northeast Specialty Insulations, Inc.
1 Watson Place
Saxonville, MA 01701
(617) 877-0721

Owens-Corning Fiberglas Corp.
Fiberglas Tower
Toledo, OH 43659
(419) 248-8102

A distributor and fabricator of all types of insulation, including urethane, fiberglass, flexible tubing and Foamglas. Insulation fabricated to customer's specifications.

Insulation SI-100
Glass fiber insulation that is unbonded and unlubricated to prevent outgassing. Standard widths: 24″, 36″, 48″ and 72″. Made to order: 40″ through 72″. Lengths: 1″ increments. Thicknesses: 1″ or 2″. Manufacturer claims R-3.7 to R-4.3 per inch at 75°F (24°C). Stable to 1000°F (538°C).

Solar Absorber Paints

Kalwall's Black Absorber Paint
Martin's Flat Black Latex
Martin's Latextra Acrylic Latex
Rust-Oleum's Flat Black
Rust-Oleum's Midnight Black

Solar Usage Now, Sun-In Absorber Paint
This graphite paint has very good heat-collecting characteristics and is reasonably priced.

☼PROJECT
Installing, Plumbing and Wiring Your Collectors

> *If there are ten ways to build a collector, there are ten hundred ways to install it. Good planning is important; correct execution is critical. Solar installing is carpentry, plumbing and electrical work all mixed together, and the following information should help to keep things from getting all mixed up.*

You probably haven't gotten this far—with your collectors ready and waiting—without having at least a general plan for how and where they're going to be installed. Unless you can hide them somewhere in the deep recesses of a flat roof, they're probably going to be put on a roof or wall or out in the yard somewhere, exposed, and it's important to consider the appearance of the installation as well as its function. Solar collectors aren't skylights or roof vents or windows or any other "standard" building component, and they're likely to stand out. They can stand out nicely or they can be visually obtrusive, poorly placed and generally out of kilter with the rest of the house. If the collectors are going to show, the installation is a home improvement as well as an energy improvement. A well-planned and well-executed installation will enhance the value of your home because of its energy benefits *and* its visual merits.

Collector placement is also involved with the design of the whole system. With some of the flat plate systems, it matters little where the collectors are placed relative to other components, but other system designs impose certain unavoidable requirements on the relationship of the collectors to the pipe runs and tank placement. It's advisable that you go ahead after reading this article and study the systems described in "A Catalog of Flat Plate Collector Systems," found later in this section, so you can make a final determination of which one you're going to

use and how it may affect collector placement. From there you should draw up a firm plan of how you're going to apply the system to your house. This will help prevent errors when it comes to doing the real work.

Installing Your Collectors

A few basic parameters and rules of thumb can help with the layout of the system. The collectors must of course be out of shade year-round. This is what makes roofs the first place to look for a mounting location. If you're uncertain as to whether or not a tree or other object is going to cast a shadow on your roof, use the calculation methods described in the previous section in "The Sun in Your Place" (figures I-29 and I-30). They will pinpoint the length of a shadow on a vertical wall and on a roof at any time of the year.

As the solar domestic hot water (DHW) heat gain calculation showed (see "Finding Your Heat Gain," in Section I), orientation can somewhat affect a collector's performance. If your pitched roof is oriented 30 degrees or more away from true south, you've got a couple of options for using that pitch to get an adequate solar fraction. Another collector module can be added to make up for the loss of gain

Photo II-4: There may be a few magic moments in the installation of a solar DHW system, and one of them is when the collector is put in its spot, from whence it will deliver your hot water in the years ahead. Note that the collector is covered to prevent it from overheating until it's filled with water.

with increased collector area. Or, you can install the collectors in a sort of "cocked" position so that they do in fact face south, and at the proper tilt. This latter option may mean some compromise in appearance, but done properly an installation like this can have a clean, if unusual, look. Metal or pipe mounting bracket systems are somewhat less obtrusive than bulky wood struts; they can be painted to match and blend with the roof color, as can the collectors. In fact, that may be the preferable way to both preserve the collector box and help a collector array "disappear" into the roof pitch. With unshaded flat roofs, collectors can be oriented and tilted just right, never to be seen from the ground.

Speaking of tilt, the general rule of thumb for collector tilt for solar DHW collectors is that the angle should equal your latitude, plus or minus 5 degrees. We say plus or minus because if your roof pitch is within that range, it's not necessary to increase or decrease the collector tilt, unless of course you're truly fanatical. But actually the truly fanatical would devise a mounting system that would allow for tilt angle adjustment at different times of the year. As the winter solstice approached, the collector would reach its steepest tilt, and as the summer solstice came closer, the collector would fall back to its shallowest angle. But again, fanatical types only need to heed this note.

Photo II-5: A strut-type mounting system solves the problem of poor orientation and a shallow roof pitch. This solution is somewhat cheaper than the other alternative: increasing the number of modules, to make up for the loss of efficiency with the collectors mounted flat to the roof.

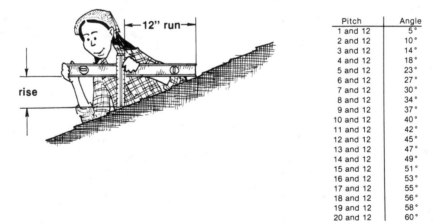

Pitch	Angle
1 and 12	5°
2 and 12	10°
3 and 12	14°
4 and 12	18°
5 and 12	23°
6 and 12	27°
7 and 12	30°
8 and 12	34°
9 and 12	37°
10 and 12	40°
11 and 12	42°
12 and 12	45°
13 and 12	47°
14 and 12	49°
15 and 12	51°
16 and 12	53°
17 and 12	55°
18 and 12	56°
19 and 12	58°
20 and 12	60°

Figure II-6: If your roof pitch is apparent inside the house (pitched ceiling or unfinished attic), you can do the same exercise without getting up on the roof. The procedure is just a little different. Lay one end of the level onto the pitched ceiling or roof rafter, find a horizontal line and then measure vertically up to find the rise.

If it does turn out that your roof pitch angle is too shallow or too steep (as it might be with an A-frame house), you can build a simple mounting rack to get just the mounting angle you need. The first step is to find out what the roof angle is. You need a level and a tape measure, and you need to be up on the roof. As figure II-6 shows, one end of the level is placed on the roof, and the other end is moved up and down until the center bubble indicates a horizontal line. At a point that is 12 inches from the roof end of the level, measure down to the roof with the tape measure, keeping the tape perpendicular to the level. Note the number of inches from the roof to the bottom edge of the level. You've just found the "rise and run" of your roof, which can be translated into a pitch angle using the table in figure II-6. Now subtract the pitch angle from your latitude or, if the pitch angle is greater, subtract latitude from pitch; the result is the angle at which the mounting rack will be built. See the box, "A Collector Mounting Rack."

Once questions concerning shadows, orientation and tilt have been resolved, there may still be a big space available on the roof for the relatively small collector area to be installed. Determining collector placement involves foreseeing the visual result and also some practical considerations. Color isn't the only way that collectors can be blended with a roof: Proper positioning can also help to quiet their visual impact. Putting an array of collectors smack dab in the middle of a roof may only serve to emphasize their presence, while putting them off to the side can diminish their contrast with other building elements. Make a quick scale drawing of the roof in question and play around with different locations, on paper and in your mind's eye, to find what seems to be the best spot.

The other consideration with collector placement has to do with the plumbing. It's always a good idea to minimize the lengths of the outside pipe runs to reduce heat loss and the cost of protecting insulated pipe from the elements. The sooner the pipes can be brought through the roof or through the side of the house, the better. Look at your system diagram to see where the pipes are going, from the collectors to the storage tank. If you're adding a tank (to be discussed shortly), it's entirely possible to put it up in the attic or in a second-floor closet to minimize the run to the collectors. In fact, if you're adding a storage tank with a built-in electric element for back-up heat, you can put the whole DHW system upstairs. You'll have to be able to bring a cold supply line up to the tank and also come away from the tank with a hot water outlet line that ties in to all the points of hot water use.

No Roof

Although we've emphasized roof mounting thus far, you don't have to pin all your hopes for a solar DHW system on the availability of a clear roof area. There are always the walls, and if they don't give you a spot, there's always the good old ground. The first question to ask about a wall mount is whether it will interfere with other solar plans you may have for the south wall of your house. If there are such plans, how can the collectors be included? A line of tilted collectors could conceivably serve as an overhang that would shade a space heating collector in summer. Collectors could also become part of a sunspace or greenhouse roof pitch. If they're protected from the inevitable flying object (with tempered glass or plastic glazing), they can be mounted low on a wall, for example, between the ground and the top of the building foundation. For mounting a collector on a wall and on the ground, Chris Fried has developed a simple bracket system using aluminum angle. (See the box, "Wall Mounts for Flat Plates.")

If the collectors are to be wall mounted, every attempt should be made to make them as much a part of the existing wall as possible. When building your collector "eave," duplicate the original siding material, trim work and color as best you can. Keep the design simple, and it won't stand out so much. Put enough effort into the finished appearance of the thing, and it will be an elegant home improvement as well as an effective energy improvement.

A Collector Mounting Rack

If your roof pitch is either too shallow or too steep, you can build this mounting rack to get your collectors at exactly the desired angle. The rack is a right triangle in which the hypotenuse (A) is formed by the collector itself. The two other sides (B and C) are cut from 2 × 4 material. Part D is a beveled strip of wood that is also cut from 2-by material. It is used as a stop at the low edge of the collector. It is screwed to part C and the collector. Metal strapping joins the collector with part B. C and B are joined with ¼ by 3½-inch bolts. Because all collectors aren't created equal (in size), there's no standard measurement for all these parts, but there is a standard formula for figuring out the correct lengths and cutting angles. You've already seen a little simple trigonometry in the first section, and here is a little more. As before, all the information you need is right here.

A = total width of the collector, in inches

Angle 1 = angle required for attaining desired collector tilt

Angle 2 = 90 degrees − angle 1; angle 2 is the cutting angle for the top end of part B

B = (A × sine of angle 1) + 1½″

C = A × sine of angle 2

Here's an example: The collector is 25 inches wide. The existing roof has a 23-degree pitch (5 and 12), but the desired collector tilt is 40 degrees. Therefore the rack needs to create an additional 17 degrees of tilt (40 − 23 = 17).

A = 25

Angle 1 = 17°

Angle 2 = 90° − 17° = 73°

B = (25 × 0.292) + 1½″ = 8.8″ = 8 13/16″

C = 25 × 0.956 = 23.9 = 23⅞″

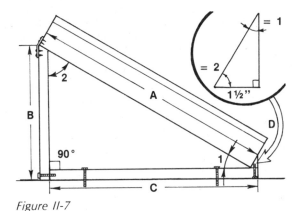

Figure II-7

No Wall

Ground mounting really ought to be your last resort simply because the whole procedure involves so much more work. The collector rack must be absolutely anchored to the ground, which means that any concrete footing has to be dug below the soil frost line. Ground mounting may put the collector a fair piece from the house, necessitating a long underground pipe run. But if it's your only option, the job is not, of course, impossible. In "A Catalog of Flat Plate Collector Systems," found later in this section, see the antifreeze fluid system for some ideas on how one ground-mount system was handled.

A Roof Installation: The Steps

1. If the roof is your spot, the first order of business is to outline the perimeter of the collector

(continued on page 178)

Deg	Sine	Deg	Sine	Deg	Sine	Deg	Sine		Deg	Sine	Deg	Sine	Deg	Sine	Deg	Sine
		27°	0.454	45°	0.707	63°	0.891		18°	0.309	36°	0.588	54°	0.809	72°	0.951
10°	0.174	28°	.470	46°	.719	64°	.899		19°	.326	37°	.602	55°	.819	73°	.956
11°	.191	29°	.485	47°	.731	65°	.906		20°	.342	38°	.616	56°	.829	74°	.961
12°	.208	30°	.500	48°	.743	66°	.914		21°	.358	39°	.629	57°	.839	75°	.966
13°	.225	31°	.515	49°	.755	67°	.921		22°	.375	40°	.643	58°	.848	76°	.970
14°	.242	32°	.530	50°	.766	68°	.927		23°	.391	41°	.658	59°	.857	77°	.974
15°	.259	33°	.545	51°	.777	69°	.934		24°	.407	42°	.669	60°	.866	78°	.978
16°	.276	34°	.559	52°	.788	70°	.940		25°	.423	43°	.682	61°	.875	79°	.982
17°	0.292	35°	0.574	53°	0.799	71°	0.946		26°	0.438	44°	0.695	62°	0.883	80°	0.985

Photo II-6: A collector mounting rack is a must for flat roofs, unless you live at the equator. To resist strong gusts the racks must be very well fastened to the roof, and the collectors must be very well fastened to the racks.

J.C.

Wall Mounts for Flat Plates

As you'll see in Chris Fried's description of the solar DHW drain-back system, the closer the collector is to the storage tank, the better. This means either raising the tank to meet the collector, or lowering the collector to meet the tank, or both. When installing this on other flat plate systems, there may simply be no feasible roof location, as when steep roof pitches face east and west, which makes a wall mount the next best choice. Chris has handled a lot of wall and wall-ground mounts in his solar DHW workshops and he's developed a simple mounting system that uses aluminum flat and angle stock, which is easily handled with a hacksaw, a drill and some elbow grease (for bending the stuff). As the following illustration shows, this bracket system can be used to get both the right collector tilt and orientation.

You can fabricate a standard wall-mount bracket from three lengths of aluminum angle bracket with holes drilled wherever bolts or lag screws are needed. The piece that is bolted to the collector is a piece of angle bent back onto itself. A face of the angle must be cut to make the bend possible. This piece is screwed to the collector frame and bolted to a small piece of angle mounted right under the bracket on the bottom of the collector. A third short piece of aluminum angle is lag-bolted to the house wall. It's important to find a stud, post, sill or top plate to bolt into to ensure a strong fastening. With masonry walls a high-quality fastener is essential. After the wall piece is in place, the collector brackets can be bolted on.

If there are no corrections to be made for collector orientation, the two wall pieces are identical, but if there is a need for an azimuth correction, one wall bracket can be made longer to hold the collector out at the right orientation. The long wall bracket shown is simply a piece of bent angle stock that's been triangulated with a piece of flat stock. In order to bend the angle stock to form a right angle, an isosceles triangle piece has to be cut from one face of the angle at the bend point. The long bracket is fastened to the same standard wall piece and bolted to the standard collector bracket. You can figure out the extension needed for the long bracket by doing some simple trigonometry. Find the angle required to get the proper orientation. Multiply the sine of that angle times the distance between the two collector brackets as they are actually mounted on the collector. The result is the distance that the long wall bracket should extend from the wall.

Brackets at the two bottom corners of the collector hold the collector at the proper tilt angle. If a long wall bracket has been used, the bottom bracket below it must also be longer than its partner on the other side. As it's shown in the illustration, the bottom bracket is made with two pieces of aluminum angle: One is screwed to the wall with lag bolts; the second piece is bolted to the wall piece and to the bottom of the collector. The tip of the horizontal piece, where it's secured to the collector, has to be bent up slightly so that it lies flat to the collector frame. To counteract side-to-side movement at the base of the collector, diagonal braces should be run from the bottom brackets back to the wall or between the two brackets, using the same angle stock.

In a ground-wall mount, when the collector bottom is fastened to the ground, support the collector with two ½- or ¾-inch galvanized steel pipe sections anchored into a concrete footing that has been dug below the soil frost line. The bottom of the collector should be at least 1 foot or more off the ground in northern locations to clear usual snow levels. The collector is fastened to the pipe by bending a piece of flat aluminum stock into a collar around the pipe and bolting it to another flat piece screwed to the collector. This setup allows you to adjust the tilt angle before you tighten the bolts on the collars.

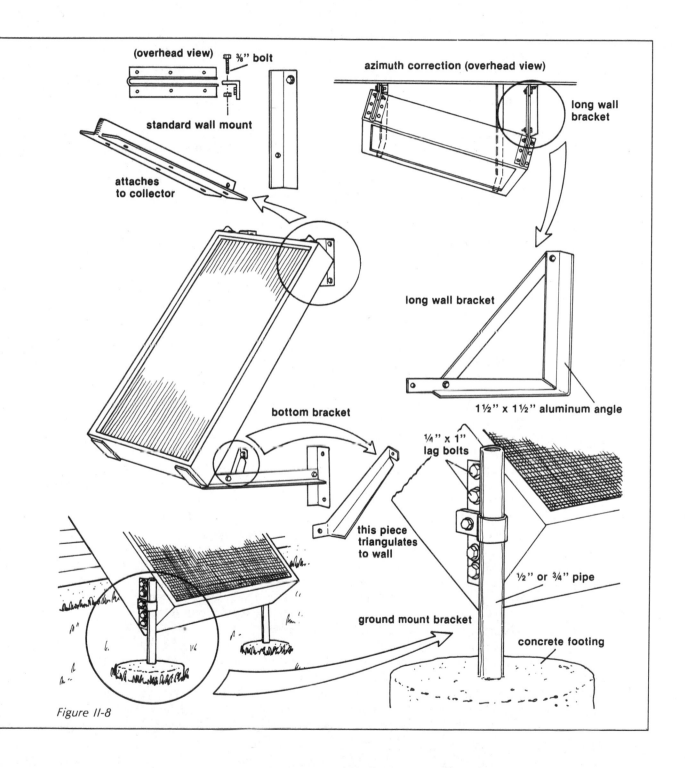

(overhead view) ⅜" bolt

standard wall mount

attaches to collector

azimuth correction (overhead view)

long wall bracket

long wall bracket

1½" x 1½" aluminum angle

bottom bracket

this piece triangulates to wall

¼" x 1" lag bolts

½" or ¾" pipe

ground mount bracket

concrete footing

Figure II-8

array. This means getting up there and working around with a tape measure and a chalk line. If the roof is steeper than a 3- or 4-and-12 pitch (steeper than 18 degrees), you'll have to set up a stage or platform to carry you and your tools. Clinging to a steep pitch while trying to accomplish something is patently unsafe, and even if you come away without injury, you waste a lot of muscle energy trying to be like Spiderman. You can rig up your own platform with scrap wood and a 2×10 or 2×12 plank, or you can rent what are called "staging brackets" or standard platform scaffolding.

2. Once you've chalked out the perimeter of the collector array, find exactly where the roof rafters run through that area. The collector mounts will be lag-bolted to the rafters. Perhaps the rafter ends protrude at the roof overhang, or perhaps they're visible from the inside of the house. Maybe you'll have to make a couple of "test bores" with a nail to find the first rafter; then you can measure across 16 or 24 or whatever inches to find the next one, and so forth. With roofs that are decked with 1- or 2-inch-thick boards, there's no need to find the rafters as the collector mounts can be bolted right to this decking. You can also add blocks between rafters and bolt into them, or you can drill right through a block and use a through-bolt (carriage or machine bolt, washer and nut). The ways are many.

3. If the collectors are to be mounted flat to the roof, 2×2 stringers should be used to raise the collectors off the roof by about 1½ inches. It's very important to raise the collectors in this way so that rain and melting snow can run under the unit. Whether the long side of the collectors are perpendicular or parallel to the rafters, the stringers are mounted perpendicular to the rafters. The stringer material should be a naturally rot-resistant wood or wood that has been thoroughly soaked with preservative. The actual pieces should be cut just as long as the collector, and instead of installing them perfectly parallel with the roof ridge, build in a slight cant so that water doesn't collect behind them.

When mounting racks are used to increase or decrease the collector tilt, they can either be bolted on top of stringers or right over the rafters, eliminating the need for stringers. The second method keeps the collectors in a lower profile on the roof. Whichever mount is used, the racks or the stringers, a ¼- or 5/16-inch lag bolt will provide adequate holding power. Before the bolt goes into the hole, fill the hole with a glob of roof tar (between the stringer and the roof).

4. When the mounts are ready, it's time to get the collectors up on the roof. This step invites all sorts of ingenuity, and depending on the situation there are probably several ways to get the same results. Start with at least two strong people or three regular people. With two people, one ladder is needed and with three people, two ladders. As photo II-7 shows, the collector is roped up and then pulled and pushed up the ladders, a simple enough task but one that should be done with great care for the safety of people and (above all!) that collector you have put so much work into.

5. After the stringers are fastened, install 2×2-inch angle brackets at the ends of the stringers in the manner shown in figure II-9. These brackets will secure the collector at four points. If two stringers are to carry just one collector, then four brackets installed as described will do the job. But if two or more collectors are installed side by side, there will be two inside edges that have to be dealt with. To secure the inside edges, the angle brackets should be prefastened to the collectors at the points where the box meets the stringers. By leaving a 2-inch gap between the collectors, the other tang of each bracket can be fastened to the stringers. When the mounting rack is used, the collector is fastened as shown in figure II-7. At the lower edge put screws through the beveled stop strip into the frame. At the top edge bend a metal strap over the collector frame and the rack upright (B) and secure it with wood screws.

6. Once the collector(s) is secured, attach some kind of opaque cover to it. Until the collector is filled and has water running through it, it can get very, very hot as an empty, "stagnating" collector, which is not good for collector health. Even with a cover it will still heat up, but not nearly so much.

Photo II-7: This simple hoisting technique will save you a lot of sweat in getting heavy, awkward collectors up on the roof.

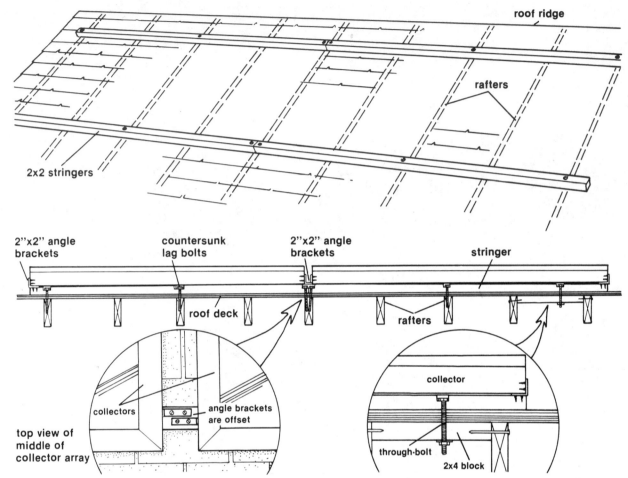

Figure II-9: This is a simple layout for the installation of two collectors mounted end to end, flat on a roof. The 2×2 stringers are placed parallel with the roof ridge, and they are spaced a few inches less than the width of the collector so as to be concealed. All the lag bolts are countersunk to be even with the surface of the stringer material, and the holes are filled with roof patch. The detail at lower right shows how to gain a fastening point if there is no rafter available. Run a 2×4 block between two rafters and use a through-bolt for the fastener. The detail at lower left simply shows how the angle brackets are offset at the middle of the array so that both can be fastened to the stringer.

Tanks for the Hot Water

The required size of the storage tank is directly dependent on collector area, and it directly affects the efficiency and usefulness of the total system. With an undersized tank the domestic water will

generally be overheated, as if the thermostat on a water heater were turned up too high. Overheated water means that the collectors run hotter and less efficiently. Heat loss from the tank and the pipe lines is also increased, more waste. With an oversized tank the collectors are indeed putting thousands of

Btu's into your domestic water, but they're dispersed into too many gallons. This limits temperature buildup in the water, which means you're still dependent on auxiliary power to get useful hot water (115 to 120°F, 46–49°C).

To avoid the pitfalls of either extreme, you can use a simple sizing guideline: For every square foot of collector glazing area, there should be 1½ to 2 gallons of storage capacity in "sunny" climates (with solar gain of more than about 550 KBtu per square foot per year, as listed in table I-8, column D). For climates that aren't so sunny, a ratio of 1 to 1½ gallons per square foot of collector is appropriate. These guidelines are clearly full of slack, but they have to be because of the vagaries of climate, the areas of standard collector modules and especially because of the sizes of standard hot water tanks. Collectors come 16, 20, 24, etc., square feet at a time, and you can't build half a collector. It's the same with tanks: They're made in standard volumes (52, 60, 80, 100, and 120 gallons), and you can't fill just part of a tank to create the right storage volume. Hence, there is no pinpoint perfect sizing parameter. (That's not 100 percent true: In one type of flat plate system, the drain-back system, you can vary the storage volume because the tank isn't pressurized by the domestic water system.)

Using these guidelines you can first figure out if you need to add a tank or just use the existing water heater for double duty: solar storage and backup. If, for example, you're installing 20 to 30 square feet of collector and you have a 40- or 52-gallon water heater in place, you've got all the stor-

age you need. If the heater is electric, you can shut off the bottom element and use just the top element for back-up heat. As figure II-10 shows, the cold supply to the tank is located down where the tank

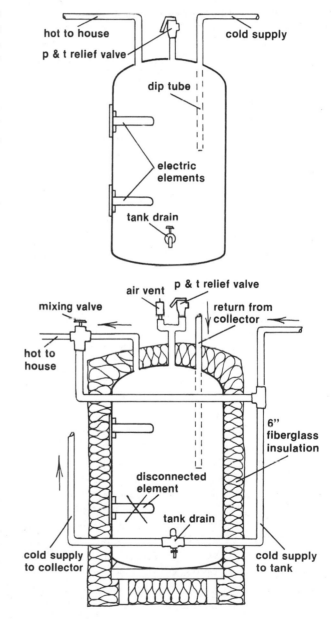

Figure II-10: This is how your old water heater can be reborn as a solar storage tank. The cold supply becomes the collector return line, and the cold line is diverted to the tank drain port, which also becomes the collector supply line. The cold line branches off before it reaches the tank, and the branch is run to the tempering valve. For electric water heaters the bottom element is disconnected, and where possible the tank is blocked up so that insulation can be stuffed under it.

drain valve is; that's also where the supply line to the collector originates. The plumbing configuration for gas heaters is the same, but you can only turn down the gas thermostat.

With larger collector arrays that would thermally overwhelm your present water heater, you must add a storage tank sized as if there were no storage at all. That is, you can't count the water heater as part of the storage system when a second tank is added. Thus 40 square feet of collector will need a 40- to 80-gallon tank depending on the climate, with a 52- or 60-gallon tank probably being the best choice. Generally, it's better to size conservatively, on the short side, to ensure an adequate temperature rise through a day's worth of solar water heating. The tank should be a glass- or stone-lined unit rated for up to 150 psi hot water service. If you foresee having to replace your present water heater in the near future, you can make your DHW system a one-tank system by installing a solar storage tank that is equipped with a back-up electric heating element. A single-tank storage system is more efficient than a two-tank setup because of reduced heat loss.

In many cases the second tank will be placed right next to the existing water heater, although sometimes putting the tank close to the collectors makes the installation easier when all the vagaries of plumbing and wiring are considered. If the tank is to be installed in the attic, it's important to make sure that the ceiling joists are strong enough to carry the concentrated load of 500 to 700 or more pounds of water. With 2×4 joists (sometimes found in old houses), the load of the tank should be spread out over three or four joists by placing 2×4's under the tank, perpendicular to the joists. Also, any tank placed over a living space should sit inside a shallow drain pan (available at plumbing supply houses) just in case the tank ever decides to spring a leak. A drain hose can lead from the pan to some appropriate drain location, such as through a small hole in an exterior wall. If you happen by the hose outlet one day and see it dripping, you'll know what's up. Finally, any tank you put anywhere should be heavily insulated, with a minimum of 6 inches of fiberglass.

All solar heating systems should be equipped with a *tempering* or *mixing valve* located at the outlet of the solar storage tank (in a one-tank system) or the outlet of the water heater (in a two-tank system). This valve both protects you and increases the overall efficiency of the DHW system. Suppose, for example, that your household were empty for a few days, and the collectors were able to raise the storage temperature to 160, 170, even 180°F (71, 77, 82°C). You might come back and flip on a hot water tap, fully expecting the normally not-too-hot water to come your way and zowee!!!, you're zapped with a splash of solar super-heated water. Ouch. What's also happening is that the super-heated water flowing through the lines is dumping a lot more heat than lower-temperature water. Waste. A tempering valve solves the problem. As figure II-10 shows, this thermostatically controlled valve mixes cold water with overheated water to send just the right hot water temperature into service. These valves are also adjustable, so you can set the hot water outlet temperature as low or as high as you want. Usually it's set just a little hotter than the thermostat setting of the back-up heat source (electricity or gas, 115 to 120°F, 46–49°C).

Laying Out the Plumbing

In some cases the pipe runs between the collectors and the storage tank are variously affected by the type of system that's being installed, by the location of the tank and by other factors that lead to the final collector-storage tank relationship. Your system plan drawing will reflect all the decisions and compromises that go into this positioning. To some degree the plumbing layout can affect where the major system components might be put. As mentioned, it's always a good idea to minimize outdoor pipe runs for reasons of expense and heat loss. Concealment should be another goal. Pipe runs that follow roof pitch angles and the building's vertical and horizontal lines will blend better than runs that are skewed. As with the collectors, color can also help in the disappearance of pipe runs. Outside, pipe that is run vertically through a 3- or 4-inch diameter white PVC pipe can look like a gutter downspout. If it's also painted to match wall or trim color, the illusion is even better. Of course, if the collector lines can be passed through the roof or wall right at or near the mounting point, there's no need for visual trickery.

What you do outside affects, and is affected by, what you want to do inside. If your collectors are going up on the roof and your tank is going down in the basement, you've got 20, 30, 40 or more feet of vertical run to deal with. With the right tools, namely long drill bit extensions, pipes can be run through interior or exterior stud walls, as long as they can still be insulated. Perhaps the pipes can more easily be

dropped through closets or inside built-in cabinets. If there is no choice except to run pipes through a living space, they can be boxed into a corner or into a jog in the wall with wood or Sheetrock and painted, papered or otherwise finished to match the existing wall. Consider all the angles; it's likely that you can come up with some ''creative'' routes to put your pipes out of sight. Do not, of course, place any control valves where they are not easily accessible.

When more than one collector module is used, it's important to connect them in *parallel* rather than in *series.* In a series connection the outlet of one collector is connected directly to the inlet of another and so forth until the pipe returns to the tank from the last collector outlet. With this arrangement the collectors will run hotter since the water in them

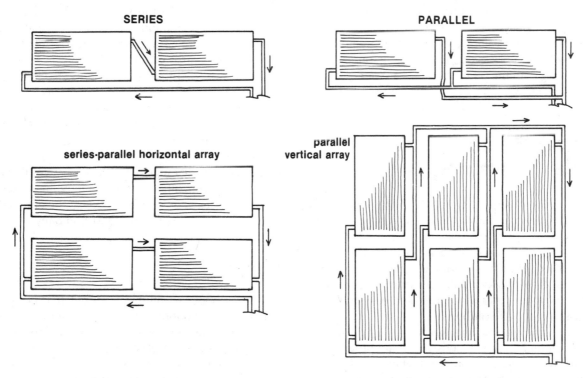

Figure II-11: For collector arrays involving two or three modules, parallel plumbing is the most efficient flow pattern, although it does use a little more pipe than a series connection. For larger arrays a parallel/series or a totally paralleled format are two options. Much depends on site conditions, especially where the pipes are coming from and where they're going relative to the positioning of the collectors.

becomes progressively hotter as it flows from one collector to the next. This makes for higher-temperature water, but also for lower system efficiency because a hotter collector loses more heat back to the atmosphere. The preferred parallel connection has the collector supply line manifolded to separate collector inlets, as shown in figure II-11. The outlets are treated the same way. The collectors run cooler; they deliver more Btu's to your DHW load, and the water temperature is usually hot enough anyway. For larger collector arrays (more than five collectors—that's a lot of hot water!), a series-parallel arrangement may be necessary for practical purposes, or more than one level of manifolding can be used to preserve the parallel flow pattern.

When you've got the collectors in place and the plumbing completely laid out, the real plumbing begins. You'll find that solar plumbing isn't just joining pipe; there's usually a little roofing or carpentry work awaiting the solar installer, and in the next section we'll cover some basic plumbing practices along with the other tasks that are necessary for proper installation.

Some Basic Solar Plumbing

Plumbing is probably the biggest part of installing your solar DHW system. It can be every bit as involved as, say, plumbing a complete bathroom, but there's no mystery to it: Solar plumbing is like any other standard plumbing job even though it involves a few tasks that make it uniquely solar. If your system design is sound, and if you proceed with care, one step at a time, all will be well.

There are several kinds of pipe available for DHW plumbing, and it's possible that you will have to deal with more than one kind while installing your system. You will be tapping into the existing pipes of your house, running new ones out to the collector and back to a storage tank, then back to the hot water distribution lines. If, for example, your storage tank

uses galvanized steel fittings, you will have to make the switch from steel to the house plumbing, which is usually copper. If you decide to run plastic pipe from the house plumbing to the collector, you will need to make the transition between plastic and copper in the house and at the collector and once again when the lines run to the storage tank.

Plastic

In recent years, plastic pipe has established itself as the material of choice in many residential applications. There is a new kind of flexible plastic tubing made of polybutylene that is well suited for solar DHW applications. Being flexible, the tubing doesn't require precise measurements and joins easily with ring-clamp or threaded compression fittings. It can be used continuously at 200°F (93°C) at 70 psi and at 180°F (82°C) at 100 psi. These fittings can withstand pressures of up to 150 psi for an hour at 210°F (99°C). The plastic components are chemically inert and they don't corrode or collect minerals. The tubing has survived freeze testing down to −50°F (−46°C) without bursting. The pipes should, however, be protected from direct sunlight to prevent ultraviolet deterioration. This plumbing system includes a complete range of fittings and valves that are compatible with other standard plumbing systems. (See the "Hardware Focus" section at the end of this article for the listing for the Qest company.)

Copper

Copper is still the most widely used material for supplying domestic water, and it's often used for solar preheaters, which are usually copper. It's more expensive than plastic pipe, but it's very durable and not hard to work with in the sizes used for water supply. The one thing that makes copper seem a little intimidating to some people is that it is usually joined by sweat soldering, which can be tricky for the beginner. A soldered joint can look good and still leak. Since most domestic water supply lines are copper, though, you'll probably have to deal with it at some point in the installation. With a little practice beforehand, handling copper won't be much more difficult than joining plastic pipe.

Copper pipe is available in rolls of bendable tubing or lengths of rigid pipe. The bendable kind, called soft temper, is easier to work with in tight spots since measurements between fittings do not have to be quite as exact as with rigid (hard temper) pipe. Bendable tubing is more expensive than hard temper, and you have to be careful not to kink it when making tight bends. It comes in 30-, 60-, and 100-foot rolls, while rigid pipe is generally sold in 10- and 20-foot lengths.

There are three pipe thicknesses to choose from: type K, L and M. Type K is thickest and not often used in residential DHW systems; M is the lightest weight, and L is in between K and M. Bendable tubing comes only in type K and L, while rigid pipe is available in all thicknesses. Generally types L and M are used with DHW systems, with M used where rigid pipe is needed and L used where bendable pipe is needed.

Sweat Soldering: No Sweat

When you first start doing it, soldering copper fittings is most intriguing. With a little flux and a little heat the solder is almost magically drawn up in between the fitting and the pipe to make a permanent

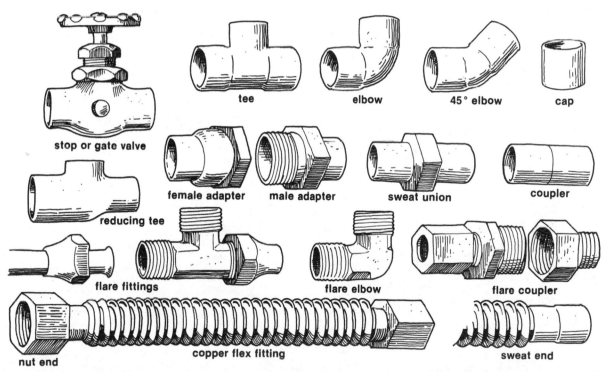

Figure II-12: This is just a small sampling of the dozens of kinds of fittings used to join copper pipe. Sweat solder fittings are the most commonly used, and flare fittings are useful at points where future disassembly may be necessary. The gasketed nut end on a copper flex fitting simplifies the transition from threaded steel to copper pipe.

metal-to-metal seal that can handle any temperature or pressure a DHW system is likely to experience. Sweat soldering demands your concentration for a fairly simple procedure and, if you stick with it, every joint should come out right. There's something very satisfying in pressurizing a complete copper system and having no gurgling leaks or needle sprays emitting from any joints. Once you've mastered sweating, you'll be able to tread fearlessly into new frontiers of plumbing.

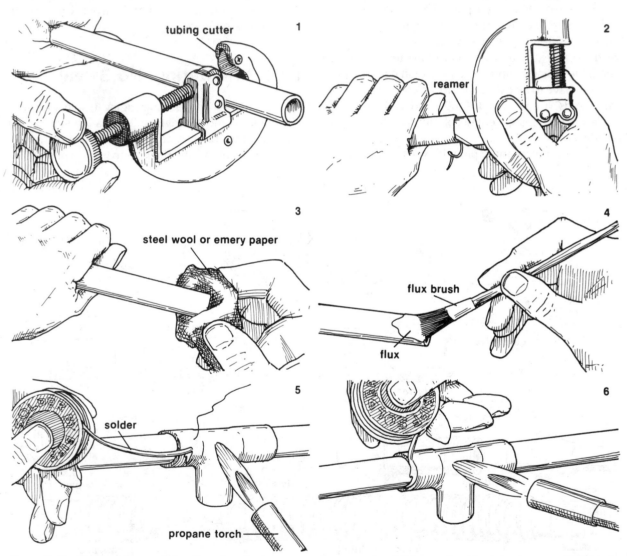

Figure II-13: After you've read the directions in the text, make a couple of "dry run" solder joints to get a feel for the procedure before doing the actual plumbing. One step this illustration doesn't show is the cleaning of the inside surfaces of female sweat fittings. The job is made easy with a small, round wire brush that quickly cleans and roughs up the copper, and it's as important as cleaning the end of the pipe.

A checklist of the tools you'll need for sweat soldering can be found in "Working" in Section I. The first step is to cut the pipe, a task made easy with a tubing cutter. Then use the reamer that comes with most tubing cutters to carve out the inside burr that remains from the cut. Rub the outside of the pipe with emery paper, fine sandpaper or coarse steel wool. Having a clean, scratched surface ensures a good bond between the copper and the solder. Give the fitting the same treatment. A ½- or ¾-inch diameter round wire brush makes fast work of preparing female fittings. For long or complicated pipe runs with several turns or tight spots, it's a good idea to put together the entire run, flux and all, rather than assembling and soldering one joint at a time.

Brush a coat of flux onto the pipe end and the inside of the fitting, and join them together. Heat the assembly with a propane torch. The flame itself should be about 1 to 1½ inches away from the metal. The inner cone of the flame is the hottest part, and the metal should be just inside it. Touch the solder to the opposite side of the assembly from the flame; if it melts, move the flame onto the pipe, about 1 inch away from the joint. (Don't keep the flame directly on the joint, or you may overheat it, making it hard to get a proper sweat.) Start feeding solder into the joint until a rim of solder appears completely around the fitting. What happened? Capillary action pulled the solder up into the joint. (So much for the magic.) The solder appears bright and shiny at first, but about 15 to 30 seconds after the flame has been removed, it becomes dull, indicating that the joint is set.

Copper conducts heat very well, and you should be careful about touching the pipe soon after it has been heated. For the same reason it will also readily lose the heat in the hot water it's carrying, making pipe insulation essential. Be careful about soldering near wood or other combustible materials. It's good practice to use a small sheet of metal or an asbestos square to keep the flame away from anything that can ignite.

There's another way to work with copper pipe that doesn't require sweat soldering. A *flare* *fitting* works by compressing a soft temper pipe end onto a brass cone. The pipe end has to be flared with, of all things, a flaring tool so it will seat properly over the cone end of the brass fitting. Flare fittings are somewhat more expensive than copper sweat fittings but they do have the advantage of permitting easy decoupling of a pipe section, which may be useful when it comes to servicing the parts of your DHW system.

Steel

Galvanized, threaded steel pipe is sometimes found in older homes. It is heavy and cumbersome to work with, and it's not advisable to plumb a new DHW system with it. If your present plumbing is steel, you can tap into it by disassembling a section, working in the new fitting (such as a tee), and reassembling the section with the aid of a *union,* which joins two facing threaded pipe ends. With threaded pipe you cannot of course unscrew a piece in the middle of a run. It has to be cut; the two "raw" ends have to be threaded (at your local hardware or plumbing store) to receive the new fitting and the union. You could forget about galvanized pipe except for the fact that many tanks suitable for hot water storage use threaded galvanized fittings (see Figure II-5).

One thing you must always do with threaded steel fittings is put pipe dope or Teflon plumber's tape on the threads before you join two pieces. Either material will keep the threads from rusting, making it easier to take the joint apart if the need arises. More importantly, dope and tape seal the joint, preventing leaks.

Transitions

A likely plumbing scenario for a solar DHW installation is that you start at your home's copper water supply and run copper or plastic tubing up to a copper flat plate collector or to a batch-type collec-

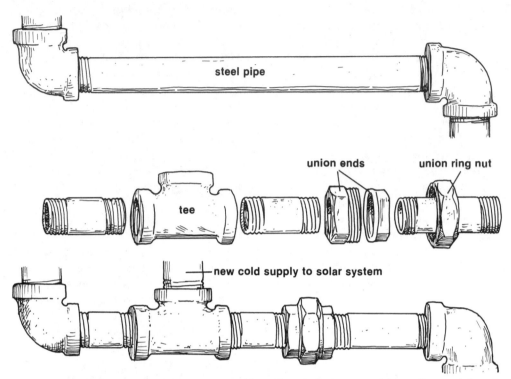

Figure II-14: Union fittings allow you to modify a threaded pipe run without disassembling the whole run. To add a tee section, the pipe is first cut and then converted into shorter pipe sections to accommodate the new fittings. The last thing to be tightened in the reassembly is the union ring nut.

tor that probably uses galvanized fittings. With flat plate systems there is usually some type of storage tank in the house, which also may use galvanized fittings and possibly a pump that may have copper sweat fittings or threaded brass or plastic fittings.

Standard plumbing equipment is designed to permit interconnections of different materials, but there are a couple of points to bear in mind. When joining dissimilar metals, such as steel and copper, you should use special fittings to prevent corrosive electrolytic action between them. In the presence of moisture, different metals in contact can create a miniature battery, causing one of the metals to corrode slowly. A watertight fitting that separates the two metals with a neoprene washer or some other nonmetallic substance is required. Brass and copper can be joined directly because they are sim.lar, and

brass can also be used as an intermediary between steel and copper.

The connection of plastic pipe to copper or steel presents a little different problem. Since it is impossible to weld or solder plastic to metal, a threaded connection is required, which usually has to be wrenched down very tight to ensure a leak-proof connection. Since threaded plastic is not as durable as copper or steel, be careful not to over-tighten the plastic during installation. There is no electrolytic action between plastic and any metal.

One fitting that will make life easier when you work with copper is called a copper flex fitting. One end is a standard female sweat fitting, which runs into a length of corrugated tubing that can be easily bent to shape. At the other end there is a neoprene washer inside a threaded compression fitting. This fitting completely separates the copper from the pipe it joins, making the flex unbeatable for copper-to-steel transitions.

Running Pipe

All pipe runs should maintain a minimum pitch of ¼ inch per foot of run with drains at low points and vent valves at high ones to allow complete drainage of the system if that becomes necessary. All discharge should be directed to an acceptable disposal location, especially nonpotable, closed loop, antifreeze liquids. Check your local codes for other requirements.

Pipe must be well supported, with allowance for expansion and contraction of the pipe. Inside the house, steel or copper should be supported every 7 to 10 feet and rigid plastic should be supported every 2 to 3 feet. Flexible plastic pipe run horizontally needs support about every 12 inches. For copper and steel, use the same metal for support as for the pipe to avoid the possibility of electrolytic action, or make sure that pipe insulation separates dissimilar metals. There are many kinds of pipe hanging hardware for indoor and outdoor use; a visit to your local plumbing supply will help you identify what you need.

You can notch joists or studs for pipe, but keep the size of the cuts to a minimum or you'll weaken the structure. For vertical studs, cut no deeper than one-half the width of the stud (¾ inch for a 2 × 4). Notches in studs should also be covered with solid metal straps to protect the pipe from future nails. If you must notch floor joists, cut no more than one-quarter of their depth, or better still, drill holes through the joist. Holes don't weaken the structure of studs or joists nearly as much as notches; thus they can be located anywhere along the length of the support member. Never locate one hole directly above another; stagger them. Outside pipe supports must be attached securely to the structure of the house and carefully waterproofed.

Any run of hot water pipe will of course lose heat, and for a solar system this can make the difference between one that performs on par with expectations and one that does not. All inside and outside hot water pipe runs must be insulated to at least R-4, including insulation wrapped around all valve bodies. It's always a good idea to try to minimize the length of pipe runs outdoors and through unheated areas.

Pipe Insulation

There are several ways to insulate pipe: with nonrigid foam insulation, with rigid foams made in whole and split sections that fit over the pipe, and with preinsulated pipe sections. Some insulating materials are degraded by ultraviolet light and should be protected from the elements when used outside. In some cases rigid insulation comes encased in a protective plastic shell; in others the best policy is to insert the pipe and insulation into a 3- or 4-inch PVC pipe. For example, ½-inch copper supply and return pipes, with insulation, can be run through a single 4-inch PVC pipe, along with the sensor wire, making the installation weather-tight and better looking than running everything separately.

When working indoors or outdoors with pipe insulation, it's important to seal all seams with glue or tape or whatever else the manufacturer provides. Loose foam means heat leaks. Flexible foam insulation tubes should be slipped onto the pipe sections *before* they're assembled. This saves the task of slitting the insulation sleeves and then gluing them back together again. To complete solders or plastic welds the foam can be held back from the work with spring clamps or Vise-Grips.

Some rigid foam insulation systems come with preformed elbow, tee and other shapes to ensure total coverage. Generally rigid (urethane) foam insulation has a higher R-value (R-7 per inch) and is recommended for very cold environments. It's more expensive than flexible foam, but it is cost-effective for solar DHW applications.

Preinsulated pipe comes in a variety of sizes and pipe materials, so you can use it with just about any pipe you have to work with. It usually comes with a complete package to allow joining and proper sealing of the exterior, ensuring proper weather protection. Preinsulated pipe is therefore a little easier to work with, though it is of course more expensive than flexible and rigid foam. Some brands are available with two pipes mounted inside a single protective pipe filled with high R-value urethane foam, perfect for solar collectors.

Roof and Wall Penetrations

Since it's best to get the insulated pipe into the house with the shortest run possible, pipes leading from roof-mounted collectors often penetrate the roof itself. The usual practice is to run the pipe through a standard vent flashing or "boot." This flashing goes under the shingles above the pipe and lies on top of the shingles below. It's sold in standard pipe widths, and it's a simple matter to match the size

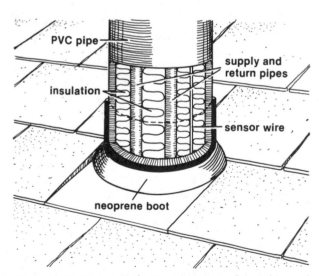

Figure II-15: Neoprene boots are commonly used to seal and flash sewer and other household vent pipes, and they are well-suited for use in weatherproofing solar plumbing. Generally roof penetrations in the vicinity of the collector should be considered ahead of long, exterior pipe runs for better looks and better economy.

to your pipe, especially when using PVC pipe as a sheathing over insulated pipe. The boot has a neoprene collar that fits around the pipe to assure a watertight seal. One brand has separate collars for the supply and return lines and a tiny collar for sensor wire. But generally a one-hole boot will do fine; they're available at plumbing and hardware stores. Don't use the boot as a pipe support. Secure the pipe as soon as possible after it enters the house.

Wall penetrations are not as exposed to rain and snow as are those in roofs, but they, too, must be well sealed to eliminate infiltration heat loss from the building. Rigid insulation can be surrounded by a plastic sleeve where it runs through the wall. The opening is then sealed with caulk. Since nonrigid insulation will not support the weight of the pipe, the pipe must be completely supported before it enters and after it leaves the wall. The same sleeve system used with rigid foam can then be used. If the insulated pipe enters a foundation wall, the insulation must be carefully protected against moisture to prevent absorption and damage to the insulation; again the plastic sleeve offers good protection.

There are many products on the market with new plumbing and insulation systems arriving all the time. Some are improvements; some are no better than the old stuff they're purported to improve on. Look over the manufacturer's specifications carefully to be sure the products you get are compatible with each other and suitable for the application you have in mind.

Filling the System

It doesn't make any real difference if you plumb your DHW system from the tank to the collector or vice versa, but when it's all hooked up, it's time to put your plumbing skills to the acid test by filling the system. Pressurized (open loop) systems are filled by simply opening the cold water supply valve to the tank.

If it's possible to run water through the pipes and bypass the tank, it's a good idea to flush

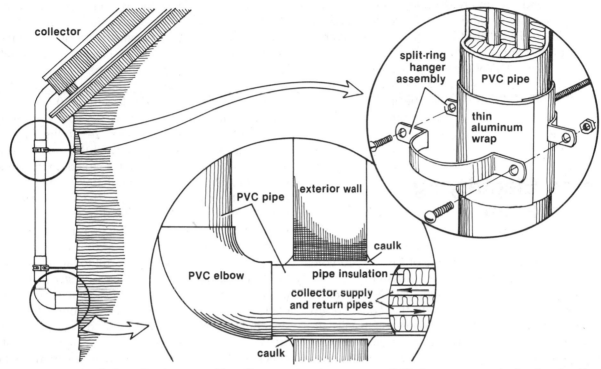

Figure II-16: A simple but effective way of handling exterior pipe runs uses PVC pipe to encase the insulated collector supply and return lines. The split-ring hanger should be bolted to a solid backing (masonry, wall stud, rim joist, etc.) where possible. Where the PVC pipe passes through the wall, use liberal doses of caulk to seal out cold air infiltration.

the lines free of metal or plastic bits and flux or pipe dope, all of which can accumulate during the installation. When the system is filling, air will be forced out of the air vents at the collector and at the tank(s), and when the hissing stops, that should be all there is to it. Hopefully you won't experience the disheartening sight or sound of drip, drip, drip, or of some fitting making like a fountain.

It's a good idea to have a friend observe parts of the plumbing you can't see when the system is filling. It can happen that one joint is "forgotten" during the plumbing, and suddenly it begins to gush forth, to everyone's surprise and dismay. If you can't see that part of the system, water could get into the wrong place before you discover the calamity.

If you do end up with a leak in a copper pipe system, that section must be drained completely

before the fitting can be resoldered. Water in the line will prevent the joint from getting hot enough to melt solder. The same is true for solvent-welded plastic pipe, but for different reasons. Compression fittings can be tightened, but be careful not to overtighten and risk damaging the fitting. Leaks in compression joints may result from uneven seating or dirt in the seal. Check these before you torque the joint to death. Threaded steel fittings can be tightened too, but if you're tightening one end of a pipe section or fitting, you're loosening the other. If it's a section you've worked on (why else would it be leaking?), there may be a union where you tapped the run. Start taking up the slack from there. If there's no union, you may have to decouple the steel section from the rest of the system and retighten everything from one end to the other. Two kinds of flat plate systems, the drain-back and the antifreeze fluid system, require different filling procedures, which are described in "A Catalog of Flat Plate Collector Systems," found later in this section.

If your DHW system is passive, with no electrical controls, it's ready for operation. Uncover the collector and let the heating begin. With active systems the electrical work begins when the plumbing is done.

Some Electrifying Information

If you are installing a solar water heater that has a pump, you'll need a source of 115- to 120-volt electricity for it and for the other system controls such as a thermostatic switch and possibly an electric valve. Hiring an electrician to do this could be an expensive move for a relatively simple job. If you can do your own plumbing, you can certainly do the wiring—as long as you take a few precautions and make sure you know what to do before you start.

The first step is to locate a source of power near where the system's electrical components will be placed. There ought to be a circuit nearby that you can tap into; you just have to be sure the circuit can handle the small additional load. The pump for a typical collector probably won't exceed 1/20 hp, or about 1.8 amps, and the thermostat load is negligible, less than 0.05 amps. Allowing the motor 25 percent extra for start-up load, the whole thing will draw less than 2.3 amps, so there's a good chance you can use any circuit. Find out what other loads are on the circuit to be sure there is enough extra capacity. If there's no circuit available, a new breaker (15 amp) must be added to your breaker panel and a 14-gauge wire run to the pump location.

To add up the loads on an existing circuit, disconnect the circuit in question by switching off the breaker or unscrewing the fuse; then see which fixtures or receptacles (sockets) have lost power. You should check everything in the house to be sure,

because a circuit may follow some unusual paths. When you know what's on the circuit, add the amperages of appliances that might be on simultaneously. If some of the loads are labeled in watts only (e.g., light bulbs), simply divide the wattage by 120 (volts) to get the amperage. If the total load, including the pump and thermostat, is less than 80 percent of the amperage rating of the *wires,* you're in good shape. Don't depend on the size of the fuse or breaker to tell you the capacity of the circuit. Look at the wire to see what gauge is marked on the cable insulation. If you can't find out that way, shut off the power to the main fuse or breaker panel, remove the black conductor and check the diameter of the wire with a wire gauge. Number 12 wire is rated for up to 20 amp service; #14 carries up to 15 amps.

Even if the wire you want to tap into passes the tests for right voltage and carrying capacity, it could still be the wrong line for your solar DHW controls. The conductors could be connected to a switch. Hook into one of these and your pump will cut out every time you turn off the switch. It pays to check out the wires thoroughly before you make the changes.

To tap into the circuit, shut off the power to the circuit. Check with a test light to be sure the power is off, then simply cut the cable and mount a junction or outlet box next to the cut. There are many kinds of boxes designed for installation to studs, in plaster and lath, or in drywall. There are bar hangers that allow you to attach a box between exposed joists or studs.

Whether you install a junction box or an outlet box with a receptacle depends on the type of thermostat. Some thermostats have plugs for the power supply and a receptacle for the pump. You'll need to have a receptacle near a plug-in type thermostat. Other thermostats have a built-in electrical box to which you fasten the electrical cable for the power supply and the pump. In this case, you'll still need a junction box when you tap into the existing circuit. Some systems use a mixture of plug-receptacle-cable connections. Whatever the wiring combination, any time there are wires spliced together or attached to terminals, they must be enclosed in a UL-approved electrical box, whether it is separate from the pump or thermostat or attached to it.

Junction boxes usually have no built-in cable clamps, so you'll need to buy some to fasten the cable to the box. Junction boxes are covered with knockout holes. Determine which ones are the most convenient for the cable entries; insert a screwdriver in the slot in the knockout and simply bend the knockout out. Strip the sheathing off each end of the cable you've previously cut. Do this very carefully because if you nick the insulation of the conductors inside the sheathing, a short circuit is possible. Very bad. Work carefully.

The idea is to strip the sheathing from the cable up to where they enter the box. Thread the wires through the knockout holes which have been fitted with cable clamps. Give yourself about 6 inches of cable inside the box so you'll have enough slack to work with. Join the new cable to the newly cut cable ends with wire nuts: black to black, white to white, ground to ground. The result will be a trio of black and a trio of white wires, all bound together with the wire nuts. The ground wire will need a fourth line because you'll need to run a short "pig-tail" from it to the box if it's metal. Make a loop at the other end of the pigtail from the wire nut and screw it to the box. Scrape the metal under the loop to ensure good contact.

You can use either Romex or BX (metal-armored) cable for your circuit tap. Electrical codes vary, but usually exposed cable should be armored.

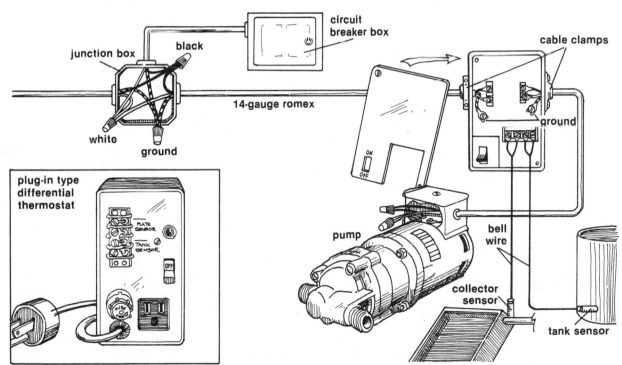

Figure II-17: This schematic shows the typical wiring for an active DHW system. The 120 VAC power is brought to the thermostat, which controls the 120 VAC pump based on the signals from the collector and tank sensors. The connections to most thermostats and pumps are standard, although some DHW control packages do have plug-in components, which can save installation time.

If you can run wire inside a wall or in between exposed joists, Romex will suffice. If you run the cable any distance to the thermostat and pump, support the cable with cable straps or staples. Don't drive staples too deep into the joists or studs or you may damage the conductors. Any cable must be supported within 12 inches of a metal box and within 8 inches of a nonmetallic one. The cable itself should be supported about every 4½ feet. You can run unprotected cable along (parallel to) studs or joists, but if it runs across them, it must be attached to a board (1 × 4 will do) to avoid possible damage. A cable needs no support where it runs inside a wall or through studs. Finally, run the cable to the thermostat. There should be a wiring diagram to tell you how to connect it.

If the thermostat doesn't have built-in plugs and receptacles, remember as you make your connections that the black (hot) wire from the thermostat to the pump is the switching line. Make sure you correctly place the black wire from the source (line) and the black wire from the pump (load) on the thermostat terminals. These terminals are often labeled "line" and "load" and should be wired accordingly. The white (neutral) wires are never switched, and the pump neutral will directly join the source neutral at a wire nut. A short jumper will also join these neutrals to a thermostat terminal, often labeled "com" (common).

Sensors

The differential thermostat is basically a "smart" switch for the pump. From it run two low-voltage sensors, one to the hot water outlet of the collector and the other to the storage tank. When the thermostat senses that the collector is 5 to 15°F (−15 to −9.4°C), depending on the model used, warmer than the storage tank, indicating that there is heat to be gained, it kicks in the pump. The thermostat operates on line voltage (115 to 120 volts AC), but the sensors are at 5 to 10 volts DC. The sensor terminals are clearly marked on the thermostat to indicate what goes where, so you should have no trouble in making the right connections.

For sensor wire you can use standard "bell wire" available at hardware stores. Bell wire comes in 2-, 3-, 4- and 6-wire configurations, with each wire having a different insulation color, so depending on the sensor arrangement your thermostat uses, you can choose a type to suit your needs. Some differential controllers, for example, require separate heat and freeze sensors placed at the collector, which makes 4-wire bell wire the right choice, combined with a 2-wire run for the tank sensor. For the sake of neatness, run the wire from the collector along the plumbing, taped onto the pipe insulation and/or run inside the protective plastic pipe that carries the insulated collector supply and return lines. Bell wire is usually a solid, not braided, copper wire that can snap easily if it's "overexercised." When connecting the sensor leads to bell wire, it's best to make a solder connection rather than a wire nut connection because it's possible for a twist connection to corrode and affect the signal from the sensor.

Sensor installation should be done with the utmost care, for in active DHW systems these little items are what give both hot water and, in some systems, critical freeze protection. Generally the collector sensor should be taped to bare pipe at the high end of the collector array. If the sensor is for heating only, it should be insulated as much as the rest of the pipe. If it serves both heating and freeze protection functions, *do not insulate it!* An insulated freeze sensor won't know what the weather's like perhaps until after the collector has begun to freeze, which could make it impossible for the system's freeze protection to come to the rescue. The tank sensor should be heavily taped to the bottom of the tank, preferably 180 degrees away from the supply line to the collector (which may also be the cold supply to the tank). The tank sensor should be as heavily insulated as the rest of the tank. All tank and collector sensors should remain somehow accessible for future testing and maintenance. Generally the manufacturer of your control system will state quite specifically where the sensors should be placed, and they say that because that's where they *must* be placed to satisfy the requirements of that particular system.

Final Check and Start-Up

Once you've made all the connections, but before replacing any cover plates, look over all your wiring to make sure everything is in order and that no bare conductors are touching other conductors or metal boxes (except for bare ground wires). When you're sure all is well and the rest of the system has been completed and charged with water or fluid, uncover the collectors and turn on the power.

The thermostat should have a manual override switch in the "off" position. If the sun is shining, turn it on to activate the system. If the collector is hot enough and the tank is cold enough, the pump ought to go on, and you can start to feel the first solar-heated water coursing through the lines. If there's not enough solar energy available, bring a glass of hot water up to the collector, immerse the heat sensor in it and have a partner yell up if the pump goes on. As long as you're going up there, you can also test the freeze protection circuit. Bring up a glass of ice water and a thermometer and mix hot and cold water until the thermometer reads the temperature at which your system's freeze protection is activated. Then dunk in the freeze sensor. If it's a drain-down system, the collectors should empty out, and if you have recirculation freeze protection, the pump should start up again. (These freeze protection systems are described more fully in "A Catalog of Flat Plate Collector Systems," found later in this section.) It's important to make the freeze check before the event really occurs, of course.

System Maintenance

The do-it-yourself solar DHW system doesn't stop being do-it-yourself once it's installed and operating. Like the other parts of your house, or your car or your land, your DHW system will need a minimum, periodic maintenance routine to keep it running properly and to ensure that the mechanical system itself (collector, pipe insulation, building penetrations, etc.) is intact and not suffering excessive deterioration.

A good place to start is to make a final plumbing and electrical schematic drawing of the complete system as it was installed, noting all plumbing and electrical parts and their specifications (sizes, voltage, etc.) and all transition points such as building penetration and pipe material changes. Plumbing lines should be marked with flow directions. Along with the schematic, which should be posted on the tank or near the thermostat, make up a simple log or sheet for keeping track of system operation, maintenance and repairs. If you're into it, keep track of your energy bills to see how much they are decreased at different times of the year. Somewhere, with the log or the schematic, collect all the manufacturer's literature in one place.

If your system is running normally, you can keep it that way with simple maintenance checks every three months, with more extensive checks made every six months. The simpler the system, of course, the less looking-over it will need. Passive DHW systems (flat plate, thermosiphon and batch-type) that don't have electrical controls need only be checked to make sure that all the hardware is intact, that all valves operate properly and that the collector glazing seals and box aren't deteriorating. Active systems need that and more. In both types the three-month check should include bleeding of all air valves. In pressurized systems, only a small amount of air should emit from the air valve before water appears. If it takes more than five seconds for water to appear, the air valve may be defective and should be replaced. After shutting down the water pressure, remove the valve and see if it's clogged somehow. Clean and replace it and check the valve again in a week. If the condition persists, the valve isn't doing its job and should be retired. To complete the three-month check, manually pop open all pressure-relief valves to make sure they aren't stuck and can close completely. Give an eye to interior and exterior pipe runs to make sure nothing unusual has happened. Check all building penetrations for possible leaks. Depending on the environment, the collector glazing may need to be cleaned.

The six-month check should include all the above tasks plus a close check on the condition of the collector box with maintenance as necessary. Unpainted collector boxes should receive regular treatments with sealer-preservative. With active systems make the same checks on the electrical controls that were made when the system was first started up. It's especially important to check the freeze protection circuit just before the onset of cold weather. If the manufacturer calls for lubrication or other service for the pump or other component, it should be done in the six-month check. Other maintenance routines that are specific to certain system designs are dis-

cussed in "A Catalog of Flat Plate Collector Systems," found later in this section.

These simple maintenance routines aren't meant to add a worry factor to your DHW system; quite the contrary. If you look after the thing just four days out of 365, you'll be able to forget about it for the other 361 and enjoy faithful operation. And part of having worry-free operation comes with choosing the right type of system for your climate, for your particular application and for your hot water needs. In "A Catalog of Flat Plate Collector Systems," which follows, we present five generic system types, each with its own advantages and appropriateness for the above-mentioned factors. After that you'll run into the batch-type solar DHW systems, which have advantages of their own. Choose carefully and install carefully, and the rewards—solar showers, shaves, baths and dishwashing—will be sweet.

John Blackford and Joe Carter

References

Plumbing

Adams, Jeannette. 1977. *Complete home plumbing and heating handbook.* New York: Arco Publishing Co.

Alth, Max. 1975. *Do-it-yourself plumbing.* New York: Harper & Row Publishers.

Day, Richard. 1969. *The practical handbook of plumbing and heating.* New York: Fawcett Publications.

Hylton, William H., ed. 1980. *Build your harvest kitchen.* Emmaus, Pa.: Rodale Press.

Massey, Howard C. 1978. *Plumbers handbook.* Solana Beach, Calif.: Craftsman Book Co.

Time-Life Books, eds. 1976. *Plumbing.* Alexandria, Va.

Wiring

Colvin, Thomas S. 1979. *Electrical wiring: residential, utility buildings and service areas.* Athens, Ga.: American Association for Vocational Instructional Materials.

Departments of the Army and Air Force. 1975. *Interior wiring.* Washington, D. C.: U.S. Government Printing Office. no. TM 5-760/AFM 85-24.

Duncan, S. Blackwell. 1977. *The complete handbook of electrical & house wiring.* Blue Ridge Summit, Pa.: Tab Books.

Hylton, William H., ed. 1980. *Build your harvest kitchen.* Emmaus, Pa.: Rodale Press.

National Fire Protection Association. 1977. *National Electrical Code, 1978.* Boston.

———. 1977. *One- and two-family residential occupancy electrical code.* Boston.

Richter, H. P. 1978. *Wiring simplified.* 32d edition. Minneapolis: Park Publishing.

Time-Life Books, eds. 1976. *Basic wiring.* Alexandria, Va.

———. 1978. *Advanced wiring.* Alexandria, Va.

Hardware Focus

A High-Pressure, High-Temperature Plastic Plumbing System

Qest
U.S. Brass Division
P. O. Box 1746
1900 W. Hively Ave.
Elkhart, IN 46515
(219) 294-7541

Qest Products

Qest manufactures a complete plumbing system for domestic water that includes flexible, hot/cold polybutylene piping in sizes ¼'' to 1'' ID and fittings and valves of Celcon, a rigid plastic. Chemically inert, the plastic components are resistant to corrosion and mineral buildup. The Qest system is suitable for applications up to 210°F (99°C) with pressures up to 150 psi (intermittent service), which is more than adequate for the average home. The system has survived testing down to −50°F (−46°C) without bursting under freezing conditions. The manufacturer notes that the tubes should be shielded from direct sunlight to prevent deterioration. The Qest fittings have standard plumbing threads so they are compatible with most other existing systems and the price list seems to show that the Qest system is cheaper than copper or steel plumbing. A catalog of Qest Products is available from the manufacturer.

Differential Thermostats for Solar Domestic Hot Water Systems

A differential thermostat is basically a "smart switch" that turns a solar DHW pump on and off depending on certain temperature conditions in the system. When the temperature of the fluid in the collector is higher than that in the storage tank, the differential thermostat senses this and activates the pump to circulate the heat to the storage tank. When the temperatures in the collector and the storage tank are nearly the same, the thermostat stops the pump because there is little or no heat to move to storage.

Some thermostats have fixed turn-on and turn-off differentials, while others can be adjusted. Most systems turn on when there is a difference of about 10 to 20°F (6–11°C) between the temperature in the collector and storage and then turn off when there is about a 3°F (1.7°C) difference.

American Solar Heat Corp.
7 National Place
Danbury, CT 06810
(204) 792-0077

Sunmeter, Model 1100
On differential 18°F (10°C). Off differential 2°F (1.1°C).

C & M Systems, Inc.
Saybrook Industrial Park
Elm St.
P. O. Box 475
Old Saybrook, CT 06475
(203) 388-3429

Solar Stat SSA
On differential 12°F $\pm$ 3°F (6.7°C $\pm$ 1.6°C). Off differential 5°F $\pm$ 3°F (2.8°C $\pm$ 1.7°C).

Dan-Mar Co., Inc.
RR 2
Box 338B
Wikel Rd.
Huron, OH 44839
(419) 433-4479

Solar Heating System Controller, TC17
On differential fixed at 12°F (6.7°C). Off differential fixed at 5°F (2.8°C). Recirculation freeze protection.

Deko-Labs
P. O. Box 12841
University Station
Gainesville, FL 32604
(904) 372-6009

Temperature Comparator, Model TC-3
On differential fixed at 10°F ± 1°F at 135°F (5.6°C ± 0.6°C at 57°C). Off differential fixed at 5°F ± 1°F at 135°F (2.8°C ± 0.6°C at 57°C). Recirculation freeze protection.

del Sol Control Corp.
11914 U.S. No. 1
Juno, FL 33408
(305) 626-6116

Differential Control, Model 02B
On differential 8°F (4.4°C). Off differential 7°F (3.9°C). Recirculation freeze protection.

Ecotronics, Inc.
7745 E. Redfield Rd.
Scottsdale, AZ 85260
(602) 948-8003

Soltroller Proportional Controls, SPC Series
Customer specifies on/off differential setting from 4 to 8°F (2.2–4.5°C). Drain-down freeze protection.

Hawthorne Industries, Inc.
Solar Energy Division
3114 Tuxedo Ave.
West Palm Beach, FL 33405
(305) 684-8400

Fixflo Control H-1507-C, H-1503-C
On differential 18°F (10°C). Off differential 6°F (3.3°C). Recirculation freeze protection optional.

Fixflo Control H-1503-B
On differential 14°F (7.8°C). Off differential 8°F (4.4°C). Recirculation freeze protection available.

Heliotrope General
3733 Kenora Dr.
Spring Valley, CA 92077
(714) 460-3930

Delta-T Differential Temperature Thermostat
Nonadjustable models on/off differentials: 9°/3°F (5°/1.7°C), 4.5°/1.5°F (2.5°/0.8°C), 15°/5°F (8.3°/2.8°C), 15°/1.5°F (8.3°/0.8°C). Adjustable models from 3 to 24°F (1.7–13.3°C) with off fixed at 1.5°F (0.8°C). Recirculation or drain-down freeze protection available.

Independent Energy
P. O. Box 732
42 Ladd St.
East Greenwich, RI 02818
(401) 884-6990

C-100, C-120
Three-digit temperature display. Standard on differential 20°F (11.1°C). Off differential 5°F (2.8°C). Drain-down or recirculation freeze protection optional.

JBJ Controls
P. O. Box 383
Idaho Falls, ID 83401
(208) 522-2200

Solar Control, Model 79F
On differential 15°F (8.3°C). Off differential less than 5°F (2.8°C). Drain-down freeze protection.

Johnson Controls, Inc.
2221 Camden Court
Oak Brook, IL 60521
(312) 654-4900

Solid State Differential Temperature Controller, Series R34D
On/off differential adjustable from 0 to 40°F (0–22.2°C).

Natural Power, Inc.
New Boston, NH 03070
(603) 487-5512

Differential Thermostat, Series S25 and S26
On differential can be set anywhere between 3 and 25°F (1.7–13.9°C). Turn off adjustable between 0 and 22°F (0–12.2°C). Drain-down or circulating freeze protection.

Pyramid Controls
P. O. Box 2211
Martinez, CA 94553
(415) 229-2470

Solar and Thermostat Control, SC-310
On differential adjustable from 1 to 20°F (0.6–11.1°C). Off differential is adjustable from 2 to 8.5°F (1.1–4.7°C). Recirculation or drain-down freeze protection.

Rho Sigma, Inc.
11922 Valerio St.
N. Hollywood, CA 91605
(213) 982-6800

Proportional Control, RS 500
Unit varies flow rate with temperature difference. Programmed to customer specifications. Drain-down freeze protection.

Richdel, Inc.
P. O. Drawer A
1851 Oregon St.
Carson City, NV 89701
(702) 882-6786

Differential Thermostat Control, Model R409
On differential 15°F ± 1°F (8.3°C ± 0.6°C). Off differential 5°F ± 1°F (2.8°C ± 0.6°C). Drain-down freeze protection.

Robertshaw Controls Co.
Temperature Controls Marketing Group
100 W. Victoria St.
Long Beach, CA 90805
(213) 636-8301

Solar Commander, SD-30
On differential 15°F ± 5°F (8.3°C ± 2.8°C). Off differential 5°F ± 3°F (2.8°C ± 1.7°C). Recirculation freeze protection optional. Adjustable on differential, 8 to 20°F (4.4–11°C), optional.

Solar Control Corp.
5595 Arapahoe Rd.
Boulder, CO 80302
(303) 449-9180

Solid-State Solar Hot Water Controller, Model 77-171
Adjustable on/off differential. On typically 20°F ± 2°F at 100°F (11.1°C ± 1.1°C at 37.8°C). Off typically 4°F ± 2°F at 100°F (2.2°C ± 1.1°C at 37.8°C). Freeze protection circuitry.

Solar Sensor System Electronic Controls
4220 Berritt St.
Fairfax, VA 22030
(703) 273-2683

Solar Sensor
On/off differential adjustable from 5 to 20°F (2.8–11.1°C).

Solarsystems Industries, Ltd.
No. 2-11771 Horseshoe Way
Richmond, B.C., Canada V7A 4S5
(604) 271-2621

Differential Thermostat
On differential adjustable from 9 to 27°F (5–15°C). Off differential adjustable from 0 to 7°F (0–4°C). Drain-down freeze protection.

TA Tour & Andersson, Inc.
652 Glenbrook Rd.
Stamford, CT 06906
(203) 324-0106

Differential Temperature Controller
On/off adjustable differential from 0 to 54°F (0–30°C).

Thanor Enterprises, Inc.
817 West St.
Wilmington, DE 19801
(302) 475-8257

Differential Thermostat, Models S, SF, SFH
On/off adjustable differential from 0 to 20°F (0–11.1°C).

West Wind Electronics, Inc.
P. O. Box 1657
Durango, CO 81301
(303) 588-2275

Temperature Differential Switch
On differential adjustable from 0 to 9°F ± 1.8°F (0–5°C ± 1°C). Off differential factory set at 1.8°F ± 9°F (1°C ± 0.5°C).

Willtronix
1927 Clifton
Royal Oak, MI 48073
(313) 399-9557

Differential Temperature Thermostat, WD-1
Operating temperature is adjustable with a constant differential of 19°F ± 4°F (10.6°C ± 2.2°C).

Wolfway Product Consultants, Inc.
R.D. 1, Box 1135
Tamaqua, PA 18252
(717) 668-4359

Solar Controller Differential Thermostat
On differential preset at 15°F (8.3°C). Off differential preset at 3°F (1.7°C). Adjustable within a wide range. Completely assembled units or kits.

Pumps

The pump product descriptions include the type of housing material and the horsepower range. Closed-loop systems using antifreeze fluids with corrosion inhibitors can use circulating pumps with iron housings, but open-loop water systems must utilize pumps made of stainless steel, bronze or plastic. The flow rate through a DHW collector loop is usually very low, so in most cases a circulator pump with a low horsepower (1/100 to 1/12 horsepower) will be adequate.

Armstrong Pumps, Inc.
93 East Ave.
North Tonawanda, NY 14120
(716) 693-8813

In-Line Mini Compact 5-Speed Circulator, S-15B
Iron or bronze. Armstrong rates pump at 80 to 100 watts; not rated for hp.

Dayton Electric Manufacturing Co.
5959 W. Howard St.
Chicago, IL 60648
(312) 647-0124

Teel Hot Water Booster Pump, Model 1P760A
Bronze. Hp: 1/100. (Available through W. W. Grainger, Inc., a national distributor. Check your phone book for an outlet in your area.)

Edwards Engineering Corp.
101 Alexander Ave.
Pompton Plains, NJ 07444
(201) 835-2808

Edwards Even-Flow Circulators
Iron or bronze. Hp: ⅛, 1/3, ½.

Grundfos Pumps Corp.
2555 Clovis Ave.
Clovis, CA 93612
(209) 299-9741

Grundfos Domestic Circulator, Models UPS 20-42, UP 26-64, UM 25-18, UP 25-42, UP 25-64
Iron or bronze. Hp: 1/12, 1/20, 1/32, 1/35, 1/64.

Hartell Division of Milton Roy Co.
70 Industrial Dr.
Ivyland, PA 18974
(215) 322-0730

CP Series
Brass or thermoplastic. Hp: 1/150, 1/50.
GPPS 45H Series
PPO thermoplastic. Hp: 1/7.

Hi-Tech, Inc.
3600 16th St.
Zion, IL 60099
(312) 746-2447

Series 908
Celcon, Ryton or ceramic. Hp: 1/100 to ⅛.

KEM Associates, Inc. (distributor)
153 East St.
New Haven, CT 06507
(203) 865-0584

Myson 'L' Series Water Circulating Pump
Iron or bronze. Hp: 1/25, 1/20.

Little Giant Pump Co.
3810 N. Tulsa St.
Oklahoma City, OK 73112
(405) 947-2511

Little Giant CMD-100 Series
Bronze or nonmetallic. Hp: 1/150, 1/125, 1/20.

March Manufacturing, Inc.
1819 Pickwick Ave.
Glenview, IL 60025
(312) 729-5300

Seal-less Pumps
Iron, bronze or plastics. Hp: 1/200, 1/100, 1/75, 1/50, 1/40, 1/35, 1/25, 1/20, ¼.

Richdel, Inc.
P. O. Drawer A
1851 Oregon St.
Carson City, NV 89701
(702) 882-6786

Solar Recirculation Pump, Model R 798
Thermoplastic. Available in CPVC, glass-filled polypropylene and glass-filled polysulfone. Hp: 1/35.

Taco, Inc.
1160 Cranston St.
Cranston, RI 02920
(401) 942-8000

Solar Cartridge Circulator, Models 008, 008V, 007, 006B, 008B
Iron or bronze. Hp: 1/40, 1/25.

Pipe Flashing

One of the trickier aspects of installing a collector system on the roof of a home is to ensure a watertight seal for the pipe and sensor wire that must pass through a hole in the roof.

Specialty Products Co.
P. O. Box 186
Stanton, CA 90680
(714) 828-9730

Poly-Flex
Flexible thermoplastic unit that fits over pipes and sensor wires to form a watertight seal. Poly-Flex can be used with any kind of roof material and at any pitch angle and is resistant to ultraviolet rays. Various models and sizes are available.

Pipe Insulation

The R-values (per inch) listed below are those claimed by the manufacturers. Actual R-values will vary according to temperature conditions and aging of the insulation.

Armstrong Cork Co.
Liberty & Charlotte Sts.
Lancaster, PA 17604
(717) 397-0611

Standard Armaflex Pipe Insulation
Flexible, elastomeric insulation in thicknesses of ⅜'', ½'', ¾'' for use with ⅜'' to 5'' diameter pipe sizes. Unslit tubular form can be slit and secured with Armstrong 520 Adhesive. Unslit, it is slipped onto piping before it's connected. Rated R-4.

Celotex Corp.
Tampa, FL 33622
(813) 871-4575

Celotemp 1500
Molded, rigid insulation of expanded perlite. Thicknesses from 1'' to 3'' for pipes ½'' to 24'' in diameter. Rated R-2.3 at 140°F (60°C).

CPR Upjohn
555 Alaska Ave.
Torrance, CA 90503
(213) 320-3550

Knauf Fiber Glass
240 Elizabeth St.
Shelbyville, IN 46176
(317) 398-4434

Northeast Specialty Insulations, Inc.
1 Watson Place
Saxonville, MA 01701
(617) 877-0721

Owens-Corning Fiberglas Corp.
Mechanical Division
Fiberglas Tower, OH 43659
(419) 248-8102

Pittsburgh Corning Corp.
800 Presque Isle Dr.
Pittsburgh, PA 15239
(412) 327-6100

Sekisui Products, Inc.
1800 W. Blancke St.
Linden, NJ 07036
(201) 862-1414

Sentinel Foam Products, Inc.
Energy Conservation Division
Hyannis, MA 02601
(617) 775-5220

Teledyne Mono-Thane
1460 Industrial Parkway
Akron, OH 44310
(216) 633-6100

Thermacor Process, Inc.
P. O. Box 4529
500 N.E. 23rd St.
Ft. Worth, TX 76106
(817) 624-1181

Urethane Molding, Inc.
RFD 3, Rte. 11
Laconia, NH 03246
(603) 524-7577

Trymer
Urethane. Rated R-7.14. Fabricated to customer's specifications.

Knauf Pipe Insulation
Glass fibers bonded to thermosetting resin. Plain or jacketed. Available in thicknesses from ½" to 6" for copper pipes ⅝" to 6⅛" in diameter. Self-sealing lap available. Rated R-4.3 at 75°F (23.8°C).

Urethane insulation with PVC jacket in thicknesses from ½" to 8" for pipes ⅝" to 12'. Rated R-7.14.

Fiberglas 25 ASJ/SSL
Available in thicknesses of ½" to 3" for use with ⅝" to 6⅛" diameter copper pipe sizes. Self-sealing lap and end joints with pressure-sealing adhesive. Rated R-4 at 75°F (23.8°C). Rated R-3.3 at 200°F (93.3°C).

FOAMGLAS Uni-Jac
White, vinyl-jacketed cellular glass. Thicknesses from 1½" to 4" for pipes ½" to 36". Rated R-2.7 per inch at 50°F (10°C) to R-2.6 at 75°F (25°C).

Eslon Zip-Sleeve
Closed-cell polyethylene foam. Wall thickness ⅜" for pipes ⅜" to 2" in diameter. Rated R-4 at 75°F (23.8°C).

Sentinel Pipe Insulation
Closed-cell polyethylene preslit foam tubes. Wall thickness ⅜" for copper pipe sizes ½", ¾", 1". Rated R-3.57 at 75°F (23.8°C). In most applications, no need for tape, clamps, or adhesives.

Foamedge Pipe Cover
Vinyl-clad polyurethane foam. For copper pipe diameters ⅜", ½", ¾", 1", 1¼" OD. Rated R-4.34 at 75°F (23.8°C). Preslit and snaps over pipe.

Thermacor C
Urethane pipe covering in thicknesses from ¾" to 4" for pipes ½" to 48" in diameter. Outer jacket of aluminum foil reinforced with fiberglass, PVC, black vinyl or other materials. Rated R-7.7.

Insuljac
Urethane foam pipe covering. Outside and core diameters: 3" × 1", 4" × 1", 4" × 2", 5" × 2", 6" × 3". Rated R-7.14. PVC exterior.

Insultek Corp.
82 Crestwood Rd.
Rockaway, NJ 07866
(201) 625-3828

Rovanco Corp.
I-55 and Frontage Rd.
Joliet, IL 60436
(815) 741-6700

Thermacor Process, Inc.
P. O. Box 4529
500 N.E. 23rd St.
Fort Worth, TX 76106
(817) 624-1181

Energy Systems, Inc.
4570 Alvarado Canyon Rd.
San Diego, CA 92120
(714) 280-6660

Gascoigne Industrial Products, Ltd.
Kee Klamp Division
P. O. Box 207
Buffalo, NY 14225
(716) 685-1250

Solar-Eye Products, Inc.
1300 N.W. McNab Rd.
Building G&H
Ft. Lauderdale, FL 33309
(305) 974-2500

Solar Warehouse, Inc.
140 Shrewsbury Ave.
P. O. Box 639
Red Bank, NJ 07701
(201) 842-2210

Sunworks
P. O. Box 3900
Somerville, NJ 08876
(201) 469-0399

Preinsulated Pipes

"R-7+"
Copper carrier tubes from ½" to 6". Isocyanurate polyurethane foam insulation. Rated R-6.69 to R-13.83 per inch at 75°F (23.8°C). PVC jacket.

Insul-8
Carrier tubes from ¼" to 32" of copper, steel, aluminum, PVC, fiberglass or stainless steel. Urethane foam insulation in thicknesses of 1" (rated R-7.7) to 5". Outer jackets of PVC, steel, coated steel, aluminum, fiberglass, stainless steel or polyethylene.

Thermacor D
Standard carrier tubes from ½" to 16" of steel, copper, PVC, fiberglass-reinforced plastic, epoxy-lined asbestos cement or other materials. Urethane foam insulation in 1" thickness (rated R-7.7) through 6" diameter pipe. Outer jacket of polyester and woven glass.

Mounting Rack Hardware

E.S.I. Bracket System
Roof mounting bracket that fits almost any collector. Kit includes ⅜" × 2" lag bolts, 5/16" × ½" bolts and washers, roof jacks (flashing), pipe supports, collector sleepers. Suitable for installations on shake, tile or composition roofs.

Kee Klamps
Malleable iron slip-on pipe fittings to be used with standard pipe (½" to 2") to fabricate mounting rack.

Universal Mounting Frame
Adjustable aluminum mounting frame kit for flat and pitched surfaces.

Sol-R-Mount
Adjustable aluminum and cadmium-plated steel mounting legs. Universal mounting bracket for any roof pitch.

Selector Rack
Adjustable aluminum mounting rack. Twenty tilt combinations possible between 0° and 49° above flat or pitched mounting surface.

Photo II-8: Collectors in the sun, a classic pose

A Catalog of Flat Plate Collector Systems

Five Systems for All Seasons

All flat plate DHW systems do the same thing: Water circulates between a collector and a storage tank, and it gets progressively hotter because of the solar energy gained by the collector. If Nature gave us only day and night and no other complications, we could get away with at most two kinds of flat plate systems: one that circulated the water by convection (thermosiphon), and one that used an electric pump to do the same thing (forced circulation). But Mother also gives us winter, when water freezes and guarantees the demise of any unprotected flat plate system. This is why we'll be looking at five, not two, different collector systems. In each the basic difference is in their approach to the problem of freeze protection. One system *(thermosiphon)* literally puts the collector and all the attendant plumbing in the house where it's protected by building heat. In another *(drain-back)* the collector and the pipe lines all empty back into the storage tank whenever the pump is off. The *recirculation* system borrows a little of the day's "solar profit" to circulate hot water from storage through the collector, keeping it from icing up. When the sensors in a *drain-down* system read a near-freezing temperature, they activate a special valve that drains a couple of pints of water from the collector and then refills it when sunshine returns. In the *antifreeze fluid* system water is replaced by a special fluid that never freezes. You can't drink that fluid (it's too expensive, and it might make you feel bad), so a *heat exchanger* is used to take heat from the fluid and put it into your domestic water.

As a group these systems give you a flat plate option for any climate and installation scenario. You'll see that some systems are best suited for certain climates and in some cases for certain kinds of installations; some systems can be used anywhere. Some are less expensive, and some are more efficient than others. Thus there will probably be more than one choice that's suitable for your application, and your choice of *the* system will ride on a mixture of factors. But whichever you select (or *if* you select; look over the batch-type DHW systems after "Catalog"), you are choosing from a collection of reliable systems, and reliability, especially in terms of freeze protection, is the first order of business for a flat plate collector.

☼ PROJECT

A Thermosiphon System for "Indoor" Collectors

This is the simplest of all open loop, flat plate DHW systems. No pump, no electrical controls, the sun is the "driver" that creates flow through the collector. Water heated in the collector becomes less dense and rises into the storage tank, while cooler, denser water falls from the tank and enters the collector, all in a continuous cycle that lasts as long as the sun delivers enough energy. With no electrical controls and no antifreeze fluid, a collector in this system has no built-in freeze protection except for the protection of installing it behind glazing within the thermal envelope of the building, where it isn't supposed to freeze. What follows is one architect's approach to solving a couple of problems with a rather different sort of collector.

The bright red wooden roll shades on our greenhouse, a landmark for solar aficionados and tourists, were fast deteriorating. They were very old, and although I had rehabilitated them once, they were finally beyond repair. Big wooden shades were no longer available and to get aluminum ones—in red—was going to cost over $1500. We reluctantly ordered them despite the cost, but the next day we were delivered from our dilemma by a brochure that arrived to announce Big Fin. I was immediately sold.

Here was a product that would shade plants in summer and heat water year-round for less than the cost of the aluminum shades. The collector loses most of its excess heat to the greenhouse, unlike exterior-mounted collectors that lose heat to the outdoors. Other positive features are no pump, no thermostat, no antifreeze fluid, no heat exchanger and no drain-down valve. But what's Big Fin?

A Big Fin module is an 8-foot long by 8-inch wide by 1/16-inch-thick aluminum extrusion painted black on one side. It becomes a collector when it's snapped onto a ¾-inch copper tube (type M). In our installation this assembly was attached directly to the inside of our greenhouse glazing support bars, held away from the bars with nonmetallic spacers to reduce conductive loss. Each fin was drilled with over-

sized holes to receive four bolts about 18 inches in from each end. I bolted all the fins into place to form a nearly level array with no high spots that might collect air bubbles; then a plumber sweat soldered all the copper tubes together. I ordered ten fins (about 53 square feet) to supply hot water for our "water-tight" family of four. Five rows of fins would also fit into the greenhouse without blocking our view to the yard. By leaving about 2 inches between the fins, dappled light reaches the rear of the greenhouse, where shade-loving plants get just what they need to prosper.

How Does It Work?

The 1-inch collector supply (downcomer) and return (riser) lines connect with two storage tanks in a small attic space about 8 feet above the greenhouse. The two tanks, plumbed in series, add up to about 63 gallons of storage. A single 66-gallon tank would have been installed vertically, but there just wasn't enough headroom, so two tanks plumbed in series became the almost-as-good compromise. The advantage of having a single vertical tank is due to the fact that the slow thermosiphon flow rate does

Photo II-9: The rows of Big Fin were fastened directly to the greenhouse structure with spacing that would still admit low-angle sunlight, which is vital to wintering plants.

not disrupt *temperature stratification* in the tank, whereby the hottest water stays at the top of the tank, and the coldest water stays at the bottom with minimal mixing in between. When a hot water tap is opened, the hottest water flows to the gas water heater located in the basement and from there to the point of use. It would also be possible to use a single DHW tank with a built-in electric back-up element, really the optimum arrangement for efficiency and cost. The element is located in the upper half of the tank to ensure that there will always be sufficient reserve of hot water.

Two valves at each end of the Big Fin array allow the water circulation pattern within the collectors to be altered, and a hose bib at the low point of the array allows for emergency or maintenance drain-down (the fins are installed at a gentle slope to help the drainage). If, for example, the greenhouse glass were broken and the collector were exposed to freezing temperatures, quick drain-down capability would be critical.

The five runs of Big Fin are joined at each end with a 1-inch manifold pipe, which is where the flow control valves are located. With the valves open, water enters the array along the bottom fin run only, then returns in parallel through the remaining four runs. Because of the placement of the valves, a more serieslike flow can be created by closing both valves, as the detail in figure II-19 shows. In this mode only the top two runs flow in parallel, resulting in a higher outlet temperature. The water gets 5 to 7°F (3–4°C) hotter this way; I've measured up to 132°F (56°C) at the outlet.

To help maintain temperature stratification in the tank, we used a "drooler" at the tank inlet. This is a simple, home-built device that permits the town water to enter the tank with minimum turbulence. The drooler is a section of drilled copper pipe that is inserted into the tank to disperse incoming cold water in a downward direction so that the stratification will not be upset. If the cold supply to the tank were allowed to rush into the tank helter-skelter, there would be much more mixing of cold and hot water, which defeats stratification. See figure II-20 to see the way in which the drooler is made and installed.

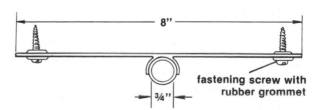

Figure II-18: Big Fin Cross Section

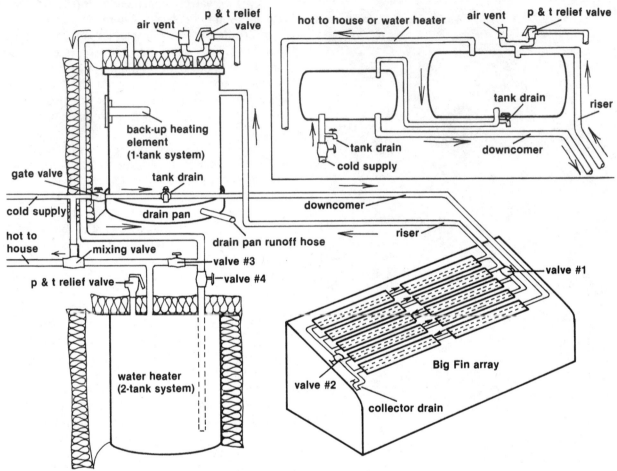

Figure II-19: Plumbing for thermosiphon flow is the simplest of all the flat plate systems. The downcomer runs to the base of the collector from the bottom of the tank, and the riser returns to the top of the tank. The flow patterns through the Big Fin in the Kelbaugh's installation are controlled by valves 1 and 2. When they are both closed, the three bottom courses have a series flow, and the top two courses have a parallel flow. Valves 3 and 4 control the water heater bypass. To bypass the heater, valve 4 is closed and 3 is open. The inset at the top shows the series plumbing for two tanks in a thermosiphon system; this is what was done at the Kelbaugh's because there wasn't enough space for a single vertical tank.

How Well Does It Work?

Our preliminary experience indicates that the system works well. It was summer when I recorded outlet water at 132°F (56°C). This was much hotter than we were used to, and we usually had to mix it down with cold water at the tap or shower head. The dishwasher and automatic washing machine, on the other hand, are quite happy to get water this hot. (Editor's note: A tempering valve can also be used to premix the hot and cold water before it reaches any points of use. Dishwashers, however, need higher temperatures.) The day that we collected 132°F water was a bright sunny day with an

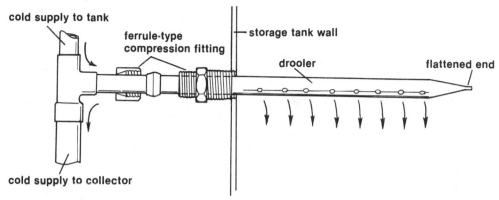

Figure II-20: The plumbing for the drooler uses standard hardware that is slightly modified. The brass ferrule fitting must be drilled out to allow the drooler pipe to pass all the way through it. Drill 1/4" holes along only one line on the drooler end of the pipe, and be sure to face the holes down toward the tank bottom when installing it.

outdoor temperature of 90°F (32°C) and a greenhouse temperature of 100°F (38°C). Performance is so good in the summer that we can literally shut off the water heater and bypass it with some simple plumbing (see figure II-19).

In winter, of course, the water doesn't get quite so hot. There is less solar energy available, and in our case the tilt angle of the collector, which conforms to the pitch of the glazed greenhouse roof (26 degrees), is a bit on the low side. Nevertheless, when checked on a cold December day, the outlet water temperature has reached 100°F.

The temperature rise through the array is usually about 15°F (8°C), indicating a relatively brisk flow rate for a thermosiphon system; I estimate the rate to be about 0.5 to 0.7 gpm, compared with 1 to 2 gpm for pumped systems. Obviously, if the water languished longer in the array, the temperature rise would be greater, but fewer Btu's would be delivered to the tank because the fins would run hotter and lose more heat by reradiation. I insulated the riser and the downcomer pipes with ½-inch foam sleeves to maintain the temperature in the pipe and thereby to help gravity circulate the water.

Insulating the storage tank is critical. When first installed, the tanks were uninsulated for two weeks. When Steve Baer, who developed Big Fin, was in town for a conference, he was quick to throw some blankets and a sleeping bag over them. Even with this improvised insulation, the tank temperature immediately climbed from about 105 to 115°F (41–46°C). It was not until we smothered the tanks with 7 to 9 inches of fiberglass that temperatures in the system began to exceed 120°F (49°C).

Deeper in the Black

During the first (summer) month of operation, the gas meter recorded usage of only 850 cubic feet instead of the typical 1800 cubic feet, a 53 percent reduction and a savings of about one million Btu. In winter I estimate that the system will deliver one-half to two-thirds as many Btu's. The angle of the greenhouse glazing is less favorable to low-angle solar gain, but shading by foliage will also be less. In any case, annual savings should be nine million Btu, averaging about 450 Btu per square foot of collector per day. This Btu saving is worth about $70 per year at the local rate for natural gas (75¢ per therm) and about $210 at the local electrical rate of 8¢ per KWH.

The system cost $1400 installed, less than new red aluminum shades. If I had located the storage tank at the top of the greenhouse, or at least nearer the greenhouse, the plumbing costs would have been considerably less. It's generally better to minimize the collector-tank separation in thermosiphon systems to reduce pipe friction, which reduces flow rate. The tank must of course be above the collector. There is every reason to believe the system would cost less than $1000 in most instances, especially in do-it-yourself installations. (Note that $375 was a labor cost.) In situations where the storage tank could be eliminated entirely (e.g., when the existing water heater is big enough and in the right place), the cost could conceivably drop to one-half of mine. The following is a breakdown of costs for the system:

ten 8-foot Big Fin sections	$ 266
1-inch pipe, fittings, gate valves, etc.	$ 475
shipping from New Mexico	$ 37
storage tanks	$ 172
rubber grommets	$ 10
tank and pipe insulation	$ 40
absorber paint	$ 24
total materials	$1024
labor (plumber)	$ 375
total job	$1399

($26 per square
foot of collector)

With a 40 percent reduction for the federal tax credit, my real cost was $840. With a 10 percent annual escalation rate in the cost of fuel, the system should pay for itself in about 7 years with gas and in 3.5 to 4 years with electricity. This payback period could be somewhat reduced with a lower-cost installation. Typically, thermosiphon systems are the least expensive of the flat plate options.

In its secondary role as a greenhouse shading device, the Big Fin isn't quite as effective as were the old wooden shades, which provided more opaque surface area for summer shading. Since the Big Fin is fixed, we couldn't add the same surface area because the collector would block too much valuable winter sunlight. The greenhouse now runs warmer in summer. Before Big Fin the greenhouse typically never got hotter than 100°F (38°C) during the summer, but on occasion it has reached 110°F (43°C)—an uncomfortable temperature for humans and ladybugs, but seemingly acceptable to my plants and parakeets. (Part of the temperature increase is also due to the fact that I added a second layer of glazing to the greenhouse. This somewhat improved winter performance in minimizing heat loss, keeping the greenhouse warmer in daytime and assuring that the temperature remains above freezing at night, which guarantees freeze protection for the collector.)

In our estimation this system compares favorably with active solar DHW systems. It preheats our domestic water and shades our plants in summer. It uses no heat exchangers, antifreeze, thermostat or pump, and it's smart enough to run itself. All of the system's thermal losses are to the greenhouse rather than back to the outside. It costs less than some active DHW systems, and it performs almost as well. In short it has a great deal to recommend itself to anyone who has or who will have an attached greenhouse. But alas, in our case there are drawbacks: The greenhouse is sometimes warmer in summer (curable with increased ventilation), and we have lost our patch of red.

Doug Kelbaugh

Other Thermosiphon Systems

A thermosiphon collector doesn't have to be mounted inside a greenhouse, but that placement does eliminate the threat of damage by freezing. Other possibilities include devoting all or part of a skylight glazing to a collector, or perhaps a clerestory opening would provide a suitable mounting location. And Big Fin isn't the only collector suitable for indoor installation, although it does present a nice appearance. You could use any standard absorber plate, and it would be a good idea to put some insulation behind the plate to reduce reradiative heat loss from that side. A standard collector box could be built without glazing and then mounted up close to the building glazing, but not sealed to it.

Collectors for thermosiphon systems can also be installed outside, if you're willing to provide

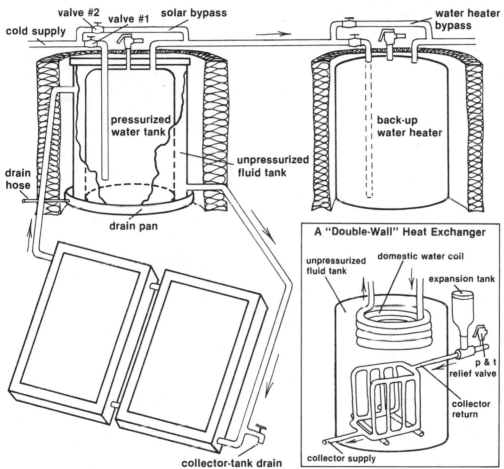

Figure II-21: This thermosiphon system uses an antifreeze fluid as the circulating medium, which takes care of freeze protection when the thermosiphon collectors are mounted outside. In the two-tank heat exchanger a pressurized water tank is immersed in the unpressurized fluid tank, and heat is conducted through the wall of the pressure tank. The double-wall exchanger has a "cage" made of 1-inch copper pipe immersed in an intermediate heat exchange fluid that is contained by an unpressurized tank. This fluid delivers heat to the domestic water coil. The latter option is useful with large collector arrays of over 80 square feet. With either heat exchanger the plumbing can be rigged to incorporate both a solar and a water heater bypass loop.
Source: Adapted from Roger Bryenton et al., The Solar Water Heater Book, *1980, with permission of Renewable Energy in Canada, Toronto.*

"manual" freeze protection, i.e., to drain the collector and all exposed plumbing (including insulated lines) whenever cold weather comes along. You lose the automatic freeze protection of the building's thermal envelope, but the system will still be a viable one. To eliminate the bother and possible disaster of an unexpectedly frozen night, you could drain the system seasonally if during winter you could use another alternative DHW heat source, such as wood or coal. But it won't make much sense to have a collector system that can only operate for 6 to 9 months out of 12. (Editor's note: There is an important limitation on the use of open loop systems like the thermosiphon and others that circulate tap water through them. In areas where the water has a high mineral content, the high temperatures of flat plate collectors can cause the minerals to precipitate onto the water tubes in the absorber, causing something akin to arteriosclerosis in people. Flow is restricted and heat transfer efficiency is reduced because the mineral acts as an insulation. In advanced cases, total blockage can occur, solar thrombosis. So if minerals are generally a problem in your area, it's probably best to use a closed loop system, such as the drain-back or the drain-back II or the antifreeze systems described in the following projects.)

Thermosiphon Plus Antifreeze

If the simplicity of the thermosiphon attracts you, but you have no freezeproof location, you might consider using a special antifreeze liquid like propylene glycol in a closed thermosiphon loop in combination with a heat exchanger in the storage tank. The heat exchanger would transfer solar heat to domestic water.

There are several ways to run a closed loop system. For an average-sized family, a tank within a tank is effective (see figure II-21). The freezeproof

Photo II-10: If the thermosiphon collector were installed high on the wall or roof of a house, there might be a problem with finding an even higher place for the storage tank. The problem was solved in this installation by building a tank rack out of 2×4's, 2×6's and 2×8's. Bolted to the rafters and the wall, a rack like this will easily support 500 to 700 pounds of filled tank with no need for any other structural changes.

liquid circulates in a large plastic tank around a smaller, pressurized steel tank that holds the domestic water. Keeping the distance between the two tank walls to about 1 inch minimizes the amount of expensive antifreeze liquid required.

For a large water heating load extra collectors and a larger storage tank would be needed. Using a tank within a tank in that case would require too much propylene glycol to be economical. An alternative is to surround a pressurized domestic water tank with copper coils rather than a larger tank. Or you can create a double-wall exchanger by building a copper pipe cage in the bottom two-thirds of a tank and forming a coil at the top of the tank with 100 feet of soft copper pipe. The copper cage, filled with freezeproof liquid, runs to the collector while the coil carries the domestic water. The tank itself holds water that heats up as the liquid from the collector circulates through the copper cage. Since the water temperature will stratify, the top third, where the domestic coil is located, is the hottest. The trade-off with using antifreeze fluid and heat exchangers is some loss of efficiency, on the order of 10 to 20 percent, compared to open loop systems. (For more information on this idea, see *The Solar Water Heater Book* by Roger Bryenton, Ken Cooper and Chris Mattock, published by Renewable Energy in Canada, 41 Riverdale Avenue, Toronto, Canada M4K 1C2.)

However you design the system, there are a few guidelines that must be adhered to. The bottom of the tank must be at least 12 inches higher than the collector outlet. All pipe runs to and from the collector must run smoothly without any dips that would compete with the gravity of falling liquid or with the force of the rising liquid. Thermosiphon flow is very sensitive to twists and turns in the pipe run. It's also sensitive to the friction of the fluid against the walls of the pipe lines. To minimize friction, the minimum pipe diameter for all runs should be 1 inch or larger. Using a collector made with ¾-inch pipes instead of ½-inch pipes also helps to promote a healthy flow rate. Also, don't spend more money using fancy globe valves in a thermosiphon system because they restrict flow more than standard gate valves. If you build a thermosiphon collector for outdoor use, it should be double glazed to reduce the heat loss from the hotter-running absorber. Basically it's recommended that you have some sort of built-in freeze protection in a thermosiphon system (and in any flat plate system for that matter), either with building envelope or antifreeze fluid protection. That way you won't have to outguess the weather, because the system will take care of itself.

John Blackford

Hardware Focus

Big Fin

Easco Aluminum
Solar Division
P. O. Box 73
North Brunswick, NJ 08902
(201) 249-6867 or toll-free number:
(800) 631-5856

Absorber Plate
Can be used for "homemade" Big Fin absorbers; aluminum; 8' long; 6" wide; compatible with ½" copper tubing.

Zomeworks
P. O. Box 712
Albuquerque, NM 87103
(505) 242-5354

Big Fin
Aluminum fins 8" × 96" long; flat black on one side; satin anodized on the other; with ¾" copper pipe press-fitted onto back.

☀PROJECT
The Drain-Back System

Empty collectors can't freeze. This unpressurized (closed loop) system is designed to automatically drain itself whenever the pump is off, without using electrically operated valves or other special components. When the differential thermostat senses that solar heat is available, the pump is switched on to draw "collector" water from an unpressurized tank and run it through the collector. When the pump stops, gravity takes over; water in the collector and the connecting pipes simply drains back into the tank. A heat exchanger carrying pressurized domestic water is immersed in the upper part of the tank. When hot water is used, the water is preheated in the exchanger before it flows into the back-up water heater and out to points of use.

Some Penn State University professors have been known to say, "If you can make a solar heater work here at State College, it will work anywhere!" That may be an exaggeration, but with 6000 degree-days and cloudy weather 40 percent of the year, the climate does pose a formidable challenge to any solar heater.

Harry and Elizabeth Brown decided to take the challenge, determined to reduce their climbing electric bills. The south side of their single-story home was only partially shaded from the sun, and it seemed feasible to mount a collector on the roof and put a storage tank in the basement. The big problem was choosing, from among several designs, the best system for their needs.

As it turned out, Harry and Elizabeth were in the right place at the right time. The first of a series of hands-on, solar DHW workshops was to be held nearby, and the weekend workshop was open to anyone interested in building a drain-back solar water heater, a reliable design sized to provide approximately 50 percent of the hot water needs of a family of four. Although they had no experience with solar and little with plumbing and carpentry, Harry and Elizabeth decided that if they were going to take the workshop and build a solar water heater, it might as well be their own! So they dove in, bought all the various and sundry materials, built the collector, and with the help of the workshop instructor and participants did the installation.

As with any solar water heater that utilizes the drain-back approach to freeze protection, the collector had to be located above the storage tank. Just as important, the pipes running to and from the collector had to be installed at a continuous downhill slope, free from any dips or bends that could trap water. Another consideration (important for this and any unpressurized, closed loop system) is the height differential between the tank and the collector. Because the pump must literally push water from the tank to the collector, the installation should be tailored to minimize that differential to avoid the need for a larger pump that would consume more electricity and reduce the overall benefit of the solar system. With a pressurized, open loop system there is little of this concern because the collector is already filled, requiring a smaller (1/100 hp) pump to push the water around. Mounting the collector on the ground is often the best way to minimize the pumping head. The top of the collector should usually sit no more than 9 feet above the water level of a basement tank (which is raised up on a platform). This 10-foot "head" is easily handled by a 1/50-hp pump and, if there is a basement window available for passing the pipes in and out, the installation is simple. The entire system takes about two man-days to complete.

In this case, however, a ground-mounted collector was out of the question because of excessive shading, but a small shed that was attached to the house's southeast side offered a suitable surface about 6 feet above the ground. With the top of the collector 6 feet above that and the tank raised 2 feet off the basement floor, the pumping head amounted to 14 feet. Since the standard 1/50-hp pump could lift water only 11 feet, a 1/20-hp unit, which can pump against 16 feet of head, was needed. The added pumping power (1/50 hp = 50 to 75 watts; 1/20 hp = 100 to 150 watts) was a small sacrifice, though, for the system arrangement was otherwise perfect. When choosing a pump, read the manufacturer's specifications for the maximum head capability. It's always better to buy a little extra horsepower to guarantee an adequate flow rate, so if your system is "borderline" vis-à-vis pump size, go with the next bigger size.

Operation

Whenever the system's differential thermostat senses a collector temperature that is a few degrees higher than the tank temperature, it activates the pump. The pump drives water from the bottom of the storage tank up to and through the collector. The pump may cycle on and off a few times in the morning and late afternoon, but if the sun is strong and the collector is unshaded, it should run continuously from about 10:00 A.M. to 3:00 P.M. With each pass through the collector, the water temperature increases a few degrees.

At the end of the day or when heavy clouds roll in, the thermostat senses no significant temperature difference and stops the pump, allowing the system to drain by gravity. The ¾-inch ID pipes have a

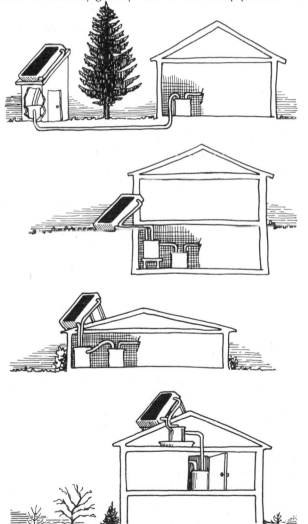

Figure II-22: In a drain-back system the collector should be located so as to minimize the vertical distance between it and the tank. In the top panel the house itself is shaded year-round by an evergreen, so the collector is placed away from the house, perhaps on an outbuilding. A more typical installation, as shown in the second panel, has the ground-mounted collector connected to a raised drain-back tank located in the basement. For homes without basements, a roof-mounted collector may not be too far from the tank, which could again be raised. The bottom panel shows how a two- or three-story house can still be compatible with a drain-back system. Note that the drain-back tank is equipped with a drain pan should the day ever come that the tank springs a leak.

big enough cross section for collector water to return to the drum, without creating an air lock that might inhibit the draining action. For this reason ¾-inch pipe is the *minimum* dimension for the collector supply and return lines in drain-back systems.

A 60-foot-long copper coil submerged in the upper (hottest) level of the storage tank is connected with the pressurized, potable domestic water line. Heat passes rapidly through the large surface area of the coil and effectively preheats the potable water before it is fed to the back-up water heater. Depending on the flow rate, the water at the outlet of the coil is quite nearly as hot as the collector water.

Performance

By late afternoon on a sunny day, the water temperature in the storage tank is between 110 and 160°F (43 and 71°C), and on the average it gets 50°F (28°C) warmer than in the morning. This 50° rise represents an average of about 20,000 Btu of collected energy per day for a conserving family of four using 12 gallons of hot water per person per day. If they use hot water at 120°F (49°C), their daily water heating energy consumption will be about 26,000 Btu. The solar preheater thus provides about 75 percent of the heat required on a sunny day, or about 400 to 450 Btu per day per square foot of collector.

The heat contributed by the preheater is worth about 26¢ per day after you deduct the cost of operating the pump (about 5¢ per day). This is where the pumping head, the distance the water must rise, is significant. A 1/50-hp pump—which can pump against heads of up to 9 feet—would have an operating cost of about 2.5¢ per day. With 60 percent of the days sunny in Pennsylvania, the Browns' preheater saves about $57 per year over the operating cost of an electric hot water heater at 5¢ per KWH for electricity.

Assuming a $650 initial outlay for materials and a 12 percent annual increase in the cost of electricity, this system will pay for itself in 5 to 6 years if the owners take the 40 percent federal tax credit (which lowers the real cost to $390), or 8 years if

they don't. If the system lasts a minimum of 15 years, the savings over the cost of the materials would be $1810 with the 1/20-hp pump or $1970 with the 1/50-hp pump, assuming the owners take the tax credit. In the long run the difference in savings between the larger and smaller pumps doesn't really amount to much.

Evaluation

With good reason the Browns have been very satisfied with the system. It cost them far less than a commercially installed system; it has reduced their hot water bills substantially; the collector and the plumbing blend well with the home's exterior, and the whole system requires little maintenance. If you have a busy schedule (like most Americans), you want a solar system to be as automatic as possible. This is not only a matter of convenience, but also of necessity, for once the newness of the solar heater has worn off, it is likely to be ignored. (When was the last time you drained sediment from your water heater tank, or cleaned the heat exchanger in your oil furnace?)

The maintenance that a system requires is dictated by: 1. Its design. (Does it have a movable reflecting/insulating collector cover, a pump, electrically operated valves, antifreeze and so forth?) 2. The way it is installed. (Is the collector securely fastened; has pipe insulation been protected from the weather; has the system been filled with the proper heat-transfer fluid?) 3. Lady Luck. (Has that 75-foot oak tree sliced the collector; has your car missed the garage but found a ground-mounted collector; has lightning fried the differential thermostat?)

The drain-back system may come the closest to providing a worry-free low-maintenance system. There are no daily chores such as manual valve operations, no electrically operated valves to jamb, and no antifreeze (which can deteriorate over a long period of time). The only moving part is the pump. If it is installed carefully and maintained properly (if it needs maintenance at all) the only other maintenance required for a drain-back system is an annual water-level check and inspection of the exterior components. If you've chosen the drain-back system because of the high mineral content of your water, you might consider filling the drain-back tank with rainwater or distilled water to keep minerals out of the collector loop altogether. Any ''topping up'' that might be needed later could also be done with mineral-free water.

Variations, Modifications and Other Options

As with most solar DHW systems, many installation variations are possible. Remember that the tank must always be lower than the collector and so should any part of the system that is exposed to weather, meaning freezing temperatures. Remember too that the altitude difference should be minimized

and that the interconnecting pipes must slope continuously downhill, no ups and downs. You can mount the collector on the ground or on a wall of the house and locate the solar tank in the basement, or for a roof mount the tank can be placed in the attic or in a second-floor closet. (Whenever a storage tank is placed over a living space, it should always be put in a drain pan to guard against the day when it might, just might, spring a leak.) The collector and tank can even be located in a separate building or shed and connected to the house by underground pipes.

The storage tank used in this system, a 55-gallon drum coated and sealed with epoxy enamel, is sized for use with about 30 to 48 square feet of collector area. If more collector area is planned, the unpressurized tank must also be increased in volume. Some of the options for larger tanks include galvanized livestock waterers (100 gallons and larger) and insulated fiberglass tanks made for solar applications

Photo II-11: This installation shows an excellent application for the drain-back system. Because the house has no south-facing roof pitch, a wall-mount was the next best alternative. This location helps to minimize the vertical distance between the collector and the storage tank.

(120 gallons and larger). Any tank you choose should of course have a cover, or be coverable, to minimize evaporation of the collector water. With more absorber area there is more pipe friction, and a slightly larger pump may be needed. Check the manufacturer's specifications for pumping capacity. With the pumping head of your system (vertical distance between collector and tank), you need to have a flow rate of about 0.5 to 0.75 gpm per square foot of collector.

A larger collector array with a bigger storage tank may also require a bigger heat exchanger to maintain efficiency. The 60-foot, ¾-inch ID copper coil used in the Browns' system has a capacity of about 1½ gallons. If the family's water-consuming habits are such that small quantities are used intermittently, then this heat exchanger works well. Another

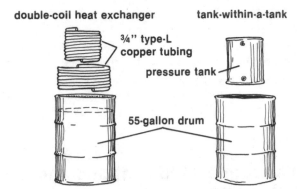

Figure II-23: Larger drain-back systems will need bigger heat exchangers. Two possibilities are shown here: The standard coil-type exchanger can simply be increased; the inner and outer coil can be plumbed in parallel or in series. The inner tank of a tank-within-a-tank option is a 5- to 10-gallon pressure tank; it should be equipped with a pressure-and-temperature relief valve and supported on blocks near the top of the drain-back tank.

Photo II-12: This is the top of the drain-back storage tank, with the water level lowered a bit to expose the coiled copper heat exchanger. Note that the tank penetrations for the collector supply and return and the heat exchanger inlet and outlet are all above the water line. The tank is shown with a polyethylene liner, which has since been replaced with epoxy paint as a means for sealing the 55-gallon drum.

desirable condition is to have water flow slowly through the coil. At a flow rate of 1 gpm or less, the solar-heated water will transfer close to 100 percent of its heat to the potable water in this coil. At 2 gpm the efficiency of heat transfer drops to about 60 percent, which is why it makes sense to install flow-restricting devices. For larger systems you can connect two 60-foot copper coils in series with the second coil wrapped smaller to fit inside the first.

You can also substitute a 5- to 20-gallon galvanized pressure tank for the copper coil. This would heat larger batches of water, but the recovery rate would be slower. The tank must be rated to withstand line pressure of up to 75 psi, and it must be rustproof. If its outlets are not copper, the fittings joining them to copper plumbing must be dielectric to prevent electrolytic corrosion.

Materials Checklist

55-gallon drum or equivalent
one pint epoxy paint
60 to 120 feet of ¾- or ½-inch ID type-L copper
    tubing

copper and gasket fittings for penetrating tank wall
pump and differential thermostat
plumbing for collector supply and return lines,
    ¾-inch ID copper or plastic pipe

Installation

The heat exchanger is prepared by forming the type-L copper tubing into a coil that fits inside the tank. Be careful to avoid kinking the line when wrapping it around the form you use. Both ends of the exchanger should come up so they can penetrate the tank wall near the top. The tank is first waterproofed with a liberal coat (or two) of epoxy enamel. After it's painted with the epoxy, four ⅞-inch holes are drilled or punched 1 inch below the top edge of the tank wall for the heat exchanger and the collector supply and return lines. All holes should be above the water level of the collector water, which should be just above the top coil of the heat exchanger. To pass the pressurized coil through the tank, use ¾-inch copper male and female adapter fittings with a gasket, as described in figure II-24. The inlet and outlet of the heat exchanger are then soldered to these assemblies, after the coil has been suspended in the tank with copper wire. These lines are then run into the cold supply to the water heater. With two gate valves you can make a simple bypass, as is shown in figure II-26.

The other two through-wall assemblies are for the collector supply and return lines, and they are handled in the same way as the heat exchanger hookup. On the outside the lines are soldered to the protruding fitting. It's important that the tank be secured to whatever it sits on so that future movement doesn't pose a risk to any of the solder joints. On the inside of the tank, the return line from the collector is plumbed with a 45-degree street elbow. This is where the hot water returns from the collector and pours into the tank, and it absolutely must be above the tank water level in order for the collector to drain every time the pump is off. The inside of the supply line to the collector is plumbed with a "dip tube" that extends to within 6 inches of the bottom of the tank. This tube supplies the coldest water in the tank to the collector for maximum heat exchange efficiency. Before filling the tank, wrap a piece of fiberglass (not aluminum) insect screen around the end of the dip tube to filter out large particles from the collector loop. When making the collector connections, it's important to ensure that both the supply and return pipe runs have a consistent slope with a drop to the tank of at least 1 inch for every 3 feet of run.

The pump is installed on the supply line to the collector, as shown in the schematic. You also have the option of using a fractional-horsepower submersible pump, which may be a little better because the waste heat from the pump motor is put directly into the collector water. With a submersible pump there is no dip tube, but it's still a good idea to attach a screen filter to the pump intake.

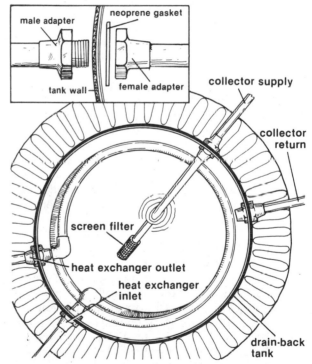

Figure II-24: To keep your pipes from getting tangled, the collector supply and return lines can be plumbed into one side of the drain-back tank, and the heat exchanger inlet and outlet can be run through the other side. The collector supply line runs to within 5 inches of the bottom of the tank, and where it terminates it's wrapped with nylon insect screen, which acts as a filter.

Drain-Back II: A Dual Pump System

One of the "sort of" drawbacks of the basic drain-back system is that at any given time there is just 1½ to 3 gallons of potable domestic water in the exchanger coil that is at the same high temperature as the collector water. Until there is a demand for hot water, the solar heat essentially remains in the collector tank. But there is a way to maintain the drain-back freeze protection feature *and* simultaneously transfer solar heat into a pressurized domestic hot water tank by using the dual pump system described here. The schematic differs little from the original drain-back system except for the fact that the storage tank for the collector water is somewhat smaller, at 7½ gallons, and there is a pump that circulates water through the domestic coil, between the collector tank and the domestic hot water tank.

The two pumps are controlled by a single, standard differential thermostat, which is possible because the pumps are wired in parallel. Thus, whenever the solar (collector) pump is switched on by the thermostat, the coil (domestic water) pump is switched on too. As solar heat is being delivered to the collector tank, it is transferred to the water flowing through the domestic water coil and hence to the domestic hot water tank.

Neat? Smart? Yes, but to get that improved performance you've got to spend a little more, of course (no free lunches, right?). The collector tank we use is an all-copper tank with a built-in heat exchanger called the Everhot, made by Everhot All-Copper, Inc., 5 Appleton Street, Boston, Massachusetts 02117. Telephone: (617) 542-1226. We use their #8 tank (7½ gallons) with a #4 coil (15½ square feet of surface area); the tank does have some insulation, but it's so small that we can add a 12-inch-thick insulation blanket all around it, using less than a roll of 6-inch fiberglass. You could use a small tank of your own finding or making, perhaps some sort of 8- to 10-gallon plastic container, which could be prepared in the same way as the larger tank described in the previous project. With a plastic container you wouldn't need a polyethylene liner, and you probably wouldn't have room enough for a submersible pump.

To create 15 to 20 square feet of heat exchanger surface area, you would have to use 65 to 90 lineal feet of ¾-inch (⅞-inch OD) type-L tubing, or 90 to 120 lineal feet of ½-inch (⅝-inch OD) tubing. The ½-inch tubing has a somewhat better surface-area-to-volume ratio, which would make for better heat transfer between the collector water and the domestic water. That much ½-inch tubing, 90 to 120 feet, only displaces about 1½ to 2 gallons in the small collector tank.

The installation of this system differs little from that of the basic drain-back system. The collector must still be above the collector tank, and the pipes must be run on a continuous slope so they can drain whenever the collector pump is off. Because the tank we use is sealed, we also install a standard sight glass on the collector supply line for easy checking of the water level in the tank (which must be just below the level of the collector return line when the system is drained). For most installations we use a 1/20 hp pump for the collector pump, though with longer runs to a higher collector (over 100 feet of pipe both ways), we use a 1/12 hp pump. The pump for the domestic coil can be a 1/100- to 1/50- hp unit. It can pump to an added solar storage tank with or without a back-up element, or to the existing water heater, if it's big enough. This dual pump system has drain-back plus improved heat transfer; it's a little better than drain-back I, and the extra cost is certainly cost-effective.

Gary Gerber

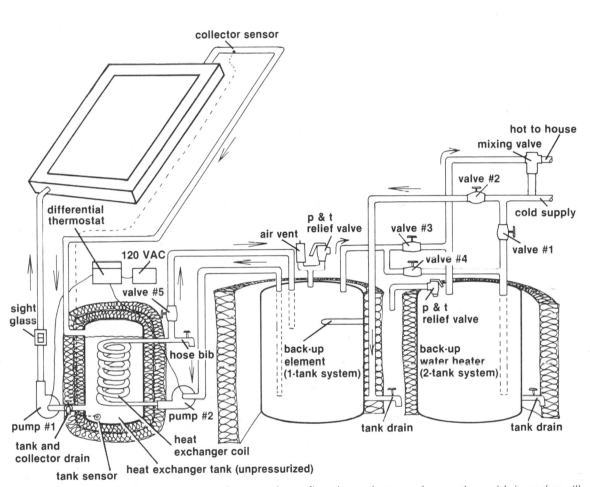

Figure II-25: Drain-back II may seem rather complex at first glance, but spend some time with it, and you'll know how it flows. The sight glass on the collector supply line is installed near the top of the heat exchanger tank, where it is used to check the tank water level. Valves 1 and 2 control an optional solar bypass when a back-up water heater is used. In normal solar operation valve 1 is closed and 2 is open. Valves 3 and 4 control the water heater bypass. When no backup is needed, valve 4 is closed, and 3 is open. In normal operation valve 5 is open. When obtaining a differential thermostat to control two pumps, make sure it's rated to carry the higher amperage.

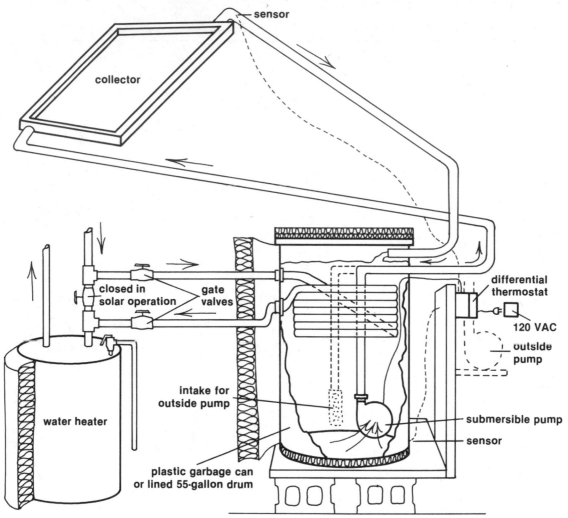

Figure II-26: Minimizing the height differential between the storage tank and the collector is an important design element in this unpressurized system. That is why lowering the collector with a wall or ground mount is desirable. The storage tank can also be raised. A water-filled 55-gallon drum weighs about 500 pounds, and building a wood or concrete block platform would not be difficult.

The tank and pump installation and the collector pipe runs are essentially what make this system different from other installations. Beyond that, you have a standard collector installation with standard plumbing and electrical practices. With a proper installation the drain-back system will provide total freeze protection, and though there is an efficiency trade-off with the heat exchanger, the simplicity of the system and the peace of mind it gives are attractive advantages.

After giving dozens of workshops for building collectors and installing the drain-back system, Chris Fried has developed a detailed installation description in his booklet, Build a Drain-Back Solar Water Heater, *which is available for $4.25, postage paid, directly from the author at R.D. 3, Box 229-G, Catawissa, Pennsylvania 17820.*

Chris Fried

PROJECT

Recirculation and Drain-Down Systems
Electrically and Mechanically Controlled Freeze Protection

You've seen freeze protection by the building thermal envelope and by the friendly force of gravity. Here are two flat plate systems that are protected with electrical and mechanical controls. The recirculation *system works by taking a little hot water from the solar storage and pumping it through the collector array. It's good for climates that only occasionally experience just-below-freezing temperatures. The* drain-down *system, not to be confused with* drain-back, *uses a special valve that closes off the pressurized lines to the collectors and allows them to drain away the 1 or 2 gallons they and the pipelines carry. Both are open loop (pressurized) systems that circulate potable water between the collectors and the storage tank. Do these controls invite proofs of Murphy's Law? Ultimately, everything breaks down, even Murphy, but when installed and used properly, these systems have proven themselves to be reliable.*

Recirculation

This is the simplest of these two systems because the differential thermostat and the solar pump double as both the primary heat transfer and the freeze protection controls. Certain models in the vast array of differential thermostats are made to accommodate a third sensor along with the usual collector (high temperature) and tank (low temperature) sensors. The third sensor is used to detect the onset of near-freezing conditions and tell the thermostat to turn the pump on. These sensors usually cause the pump to operate in the 36 to 40°F (2–4°C) range, far enough above the deadly 32°F to ensure safe passage for the collector through the cold night. The type of sensor that is used depends on the type of differential thermostat you buy. One type is a *thermistor,* which sends a variable resistance signal to the thermostat, which varies with the temperature of the pipe, and at a certain resistance (which increases

as the temperature drops) the thermostat tells the pump to go on. The other is a switch-type sensor (bimetallic) that opens or closes at a present low temperature and causes the pump to go on.

The freeze sensor is placed on the collector inlet right before the pipe passes into the collector. This short pipe section, just ½ inch or so long, should *not* be insulated where the sensor touches it. If it were insulated, the sensor might not know what the true atmospheric conditions were, especially if there were a sudden drop in air temperature. Leaving it exposed guarantees that there won't be any lag time between the onset of cold weather and the switch-on of the pump. Simply tape the freeze sensor to the inlet, and that'll be that. These sensors are so small that you will be able to insulate around it and cover most of that ½-inch pipe section to minimize pipe heat loss during normal solar operation.

Differential thermostats are generally solid-state devices, and as such they have a minuscule failure rate. But Murphy can sneak in at the slightest

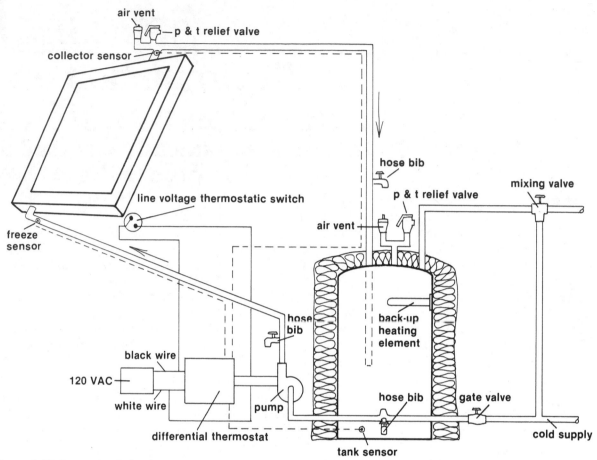

Figure II-27: In a recirculation freeze protection system, it's the differential thermostat that provides the logic for turning on the pump when the freeze sensor reads air temperature that's too cold, usually in the 36 to 40°F (2–4°C) range. The line-voltage thermostatic switch backs up the sensor system in case it fails. It is wired in parallel with the differential thermostat and installed next to the freeze sensor on the collector supply line. Thus when the recirculation water heats up the collector, the switch opens and shuts off the pump.

opportunity (neither mankind nor solid state is perfect!), so it's a good idea to have a back-up freeze protection control just in case the thermostat goes awry. This back-up is not a difficult thing to provide.

The second freeze sensor is connected directly to the solar pump, completely bypassing, and thus independent of, the differential thermostat. This sensor is connected to a 120 VAC line voltage circuit between the 120 VAC source and the differential thermostat, as shown in the schematic (figure II-27). Thus if the thermostat failed, and the 120 VAC

source were still alive, the second freeze sensor would be available to turn on the pump with the approach of freezing temperatures.

Several types of sensors are available for this function. To begin with, the sensor must be rated for 120 VAC service; it might be called a "line-voltage" device, which means it's made for a 120 VAC circuit. One of the fancier versions of these controls is known as a "remote bulb outdoor thermostat," which has an adjustment control so you can set the temperature at which the pump will go on. That setting could be one or two degrees below the temperature at which the differential thermostat initiates its own freeze protection mode. The remote bulb is the actual sensing device, and it is located on the collector inlet next to the freeze sensor. The body of this thermostat should be mounted in a waterproof electrical box, which should also be placed out of the weather, perhaps behind the collector. Then 14-gauge wire is run down to the pump, possibly through the PVC pipe that carries the insulated collector supply and return lines and the other sensor wires.

Another type of thermostatic switch is called a "limit switch," which also has a variable-temperature on-off function. You must buy one with an adjustment range that covers the 35 to 45°F (2–7°C) range. The simplest and cheapest of all these thermostatic switches is the "snap-disc." It doesn't have a variable control, so you must buy a snap-disc with a precise temperature rating (they are available in the desired freeze protection temperature range). You'll be able to find any of these items at an outlet that supplies the heating, ventilation and air conditioning trades (HVAC). All of these switches operate essentially the same way. When the temperature drops to the set point of the switch, it closes, or goes "on," which activates the pump. If the differential thermostat dies, you've got a most reliable backup with one of these switches.

The above sensor handling and wiring instructions are really the only special installation procedures associated with recirculation systems. Beyond that it's basically a regular active flat plate system with a pump and a thermostat and a few manual valves placed here and there. The collectors can be in any position relative to the location of the storage tank; sizing parameters are the same, and because there's no concern for the pipe friction of thermosiphon systems or the draining ability of drain-back systems, you can use ½-inch pipe for the collector supply and return lines instead of 1- or ¾-inch pipe. There is also a performance advantage for open loop systems over heat exchanger systems, which have some energy loss in the transfer of heat from collector water to potable water.

But for all its simplicity and good performance a recirculation system is not a flat plate DHW option for all climate zones. Put the system up in a Wisconsin winter, and it'll be spending most of its time in the freeze protection mode. In a very cold climate this type of freeze protection would use up too much of the stored solar gain in the battle against ice. In other words, too many solar Btu's would be dumped into the atmosphere at night relative to the number that were collected during the day. So this

Photo II-13: In this installation the collector supply and return lines run into the drain-down valve (black box) just above the man's arm. When the valve closes, the drain water flows through a small plastic tube to the soil pipe at left. The finishing touch for this installation is the tank insulation, no less than 6 inches of fiberglass secured with plastic straps.

system is really best suited for climates where freezing in winter is more the exception than the rule, where the coldest nighttime temperatures are usually in the high 20s to the low 30s (°F). As a rule of thumb climates that are suitable for recirculation freeze protection should have no more than about 3000 heating degree-days during the heating season (see table I-8, column C). If you know more about your lowest winter temperatures, the system may be suitable for your locale even if it tallies somewhat more than 3000 heating degree-days. But above all else this system is meant for milder climates. Because there is a failure mode in this system, loss of electrical power, it simply and absolutely must not be used in climates where freezing is common, unless there is also some sort of alarm system in place to warn you of a loss of power during cold weather.

Drain-Down Systems

Drain-down freeze protection is a recent and much needed development for open loop systems. It provides an essentially fail-safe solution to the freeze vulnerability of water-filled metal collectors and exposed pipe runs, because the special valves used in these systems (there are currently three on the market) are both electrically and mechanically controlled. The freeze protection mode of these valves is initiated by mechanical control, and therein lies the fail-safe feature. It's much harder for Murphy to fiddle around with mechanical systems than electrical systems. Let us explain further.

Drain-down valves have two operating modes: The electrically controlled "fill" mode (valve open) fills the collectors with line-pressure domestic water. This mode is for normal solar operation (pump on) and for when the pump is idle (no solar input, temperature above the freeze protection set point of the differential thermostat). The mechanically controlled "drain" mode (valve closed) cuts off the water pressure to the collectors and drains them and all exposed plumbing (or all the plumbing that is

above the valve). This is a "power off" mode in which the differential thermostat switches off the drain-down valve, and a strong spring returns the valve to the normally closed position. ("Normal" describes the condition of the valve when it's "off," or when there is no electrical power applied to it; other kinds of electric valves are designed to be normally open, i.e., they close when power is applied.)

There are three conditions that cause the drain-down valve to close and drain the system. 1. If the collector sensor senses an overheat condition, the differential thermostat will cut power to the valve. 2. The same thing occurs if the freeze sensor senses a temperature at the freeze protection set point. 3. If there is a total loss of electrical power in the differential thermostat, the collectors will drain because as far as the valve is concerned, the differential thermostat has cut power to it. We also specify a back-up freeze protection circuit that is similar to the one used in the recirculation system. The difference in the drain-down system is that the 120 VAC thermostatic switch is wired to both the drain-down valve and the pump (see figure II-28).

We know of one differential thermostat that satisfies all our requirements for drain-down system control. The Hawthorne model number 1506 is designed for drain-down systems and has all the right logic and override functions in a three-sensor control unit. For standard two- or three-sensor units a 120 VAC backup should be provided for total freeze protection reliability. As the schematic shows the line voltage sensor is a single-pole, double-throw (SPDT) remote bulb thermostat. The remote bulb is taped to the collector inlet just outboard of the collector, and the switch itself is placed on the back of the collector in a weather-tight enclosure. The switch will have an adjustable temperature control, which can be set just below the freeze protection set point of the differential thermostat. For outdoor wiring be sure to use a 14-gauge cable rated for that kind of service. The heat tape that wraps around the air vent is wired directly from a 120 VAC source.

If you haven't guessed it already, there is a possible failure mode in this freeze protection plan.

If the valve sticks in the "fill" (open) position, the collectors can't drain. Hello, Mr. Murphy. But with proper installation and maintenance you can all but eliminate the possibility of this kind of valve failure.

Drain-down systems are not inherently fraught with failure points; quite the contrary. The valves have been tested and tested and tested; they have cycle lives (fill-drain cycles) that show many years of useful life before major maintenance or replacement is necessary. The rest of the system has just a few installation requirements that will also help to ensure proper operation and long life. The drain-down system can be used in much colder climates than would be feasible for the recirculation system. Because the drain-down mode completely eliminates water from the exposed parts of the system, there really isn't any climate that would overcome the capability of this system to protect itself. Nor does the drain-down function use any of the stored solar energy for freeze protection. With all things considered this system can be used in any climate.

Since this is a self-draining type system, the collectors and the supply and return must be placed above the drain-down valve. In the drain-back system the collectors had to be above the storage tank, but this is not the case with this system. Everything on the collector side of the drain-down valve will be emptied in the drain mode. Therefore the valve and everything on the other side of the valve remain filled with line-pressure water and must be placed in a location that will never experience below-freezing temperatures. As the schematic shows, the collector array must be supplied with an air vent/vacuum breaker at the highest point in the array. This device admits air to the collectors when they are draining and purges air when they are filling. Air vent/vacuum breakers should be fully insulated, but not so much as to block the passage of air. Our experience shows that they should also be freeze-protected with electric heating cable so they don't ice up and become jammed in either the venting or vacuum-breaking position. The collectors themselves must be installed with a slight cant in the risers (the collector tubing between the headers) if they are horizontal. If the risers are installed vertically, then the headers will be horizontal, and the collectors should be installed so that the headers are slightly canted. This canting ensures that there won't be any chance of standing water in the array after the collectors are presumably drained. The collector supply and return lines must also be installed on a minimum slope of at least ¼ to ½ inch of drop per lineal foot of run, and as with the drain-back system there should be no built-in ups and downs in the lines that might trap water.

The manufacturers' literature for the various drain-down valves now available all state that they are rated for use at line pressures up to 125 psi. Well, they don't blow up at 125 psi, but our research has shown that their performance is impaired at high water pressure. Of course 125 psi is unlikely for municipal or well-pump line pressure, but even our test conditions (70 to 80 psi) seemed to challenge these valves. For that reason we recommend a maximum operating pressure of no more than 40 to 50 psi. If your water pressure is higher than that, you can reduce it by installing the pressure-reducing valve described in "Lightening Your DHW Load," found earlier in this section.

Another valve specification that's unique to this system is the water strainer that is shown installed where the cold main breaks off to the hot water plumbing. This is a standard fitting available in a bronze (corrosion-proof) casting from plumbing supply outlets. The strainer helps to protect the drain-down valve by collecting any particulates that might slip into the water line from the water main. If your system is prone to excessive particulate content, the seals in the drain-down valve could, over time, be worn down to the point where a total seal could not be achieved when the valve is closed. The worst (and most unlikely) possibility is that a particle might cause the valve to stick in the open position (but it'd have to be a pretty darn tough particle to do that).

The minimum maintenance routines described in "Installing, Plumbing and Wiring Your Collectors," found earlier in this section, should of course be followed for drain-down systems. Especially important is the periodic testing of the freeze

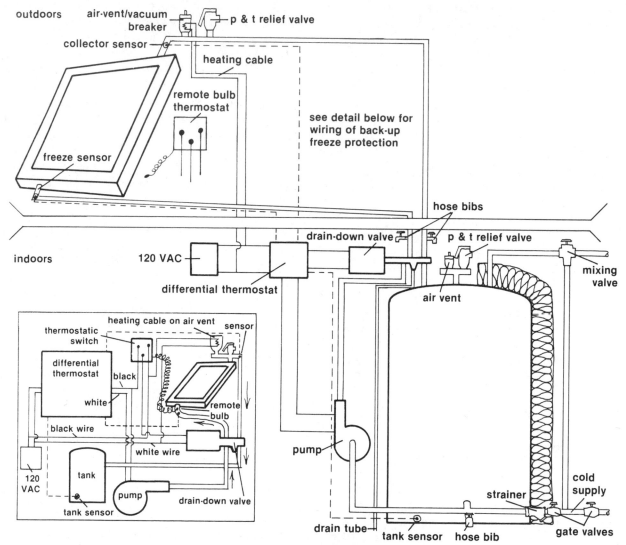

Figure II-28: This is a complicated schematic, but give it a long look and everything will fall into place. For starters, everything above the dividing line can be exposed to freezing temperatures, and everything below the line can't. At the top of the system, the heating cable is wrapped around the air vent/vacuum breaker and wired directly to a 120-VAC source. The wiring detail at lower left shows how the remote bulb thermostat (single-pole, double-throw) is wired into the drain-down control circuitry. With this wiring the remote bulb thermostat will not interfere with the normal operation of the drain-down system. It only comes into play if there is an unusual failure in the drain-down circuit. For manual draining of the system, two hose bibs are installed just below the drain-down valve, indoors of course. The drain tube running from the drain-down valve should terminate at a drain, and there should be no major convolutions in the line that might interfere with proper draining.

protection mode, in which the freeze sensor is immersed in ice water. If there is a line-voltage back-up freeze sensor, disconnect the 120 VAC power to the differential thermostat, put the sensor in a plastic bag and immerse it. Use a battery-powered continuity tester or an ohmmeter to see when the switch opens. It should open on temperature drop and close on temperature rise. The drain-down valve should also be checked internally from time to time. This will be especially important in areas with poor water quality (high mineral content). Combining high, solar-heated water temperatures with high mineral content in the water creates excellent conditions for mineral deposition on pipes and inside valves, especially metal valves. Plastic valves don't tend to collect nearly as much as their metal counterparts, but still they should be checked. Every 12 months, just before the cold weather comes in, look inside your drain-down valve for mineral deposits and general wear and tear. If there seems to be excessive wear on the sealing surfaces (usually consisting of O-rings or some other replaceable gasket), remove them and put in new ones identical to the manufacturer's specifications.

With faithful maintenance the drain-down system is a reliable solar DHW option; it's been proven. It has the same performance pluses as found with the recirculation system: direct circulation of domestic water through the collector and no heat exchanger. If you're contemplating an open loop system for a cold climate, this is the one.

Joe Carter
Bob Flower
Harry Wohlbach

Hardware Focus

Drain-Down Components

Richdel, Inc.
P. O. Drawer A
1851 Oregon St.
Carson City, NV 89701
(702) 882-6786

Solar Hot Water Control Valve, Model R797
120 VAC 60 Hz electric motor. Pressure rating: 100 psi. Accepts ½" pipe.

Solar Dynamics of Arizona
P. O. Box 647
1100 N. Lake Havasu Ave.
Lake Havasu City, AZ 86403
(602) 855-7555

Solar Dyne Sola-Sentry
12-volt charging system. Pressure rating: 150 psi. Accepts ½" pipe.

Sunspool Corp.
439 Tasso St.
Palo Alto, CA 94301
(415) 324-2022

Sunspool Collector Drain-Down Valve
12, 24 or 110 VAC models. Pressure rating: 125 psi. Accepts ½" pipe.

Differential Thermostat

Hawthorne Industries, Inc.
3114 Tuxedo Ave.
West Palm Beach, FL 33405
(305) 684-8400 or toll-free: (800) 327-7300 (except Alaska, Florida, Hawaii)

Fixflo Control H-1506-C
On differential 16°F (9°C). Off differential 6°F (3.3°C). Drain-down freeze protection.

PROJECT
A Fluid Freeze Protection System
Absolute Antifreeze

By taking the water out of a flat plate collector loop and replacing it with an antifreeze fluid, total, unquestionable freeze protection is achieved. Murphy doesn't have a chance. This is naturally a closed loop system since you can't consume antifreeze fluids, and like the drain-back loop a heat exchanger is required (with some loss of performance relative to open loop systems). Probably the most important factor in setting up this system is choosing the right fluid. Some types are poisonous, some will eat away at your copper collector, and some are inert on both counts—those are the ones to use. Antifreeze fluids also cause some loss of performance because they aren't as good a heat transfer agent as water is. But in the no-free-lunch line of thinking, that is the price for eliminating the frost factor.

More than 25 years as an industrial arts teacher turned Andrew Brock into an incurable do-it-yourselfer, and although he's retired now, he's as busy as he's ever been. In the past several years he's built several additions to his comfortable ranch house in Nazareth, Pennsylvania, put up an attached greenhouse, developed an orchard of peach and apple trees and converted his fireplace into a circulating hot water space heater that warms the entire first floor of the house. But the project he's most proud of is his solar water heater. Built from scratch using Brock's own design, the system supplies him and his wife with all of their hot water from March to October and supplements their winter supply. The Brocks figure their savings at about $200 a year compared with the days when they heated water with their coal-fired boiler.

Although Brock had been studying solar applications for the better part of 20 years, he didn't get around to building his own system until 1977. His problem was in coming up with a system that was fairly simple and maintenance-free, and one that

would provide absolute freeze protection through a Pennsylvania winter. To get all that he settled on a closed loop, forced circulation system that used antifreeze fluid instead of water for solar heat transfer.

As was described in the drain-back system, a closed loop system has all the basic solar DHW parts (the collector, a pump, a differential thermostat and a storage tank) plus a heat exchanger and, in this system, a special fluid. The electrical controls work the same as those in any forced circulation system. The differential thermostat reads the temperature difference between the collector and the bottom of the storage tank, and if the former is hotter than the latter, the pump goes on. Unlike the drain-back system, the heat exchanger carries low-pressure closed loop fluid. (In the drain-back system the exchanger carried the line-pressure domestic water.) Because of this difference, the heat exchanger in this system is usually immersed inside the domestic hot water storage tank (which is usually equipped with a back-up electric heating element so that the system needs only one tank). Some "heat exchanger tanks" are available with the exchangers wrapped around the outside of the tank, but that kind of arrangement is a little less efficient than the immersed exchanger configuration. Another difference with this system is in the kind of absorber plate that can be used. Because the closed loop operates at less than 30 psi, it's possible to use lower-cost, low-pressure absorber plates in the collector (including SolaRoll).

The presence of the antifreeze fluid requires some different plumbing hardware. When the fluid gets hot, it expands, just like the water in a closed loop hot water radiator system. Since the fluid can't be compressed, it needs an expansion tank to take up the extra volume of hot antifreeze. It also needs an overflow valve (pressure-relief) to back up the expansion tank in case its limits are exceeded. Another added piece of hardware is a standard air purging unit, which sits over the expansion tank. This is also found in modern hot water radiator systems, and in both systems it serves to pick up air from the collector loop and purge it to the atmosphere. Without these safeguards for fluid expansion, there is a

risk of damage to the system from overpressurization. The system is of course equipped with pressure-relief valves, but to use them for pressure relief without the other components would mean dumping expensive antifreeze fluid every time the pressure was too high.

For his antifreeze fluid, Brock made up a 50-50 mix of water and ethylene glycol (ordinary automotive antifreeze). This mixture provides freeze protection down to about −30°F (−34°C). However, ethylene glycol is highly toxic and should be used only in systems with double-wall heat exchangers to ensure that potable water won't be contaminated. Also, double-wall heat exchange systems cost more than single wall, and there is some performance reduction because of reduced heat exchange efficiency. A preferable fluid used by many solar installers is nontoxic propylene glycol. It's a little harder to find and somewhat more expensive at about $15 a gallon, but it's a safe fluid. One source is travel trailer and camper suppliers; people pour propylene glycol into their camper toilets and sink

Photo II-14: Since his greenhouse was "in the way," Andrew Brock found his collector spot farther out in the yard. From there the heavily insulated pipes run underground through the foundation and into the basement.

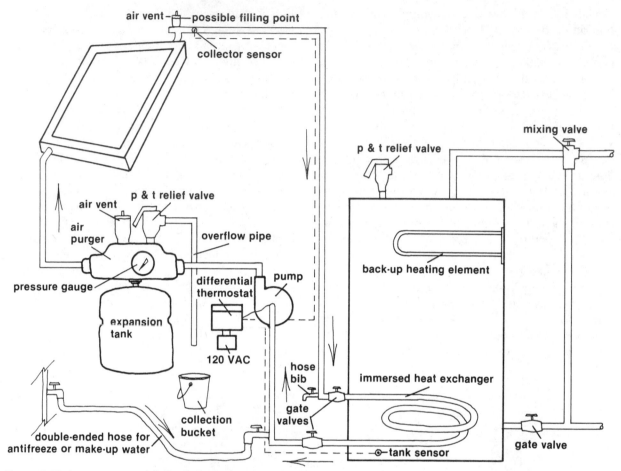

Figure II-29: In some ways the plumbing and wiring for the antifreeze system is the simplest because there's no concern about freeze protection. The expansion tank-air purge assembly is available at a plumbing supplier. The fill line shown at lower left can be used to supply the water for the antifreeze-water mixture or, as Mr. Brock did, it can be connected to a pressurized tank containing the antifreeze. The other possible fill point for the antifreeze is way up at the top of the system at the air vent, which can be unscrewed to make way for a funnel or a fill tube.

drains to prevent damage to them in below-freezing weather. Be sure to buy the glycol with corrosion inhibitors to protect metal pipes. When the glycol begins to break down (a single filling should last five years or more if the system operates properly and doesn't chronically overheat), it becomes acidic. The corrosion inhibitors counteract this acidity. It's important to check the pH level of the liquid once or twice a year using standard litmus paper. If the paper

indicates a high acid level, it's time to replace the fluid or add more corrosion inhibitor.

A Ground Mount Installation with Underground Plumbing

After the system was designed, Brock was faced with another problem: where to mount the

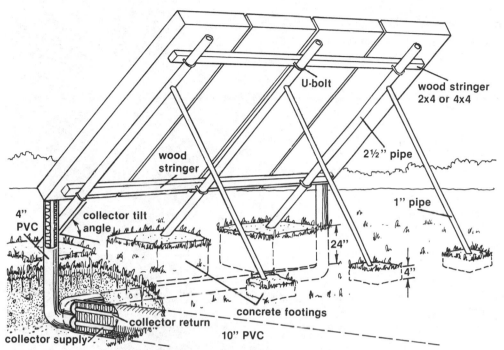

Figure II-30: A freestanding ground mount needs a triangulated support structure primarily to stand up to wind loads. In this version 2 1/2-inch pipes are sunk into concrete footings and then braced back to the ground with 1-inch pipe. When collectors are this exposed to cold weather, it's a good idea to increase the amount of insulation behind the absorber plate. The underground pipe run contains both the supply and return lines in 10-inch PVC that's filled with urethane foam. Maximum insulation is necessary to minimize pipe heat loss between the collectors and the storage tank.

four 4 by 9-foot collectors he had built the previous winter. "My roof sloped the wrong way, and it's an A-1 slate roof," he says. "I didn't want anybody walking around up there and punching holes. I have plenty of lawn, and I figured if I put the collectors out there I could get rid of a little mowing." The greenhouse had taken up the only available south wall area.

To carry the collectors at the desired angle, Brock set three 2½-inch diameter metal pipes into 2-foot-deep concrete footings. As figure II-30 shows, the pipes were set at 52 degrees, and 4 × 4 cross-ties were bolted to the pipes to form a suitable mounting platform for the collectors. Later, he discovered that

the panels were shifting because of wind and freezing around the footings, and that the movement was causing ruptures in the plumbing. To stop the movement he added supports in the form of 1-inch pipes run from the main pipes to small footings in the ground.

There was still another problem to overcome in this installation. The spot he had chosen for the collectors was nearly 60 feet from the basement location of the storage tank. Even buried underground, there would be a great deal of pipe heat loss from the collector supply and return lines with such a long trip between the tank and the collectors. Brock's solution was to suspend the lines in the mid-

dle of a 10-inch PVC pipe, and then pump in urethane foam insulation. The pipe was then buried 2
feet underground. After installing the collectors, he
dug a 2-foot trench to the spot where the pipes were
to enter his basement.

Twenty-foot lengths of pipe were used so
as to have as few underground joints as possible. The
assembly of long insulated pipe sections began with
wrapping fiberglass insulation around the 1-inch copper supply and return lines. ''I wrapped them with
fiberglass because I didn't think the foam insulation
could withstand the high temperatures of the solar-
heated fluid,'' Brock noted. (Urethane is rated for up
to 200°F, 93°C, service.) To support the pipes dead-
center inside the 10-inch PVC, Brock made wooden
crosspieces, one for every 2 feet of pipe run, and
secured them to the pipes. Then the entire assembly
was slipped into the PVC. Before they went in, Brock
pressure-tested all of his solder joints with compressed air at 100 psi. If the pressure gauge showed
any drop, it meant there was, alas, a leak somewhere. Brock's line had a couple, and once they
were fixed the pipes were slid into the PVC, and the
assembly was made ready for shooting in the foam.
To ensure that the wet foam would totally fill the PVC
before it set up, Brock drilled 1¼-inch holes into the
PVC at 18-inch intervals. A local contractor was
hired to inject the foam. With about 4 inches of high
grade foam surrounding the collector lines, Brock
created a very effective R-28 thermal barrier against
pipe heat loss to the ground.

After the underground work was done,
Brock broke a hole through his basement wall and
ran the plumbing through it. Just inside the wall, on
the hot water return pipe from the collector, he installed the expansion tank-air vent assembly. The
tank contains a rubber diaphragm with pressurized
air on one side and the system's hot water on the
other. As the hot water expands, the diaphragm
compresses the air to make room for the liquid. Just
behind the expansion tank, Brock installed a 30 psi
pressure-relief valve to serve as the overflow control.
A hose placed on the outlet of the valve drains the
fluid into a bucket on the floor in the event of a
pressure release. This fluid can be returned to the

Photo II-15: The antifreeze loop runs into the basement
and into the storage tank by way of the immersed heat
exchanger. The little tank in the foreground is the compressed air tank that Brock uses to fill and add antifreeze
to the system. The white pot collects any overflow that
might occur when the fluid is heated and expands.

system during normal maintenance.

Brock used a 1/12-hp water pump, which
he now feels is a little too large for the job. In practice, the pump size will vary according to the horizontal and vertical distance to the collector. To get
a good circulation rate (1½ to 2½ gpm) in an un-

pressurized system, you'll need between a 1/30- and 1/12-hp pump; 1/20-hp pumps are commonly used. The pump should be able to handle the antifreeze solution without corroding—stainless steel or inert plastic innards are recommended.

On the collector supply line Brock added a spigot (hose bib) that he could use for charging the system. A hose runs from the spigot to a 16-gallon aluminum water tank that Brock salvaged from a mobile home and fitted out with a fill valve and an air valve. To charge the system he simply fills the tank with the antifreeze/water mixture and hooks the air compressor to the air valve. When the compressed air pressure is applied, the liquid is forced into the pipes.

Of course not everybody has an air compressor lying around, so there is a way to fill the system just by pouring. The first step would be to fill the system with water to check for any leaks. The air vent at the top of the collector array is a good filling point, or you can use a double-ended hose to connect the collector drain valve (hose bib) with a line-pressure spigot and fill the system that way. Once you're certain there are no leaks, drain all the water from the collectors and the pipe runs, but don't just chuck the water; instead collect it to find out how many gallons of fluid you'll need to buy. Pour in the fluid through the collector air vent (after removing the air vent). If you bought a premixed glycol/water fluid, you'll want to fill the system to the brim. If you have enough straight glycol for about half the required volume, use distilled water to make up the other half. When the system is filled as much as it can be, you can use the double-ended hose connection to push in a little more water to bring the system up to pressure (10 to 20 psi, as indicated on the pressure gauge).

The storage tank is a commercial model designed for systems such as this, with a heat exchanger coil running around the bottom. To install it, he had only to hook up the exit and return lines from the collector to the heat exchanger, run the house's cold water supply line to the bottom of the tank and connect the tank outlet to his hot water main, including a standard tempering valve.

In retrospect, Brock figures it would have been better to have a tank that included a back-up heating element. Presently he can bypass the back-up water heater when the sun is shining, using manual controls. For cloudy periods he turns the water heater back on, again manually. It's not a big job, but it is inconvenient; clearly there is an advantage to the single-tank-with-element arrangement. If he had to build the system over again, Brock would use CPVC pipe for the collector supply and return lines instead of the expensive copper. It should be noted that zinc galvanized pipe should not be used with antifreeze fluid systems. The corrosion inhibitors in the antifreeze react with the zinc coating and would ultimately rust out the pipe.

While this system is somewhat more complex and costly than other flat plate systems, it does have an undefeatable freeze protection feature that Murphy's Law can't even touch. Given the reliability of other flat plate systems, however, this system should mostly be considered for use in very cold climates, although it can certainly be used anywhere.

Mike LaFavore

With this system we're at the end of the catalog of flat plate systems, but the options for solar domestic water heating don't end here. The next series of projects deal with *batch-type* solar water heaters, wherein a pressurized tank is painted black and put in the sun inside an insulated box with double or triple glazing. In a nutshell that's the basic system; the options come mostly in the area of design. The box for a batch-type system can be square, flat, triangular, curved, wall-hung, roof-mounted and put out in the yard. They don't have the sleekness of slim-jim flat plates, but they can be integrated with the existing house and look good. They're also a good bit cheaper than flat plate systems. In the projects that follow we've got some nice examples for you.

Hardware Focus

Storage Tanks with Internal Heat Exchangers

Aquatherm
541 Main St.
South Weymouth, MA 02190
(617) 331-6700

Aquatherm Tank
Steel exterior. Stainless-steel liner. Two heat exchanger coils, each 30' of ⅝'' finned copper pipe. 182 or 273 gal capacity. Polyurethane foam insulation (R-6.5 to R-9).

The Electric Heater Company
P. O. Box 288
45 Seymour St.
Stratford, CT 06497
(203) 378-2659

Hubbell Hot Water Storage Systems
Steel exterior. Stone-lined. Removable heat exchanger coil with finned copper tubing. ½'' MPS connections. 80, 120 gal capacity. 2'' of fiberglass insulation (R-8.4).

Ford Products Corp.
Ford Products Rd.
Valley Cottage, NY 10989
(914) 358-8282

Ford Solar Water Heater
Steel jacket. Stone-lined. Single-wall or double-wall heat exchangers with leak detector. 40, 65, 80 or 120 gal capacity. Fiberglass insulation (R-8.6).

General Energy Devices
1753 Ensley
Clearwater, FL 33516
(813) 461-2557

Counter-Flow Heat Exchanger Tank
Steel exterior. Glass-lined. Counter-flow, double-walled copper heat exchanger. 80, 120 gal. Urethane foam insulation (R-17).

Resource Technology Corp.
1 Alcap Ridge
Cromwell, CT 06416
(203) 635-0266

Sun-Bank Solar Storage Tank
Steel or fiberglass exterior. Stainless steel-lined. Two internal copper heat exchangers, ¾'' pipe, 100 lineal ft each. Smallest capacity 200 gal. Urethane foam insulation (R-30).

A. O. Smith
P. O. Box 28
Kankakee, IL 60901
(815) 933-8241

Conservationist, Sun Model
Steel exterior. Glass-lined. Double-wall copper heat exchanger. 66, 82, 100, 120 gal. 3'' fiberglass insulation (R-12).

State Industries
Ashland, TN 37015
(615) 792-4371

Solarcraft
Steel exterior. Glass-lined. Four internal heat exchange chambers. 82 or 120 gal. Fiberglass insulation (R-8.33).

Vaughn Corp.
386 Elm St.
Salisbury, MA 01950
(617) 462-6683

Sepco SNR Series
Steel exterior. Stone-lined. Heat exchanger: 10, 15, 20 or 40 sq ft sizes. 66, 80, 100 or 120 gal. 2'' fiberglass insulation.

Heat Transfer Fluids

The descriptions of the heat transfer fluids listed below include the product name, the generic name (if available) and either the operating temperature range or the temperature to which freeze protection is offered. The manufacturers claim that the products are noncorrosive and either nontoxic or low in toxicity.

Bray Oil Co., Inc.
9550 Flair Dr.
Suite 301
El Monte, CA 91731
(213) 575-1212

Brayco 888 & 888HF
Synthetic oil base. 888 fluid from −100 to 550°F (−73 to 288°C). 888HF fluid from −80 to 680°F (−62.2 to 360°C).

Camco Manufacturing, Inc.
2804 Patterson St.
Greensboro, NC 27407
(919) 292-4906 or toll-free number:
      (800) 334-6960

Solar Winter Ban
Propylene glycol base. Undiluted, it offers freeze protection to −55°F (−48°C).

Dow Chemical U.S.A.
Midland, MI 48640
(517) 636-1000

Dowfrost
Propylene glycol. Freeze protection to −28°F (−33°C).
Syltherm
Silicone liquid. Operating temperature range: −50 to 400°F (−46 to 204°C).

Exxon Co., U.S.A.
P. O. Box 2180
Houston, TX 77001
(713) 656-0370

Caloria HT 43
Petroleum base. Operating range: 15 to 600°F (−9 to 316°C).

Mark Enterprises, Inc.
P. O. Box 3659
30 Hazel Terrace
Woodbridge, CT 06525
(203) 389-5598

H-30 Fluids
Synthetic hydrocarbon. Freeze protection to −40°F (−40°C).

Resource Technology Corp.
1 Alcap Ridge
Cromwell, CT 06416
(203) 635-0267

Sun-Temp Solar Collector Fluid
Freeze protection to −40°F (−40°C).

Solar Alternative, Inc.
22 S. Main St.
Brattleboro, VT 05301
(802) 254-6668

Solar Winter Ban
Propylene glycol, dipotassium phosphate and water with food coloring. Undiluted, it freezes at −55°F (−48°C).

Sunworks
P. O. Box 3900
Somerville, NJ 08876
(201) 469-0399

Sunsol 60
Propylene glycol base. Undiluted, it freezes at −55°F (−48°C).

PROJECT
The Breadbox
A Passive Solar Water Heater

This two-tank batch-type system provides 60 percent of the annual hot water needs of a family of three at a total cost for materials of about $400. Pressurized domestic water flows through the tanks when hot water is used. As it stands in the "collector" tanks, the water is preheated before entering the back-up water heater. In summer, solar energy can do it all, and the water heater is shut down and bypassed. In winter, freeze protection is less critical than with flat plate collectors, and there are a number of ways to "passively" guarantee no-fault, no-frost operation.

The Reiss house is located in Sacramento, California, where winter temperatures are in the 50s and 60s during the day and in the 30s and low 40s at night. Frost is unusual in this region, which averages a mild 2800 heating degree-days. Summer temperatures are usually in the 90 to 100°$^+$ range. As well as being a mild climate, Sacramento has a healthy solar resource of 79 percent of possible sunshine. Occasionally in winter, however, the thick "Tule" fogs roll in. These dense ground fogs can last four or five days at a stretch and reduce solar gain to near zero. Nevertheless we expected a passive solar water heater to perform well in this clear and mild climate.

The Site, the System and the Box

Our first retrofit was a solar greenhouse that we built onto the south wall of our California bungalow. Before installing the water heater we built a redwood deck just east of the greenhouse, and now it helps to connect the rather bold shapes of the greenhouse and the breadbox. Needless to say, the south wall has been transformed.

The water heater sits on a platform 8 feet above the deck on two stilts that are part of the deck foundation. This "aerial" positioning above the

shade line of backyard trees and other plantings gives the collector a clear view of the sun through most of the year. The breadbox also provides summer shading and some rain protection. It's nice to have shaded walls with our brand of summer (very hot).

We were careful to site the collector relatively close to the back-up water heater. During full (no back-up) solar operation, solar-heated water is supplied directly to various points of use, bypassing the water heater. We sized the system to be large enough to meet our total daily demand. We based our estimates of daily demand on the average American use of about 20 gallons of hot water per person per day, knowing that we didn't use that much ourselves. Our collector tanks provide 80 gallons of fully heated water on sunny summer days and have proved more than adequate for our household of three. The glazed area of 42 square feet falls within the recommended range of 0.4 to 0.6 square feet of glazing per gallon of collector tank. Another rule of thumb is to provide about 20 gallons of tank volume for each member of the household.

The two tanks are spaced far enough apart so as not to shade the other too much in the morning or afternoon. The glazing is placed as close as possible to the tanks to ensure maximum solar gain and minimum shading from the opaque side walls of the breadbox.

The success of our breadbox system is due

in part to a few conservation improvements to the existing DHW system. We wrapped 6 inches of fiberglass insulation around the back-up water heater and installed low-flow devices on all hot water outlets. The water heater thermostat is set down to 115°F (46°C), which is adequate for domestic use. With a passive solar water heater, the pattern of hot water use has a greater effect on overall performance than with other types of solar water heaters. Since the most hot water is available in the afternoon and early evening, we've adjusted our use patterns to coincide with these periods. That keeps the use of the gas heater down to a minimum.

Construction

Once the system design was worked out, we went shopping for building materials and plumbing hardware. The plumbing is done with regular stuff: dielectric unions at all copper-to-steel pipe junctions, a pressure-and-temperature relief valve, a few gate valves. Finding the glass-lined steel tanks we wanted was a bit more difficult. We went to the dump and "checked out" with a couple of discarded gas water heaters. We stripped them down to the naked tank and carefully cleaned them, looking for leaks and making sure that their glass linings were intact. The anticorrosion magnesium anode rods were removed, cleaned and put back into the tanks, but putting in new rods is a better idea, if they're available (usually at plumbing suppliers). We bought enough fiberglass-reinforced plastic (FRP) to make a double glazing. FRP glazing can withstand the high temperatures generated inside the collector and has a good lifetime in the sun. Double and even triple glazing is important for reducing reradiative heat loss from the collector tanks.

Construction began with the deck. Since the water heater weighs around 1000 pounds, we had to have a "sturdy" plan for how it would be supported. Needless to say, the building inspector gave our plans close scrutiny. Once we'd built the deck, put up the support posts and plates and tied everything into the south wall, the way was clear to

Photo II-16: The south wall of the Reiss house had to make room for a greenhouse and a deck as well as the breadbox, and the breadbox itself had to rise above the shadow cast by a tree. Ergo, the airborne breadbox. With a remote rope-and-pulley system an insulating shutter reflector system could be installed to boost breadbox performance.

assemble and install the floor of the water heater. The rest of the breadbox was built onto this platform. After the tanks were installed and plumbed, we filled the system and checked for leaks before closing it all in with the glazing. We were very careful when building the double glazing to keep everything clean and to apply the FRP flat to the glazing frame without bends or buckles. Then we insulated the exposed hot water pipes, weather sealed the collector box, installed the glazing frame and started taking solar-heated showers.

Operation and Performance

The water heater has two seasonal modes of operation: full solar in summer and solar preheat in winter. During the summer we bypass the gas

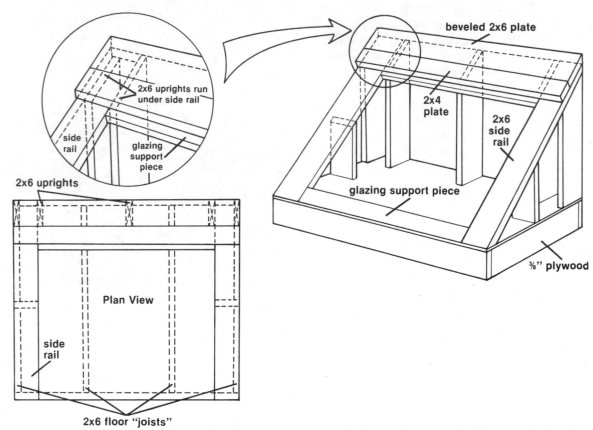

Figure II-31: Detail shows how the back wall 2×6's are arranged. The outside corner 2×6's go under the side rails and thus are 1 1/2 inches shorter than the other three back wall studs, which run up under the top plates, flush with the top faces of the side rails. With this arrangement the glazing support piece can be nailed to the three inside studs. Snap a taut chalk line across these studs to establish a straight line for nailing on the support piece. When all this framing is done, you're going to have yourself a pretty heavy item to move around, so if possible build the thing close to where it's going to end up.

water heater by closing one valve and opening another. In the winter, the valves are reversed and the water heater is restarted to provide additional heating when necessary. Full solar operation usually goes from late spring to early fall, when increasingly cooler showers tell us to bring on some natural gas.

The breadbox is providing at least 60 percent of our annual hot water load. At times the water temperature has reached over 160°F (71°C), which makes a tempering valve a good idea, especially in the bypass mode. Good thermal stratification is provided by plumbing the tilted (to a 45-degree pitch)

collector tanks in series. We can use about 60 of the 80 gallons in the collector before the incoming cold water finally causes a temperature drop. On a summer day we've found that the collector can heat 80 gallons of water twice to approximately 120°F (49°C), if we draw the first 80 gallons by midday. On clear winter days the collector is able to heat the water once during the day to between 100 and 120°F (37 and 49°C).

A computer model of this system was run for us by Davis, California, Alternative Technology Associates, and it came up with an estimated annual

Photo II-17: The tanks are mounted in the breadbox at the same angles as the glazing, and ideally they should almost be as close to the glazing as flat absorber plates are, with a gap of about 1 to 2 inches.

solar fraction of 49 percent for an average family of four. Of course, "average" consumption doesn't usually take conservation into account.

Money

The system can be built reasonably cheaply. We've proved it. The total cost of materials came to just $400: $50 for glazing, $150 for plumbing supplies, and $200 for lumber and supplies for the collector box. The junkyard tanks were free.

In the summer of 1978, we were using about 20 therms per month of natural gas. In the summer of 1979, when our hot water was entirely solar heated, we used about 8 therms a month, just for cooking. The difference shows that we had been using about 12 therms every month for water heating before we built the collector. Average consumption per house in California is estimated (by the Pacific

Gas and Electric Company) to be 30 therms per month for water heating, but our energy-conserving ways were well established before we added solar water heating. By adding the breadbox we were simply taking the next step toward self-sufficiency, reducing our gas use for water heating from 12 to about 5 therms per month (annual average). Gas, however, is a relatively cheap fuel; at about 30¢ a therm it's cheaper than fuel oil or electricity. We're saving about $25 a year. But if we were replacing electric power with solar power, we'd be saving about $127 a year (at 5¢/KWH). The 55 percent state solar income tax credit reduces the real cost of the system to $180, resulting in a six-year payback (on the cost of gas).

Building the collector was a valuable learning experience, particularly because it taught us what not to do next time. Some suggestions follow, based on our experience, and the perfection of 20/20 hindsight.

We weren't really pleased with the reflector we made. It was hard to build and waterproof, and it seemed to only slightly increase solar gain onto the tanks. Thus we've eliminated it from the construction steps. It is possible, however, to build useful reflector insulating shutters that are manually operated, and this is discussed later. When sheathing the outside of the box, make sure to use a siding-grade exterior plywood or solid wood for long life. It's important to build the box as small as possible without shading the tanks, to insulate it well (R-14 to R-19) and to seal any and all cracks in the outer "skin" to eliminate air leaks.

The 0.025-inch-thick FRP had a tendency to buckle and was difficult to apply flat to the glazing frame. Using heavier 0.04-inch-thick FRP would have made the job easier.

We were not sufficiently familiar with the characteristics of the trees on our site. One deciduous tree did not begin to lose its leaves until December, much later than expected. To our dismay, it was shading the glazing from 12:00 to 1:45 P.M. all through autumn. Know thy trees.

Because we wanted to demonstrate that double glazing alone would be sufficient for good performance, we didn't include movable glazing insulation in the original installation. Our payback calculations show that a double-glazed collector does indeed perform well on its own, and now we're ready to improve its performance. We're planning to add a curtain made with insulating pool cover material that can be rolled up and down in the morning and evening by means of a simple remote-controlled pulley-and-track arrangement. Doing this will greatly reduce nighttime heat loss. In regions with severe winters, nighttime heat loss through the glazing can lower the efficiency of the collector somewhat, and freezing of the tank is possible. In cold climates where below zero temperatures are frequent, movable insulation is essential, along with heavier (R-25 to R-30) insulation for the sides and base of the box.

Freeze protection is important for batch-type heaters, primarily for the safety of the exposed pipe runs. The tanks themselves contain so much thermal mass that they're unlikely to become a victim of Jack Frost. The whole system can be protected with movable insulation. The pipes should be wrapped with a special kind of heating cable known as Frost-Tex (made by RayChem of Redwood City, California) which gets hotter as the air temperature gets colder and doesn't use a thermostatic switch, which could fail. Another strategy is to put a little house heat into the breadbox. If the unit is attached to the house, a small operable vent can be installed to join the box with heated living space. On particularly cold nights the vent is opened, and the breadbox essentially becomes part of the heated living space.

Our collector was put up on stilts to be in the sun and out of the way, though mounting it on the ground is certainly an easier approach. To put it on the roof, you'll probably have to strengthen the rafters. Of course, access to sunshine should be your primary criterion for deciding where to install. After that, try to select a location that is the minimum possible distance to the existing water heater (or, move the water heater).

The plumbing schematic calls for a drain (hose bib) valve and a pressure-and-temperature re-

lief valve. The drain valve should be placed on the supply line, anywhere between the collector and a shutoff (gate) valve. The drain line from the relief valve must be run outside to within 6 inches of the ground to comply with most code requirements. It's okay to vent onto the roof if gutters will direct the water to the ground.

Finally, if you're getting a permit for the installation, put together a set of drawings when it's time to visit the inspector. We showed our drawings to the building inspector and used his suggestions to cut our construction costs by about 15 percent.

Construction Steps

While the heart of this project is the breadbox itself, you've still got to put it somewhere. The first few steps cover our "elevating" experience of putting the box up on stilts.

Post Footings

After determining where the post footings should go, dig a 6 by 6-inch hole at least 6 inches below the frost line. Pour in the concrete and set a $2 \times 4 \times 4$-inch block into the wet cement. This is the nailer for securing the base of the post. Level and center the block. Do this about three days ahead of the other work to allow some time for the cement to cure.

Ledger

The ledger ties the box into the building structure. We fastened an 8-foot 2×6 to the south wall with four $5/8$ by 6-inch lag screws. Be sure that you hit studs with the lags, if it's a wood-frame building. Set the ledger at the desired height (even with the top of the post header) and make sure it's level.

Posts

The 4×4 posts are nailed into the block in the footing. First, set and plumb the posts and then

Materials Checklist

2×6 lumber	4-foot rolls of FRP glazing
⅜-inch exterior plywood	silicone caulk
4×4 redwood (or other) posts for platform	2½×2½-inch aluminum corner flashing
1×2 clear, kiln-dried lumber for glazing frame	12-inch-wide aluminum flashing
2×4 lumber	copper flex fittings (copper to galvanized steel)
concrete (if posts are to be mounted in the ground)	copper tees, elbows, and ½-inch tubing
⅝×6-inch lag screws	hose bib
16d, 8d, 6d galvanized nails	galvanized steel plugs and gaskets
¾-inch ring-shank nails	pressure-and-temperature relief valve
¾-inch aluminum screws	galvanized "cross" or 4-way fitting
23-inch-wide, R-19 fiberglass batts, foil-faced	two 40-gallon tanks (such as hot water heater tanks)
foam weather stripping	heating cable
⅜-inch staples	black paint
15-pound roofing paper (tar paper)	

brace them with diagonal pieces of wood secured to the ground. Run a level line from the top edge of the ledger and mark the cutting line, which is the height of the level line minus the width of the post header. (We used a 4×6 for the post header.) Cut the posts, and nail them to the footings. Again, plumb and brace them. Then nail the header in place. The top of the ledger and the header will carry the floor of the breadbox.

Box Construction

The Floor

We laid our 2×6 floor joists on 2-foot centers across the header and ledger and secured them with 16d nails. The joists extend about 18 inches past the beam, providing a place to stand when working on the collector. If you don't need a working platform, the joists can be cut shorter and faced with 1×6 board. The box floor is made with ⅜-inch exterior plywood nailed with 6d nails to the underside of the joists. Cut pieces of 6-inch fiberglass insulation to fit between the joists, reflective or paper face up, and staple them in place. Nail more ⅜-inch plywood to the top of the floor joists to finish the floor.

The Back

Cut 2×6 studs to frame the back of the box. The top of each stud is cut at a 45-degree angle, as shown. The 2×6 top plate gets a 45-degree angle cut along the top edge, and the 2×4 plate is butted up to the lower edge of the 2×6. Assemble the back like a little stud wall with 2×6's 24 inches on center nailed to the floor of the box using 16 and 8d nails. Nail on the top plate pieces. The 2×6 plate piece gets a 45-degree bevel cut along the top edge, and the 2×4 plate is butted up to the lower edge of the 2×6. Sheathe the back with more ⅜-inch exterior plywood.

The Sides

Square the back to the floor and measure for lengths of the two side rails. Cut these pieces from 2×6 stock with 45-degree bevel cuts at each end, and nail them in place with 10d nails flush with the side and front edges of the box floor. Measure, cut

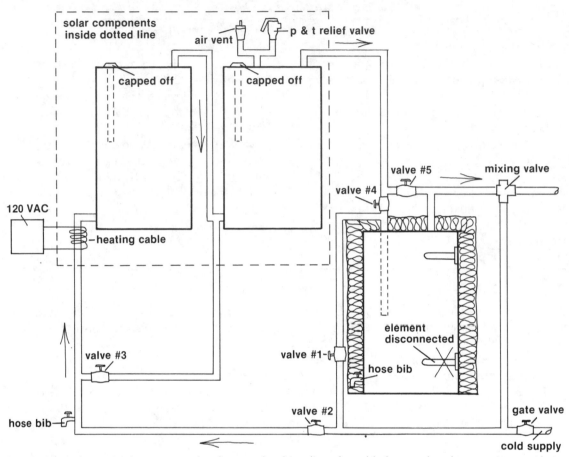

Figure II-32: This schematic shows a couple of extra plumbing lines for added control and convenience. Along with the water heater bypass (valves 4 & 5), valves 1 and 2 control a breadbox bypass, and valve 3 controls the draining of the right-hand breadbox tank. The heating cable should be used in cold climates on all pipe runs that are outside of the house. The control modes for the various valves are as follows: (1) closed for solar operation; open for breadbox bypass, (2) open for solar operation; closed for breadbox bypass, (3) open only for tank draining, (4) open for water heater backup; closed for water heater bypass, (5) closed for water heater backup; open for water heater bypass.

and install one vertical stud for each side with 45-degree bevel cuts at the top so they can fit under the side rails. Sheath the sides with more ⅜-inch plywood or siding. Make a glazing support piece by using the beveled piece that was cut from the 2×6 top plate. Cut it to fit tight between the side rails and nail it to the back wall studs so that it's flush with the top faces of the side rails. Make another beveled support strip and nail it to the floor at the front edge of the box, again flush with the top faces of the side rails. Cut pieces of 6-inch fiberglass to fit into all spaces in the back and side walls, reflective or paper face inward.

Tar Paper Floor

Cut pieces of 15-pound roofing paper (tar paper) with a utility knife, and cover the floor with it, running it about 9 inches up each side and the back wall. Crease the paper so it fits tight in all the corners.

Staple the tar paper to the floor and to the wall studs, and spray all the foil or paper insulation backing and exposed wood with flat black paint.

Glazing Frame

Use straight, untwisted, kiln-dried 1×2's that are free of large knots to make the frame. The frame dimension is equal to the total width of the box (side rail to side rail) by ¼ inch more than the distance between the two glazing support pieces. Find a clean flat surface to work on (a cement slab or wood floor will do), and sweep away all dust and debris. Lay out the frame pieces and screw and glue them together carefully so that all the surfaces at the joints are flush. When using 4-foot-wide FRP glazing,

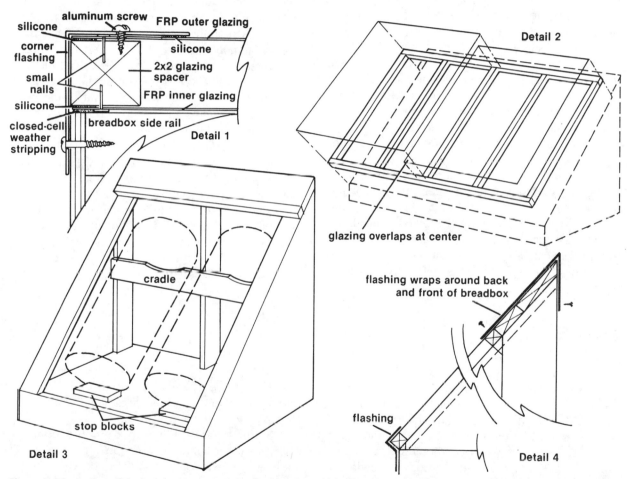

Figure II-33: In Detail 1 the aluminum corner flashing is used both to weatherproof the glazing frame and to secure it to the breadbox with the screws at the side of the flashing. The closed-cell foam ensures a virtually airtight seal. Detail 2 shows the placement of the glazing support pieces relative to the glazing frame and the side rails. It's important that these pieces be dead flush with the side rails and that they be straight and true so the glazing frame will seal properly to the breadbox. As shown in Detail 3, the tanks are set onto a 2×4 cradle that has been saber-sawed in two places with arcs that conform to the tank circumference. Then the tanks are secured with pipe strap before they're plumbed. Detail 4 shows how flashing is used along the top and bottom edges of the breadbox and the glazing frame.

place the three vertical crosspieces so as to divide the frame into four rectangles of equal width. Square the frame and brace it by tacking a scrap piece of plywood to one side. Cut two pieces of 4-foot-wide FRP glazing. Each piece will have to be trimmed to the proper width. The inside of the roll is treated with an ultraviolet inhibitor; that's the side that should face the sunlight. For a reminder put a small piece of tape on the ultraviolet-treated side of each piece. The glazing is installed over a perimeter bead of silicone. All the nailing points in the glazing (every 4 inches) are predrilled larger than the diameter of the ring-shank nail. Start fastening at a corner and work along one edge at a time. The two glazing pieces overlap 1½ inches in the middle of the frame, sandwiching a bead of silicone. Remove the plywood brace and repeat the glazing operation, facing the ultraviolet-treated side of the FRP to the inside face of the outer glazing.

Cut 2½ × 2½-inch aluminum corner flashing for the two sides and the bottom edge of the glazing frame. These pieces are fastened along the top faces with ¾-inch aluminum screws, and a bead of silicone is run between the top face of the flashing and the glazing. The top edge of the frame will be flashed by the box flashing after the glazing frame is installed.

Box Weatherproofing

Apply waterproofing stain or paint to the exterior siding of the box. Nail corner flashing to all box corners with caulk on each side of the corners. For the top flashing, cut a length of 12-inch-wide aluminum flashing 4 inches wider than the width of the box and fasten it to the box as shown in figure II-34. Run perimeter weather stripping of self-adhesive foam (about ½ by ¼ inch) around the collector aperture. The glazing frame is fastened to the box with more ¾-inch aluminum screws driven through the side of the corner flashing. The top flashing lies over the glazing frame and is fastened with aluminum screws, sealed with silicone. Of course the tanks must first be set and plumbed.

Tank Preparation and Installation

Clean the tanks thoroughly. If you're using an electric tank, install gaskets and covers over the heating element holes. Visualize how the tanks will fit into the box and screw in pipe nipples for tank inlets and outlets at the appropriate locations. The supply water enters the bottom of the number one tank, exits at the top, then enters the bottom of the second tank, finally exiting at the top of the second tank. Place a galvanized 4-way "cross" fitting (4 female threads) on this final outlet for installing a pressure-and-temperature relief valve and an air vent. The pressure-and-temperature relief valve drain line must lead outside the box. The air vent can be a standard float type. The fourth thread of the cross receives another nipple and then a copper flex fitting. Plug all other holes in the tanks that won't be used.

Cut tank supports from 2 × 4 stock. Place one tank into the box at its proper 45-degree slant and mark where the supports should go—one at the tank's bottom to keep it from sliding forward and the other under the top of the tank, against the back of the collector. Trace the curve of the tank in two places on the top support and cut along these lines to form a cradle for the tanks. Nail on the upper and lower supports. Secure the tanks with pipe strapping.

Plumbing

With the tanks in position, bend all flex connectors and measure the lengths of copper tubing needed to run the inlet and outlet lines outside the box. At the three pipe entry points (inlet, outlet, pressure-and-temperature relief valve drain), seal around each pipe with silicone.

In planning the run from the domestic water line up or out to the breadbox, try to minimize the length of exterior runs. When you know where you're going to cut the cold line, shut off the house's water main, make the cut and put in the tee fitting, or whatever is required. Do the same for the outlet line. Cut into the cold supply to the back-up water heater and add a tee fitting. Then plumb both lines to the breadbox. Insulate all the new lines. Make sure to provide a bypass at the water heater and a hose bib at low points in the collector inlet and outlet for ease in draining the system.

When all the connections are made, fill the system and check for leaks. Be sure all plumbing has been properly secured and supported. If any leaks appear (curses!), drain the lines, fix the leaks and test again. When the system is filled and leak-free, wrap the pipe lines that are exposed (outside the building thermal envelope), both inside and outside the breadbox with the heating cable. The heat tape should lead all the way into the house and plug into the nearest outlet or extension cord. In very cold climates use double or triple layers of pipe insulation

around all exposed pipe and protect the insulation with PVC (outside the box) and aluminum foil tape (inside the box).

Once all the plumbing has been installed and tested, touch up all spots on the box interior with black spray paint. Install the weather stripping along the collector box edges where the glazing frame will rest to completely seal the box. Lift the glazing frame onto the collector box and secure the glazing frame with the aluminum screws. Call in some sunshine and your breadbox will cook.

Jeff Reiss and John Burton

References

Baer, S. 1975. *Breadbox plans.* Albuquerque: Zomeworks Corporation.

Bainbridge, David S. 1981. *The integral passive solar water heater book.* Available for $11.50 postpaid from Passive Solar Institute, P. O. Box 722, Davis, CA 95617.

Bainbridge, David, and Reiss, Jeff. Breadbox designs. *Alternative Sources of Energy,* October 1978.

Burton, John. 1979. *Two tank vertical and horizontal passive solar water heater plans.* Available from Integral Design, 3825 Sebastapol Road, Santa Rosa, CA 95401.

Burton, John, and Reiss, Jeff. 1979. Passive solar retrofit, downtown Sacramento, California. In *Proceedings of the 2nd National Passive Solar Conference,* ed. Don Prowler. Philadelphia: Mid-Atlantic Solar Energy Association.

Butti, Ken, and Perlin, John. Solar water heaters in California 1891–1930. *CoEvolution Quarterly,* Fall 1977.

Davis, W. Douglas. 1978. The climax-cusp solar water heater. In *Proceedings of the 2nd National Passive Solar Conference,* ed. Don Prowler. Philadelphia: Mid-Atlantic Solar Energy Association.

Golder, John C. 1979. *Capsule collector plans.* Available from Santa Cruz Alternative Energy Co-op, #6 Old Sash Mill, 303 Potrero Street, Santa Cruz, CA 95060.

Starr, Gary, and Melzer, Bruce. 1978. An evaluation of two breadboxes. In *Proceedings of the 2nd National Passive Solar Conference,* ed. Don Prowler. Philadelphia: Mid-Atlantic Solar Energy Association.

Sunset Magazine. Breadbox water heaters. *Sunset Magazine,* September 1979.

Zweig, Peter. 1978. *Inexpensive do-it-yourself solar water heaters or preheaters.* Available from Farallones Institute, 15290 Coleman Valley Road, Occidental, CA 95465.

PROJECT
Some Breadbox Design Variations

Since a breadbox water heater is just a black tank inside a glazed and insulated box, there are really any number of ways to build one and attach it to your house. The following variations on the breadbox theme show how these collectors can be the simplest tilted box and how they can be rather elegant, integral building mini-additions. All breadboxes need not be created equal; they can be built to match different buildings, budgets and tastes.

This first variation couples the box with the building at a basement window. With the window open the breadbox is totally protected against freezing. Close attention was also paid to blending this mini-addition with the look of the house, and the result is passive solar water heating that looks downright elegant.

A Breadbox for Your Basement Window

Pity the poor breadbox heater. While you're inside sloshing around in solar-heated bath water, it's outside freezing its buns off. And if the worst happens, one of its water lines could ice up and burst, no matter how much insulation you have wrapped around them. Be nice to your breadbox by making it integral with your house's thermal envelope—built over a south-facing basement window. You can run the pipes through the window without exposing them to the winter, and the temperate air in the basement will circulate into the collector box and prevent a pipe freeze. Building heat loss is slightly increased, although heat loss from a basement, especially one that's unheated, shouldn't affect the comfort of the living space.

Construction

A 40-gallon tank filled with water weighs about 350 pounds. This corresponds to about 30

Photo II-18: With the right finishing treatments the attached breadbox looks like it was built when the house was built. The ground-level location makes this addition an excellent candidate for a movable insulation panel.

Materials Checklist

2×4 lumber	copper flex fittings for copper-to-threaded steel
2×2 lumber	connection
exterior siding to match house siding	dielectric unions or copper flex fittings
two double-glazed tempered glass units	hose bib
butyl glazing tape	three gate valves
foil-faced rigid insulation	pressure-and-temperature relief valve
bricks or concrete blocks and mortar for the	40-gallon glass-lined tank
foundation	clear silicone caulk
½-inch copper tubing with fittings	heating cable

pounds per square foot of floor area for a typical unit. Because of this and the importance of having a stable base for the glazing, the construction of the foundation and floor should be of the same quality used for houses. This is after all a building addition, albeit a small one. A good foundation in freezing climates requires a standard frost footing on firm soil. The foundation should be roughly half again as long (east to west) as the tank to allow the box to admit sunlight throughout the day. In this design the tank is set horizontally, and it rests on a small platform. Secure rot-resistant 2×4 sills on the foundation and set 2×4 joists on them. Run beads of caulking between all framing members and the building and foundation, anywhere that could be an opening for cold air infiltration into the collector airspace.

Before you fasten the joists to the sills on the house wall, fasten chicken wire or some other mesh or screen material to the bottom of the joists to support the fiberglass insulation. Then toenail the joists to the sills. Staple the insulation, foil or paper side up, to the top of the joists, and nail ⅜-inch plywood over the joists for flooring. Build a frame of 2×2's on top of the floor for the side walls and glazing frame. The side walls will be insulated with rigid foam and sheathed with ⅜-inch plywood. In this design there is vertical and pitched glazing, although it's certainly possible to design for a one-piece tilted glazing.

Mount a beveled 2×4 ledger to the house wall above the basement window with lag screws

and a bead of caulk. The top edge of the header should be beveled to the angle of the glazing tilt. Run two 2×2 side rails to the junction of the vertical 2×2 corner studs and the horizontal 2×2 glazing support piece (see figure II-34). Run more vertical 2×2's from each end of the 2×4 header down along the house wall to the back floor corners. If the house has lap siding, you'll have to cut it down to the sheathing to get a flat surface. Lay a bead of caulk under the two vertical 2×2's, and fasten them to the house wall with lag screws. Now nail ⅜-inch plywood over both sides of the box. It should extend over the sill and be sealed at the bottom edge with silicone caulk. Fill the side walls with 2 to 3 inches of rigid foam.

Before installing the glazing build a tank stand with 2×4's. Set the tank in place and complete and test all plumbing connections and valves in the collector box and in the basement.

Depending on what look you prefer, the interior of the box can be painted black or dark brown or just about any color. (The tank should of course be black.) The interior can also be totally reflective to direct more sunshine back onto the tank. Your choice.

The glazing used in this design is double glass in the form of standard ready-made insulating units. Building codes will require that glass units placed at low levels must be tempered, so you will have to investigate what standard tempered sizes you can get locally. (Custom-size tempered units are very expensive.) Basically the dimensions of the box all depend on the size of the glass. Plastic glazings can of course be cut to size. With your glazing units in hand, install the top unit first (in a two-piece design). As the detail in figure II-34 shows, the glass is held in place along the bottom edge with three J-shaped metal straps that are cushioned with neoprene setting blocks. Note that the top glazing runs out over the vertical glazing so that rainwater will simply run off. To seal the glazing to the frame, use standard butyl glazing tape. Use 2×2-inch aluminum angle stock to fasten the glazing to the frame side walls. Butyl tape is also run between the aluminum and the glass.

The vertical front glazing is installed in essentially the same way. The top edge of this piece should butt up tight to the underside of the top glazing, which means that the bottom edge of the front piece will have to be blocked up and cushioned with the neoprene blocks. When the front piece is

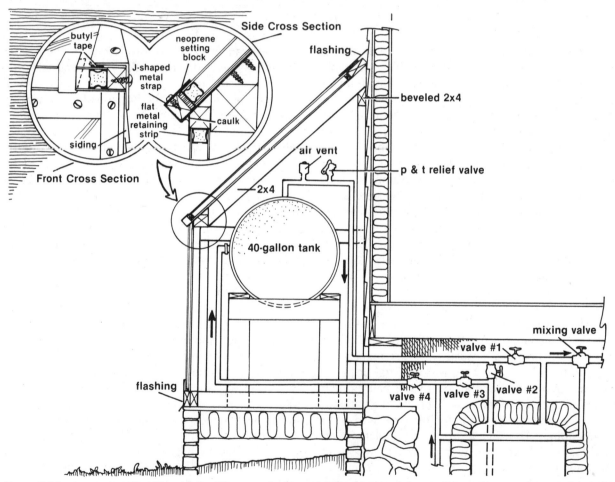

Figure II-34: A cross section like this could go on forever; that is, it's certainly possible to create a two-tank system with a longer version of the same thing. With a taller version you could stack two horizontal tanks, one atop the other. Although it's not shown here, the drain line from the pressure and temperature relief valve should be run outside the collector box, either outdoors or to a drain in the basement.

secured, run a bead of silicone along the seam where the two pieces meet. The top edge of the top piece is flashed with a strip of aluminum that has been slipped up under the exterior siding above the box, or otherwise sealed to the wall. Extend the flashing a couple of inches beyond both sides of the box to help keep water away from the back corners of the box where it meets the wall.

The siding used to finish this box exactly matches the original wall siding of the house. As much as anything else this is probably the key to achieving the well-blended breadbox, so choose your finish materials and colors carefully. Your only involvement after the box is finished will be the opening and closing of the basement window. (It doesn't have to be a basement window; cut a small hole in the wall and put in a little door or operable vent for the pipe runs and the freeze protection.) The opening and closing of the vent can be done on a seasonal basis. The window probably doesn't have to be wide open, maybe one-half or one-third open. Put a standard thermometer inside the box so you can check the air temperature on cold mornings before the sun hits the glazing. You can also use heating cable around the exposed pipes and reduce the vent opening. Using movable insulation over the glazing sections will greatly improve the performance of the system by minimizing night heat loss from the box, which means the tank will stay warmer. These could be rigid insulating shutters that could also double as reflectors by day. For some ideas on how this might be done, see the article, "Exterior Shutter-Reflectors for a Greenhouse" in Section III-F.

Performance

The bills for a four-month period during the summer after the breadbox was added show an energy savings of eight million Btu. A small part of the savings is due to the fact that the gas water heater was moved across the basement to bring it closer to the breadbox and to the kitchen and bathrooms, but other factors such as water usage and water heater temperature setting were unchanged. The owner observed that the burner of the hot water heater did not come on during his morning shower if the previous day was sunny, indicating that the preheat tank must have been warmer than the hot water heater thermostat setting, even after cooling down all night.

A conservative estimate is that the system provides about one-half of this household's hot water requirements. Using a solar fraction of 50 percent, a system materials' cost of $181.05, a natural gas cost of $2.30 per million Btu, the federal tax credit of 40 percent, the state (Missouri) tax credit of 25 percent, a general inflation rate of 7 percent and a fuel cost inflation rate of 10 percent, the economic performance is very good: The system pays for itself in two years. To get the same return from a savings account, the interest rate would have to be almost 30 percent per year compounded monthly. The system is thermally effective, inexpensive, virtually maintenance-free and simple to build. The question is: How many houses have a basement window on the south side. Millions!

Ronald Wantoch

Your Basic Breadbox

A solar water preheater box needn't be a collection of bevel and angle cuts. This one is like a flat plate collector that got fat. It consists of a simple rectangular frame made of 2×2's and 2×4's, which are filled with insulation and sheathed with plywood. The entire box can be tilted to the proper solar DHW angle and held there by a small wedge-shaped box through which the plumbing runs. This little box is also insulated along with all the exposed pipes, which are wrapped with heating cable. Like the first breadbox project, this is a two-tanks-in-series system. The 40-gallon tanks can be placed vertically or horizontally in the box.

The box itself is made with a rectangle of 2×4's for the base and 2×4 posts nailed vertically at each corner of the base. On top of this is set another rectangle the same size as the base, but made of 2×2's that are nailed to the 2×4 posts. Rigid, foil-faced foam insulation goes inside the framing, and the outside bottom and sides are sheathed with ⅜-inch exterior plywood. To support the tanks vertically mount two 2×4's across the frame base and attach cradle blocks as shown in figure II-35. Use plumber's strapping to secure the tanks. For horizontal mounting run two 2×4's across the full width of the frame and then install the cradle blocks and plumber's straps. Plumb the tanks in series with the air vent and the pressure-and-temperature relief valve and run the inlet and outlet together through the back of the box. The inside of the box can be black or reflective.

The wedge-shaped support box consists of

Photo II-19: This basic breadbox uses three vertical tanks that put about 120 gallons of domestic water in the sun. The inside of the box is painted white to make a better match with the house and to bounce a little more light onto the tanks.

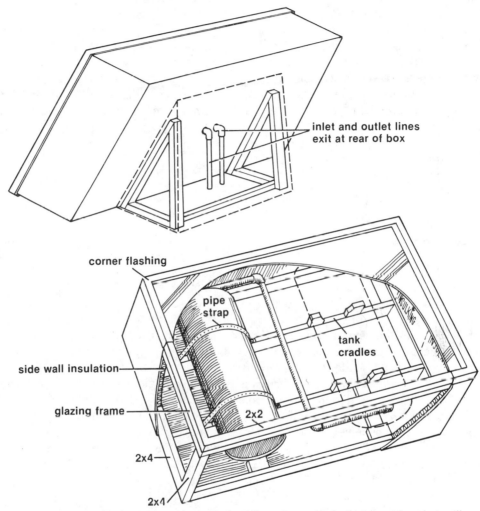

Figure II-35: You can't get any more basic than the basic breadbox. It can be built to any size that will accommodate small or large tanks placed vertically or horizontally. If the box is to be placed at ground level or otherwise within your reach, you could consider making some insulating shutters for use at night and during sunless days. The inside of the shutters could be reflective to bounce more sunlight into the box.

two triangular frames, each side made with 2×4's. These are insulated and covered with plywood after the plumbing connections are done. Weatherproof the box with paint or stain/preservative and seal all seams with exterior caulking. There are several ways to mount the box—on the ground, raised on a framed platform, or mounted on a flat roof. If you locate at ground level, place the box on concrete blocks or pressure-treated wood to keep it out of direct contact with the ground. Supports under the front of the box will prevent a strong gust of wind from tipping the whole thing forward.

The glazing can be plastic or glass. When using FRP, follow the directions in the first breadbox project. When using double-pane glass or rigid double-wall acrylic (Exolite), use the methods described in the previous project. Be sure to test all plumbing connections before putting down any glazing. Here again you have the opportunity to use movable insulation in the form of rigid shutter or roll-up blanket to boost performance.

Outside of a water-filled bucket sitting in the sunshine, this may be the simplest solar water heater you can build. With planning it shouldn't take more than a day for one person to build and maybe a half day for the plumbing. As for cost, this is another one of those $200-or-less deals that will perform as well as most any other in the breadbox line.

John Blackford

An Overhang Water Heater

If you were thinking of building a south wall solar greenhouse or sunspace or some kind of solar addition, or if you were looking at adding a lot of south glass for direct gain space heating, it would be nice to have an overhang to shade the wall or part of a solar room during warm weather. The overhang is a vital element in passive solar design. Without one you would be as victimized in summer by all that south glazing as you were benefited by it in winter. Typically an overhang means "roof," but it can also be a collector for heating water. It could be a flat plate system or a batch-type system like the one described here. For a little extra investment you get a lot more energy benefit. The overhang helps to keep things cool in summer while the collector heats your water year-round—a nice deal.

This last batch or breadbox variation uses two 30-gallon tanks placed horizontally in a triangular enclosure. As always the tanks are plumbed in series so that the final outlet water is always the hottest in the system. And as always the plumbing system simply interrupts the cold supply to the existing water heater, and a water heater bypass is installed for full solar water heating in summer.

What is really most important here is the way the enclosure is built onto the house, and the way it's finished and trimmed to match the existing siding or to somehow "disappear" to be as visually unobtrusive as possible. The overhang water heater could be the first step in a building plan that would ultimately lead up to the aforementioned solar addition, be it a greenhouse or whatever. As part of a larger solar structure it would probably blend very well. As a solo overhang it would be somewhat more apparent, but look at the photograph that shows this style; it certainly doesn't dominate the wall. The builders of this retrofit were very careful to use the same siding that was used on the house, to make the lap lines even and generally to have it look like "original equipment."

Construction

The overhang water heater, like the basement breadbox, already has an insulated back wall, your house, and there remains to be built the insulated sides and floor of the enclosure plus the plumbing and the glazing. The framing is done mostly with 2 × 4's and one 2 × 6. It was found by the workshop group that installed the pictured system that the entire enclosure could be framed on the ground and

Photo II-20: This 12-foot-wide overhang water heater is an effective energy system and possibly the beginning of a solar greenhouse or sunspace.

then braced in place with posts and fastened to the building as a nearly finished unit. The enclosure is assembled mostly with 16d nails and a good construction adhesive such as Max Bond. At a couple of critical points lag bolts were used, as shown in figure II-36. To accommodate the two 30-gallon tanks the floor of the enclosure measures 28 by 144 inches, and the height of the back measures about 42 inches. These dimensions approximate a 3-4-5 right triangle, which produces a glazing tilt of 53 degrees. For lower latitudes an isosceles right triangle would put the glazing tilt at 45 degrees.

Before installing the assembled frame, the perimeter dimensions are traced out on the building and all the siding inside that perimeter is cut away. This siding can then be used to finish the underside and sides of the enclosure. The frame is put up to the wall and held there with temporary posts while $3/8 \times 5$-inch lag screws are driven through the upper and lower ledgers into the house wall studs. To eliminate air leakage a bead of construction adhesive is run around the perimeter of the frame before it's hung on the wall.

Once the frame is hung, the bottom and one side are sheathed with 3/8-inch exterior plywood fastened with nails and construction adhesive. The other side must remain open until the bottom insulation is in place and the tanks have been slid in through the open end. The tanks are preplumbed with the major fittings, then cleaned and painted before they're placed in the enclosure. Once the tanks are in place, the end is sheathed and insulated. The insulation used is a 2-inch-thick rigid foam, the kind that has a reflective foil face. The shiny foil is left exposed to direct more solar energy onto the tanks, and the edges of the foam are covered with aluminum foil duct tape to hold the foam in place and protect the edges from degradation by sunlight. The exposed interior framing members can be painted white or silver.

At this point two 6 by 14-inch vent holes are cut into the wall. One is at the lower left-hand

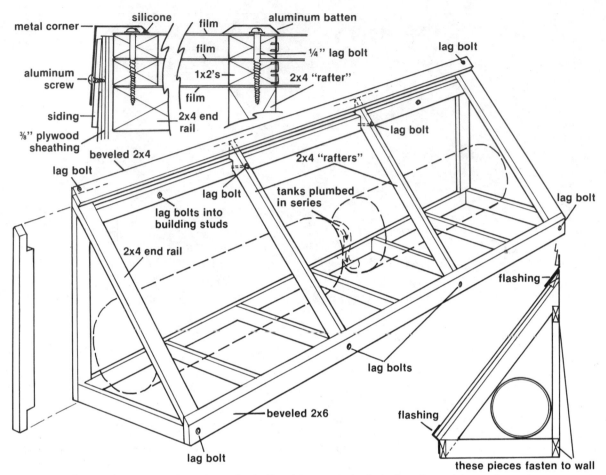

Figure II-36: The overhang water heater can be built with any desired glazing tilt, and it can be as long or as short as needs be to make it fit in with the wall in question. If this unit were being designed to become part of the opaque roof section of a solar greenhouse, the base could later on be extended over the greenhouse rafters to form the rest of the roof deck.

side of the enclosure, and the other is at the upper right. This is the coupling to the building that provides positive freeze protection for the pipes. The vents allow some warm air from the house to enter the overhang. (The pipes will also be wrapped with heating cable and insulation.) In this particular installation the inlet and outlet pipes for the tanks also pass through the lower vent to the water heater. The upper vent is equipped with an operable louver (with a piece of rigid insulation for a "flapper" type door). The tanks are plumbed in series and fixed up with the

requisite pressure-and-temperature relief valve and air vent or vacuum breaker valve. The Frostex heating cable is wrapped around the pipes, followed by pipe insulation that is a minimum ½ inch thick. The inlet and outlet lines are plumbed into the cold supply to the water heater with a standard bypass configuration. At this point the plumbing system is pressure-tested.

For the glazing we will introduce yet another configuration: triple glazing using three layers of thin solar glazing film. For this project we used three layers of Flexiguard, an acrylic/polyester film that is made in 5- and 8-mil thicknesses. It has a solar transmittance of 90 percent per layer, which yields a net transmittance for three layers of 73 percent. The third layer reduces solar gain by about 10 percent compared with two layers, but it also reduces reradiative heat loss from the tanks and the enclosure during the long, cold night. It is felt that this heat loss reduction is a greater benefit than the penalty of reduced solar transmittance, particularly in colder climates (above 5000 to 6000 heating degree-days). In climates with more than 7000 heating degree-days, some kind of movable insulation should be used over the glazing as well.

The glazing frame consists of a sandwich of the three film layers fastened to 1×2's. Use kiln-dried lumber for the glazing frame to minimize condensation between the glazing layers when the sun heats up the assembly. Drill ¼-inch weep holes into the frame members at the upper end and lower corners to further ensure against condensation buildup. The frame size conforms to the perimeter of the enclosure, with two crosspieces that align with the two "rafters" on the enclosure. As figure II-36 shows, one of the 1×2's in the frame cross section carries two glazing layers. All fastening points are on the sides of the 1×2's, not the upper or lower faces, and the fastening is done with staples. First the edges of the film are strengthened with fiber-reinforced packing tape or fabric duct tape. One edge is fastened at a time, and the opposite edge is fastened next. The film should be pulled snug to be as wrinkle-free as possible. The two frame layers are essentially independent of each other, but they are installed simultaneously. The perimeter strip of self-adhesive foam weather stripping is applied to the enclosure framing and between the two frame layers to ensure a good seal.

Then both frame layers are set in place. Aluminum battens (⅛ × ¾ inch) are laid over the side and the inside crosspieces of the frame, and the whole sandwich is drilled for ¼ × 4-inch lag bolts, placed every 8 inches.

The two sides of the collector are flashed with aluminum corner flashing, also called L-channel. If the side sheathing and the siding are flush with the edge of the side rails, the corner flashing will have to be at least 4 inches on the side face and 2 inches on the top face. You can get a cleaner look by bringing the side wall sheathing and siding about 2½ inches above the side rails. Then the flashing could be a standard 2×2. In either case it's fastened to the glazing frame and the side of the collector with small aluminum screws.

The top edge of the top flashing (8 to 12-inch-wide aluminum) is tucked under the closest lap of the wall siding. If the siding is solid plywood or vertical boards, you'll have to carefully seal the flashing to the wall. This flashing simply lies over the top of the glazing frame and is fastened with small aluminum screws. For good appearance the bottom edge of the collector glazing is also trimmed with aluminum. Before the glazing frame is fastened down, a strip of aluminum is fastened to the enclosure framing and bent down. After the glazing frames are secured, a piece of aluminum trim is fastened across the front of the collector and sealed to the glazing with silicone. Because of the odd shapes involved, some of the trim pieces will have to be custom bent, especially when aluminum siding is used. Once you've got it all together, the system will of course run itself.

Bob Flower
Mic Curd
Joe Carter

Hardware Focus

Overhang Water Heater

Conservation Concepts, Inc.
484 Middletown Rd.
Hummelstown, PA 17036
(717) 566-6360

Passive Solar Water Heater
A highly detailed set of plans for a triangular twin-tank water heater is available from Conservation Concepts for $10.00.

III SPACE HEATING

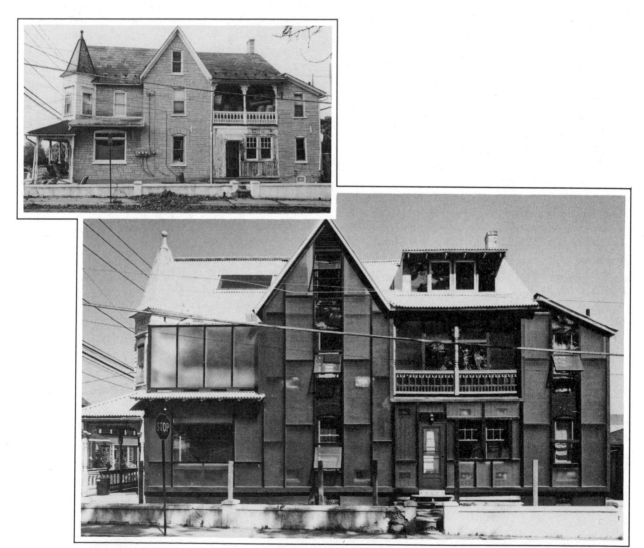

Photo III-1: Before this house was upgraded, it was a true energy disaster, but after extensive conservation and solar improvements its energy use was cut by over 80 percent. The solar features include a Trombe-type wall, an air heating collector, a solarized porch, a greenhouse (upper left) and a dormer.

One Goal, Many Paths

You want solar heating; it's not an elusive goal, but how do you get there? If you liken the possibilities for a space heating retrofit to a map, you will find that there are many routes to follow. The one you choose may follow a standard route, or it may cut some new lines onto the map. But one way or another, all routes lead to the sun. The variety of possibilities covered in the following space heating sections is proof that every home does indeed have solar potential. Building foundations, frame walls, masonry walls, interior walls, windows, doors, porches, roof pitches, flat roofs, building additions —there is room for the sun in each of these building elements. And the beauty of the variety is that there will often be more than one option for solarizing different parts of your house, and there will certainly be many options to consider if you're aiming to solarize your whole house.

From the simplest south glass installation to the full-fledged solar addition, space heating with the sun can bring quite a transformation in the workings of your house. So many of the contributors to the book have become more sensitive to their home's improved thermal processes, and also to nature's processes. Sunny, cloudy, hot, cold, breezy, windy, stormy, rainy days are all part of the life of the climate-responsive home, and as its manager you become involved in the interplay. "Adding solar" doesn't just add another chore to the list of home management duties. Rather, solar "responsibilities," if there are any, become more like pleasurable routines, not thankless tasks, because you're rewarded repeatedly by the benefits that come from solar gain. Energy benefits, cost-benefits, payback are the most commonly touted incentives for solar improvements, and they are undeniably important. But there is also a growing awareness of other benefits, benefits that can't be tagged with Btu's and dollar signs. There is a uniquely satisfying and good feeling that comes with making a stronger link with nature as we use solar design to gently tap the sun's gifts.

WINDOWS, SKYLIGHTS,
A. CLERESTORIES

Photo III-2: Direct gain feels good.

Getting the Sun Inside

Without exception, the primary element of any solar retrofit is glazing. Be it rigid glass or plastic or featherweight film, glazing is the thin skin that does so much simply by admitting sunlight and trapping its heat inside a box, a room or a building. If glazing weren't so common, it might seem miraculous for all the ways it can utilize solar gain. I ay down a piece in the desert and it will draw water from waterless ground. Put it over a pool of impure water, and it will distill it to a quality you could imbibe. It can be used to dry, preserve and cook food. It can concentrate the sun's rays enough to make fire. Glazing is the simplest solar collector, and the simplest way of using it for space heating is in *direct gain* systems.

Direct gain is the passage of solar radiation through glazing directly into the living space. When the beams of light strike walls, floors, furniture and people, their radiant energy becomes heat. Building insulation, multiple glazing layers and glazing insulation (used at night) help to keep the heat in. Little fans and interior vents can be used to distribute the heat gain to other parts of the house.

The ways of handling solar heat are only outnumbered by the ways glazing can be used to get it inside. For virtually every opportunity or limitation presented by a building or the site around it, there is a solar glazing application that can be retrofitted using common materials and techniques. Windows, skylights and clerestories—these are the basic components of direct gain systems. The variety gives freedom by allowing not only the south side of a building to be solarized, but also north, east and west sides as the need arises. These simple, passive systems will bring heating and daylighting, and the beauty of these solar improvements means that they are also valuable as home improvements. This point will be made again and again simply by what you will see in the following essays and projects. And hopefully another point will come through all the numbers and details: that getting the sun inside is good for people.

All about Glazing

There are several varieties to choose from and many ways to use them

In the simplest sense, "glazing" is any of a variety of materials that transmits radiant energy (heat and light). Conservation and solar-related applications have in recent years placed more emphasis on the proper use of glazing materials, but in different forms glazing has been in use for centuries. Some of the earliest glazing materials included oiled animal skins and various naturally transparent or translucent silicate minerals such as talc, soapstone, and mica. Paper has also been used as glazing in both European and Asian cultures. Glass, probably the most familiar and in many ways the best glazing material, has been used in buildings since the 17th century. In the past couple of decades, various plastics with characteristics similar to glass have expanded the list of useful glazing materials from one to many.

The concept of "solar" glazing, whose primary function is the control of heat from the sun, is by no means new; in ancient times Romans glazed large areas of south-facing walls to provide warmth in public baths. But until production technologies became more advanced, good quality glazing materials were generally scarce and costly. Durable mineral glazings were difficult to find, quarry, and split or cut into usable panes. Glassmaking required special knowledge and skills and high processing temperatures, which were difficult to maintain.

Until the 19th century, the use of glass was limited to expensive and special structures such as cathedrals, palaces and other exceptional buildings. New concentrations of wealth broadened the range of glazing applications during the 1800s to include large greenhouses (for exhibition and for growing exotic plants), factories (to provide daylighting) and mansions (to provide light and view). Some of these buildings, which were constructed prior to our era of cheap energy and automated heating and cooling equipment, are increasingly recognized among solar designers as a valuable source of ideas and techniques for the ways in which glazing was used. The classic example of this is Thomas Jefferson's plantation home, Monticello, near Charlottesville, Virginia.

Monticello is a remarkable example of energy-efficient design, especially because it was built in a time when there were no contemporary models nor much experience to draw upon. Jefferson was well-versed in architectural design and in building construction, but he also had a strong intuitive sense that he translated into some very effective uses of glazing and other materials. Some windows were double glazed for reduced heat loss, and many large windows were operable to allow cooling breezes to flow through the hilltop house. Deciduous trees shade many windows from direct gain in summer while allowing sunlight to pass in winter. Operable skylights were also used for daylighting and summer ventilation. In fact the building is probably most remarkable for its self-cooling capability, which made it a haven of comfort in the hot and humid Virginia summer.

Photo III-3: A couple of centuries before it had a name, Thomas Jefferson made energy efficiency an important part of his design of Monticello. Some of his ideas still represent the current thinking in energy-efficient design.

But despite Jefferson's good ideas, the cost and relative scarcity of glass meant that most homes had little or no glazing. Crude wall openings covered with oiled skins or parchment admitted some light while excluding cold, wind and rain, but provided no view to the outside. Instead of glazing, solid shutters were more commonplace. They could be moved daily or seasonally according to the weather and to the need for light or fresh air. They helped keep heat inside and were relatively cheap and easy to build and install. But unlike glazing materials, shutters couldn't provide daylighting *and* retain heat at the same time. So except in the mildest climates, compromises were inevitable: You could have light and be cold, or you could sacrifice light and be warm. The choice was clearly drawn, and decisions were usually made in favor of warmth, which meant that interiors were generally dark and close.

Cheap Energy, More Glass

Considering this history, the degree to which we now take glazing for granted is remarkable. Only in the present century have people become accustomed to large window areas. Nowadays most of our dwellings have windows in nearly every room. Daylight and view, once luxuries, have become the norm. Industrialization of the glass-making process resulted in lower costs, improved quality, and an increasing variety of types and sizes. When plastic glazings became available, they greatly broadened the range of choices. New structural systems and mounting details were devised to make use of glazing in ways that were totally new and exciting. It's clear that a glazing revolution took place, and the picture windows in our homes and the glass-skinned walls of office buildings are its legacy. But one of the most important influences on the explosion in the use of glazing was simply the availability of cheap energy.

For one thing, cheap energy helped keep manufacturing costs low. Glass is formed at very high temperatures using large quantities of high-grade heat. Petroleum is the raw material for plastics. Thus, all modern glazings are said to have a high "energy embodiment," meaning that a relatively large amount of energy is consumed as a necessary part of their production. In these days of rapidly increasing energy costs, that's bad, but prior to the energy crunch, the energy cost component in the overall cost of glazing materials wasn't nearly so great a factor as it is today.

Yet the influence of cheap energy on the use of glazing is more sharply defined with respect to the energy needed to heat and cool a building. The development of modern, expansive glazing systems has until recently been relatively insensitive to the effect they have on a building's heating and cooling loads. Now that the costs for meeting those loads have gone so high, it's clear that our conventional approaches to glazing raise the same problems of the old light versus warmth dilemma: Glazings gain heat quickly and effectively, compared with insulated walls, but they also lose heat quickly and effectively. That old dilemma only seemed to be resolved by progress in glazing technology, but in fact we were fooling ourselves with cheap energy. When energy became expensive, something had to change: The use of glazing needed to be brought into step with economic realities.

One of the first responses to rising fuel prices was to reduce the amount of glass in our buildings. But this response essentially reverts back to the standoff of the old choice: We can have light and view *or* we can stay warm at reasonable cost, but we can't have both. It doesn't have to be that way, though. Sacrificing view, light and air in order to achieve lower fuel bills is an unnecessary compromise. The old choice fails to account for energy benefits that can be achieved by proper use of glazing. It's obviously time to redefine the problem.

Solar Energy, More Glass

The glazing dilemma isn't resolved by an exclusive choice between heat gain and loss, but rather by achieving an appropriate and dynamic balance between the two, a balance that takes into account climate, seasonal change and the design of the building. In cold weather, for example, the design should of course favor heat gain and heat retention, while in warm weather the design must be flexible enough to resist heat gain and promote ventilation. Making glazing work has naturally become a major issue, and a creative opportunity, in solar building design, and much progress has been made. The advances in glazing design are readily applied to the problems of solar retrofitting because the goals are essentially the same as those for new construction. Since most people live where there is both summer heat and winter cold, designs for retrofitting must also be flexible enough to avoid letting heating and cooling goals conflict with each other. In the recent development of passive solar heating and natural cooling techniques, designers have quickly learned the danger of not working for that balance. Going too far in one direction, such as adding a lot of glass to increase solar heat gains, can lead to serious overheating problems during summer and during marginal space heating periods (early to mid fall, mid to late spring), as well as increasing building heat losses in winter.

On a purely technical level, glazing design does indeed present some rather complex problems, but in chasing down the solutions, researchers have distilled a great deal of basic, useful information that anyone can use. You needn't be frightened by the prospect of optimizing the gain-loss relationship, and you won't need the expertise of a solar consultant to do a good job. Armed with the right information, you can choose glazing materials and accessories as well as a professional designer, and you can come up with the right design for energy benefits and good appearance. One advantage of working with a designer is that he or she can give you suggestions based on experience.

A good designer will respond to what you want, and if you're not sure what you want, a designer can help you clarify your ideas.

If you're doing it all yourself, there is plenty of literature available from manufacturers and other sources to help in your planning. For starters, the ideas and data in this essay present some basic information and references you'll need to gain a good understanding of how to use glazing in any of the retrofit options discussed in this book. You'll probably find that you have a fair amount of latitude in glazing design, and there are usually several ways to do the same thing right. Glazing for the sun is by no means a risky venture; it has a history of centuries, and now that it's understood better than ever before, you can begin with the certainty that additional glazing will benefit both your home's appearance and its thermal performance.

Heat Transfer and Glazing

The main influences on heat gain and loss revolve around two basic physical properties of glazing materials: *transmittance* and *thermal conductivity*. There are also two lesser influences on heat loss that are more closely related to the glazing system design and installation: convection and infiltration.

Transmittance is a measure of the amount of solar energy that passes through a glazing material, compared with the amount that reaches its surface. This is usually expressed as a percentage. For example, a single pane of standard window glass has a transmittance of 84 percent when the sun's rays are perpendicular to the plane of the glass (0-degree angle of incidence). However, actual transmittance varies dramatically with the angle of incidence. This is one reason why the feasibility of solar retrofit is greatly diminished outside of 30 to 40 degrees east or west of true south. It's also why vertical glazing is more effective for winter space heating than is glazing in a flat- or shallow-pitched roof. As either the azimuth or altitude angle of incidence increases, the surface of the glazing reflects away a greater and greater percentage of the available sunlight, which naturally reduces heat gain compared to the gain for glazing that is properly oriented.

Transmittance is also affected by the thickness of the glazed material and by the number of layers used. Thicker glass (¼ inch), for instance, transmits a little less sunlight (about 3 percent less) than does regular single-strength glass (1/16 inch). Multiple layers both absorb and reflect away more sunlight than does a single layer. The transmittance of an assembly of multiple glazings is approximately the product of the transmittances of the individual layers. Thus, double-pane regular glass transmits about 71 percent of available light, compared with a transmittance of 84 percent for single-pane ($0.84 \times 0.84 = 0.706$). With three layers, however, transmittance falls to 60 percent ($0.84^3 = 0.59$). That much of a reduction in transmission can have a significant effect on heat gain. Note that the measurements given in table III-1 refer to single layer configurations receiving light at 0-degree angle of incidence; thus they don't give a completely accurate picture of glazing performance, but they do provide an index for comparison that is sufficient for most decision making in selecting solar glazings.

Sellers of glazing don't always know the transmittance of materials they carry, mainly because they haven't been asked often enough. They should, however, be able to get manufacturer's specifications for you. Manufacturer's literature usually gives at least one figure, the normal or average transmittance, which can be used for comparison purposes. If you need to know more, get in touch with the manufacturer's technical department.

Transmittance influences not only solar heat gain, but also building heat loss. Warm surfaces and objects (including people) can radiate heat to cold, exposed glazing surfaces, which in turn radiate heat to the sky. This factor is sometimes proffered as a reason to use glass rather than plastic glazings, because glass has a much lower radiative (infrared) transmittance than most plastics. The resistance of any glazing to conductive, convective and radiative heat loss accounts for what is commonly called the "greenhouse effect" of heat buildup inside a solar room. With the large surface area of solar "window walls," the radiative component of heat loss through the glazing can be a major factor along with conductive and convective heat loss. Because of this, reducing radiative loss has been the subject of much recent research. Special glazing products have been developed from this research and are beginning to appear on the market. One new product, a thin film called Heat Mirror, is much better at blocking infrared radiation than glass, although its transmittance (73 percent) is lower.

The thermal conductivity of glazing is not a major deciding factor in the selec-

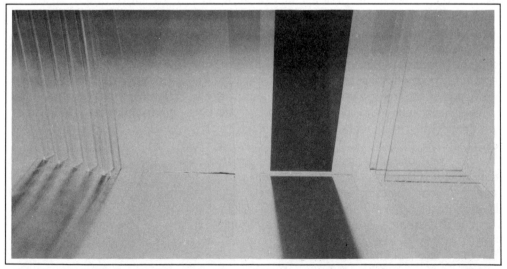

Photo III-4: With the unaided eye it's difficult to detect significant differences in transmission. When transmission through a window is reduced by 25 or even 50 percent, the apparent illumination doesn't change that much. But the energy transmission does vary greatly with the number of glazing layers, the angle of incidence and the color of the glass. The shadow differences on these samples show how these factors affect transmission.

tion of glazings. As table III-1 shows, the conductivities of glass and plastics differ only slightly, and the range of thicknesses used in glazing units is likewise small. However, thermal conductivity is a very important factor with regard to surface area and the number of glazing layers. Double the surface area, and you double the potential rate of heat loss.

But going from a single- to a double-layer glazing system is another matter altogether. Two glazing layers separated by a dead airspace have a conductive heat loss rate that is about one-half that of a single layer. Adding a third layer with two airspaces reduces conduction to about one-third of the single-layer rate. Thus the greatest benefit comes with the second layer (50 percent less conductive loss), while the third layer has a conductive loss that is 67 percent less than the single layer rate. It's also true that three layers of glass transmit about 16 percent less sunlight than two layers. Both of these factors tend to favor double glazing for south-facing applications, while triple glazing makes sense for north, east and west-facing walls.

The third source of glazing heat loss, convective heat transfer, occurs as a result of moving air "rubbing" against the glazing surface. The air picks up or releases heat by

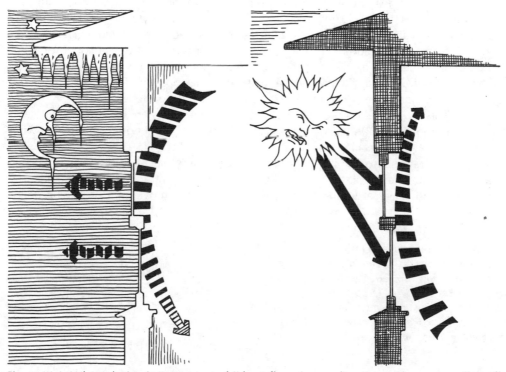

Figure III-1: When glazing is "unprotected," heat flows in any direction it chooses, cooling off warm spaces in winter and heating up cool spaces in summer. Radiative and convective heat transfer are the primary culprits, but it's not difficult to arrest them with any of a variety of controls: multiple glazing, shading, window insulation.

contact with the hotter or colder glazing. At night, for example, warm room air flowing past cold glass gives up heat, which the glass then conducts to the outside. In summer, a hot glass surface will warm cooler room air. Convective transfer is more dependent on external factors and isn't closely related to the glazing material itself. So while it is not an important factor in material selection, it is important in the overall design of a glazing system with regard to convection-reducing elements such as exterior sun and wind screening and interior movable insulation, which are discussed later in this section.

It's also possible for convective loops to develop *between* two glazing layers if they are too widely spaced. Normally the sealed gap between layers ranges up to 2 inches, and it is this dead airspace that increases the insulating value of multiple-layer glazing systems. The air gap is "dead" because it's too narrow to allow significant air movement. Air is a fluid, and it needs room to move around. A narrow gap presents too much friction for air movement, ergo, dead air. But with wider gaps air movement is possible, and it's driven by the temperature difference between the inner (warm) glazing and the outer (cold) glazing. There is, then, a limit to width of the glazing air gap as it relates to reducing conductive loss. Beyond that limit conduction is no longer decreased significantly, but convection is increased.

The final heat transfer mechanism is air infiltration. Air infiltration is entirely a system concern, since all glazing materials are of course impervious to air flow. Infiltration occurs with air leaks around the glazing—through cracks between the glazing edges and their frames or supports. It is a very important concern for it can account for 30 to 50 percent of the total heat loss in small buildings, and much of that loss is associated with windows. Minimizing infiltration is a matter of using appropriate frame materials and configurations and appropriate caulking, gaskets, seals and fasteners.

The glazing details presented in forthcoming projects are all designed to eliminate infiltration losses. On paper, that's great, but proper execution in the field is what really makes a good detail good.

Infiltration also involves *thermal expansion,* another basic property of glazing materials. Both glass and plastic expand when heated and contract when cooled. Diurnal (day-night) and seasonal temperature swings result in the movement of glazing panes

Figure III-2: A convective loop can occur between two glazing layers that are too widely spaced, wherein the radiative loss from the inner glazing is transferred to the outer glazing, which then radiates heat to the outside. When designing your multiple glazing, remember that the insulating value of the dead airspace doesn't significantly increase beyond a gap of 1 or 2 inches.

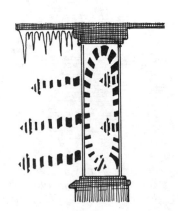

relative to their supports, and this subjects the seals (caulking, sealants or gaskets) to stresses that can cause cracks for air and water leakage. Glass expands and contracts less than plastics, which gives the former some advantage in maintaining seals, although sealing systems for plastics are quite effective. Thermal expansion is also related to the sizes of panes. The expansion property is expressed as unit expansion (change in length) per unit of original length, per degree of temperature change (see table III-1). Thus large panes will have more dimensional change than a small pane experiencing the same change in temperature. From this viewpoint, smaller panes, particularly when plastics are used, may be desirable for preventing wrinkling or outright buckling.

The difference between fixed and operable glazing can also have significant impact on air infiltration. In general, windows that you can open will sooner or later be leakier than properly installed fixed glazing, but I'm not a strong advocate of fixed glass. All too often fixed glazing systems mean stuffy or overheated rooms, and in many climates simple natural ventilation could greatly reduce energy consumed by mechanical cooling systems. Common sense favors systems consisting primarily of fixed glazing with some operable windows and/or opaque vents that are carefully selected to give the tightest possible closure with high quality weather stripping.

If these factors seem to add up only to more complication, fear not. It's important to know some basic properties of glazing and glazing systems as an aid to tackling special problems you may be facing. Enough experience has been accrued with solar glazing so that all the vagaries of handling various materials have been boiled down to some simple guidelines and basic techniques. Among the common solar retrofit options, you are unlikely to encounter questions you can't answer or problems that can't be solved for your own applications. Keep in mind that the purpose of proper design is to be in control of the heat gain and heat loss relationship. Many design decisions involve plain common sense, and from the above discussion, you already have some of the most basic decision-making criteria.

Solving Gain/Loss Dilemma

Whether you are trying to correct mistakes that were designed into your house in the days of cheap energy, planning a solar addition or dreaming of a new house, glazing will of course be foremost in your plans. As mentioned earlier, the key to solar glazing design for cold weather is to maximize heat gain while minimizing loss. In warm· weather the goal is just the opposite: Reduce unwanted gain and encourage heat loss. When the temperature shifts drastically between day and night, these conflicting goals must be pursued daily. The trick is to design your glazing to respond to the whole range of possibilities for your location, which is not a difficult thing to do.

Solutions to the gain/loss dilemma fall into two basic categories: static or dynamic. Static solutions (such as fixed multiple glazing or fixed shading devices) remain in place from season to season, while dynamic solutions (such as storm windows or movable insulation) can respond to changes in temperature and sun angle. As a general rule, static solutions are less able to respond to changes in temperature or sun angle than dynamic ones. Fixed multiple glazings, for example, are at their best in climates with small average temperature variations. Movable insulation, on the other hand, exemplifies a dynamic solution that can be useful in cold weather (as insulation) and in warm weather (as shading). Each category has its advantages and limitations, but they're not mutually exclusive. The two can often be combined effectively into a whole system that is able to respond to changing conditions without requiring excessive complexity or expense.

Static Solutions

As noted, fixed multiple glazing is an excellent means of controlling heat transfer. It can be effective in limiting either loss or gain, though generally not both at once. When it *is* important to achieve solar gain, however, anything beyond double glazing may not be the best choice. Heat losses are usually lower during sunlight hours because the temperature difference between inside and outside is not so great as at night. As a result, the addition of a third layer to south-facing glazing can reduce transmission (gain) by more than it reduces heat loss. In that case, movable insulation would be preferable to an additional glazing layer.

Special coatings offer several kinds of static solutions to the gain/loss dilemma. The flashiest example is *reflective glass* which reduces daytime heat gain. It is produced by vacuum deposition of metal particles on a glass or plastic surface, creating a partial mirror effect that reduces transmittance but preserves visual clarity. Reflective plastic films can also be applied to the inside surface of existing glazing when it's impractical to add new glass. Reflective glazing is useful if the primary goal is to reduce solar gain, but it's generally inappropriate for south glazing in cold climates where solar gain is also desirable, since transmittance is low. These films can be useful on east- and west-facing glazing, which is more subject to summer heat gain. Where requirements switch seasonally or daily between heating and cooling, dynamic solutions are preferable.

Another type of coating produced by vacuum deposition reduces transmittance of infrared radiation (heat). Its purpose is to lower radiative heat loss from building interiors. This type of static control is most appropriate in northern climates where heating is the primary need, or in cooling situations where infrared radiation from the ground or adjacent buildings is a major part of total air conditioning load. Although such coatings have been successfully applied to both glass and plastics, the only commercial product currently available (Heat Mirror) is bonded to a plastic film base, which limits its use to situations where the film won't overheat, such as a third layer between two conventional glazing layers. (Heat Mirror is not currently available, as of late 1981, to do-it-yourselfers; it is sold only to manufacturers of insulating glass.)

Special coatings share one major drawback. They are produced by a carefully controlled high-technology process and thus tend to be fairly expensive. There is reason to hope that manufacturing improvements will reduce the costs of some of these coatings, which, combined with the effects of accelerating energy costs, could soon make such materials quite attractive. Meanwhile, the benefits should be weighed carefully against the cost.

Tinted glazings are useful in controlling glare, and their absorptive properties can reduce cooling loads. As a means of limiting heat gain, however, absorptive glazings are less effective than reflective glazings because their heat absorption causes an increase in temperature of the glazing itself. Much of that heat winds up inside anyway by convection and radiation from the hot glass surface. The only reason for selecting heat-absorbing glass in this situation is that it is less costly than reflective. Small glass area, fixed overhangs or solar screens should be considered before tinted glazing as a means of reducing gain. After all, if cooling costs are increased, the lower first cost of tinted glazing isn't really an advantage. (See "Air Conditioning without Air Conditioners" in Section V for information on controlling heat gain through windows.)

Still, there are many older buildings with heat-absorbing glass that present an interesting retrofit possibility in climates with a substantial heating load. Adding high-transmittance glazing layers outside of south-facing absorptive glass can raise the winter daytime temperature of the tinted glass while reducing convective loss at night. In other than this special situation, which creates a kind of low-performance solar collector, heat-absorbing glazings are not useful in solar glazing systems. In fact, you should be careful in ordering "solar" glazing materials, because the term solar has been used by manufacturers to describe products whose main purpose is excluding, not collecting, solar energy.

Fixed shading can be beneficial in resolving the gain/loss dilemma in almost any climate. Its simplest manifestation is an overhang above a window which must be sized according to glazing orientation (azimuth) and tilt, latitude, seasonal shading of the site (by trees, etc.) and the dimensions of the glazed opening. Complicated situations may call for both horizontal and vertical projections (wing walls); the horizontal ones shade with respect to solar altitude and the vertical ones respond to solar azimuth. West-facing glazing can be particularly difficult because warmer afternoon temperatures combined with low-sun angles can cause overheating, and fixed overhangs can't effectively block the low sun. That situation requires some kind of movable shading, a dynamic solution. Still, overhangs and wing walls can be useful in temperature latitudes where both heating and cooling are essential: High-transmittance glass can be used to maximize winter gain while shading prevents excessive gain from the high-angle summer sun.

Shading devices can influence glazing selection. In fact, the cost of a shading device with low-cost regular glass can be less than high-cost reflective glass without the shade, with similar or better thermal performance. But in all applications the cost-benefit trade-offs among several alternatives should be carefully examined before you invest.

In climates dominated by a demand for cooling, fixed metal or woven fiberglass screens are often employed to reduce solar gain. They perform well in all sun angles and cost less than the more elaborate projecting shades discussed above.

Fixed reflectors are static devices intended to increase the solar radiation on a glazing surface. In some cases a shading device can also be a reflector. For example, reflective material can be installed on the underside of a clerestory overhang where it can bounce in low-angle winter sunlight.

Reflectors in passive solar designs are used to improve winter heating performance; the potential for unwanted summer gain should be avoided by careful design. Systems for solar hot water or other devices requiring year-round solar gain can always

Photo III-5: The most common form of fixed shading is the roof overhang. The overhang on the upper clerestory doesn't shade the windows from the low-angle winter sun because the length of the overhang was calculated correctly. (See "Air Conditioning without Air Conditioners" in Section V.) On the lower run of glass the overhang is abbreviated, but the ceiling joists were extended to make an arbor for a grapevine, whose broad leaves will become "seasonal" shading.

Photo III-6: When solar glazing is installed over a roof section, total transmission can be increased by 10 to 20 percent with a fixed reflective roof surface.

benefit from static reflectors. The possibility of achieving higher temperatures in reflector-augmented active collectors, however, could adversely affect some of the less heat-stable plastic glazings, and maximum temperature should be matched with the capabilities of the glazing material.

Dynamic Solutions

Movable multiple glazing is most common in the form of storm windows, which are an excellent means of limiting heat loss through glazing in winter while permitting summer ventilation. Storms used as a second glazing over single-pane windows can dramatically reduce conductive and convective heat loss without blocking a high percentage of radiative solar gain. As a third glazing layer, however, transmittance loss may limit overall performance if the primary goal is solar gain. Conventional glass storms are usually glazed with lower transmittance, high-iron-content glass, which has a transmittance for three layers of about 60 percent. A storm that is properly weather-stripped can also provide an excellent barrier to infiltration heat loss, perhaps better than double- or triple-glazed replacement windows.

The most convenient type of storm window is the common triple track. They are usually aluminum framed, and operate almost like double-hung windows. The third track holds a framed insect screen, which can be interchanged with the lower glass panel to permit either ventilation or insulation. (A simpler version of this arrangement is the double-track storm, which is a bit less convenient in terms of making the glass-screen switch.) Most triple tracks don't ventilate quite as effectively as a double-hung window because the top glass pane is usually fixed, although there may be some top-screen models. Bottom-screen triple tracks compromise the advantage of double-hung windows, which can be opened both top and bottom to vent hotter air near the ceiling. Another disadvantage where solar gain is desirable is that the insect screen is always in place, blocking 30 percent or more of the gain over half the total glazed area. Triple tracks are easily controlled from the inside, and this is their chief advantage. They can be easily retrofitted because they mount to the outside face of existing window frames, a simple job in most cases. When you're looking to buy, choose units that have tight seals for the fixed and movable window panes. When installing triple tracks, run a bead of caulk along the inside face of the nailing flange to ensure a good seal to the window frame.

Casement windows allow better ventilation because the whole window opens, not half, but they can't be used with triple tracks, because there's no way to open the casement sash with a storm in place. The solution is to have seasonally deployed storm sashes on the outside, or the inside. Seasonal storms do require fall and spring mounting and demounting. Storm panels mounted separately to each casement sash, instead of to the window frame, are another possibility. Panels don't require demounting except to clean the inside surfaces every few years. Sash glazing won't retain heat quite as well as panels over the entire frame because in the latter frame coverage is more complete and can help to reduce infiltration.

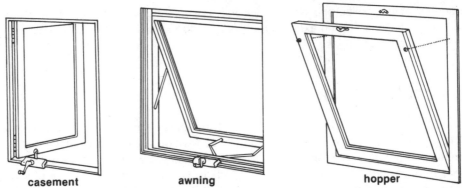

casement   **awning**   **hopper**

Figure III-3: For the sake of a little terminology, there are three basic types of hinged windows, as shown above. The nice thing about all hinged windows is that when they're properly weather-stripped, they seal very tightly when closed. Casements can usually be swung open very wide for maximum ventilation and/or catching of cooling breezes, when they're hinged to open into prevailing summer winds. Awning types provide ventilation and weather protection for walls that catch a lot of rain but have no roof overhang to keep it away. Hopper windows will provide maximum convective ventilation, but they must of course have some overhang above so they don't divert rainwater indoors. There's a simplicity to hinged windows that isn't shared by standard double-hung units, which makes them prime candidates for home building.

Some inexpensive interior storm glazings, such as acrylic plastic panels, are very attractive for their low cost and ease of installation. These panels are about 1/16 inch thick and can be easily cut to size. Special plastic mounting frames usually attach by pressure-sensitive adhesive to the inside face of the window frame, and the acrylic panels snap in or out for easy cleaning or summer storage.

Cheapest of all is a storm window made by attaching plastic film to the inside of a window frame. Groove-and-spline plastic extrusions are often sold with a roll of plastic to hold it in place. Pressure-sensitive tape works equally well, but can strip paint when it's removed. The plastic film you use can be a product sold specifically for the purpose or regular polyethylene, which isn't totally clear, as are other types. Both plastic-film and semirigid storms can be taken down and stored for another season, but they tend to suffer from wear and tear, and many have to be discarded and replaced every one to three years, depending on the care with which they're handled.

Both acrylic panels and plastic films are suitable for do-it-yourself projects. In stopping conductive, convective and infiltration heat loss, they perform about as well as the most expensive storm glazing you could install, and as a second glazing they don't greatly reduce solar gain. Their temperature limitations, however, may prohibit their use in active solar collectors. Acrylic and polyethylene films are also flammable and should never be used near kitchen stoves, woodstoves or other high-temperature areas.

Window insulation (movable insulation) is an excellent dynamic means of reducing heat loss. In fact, to get the most benefit from a passive solar installation, some kind of movable insulation is essential for preventing excessive heat loss at night. Triple

Photo III-7: Window insulation cures the chills that large areas of glazing can create. Roll-up (or down) shades, hinged shutters and mobile panels offer a variety of insulation options to meet most any situation.

glazing (R-2.7), for example, is not as effective as single or double glazing with 1 inch of movable Styrofoam (R-6.3 to R-7.2). Some of the options for movable insulation are described in Section III-B.

 Movable shading and reflecting devices perform functions similar to static shades and reflectors, but with the added flexibility of being operable. A rigid movable shade or reflector can almost always perform better thermally than a static one, because it can respond to changing solar angles. In some instances a single device such as a rigid foam panel can serve shading, reflecting and insulating purposes. Combining functions can be a way to greatly reduce costs.

 There are also several brands of solar control (transmission-reducing) films that are used with standard shade rollers. Other brands are temporarily or seasonally bonded to the inside surface of the glazing. These films can be especially useful for "shading" south-facing glazing in summer. When the heating season comes around again, the film is simply peeled off and stored.

 Automatic or remote control devices are both desirable but expensive in glazing systems. There are remote controls for glazings in special locations such as high clerestory windows, but there isn't very much in the way of off-the-shelf remote controls for external shades or reflectors or movable insulation. (Some home-built, pole-operated

Photo III-8: Movable shade/reflector devices control glazing heat loss and gain from the outside. This hinged panel boosts gain through the glazing in winter, and when fully or partially raised in summer, it resists unwanted gain. It's also a window insulation device because the core of each panel is filled with rigid foam.

and rope-and-pulley controls are shown in this and later sections.) Motor-operated drapery or shade systems have been used inside large buildings, but on the whole, motorized systems are still too costly to be feasible for residences.

There is one type of automatic device that is being used as a passive venting device in greenhouses. These vents are controlled with paraffin-cylinder "heat motors" that expand and contract with changes in temperature. This movement causes a vent to open when the interior air is too hot, and then to close when the inside cools down. They are effective in greenhouses, but neither the technology of heat motors nor passive home design are as yet fine-tuned enough to make them useful in controlling residential glazing systems.

But the words "remote" and "automatic" are entering the realm of solar glazing, largely because of the remarkable inventiveness of Steve Baer, who runs the

Zomeworks company in New Mexico. One of his products, Beadwall, was invented by a Zomeworks employee, David Harrison, and it's a truly unique idea. Beadwall is an automatic window insulation system in which a layer of Styrofoam beads is blown between double glazing that is separated by a 3- to 5-inch air gap. When the sun comes up, the beads are vacuumed back into a storage container. The system is quite effective especially for large expanses of south glass. Another Baer invention, Skylid, is an automatic skylight and clerestory movable insulation and shading system. Two connected canisters of Freon control the opening and closing of Skylid, depending on the availability of sunshine. One canister is exposed to sunlight, and when it heats up, the Freon is forced into the second canister, which gets heavier and causes the closely balanced Skylid array to open. When the inside canister gets warmer than the outside canister (sundown, room overheating), the Freon flows back onto the latter canister, shifting the balance and causing Skylid to close. (See the box "Skylights for Space Heating" in "Skylights," found later in this section.) It's likely that other automatic and remote control devices will emerge when there is sufficient demand to reduce costs and when available or new technologies can be sufficiently fine-tuned to respond to temperature and sunlight changes that affect people's sense of comfort.

Materials: A Solar Glazing Catalog

There are several generic types of commercially available solar glazing materials, each having advantages and limitations. Included here is a review of a few salient characteristics to help with the selection of glazings for particular uses. These glazings are also represented in table III-1, which gives more of the "vital statistics" of glazing materials.

Glass

Glass is a highly effective solar glazing. It is durable, suffers no loss of transmittance from aging or weathering, and in most situations it is easily cleaned. Under normal conditions (except sandstorms) its surface is abrasion-proof. Thermal expansion is modest, though it must be allowed for. Glass transmits radiant energy selectively, passing light but blocking radiant (long wave infrared) heat, a characteristic responsible for the "greenhouse effect" that is important in solar heating. Some types of glass can be more expensive than some plastic glazings, but glass has a much longer useful life than plastics. On the minus side, glass breaks, and be it tempered or regular annealed glass, a broken glass pane is a total loss. Plastics have much more impact resistance.

Sheet glass is the least expensive type, used for windows in most small buildings. You fix windows with sheet glass from the hardware store. Scavenged materials (used windows or dismantled greenhouses) are an excellent source for low-cost sheet glass. It is made in single-strength (about 1/16 inch thick), double-strength (⅛ inch thick),

(continued on page 289)

TABLE III-1 Solar Glazings

Material Type and Form	Trade Name (manufacturer*)	Comments	Installation Suggestions	Suitable Applications†			
				Flat Plate Collectors & Mass Walls		Direct-Gain Systems: Windows, Skylights	
				outer glazing	inner glazing	outer glazing	inner glazing
GLASS rigid, flat sheet tempered and annealed (window glass) tempered and annealed low-iron (high transmittance)	many Clearlite (AFG) Sunadex (AFG) Heliolite (C-E Glass) SolaKleer (General Glass) plus many others	(Some brands of low-iron glass have textured, light-diffusing transmission.) -excellent transparency and appearance -excellent resistance to UV, weather and high heat -low thermal expansion/contraction -readily available -noncombustible, chemically inert -low impact resistance; potential for breakage a safety hazard in some areas -heavy, requires strong supports -hard to handle on site	Bottom edges must be supported at quarter-points on neoprene "setting blocks." Butyl tape ("glazing tape") is a common gasket and sealant for glass.	yes	yes	yes	yes
ACRYLIC rigid, flat sheet double-wall extrusion	Plexiglas G (Rohm and Haas) Lucite L (Du Pont Co.) Acrylite GP (CY/RO) Acry-Pane (Sheffield) Exolite Acrylic (CY/RO)	(Double-wall extrusion is semitransparent.) -good transparency and appearance -good UV and weather resistance -lightweight -readily available in many sizes and thicknesses -easy to site fabricate -high thermal expansion/contraction -susceptible to abrasion (surface scratches) -softens under moderate heat	Use mechanical compression fastening around all four edges. Mounting must be resilient to allow thermal expansion/contraction. Thru-bolting or nailing not recommended. For double-wall extrusion: Special mounting hardware is available from manufacturers. Ends of open channels should be sealed.	yes	doubtful	yes	yes

*See list of manufacturers at the end of table.

† In making these recommendations for "Suitable Applications," we have weighed the factors (namely cost, performance, and long-term durability) that would be important in a typical home solar retrofitting project.
"Yes" means that the application has a very good chance of success.
"Maybe" means that the application is possible, but will require special care in design or construction to have greater than a 50/50 chance of success.
"Doubtful" means that the application might be possible, but only under very unusual combinations of cost, design, or construction requirements.
"No" means that the application is almost certainly not feasible.

Suitable Applications†		Available Thickness and Sizes (weight, lbs/ft²)	Typical Retail Price Range ($/ft²)‡	Estimated Service Life (years)	Solar Transmittance, % (Infrared Transmittance,%) §	Maximum Continuous Service Temperature (°F)	Thermal Expansion Coefficient (10⁻⁵ inch/inch/°F)	Impact Strength (tensile strength, psi)
Agricultural Greenhouses	Curved Glazings (cold bending)							
yes	no	⅛" thick (1.6) ³⁄₁₆" thick (2.5) tempered glass commonly available in: 34" × 76" 34" × 96" 46" × 76" 46" × 96" and other sizes	$0.60 to $1.60 $0.90 to $1.80	50+ barring breakage	regular glass: 82–84 (less than 2) low-iron glass: 90–91 (less than 2)	400–600°	0.47–0.51	very low (10,000–20,000)
yes	yes	flat sheets: ⅛" thick (0.70) ¼" thick (1.56) double-wall extrusion: ⅝" thick (1.0)	$1.50 to $2.00 $2.00 to $2.65 $2.50 to $3.50	25+	flat sheets: 89 (less than 6) double-wall extrusion: 83 (less than 6)	160–200°	3.4–4.0	medium (10,000)

‡ Prices shown here are for a typical 50- to 300-square-foot quantity and do not include mounting hardware, sealing gaskets, etc., which combined with shipping and handling charges can add up to 30 percent onto the glazing's stated price per square foot. Prices start to decrease drastically at quantities larger than about 500 square feet. Prices also depend greatly on the manufacturer's marketing scheme and for a given product and quantity can vary by a factor of 2 depending on the company from whom you buy.

§ "Solar Transmittance" refers to the fraction of thermal radiant energy within the solar spectrum (typically 0.4 to 2.5 microns) which is transmitted by the glazing. "IR Transmittance" pertains to the far infrared band (typically 5 to 50 microns), which represents the thermal radiation from a room-temperature (70°F) body.

TABLE III-1—Continued

Material Type and Form	Trade Name (manufacturer*)	Comments	Installation Suggestions	Suitable Applications†			
				Flat Plate Collectors & Mass Walls		Direct-Gain Systems: Windows, Skylights	
				outer glazing	inner glazing	outer glazing	inner glazing
POLYCARBONATE rigid, flat sheet double-wall extrusion (thick wall) double-wall extrusion (thin wall)	Tuffak A (Rohm and Haas) Lexan 9300 (G.E.) Poly-Glaz (Sheffield) Glazing A2 (Park Energy) Exolite Polycarbonate (CY/RO) Tuffak-Twinwal (Rohm and Haas) Qualex (Structured Sheets)	(Double-wall extrusion is semi-transparent.) -good transparency and appearance -high impact resistance -lightweight -questionable resistance to UV and abrasion -high thermal expansion/contraction	Use mechanical compression fastening around all four edges. Mounting must be resilient to allow thermal expansion/contraction. Thru-bolting or nailing not recommended. For double-wall extrusion: Special mounting hardware is available from manufacturers. Ends of open channels should be sealed.	maybe	doubtful	yes	yes
FIBERGLASS REINFORCED POLYESTER (FRP) flexible, flat thin sheet corrugated and shiplap configurations also available	Sunlite Premium II (Kalwall Corp.) Solar Plate/15 Types 546 & 556 (Filon Div.) Crystalite-T (Lasco Ind.) Glasteel (Glasteel, Inc.)	-translucent, diffused light transmission -very lightweight -good impact resistance -easy installation and mounting -readily available -questionable resistance to UV, surface erosion and high heat -requires occasional surface recoating -high thermal expansion/contraction -hard to eliminate wavy appearance of flat sheets -low solar transmittance at oblique incidence angles	Overlap at least ¾″ at edges of sheet onto support frame. Sheet should be predrilled for fasteners; holes should be at least ⅛″ oversize. Seal with silicone. Special mounting hardware for flat and corrugated fiberglass is available.	maybe	maybe	yes	yes

Suitable Applications†		Available Thickness and Sizes (weight, lbs/ft²)	Typical Retail Price Range ($/ft²)‡	Estimated Service Life (years)	Solar Transmittance, % (Infrared Transmittance,%) §	Maximum Continuous Service Temperature (°F)	Thermal Expansion Coefficient (10⁻⁵ inch/inch/°F)	Impact Strength (tensile strength, psi)
Agricultural Greenhouses	Curved Glazings (cold bending)							
yes	yes	flat sheets: ⅛″ thick (0.78)	$2 to $4	10 to 15	flat sheets: 86 (less than 6)	200–260°	3.3–4.0	high (9,500)
		¼″ thick (1.56)	$4 to $6					
		double-wall extrusion: ¼″–⅝″ thick (0.25)	$1.20 to $2.20		double-wall extrusion: 74–77 (less than 6)			
yes	yes	0.025″ thick available widths: 24″, 36″, 48″, 49½″, 60″ available lengths: 8′, 10′, 25′, 50′ (0.25)	$0.78 to $1.03	25+ 8 to 12	0.025″ thickness: 87 (10–12; 5–50 micron band)	200°	1.36	medium (10,000)
		0.040″ thick other dimensions as above (0.31)	$0.96 to $1.22		0.040″ thickness 85 (5–6; 5–50 micron band)			
		0.060″ thick other dimensions as above (0.50)	$1.29 to $1.55		0.060 thickness: 72 (less than 2)			
		other thicknesses: 4 oz/ft²=0.030″ 5 oz/ft²=0.037″						

TABLE III-1—Continued

Material Type and Form	Trade Name (manufacturer*)	Comments	Installation Suggestions	Suitable Applications†			
				Flat Plate Collectors & Mass Walls		Direct-Gain Systems: Windows, Skylights	
				outer glazing	inner glazing	outer glazing	inner glazing
FLUORINATED ETHYLENE PROPYLENE (FEP) thin film	Teflon Type 100 A (Du Pont Co.) Type A: general purpose Type C: cementable with adhesives Type L: greater flexibility for environmental extremes (to 0.090″)	-totally transparent -very high solar transmittance -superior resistance to UV and high heat -low cost -very lightweight -chemically inert, noncombustible -suitable only for inner glazings -high thermal expansion/contraction (can sag at high temperatures) -not readily available in small quantities -hard to eliminate wrinkles in installation -high infrared transmittance -easily torn or punctured	To prevent sag, should be stretched app. 1% during mounting, and support wires used under long spans. Use adhesives and mechanical clamping along edges. Nails and staples not recommended.	no	yes	no	maybe
POLYVINYL FLUORIDE (PVF) thin film	Tedlar Type 400XRB160SE (Du Pont Co.)	-almost transparent -very lightweight -high solar transmittance -low cost -high tensile strength -questionable resistance to UV and weathering -hard to eliminate wrinkles in installation (should be shrink mounted) -high infrared transmittance -embrittlement at prolonged high temperatures -not recommended for inner glazings -unknown long-term durability -can be torn or punctured under impact	Gentle heat shrinking prior to mounting will provide taut surface. Heat sealing or adhesive bonding is preferred to nailing or stapling.	yes	no	maybe	maybe

Suitable Applications†		Available Thickness and Sizes (weight, lbs/ft²)	Typical Retail Price Range ($/ft²)‡	Estimated Service Life (years)	Solar Transmittance, % (Infrared Transmittance, %) §	Maximum Continuous Service Temperature (°F)	Thermal Expansion Coefficient (10⁻⁵ inch/inch/°F)	Impact Strength (tensile strength, psi)
Agricultural Greenhouses	Curved Glazings (cold bending)							
doubtful	no	Type 100A (1 mil thick) 1 mil thick common. Thickness to 20 mil also available. 50″ width common. Other widths also available. 50′ length common. Up to 300′ available. (0.011)	$0.39 to $0.69	up to 20	96 (58; 3–50 micron band)	400°	9.0	very low (3,000)
maybe	no	All 4 mil thick. available widths: 25″ and 50″ available lengths: 25′, 50′ and 100′. (0.029)	$0.60 to $0.77	5 to 10	90 (greater than 50)	225°	2.8	low (12,000)

TABLE III-1—Continued

Material Type and Form	Trade Name (manufacturer*)	Comments	Installation Suggestions	Suitable Applications†			
				Flat Plate Collectors & Mass Walls		Direct-Gain Systems: Windows, Skylights	
				outer glazing	inner glazing	outer glazing	inner glazing
LAMINATED ACRYLIC/ POLYESTER thin film	Flexigard #7410 (3M Co.)	-totally transparent -combines weatherability of acrylics with strength of polyester -very lightweight -relatively low infrared transmittance -good solar transmittance -does not sag at high temperatures -unknown long-term durability -hard to eliminate wrinkles in installation -nonreversible -susceptible to wind flapping -can be torn or punctured under impact	Outside of roll is side of film to be exposed.	maybe	maybe	yes	yes
WEATHERABLE POLYESTER thin film	Llumar (Martin Processing, Inc.)	-totally transparent -very lightweight -high strength for a thin film -transmits almost no UV -does not sag -UV stabilizers are integral with film, not damaged by scratch or abrasion -unknown long-term durability -hard to eliminate wrinkles in installation -susceptible to wind flapping -can be torn or punctured under impact	Helpful to wrap around frame members before fastening.	maybe	maybe	yes	yes

Suitable Applications†		Available Thickness and Sizes (weight, lbs/ft²)	Typical Retail Price Range ($/ft²)‡	Estimated Service Life (years)	Solar Transmittance, % (Infrared Transmittance,%) §	Maximum Continuous Service Temperature (°F)	Thermal Expansion Coefficient (10⁻⁵ inch/inch/°F)	Impact Strength (tensile strength, psi)
Agricultural Greenhouses	Curved Glazings (cold bending)							
maybe	no	0.007″ thick 4′ × 20′ rolls and 4′ × 150′ rolls (0.052)	$0.42 to $0.65	7 to 10	89 (9.5; 6–50 micron band)	275°	2.7–5.0	low (14,000)
yes	no	0.005″ thick available widths: from 26″ to 60″ available lengths: from 50′ to 300′ (0.03) 0.007″ thick in 48″ width and above lengths	$0.46 to $0.69 $0.64 to $0.94	7 to 10	88	300°	(not available)	low (25,000)

TABLE III-1—Continued

Material Type and Form	Trade Name (manufacturer*)	Comments	Installation Suggestions	Suitable Applications† Flat Plate Collectors & Mass Walls — outer glazing	inner glazing	Direct-Gain Systems: Windows, Skylights — outer glazing	inner glazing
ANTIREFLECTIVE COATED POLYESTER thin film	3M high transmission film (3M Co.)	-totally transparent -antistatic; will not cling or attract dust -very lightweight -unknown long-term durability -suitable for inner glazing only -relatively high cost -special surface coating easily damaged by scratch or abrasion -can be torn or punctured under impact	Surface should not be rubbed or abraded. Handle film by edges only and wear clean cloth gloves. Adhesive mounting is recommended (e.g., Scotch brand 838 weather-resistant tape).	no	maybe	no	yes
POLYETHYLENE thin film	Visqueen (Ethyl Corp.) Monsanto 602, UV resistant (Monsanto Co.) many others	-almost transparent -very inexpensive -easy to install -readily available in many sizes -very lightweight -permeable to CO_2 for greenhouse application -very short service life—poor weather and UV resistance -melts under moderate heat -possible fire hazard -easily torn or punctured -very high infrared transmittance	Helpful to wrap around frame members before fastening. Special mounting hardware available from greenhouse suppliers.	no	no	doubtful	doubtful

GLAZING MATERIALS MANUFACTURERS

AFG Industries, Inc.   glass
1400 Lincoln St.
Kingsport, TN 37662
(615) 245-0211 or toll-free:
  (800) 251-0441

C-E Glass   glass
Division of Combustion
  Engineering, Inc.
825 Hylton Rd.
Pennsauken, NJ 08110
(609) 662-0400

CY/RO Industries   acrylic and
P.O. Box 1779   polycar-
697 Rt. 46   bonate sheet
Clifton, NJ 07015
(201) 560-0485

Du Pont Co.   Lucite acrylic
Plastic Products and Resins Dept.
Lucite Sheet Products/Architectural
  Group
Wilmington, DE 19898
(302) 774-2629

Suitable Applications†		Available Thickness and Sizes (weight, lbs/ft²)	Typical Retail Price Range ($/ft²)‡	Estimated Service Life (years)	Solar Transmittance, % (Infrared Transmittance,%) §	Maximum Continuous Service Temperature (°F)	Thermal Expansion Coefficient (10⁻⁵ inch/inch/°F)	Impact Strength (tensile strength, psi)
Agricultural Greenhouses	Curved Glazings (cold bending)							
doubtful	no	0.004" thick 51" × 50' rolls and 51" × 300' rolls (0.029)	$0.68 to $1.00	10+	93 (not available)	300°	3.06	low (25,000)
yes	no	0.004" thick available widths: from 10' to 42'; various lengths up to 100' (app. 0.023) 0.006" thick available widths: from 12' to 42'; various lengths up to 150' (app. 0.032) Thickness up to 0.020" available	$0.03 $0.04 to $0.05	untreated, app. 8 months UV resistant, 1 to 3 years	90 (app. 80)	120° less than 160°	30.0	very low (2,000)

GLAZING MATERIALS MANUFACTURERS—CONTINUED

Du Pont Co.
Polymer Products Dept.
Talley Bldg., Concord Plaza
Wilmington, DE 19898
(302) 772-5880

Tedlar PVF film

Du Pont Co.
Polymer Products Dept.
Technical Service Laboratory, Chestnut Run
Wilmington, DE 19898
(302) 999-3456

Teflon film

Ethyl Corp.
330 S. 4th
Richmond, VA 23219
(804) 644-6081

Filon Division
Vistron Corp.
12333 Van Ness Ave.
Hawthorne, CA 90250
(213) 757-5141

Visqueen polyethylene

FRP

Table III-1—Continued

GLAZING MATERIALS MANUFACTURERS—CONTINUED

General Electric Co.
Plastics Division
Specialty Plastics Dept.
Sheet Products Section
One Plastics Avenue
Pittsfield, MA 01201
(413) 494-1110
— Lexan polycarbonate sheet

General Glass International Corp.
270 North Ave.
New Rochelle, NY 10801
(914) 235-5900
— glass

Glasteel, Inc.
1727 Buena Vista
Duarte, CA 91010
(213) 357-3321
— FRP

Kalwall Corp.
P. O. Box 237
Manchester, NH 03105
(603) 668-8186
— FRP

Lasco Industries
3255 Miraloma Ave.
Anaheim, CA 92806
(714) 993-1220
— FRP

Martin Processing, Inc.
Film Division
P. O. Box 5068
Martinsville, VA 24112
(703) 624-1711
— Llumar polyester

Monsanto Co.
800 N. Lindbergh Blvd.
St. Louis, MO 63166
(314) 694-1000
— type 602 polyethylene

Park Energy Co.
Star Route, Box 9
Jackson, WY 83001
(307) 733-4950
— double-wall polycarbonate sheet

Rohm and Haas Co.
Independence Mall West
Philadelphia, PA 19105
(215) 592-3460
— acrylic and polycarbonate sheet

Sheffield Plastics, Inc.
P. O. Box 248
Salisbury Rd.
Sheffield, MA 01257
(413) 229-8711
— Acry-Pane acrylic and Polt-Glaz polycarbonate sheet

Structured Sheets, Inc.
196 E. Camp Ave.
Merrick, NY 11566
(516) 546-4868
— Qualex polycarbonate sheet

3M Co.
New Business Venture Division
53-3 3M Center
St. Paul, MN 55144
(612) 733-1110 or toll-free:
(800) 328-1300
— high-transmission film

3M Co.
3M Center
Building 223-2-2W
St. Paul, MN 55101
(612) 733-0306
— Flexigard #7410

SOURCES:

1. E.J. Clark et al., *Standards for Cover Plates for Flat Plate Solar Collectors: NBS Technical Note 1132* (Washington, D. C.: U.S. Government Printing Office, 1980).

2. L.C. Godbey, T.E. Bond, and H.F. Zornig, ''Transmission of solar and long-wave length energy by materials used as covers for solar collectors and greenhouses.'' In *Transactions of the American Society of Agricultural Engineers,* ASAE paper #77-4013, vol. 22, no. 5 (St. Joseph, Mich.: ASAE, 1979) pp. 1137–1144.

3. H. Hartman and J. Whitridge, ''Fiberglass Reinforced Panels for Solar Collector Glazing.'' Paper presented at the American Society of Agricultural Engineers Winter Meeting, 18–20 December 1978. ASAE paper #78-4527. (St. Joseph, Mich.: ASAE).

4. Mid-Atlantic Solar Energy Association. *Proceedings of 1979 MASEA Topical Conference on Solar Glazings* (Philadelphia: MASEA, 1979). Available from MASEA, 2233 Grey's Ferry Ave., Philadelphia, PA 19146.

NOTES:

1. Prepared by David Sellers and Bob Flower of Rodale's Product Testing Department.

2. Some items in this table are based not only on data furnished by manufacturers or independent testing, but also on our subjective evaluation and judgment. This is necessary because, for characteristics such as ''Service Life,'' manufacturers are totally unwilling to commit themselves to a guaranteed specification value.

3. CAUTION: Most plastic glazings are combustible, although in our opinion they do not pose a fire hazard any more serious than other plastic furnishings in the home (e.g., draperies, wall coverings, furniture, etc.). Some building codes may restrict or impose special conditions on the use of plastic glazings which are physically part of the building's roof or wall. Building codes generally do not apply to plastic glazings that are not part of the building (e.g., glazings on freestanding solar collectors).

4. Some manufacturers produce lower quality utility-grade glazings for less demanding nonsolar applications. Examples: Plexiglas MC, Kalwall Sunlite Regular, and Filon Types 740 and 750 (also called ''Solar-Gro Home Greenhouse Panels''). Such products look (to the eye) like their higher quality counterparts, but their long-term performance will be significantly worse. Buyers should always make sure of the exact product name and type designation before placing an order to avoid being surprised by receiving the wrong material.

5. Many manufacturers don't sell small quantities of their products at retail, but they should be able to direct you to retail and wholesale sources of supply in your area.

triple-strength (3/16 inch thick) and heavier thicknesses. With its slow-impact strength it isn't suitable for spanning large areas or for use where high wind loads are anticipated. Nor should it be used in overhead applications such as in skylights. Sheet glass usually has a high-iron content, and thus a lower transmittance (84 percent) than low-iron varieties (91 percent). But sheet glass is still one of the most common glazings for many solar applications because of its relatively low cost.

Plate glass is cast over a smooth mold, then ground and polished to eliminate surface irregularities, producing glazing with virtually no visual defects, even in very large panes. Plate glass is usually made in ¼- to ¾-inch thicknesses and is somewhat more costly than sheet glass. In applications where distortion-free view is of no concern (mass wall glazing), the extra cost of plate glass is not warranted. "Salvage plate" is used plate glass that is usually available at a somewhat lower cost.

Float glass is made by a process developed in England in which glass is cast on molten metal. This produces surfaces nearly as perfect as those of plate glass, but without the additional manufacturing step of polishing. Float glass is thus cheaper than plate and a good compromise between cost and clarity, between sheet and plate.

Sheet, plate and float glass are available in tinted, patterned and reflective varieties. As has been discussed, tints and reflective coatings should usually be avoided where solar gain is important. (Their proper use is in limiting gain where space cooling is the main problem.) Some colored glass, especially a white or "opal" variety, is used in lighting for its diffusing quality, as is patterned and frosted glass. Diffusing glass also provides visual isolation without sacrificing natural light and has been used for many years in bathroom windows, office partitions and other places meant for privacy.

A recent development for solar applications is low-iron (high transmittance) glass with a low reflectance surface treatment. One side of the glass (the sun side) has an etched or almost stippled appearance and is thus not perfectly transparent. This treatment reduces surface reflection, which can be significant at high incidence angles. Lower reflectance means increased transmission of solar energy.

Glass Strength

Glass is strengthened by several methods of heat curing in the final stages of the manufacturing process. Annealing is simply the cooling of hot glass in a slow, controlled manner. It improves strength mainly by improving material homogeneity and surface regularity. Structural failure in glass always starts at weak spots and at defects that are sometimes invisible. By reducing imperfections, annealing reduces the likelihood of breakage, for reasons other than point impact, but it's still the weakest of all glass types.

Heat strengthening involves rapid heating and cooling at the end of the forming process. Heat-strengthened glass is about twice as strong as annealed and should be specified for large panes or where wind or other load conditions risk breakage. Heat-strengthened glass breaks in the same dangerous way as annealed, in long shards. Although its added strength may reduce the chances of breakage, it should never be used as a hedge against injury involving broken skylights or glass doors.

The strongest glass that is most commonly used for solar glazing is tempered. Tempering is a process in which a sheet of glass is heated to a very high temperature and then cooled very quickly. This results in glass that is up to five times stronger than annealed glass. Tempering also changes the breaking characteristics in a remarkable way: Tempered glass shatters into small, blunt-edged particles rather than the large, dangerous shards characteristic of annealed or heat-strengthened glass. For this reason most building codes require tempered glass in and around doors and other high-risk building locations. Tempered glass performs well under wide temperature swings, which makes it useful in high-temperature solar applications. Most high-transmittance (low-iron) solar glazing is tempered. Tempered glass is more expensive than other glass types, and if you buy some, you'll see it has another important limitation: The glass is cut to a few standard sizes before tempering, and it can't be cut afterward. Your plans must conform to these standard sizes or you'll be buying much costlier *custom tempered* sizes.

Two other strengthening methods are worth mentioning, although their use in solar glazing systems is much less common. *Wired glass* is manufactured with embedded wire reinforcement and is sometimes required by building codes in skylights and other overhead glazing. *Laminated glass* is a sandwich of a plastic film between two layers of glass. It is used in vehicle windshields and in thick, multiple laminations for bulletproof glazing. The plastic holds glass fragments in place in the event of breakage, minimizing the danger of flying particles.

Plastics

Plastic glazings offer several alternatives to glass. Some varieties are lower in cost, and some have certain performance advantages. They are sometimes referred to as *thermoplastics,* and they are formed by casting, extrusion or other methods. While plastics offer some unique design possibilities, such as the fabrication of bent or curved panes, there is also the possibility that plastic glazing will deform at temperatures that can occur in some solar applications. Thus plastic glazings will generally require more structural support than will glass.

Another important physical characteristic is thermal expansion. The coefficient of thermal expansion for common plastic sheet glazings is about eight times that of plate glass, so extra care must be taken in the design of supports to accommodate this movement. Another important factor in the use of plastic glazing is fire safety. Most thermoplastics are flammable, and while this is no cause for concern with solar heating systems, it is a factor in building fire safety. Plastics should never be used in fire-hazardous areas; for instance, near heating equipment or stoves. Some building codes also limit the allowable area of continuous plastic glazings in walls or skylights. Check local codes before you buy your materials.

When they're exposed to high temperatures, plastic glazings are subject to "bowing," and this is sometimes cited as an aesthetic objection to plastic-glazed flat plate collectors (although one major manufacturer intentionally bows its collector cover to increase strength, and the same bowed configuration has been used effectively in site-glazed systems). Unintentional bowing can occur where there are significant humidity differences between inside and outside air, because some plastics absorb small amounts of water and swell slightly. Generally, though, this isn't a serious design problem.

Acrylic and polycarbonate sheets are probably the best-known plastic glazings. These rigid glazings have the appearance of glass and can be used like glass, although there is a basic difference in performance. Acrylics and polycarbonates do not block heat (longwave infrared) radiation as well as glass. This is a fact of minor importance in the overall heat balance of direct-gain systems where windows are relatively small. However, where glazed walls, Trombe walls, greenhouses or other glazing-thermal mass systems are involved, losses by infrared radiation through plastics may be significant. This loss can be reduced by adding a second layer of glazing to reduce conductive and convective heat loss.

With acrylic sheet, this can be done without greatly reducing solar gain. One of the special advantages of clear acrylic sheet is that its transmittance (85 to 92 percent, depending on thickness) is about as good as that of the most expensive low-iron glass. Acrylic sheet is also very stable; its transmittance is virtually unaffected by age or weathering. When it was first developed for use in aircraft canopies, acrylic glazing was subjected to years of desert exposure, and after polishing to eliminate surface abrasion from blowing sand, it proved to have a negligible loss of transmissivity. With its high transmittance, long life and relatively low cost, acrylic glazing is most attractive in solar applications.

Polycarbonate is more impact-resistant and more temperature-resistant than acrylic, but it has significantly lower solar transmittance (74 to 85 percent). It also discolors and loses transmittance as it ages, so it's generally less appropriate for solar applications, except where very high temperatures are anticipated, or where breakage (vandalism) is a factor, or when it's used as an inner glazing.

Double-layer extruded plastic is available in both polycarbonate and acrylic and has two major advantages over single glazing: greater rigidity (due to its ribbed channel structure) and lower thermal conductivity. Double-layer acrylic products, because of their high transmittance (84 percent) and low cost (relative to double-layer glass), are well-suited to solar applications. They do, however, have a quality that may either be an advantage or a disadvantage: Their channeled structure and their extrusion marks obscure vision. Thus they don't have the clarity of glass or regular single-layer acrylics.

Both acrylic and polycarbonate sheets are suitable for curved glazing. Plastic sheets can be cold formed in a curve, allowing a broad range of possibilities that would be too costly to consider with glass. They can also be cut and handled on-site with relative ease and safety, although great care must be taken to avoid scratching or abrading the surfaces.

Plastic film glazings are very thin, generally less than a hundredth of an inch thick. Some varieties have high transmittance in part because they are so thin. Thin films are very lightweight; large areas can be glazed with minimal supporting structure. The huge sizes that are available also make it easier to eliminate air infiltration.

Clear *polyethylene* has many limitations as a glazing material, but one major advantage: It's very cheap. In its most common solar glazing application, greenhouses, it is generally recognized as a temporary glazing because it degrades rapidly in transmit-

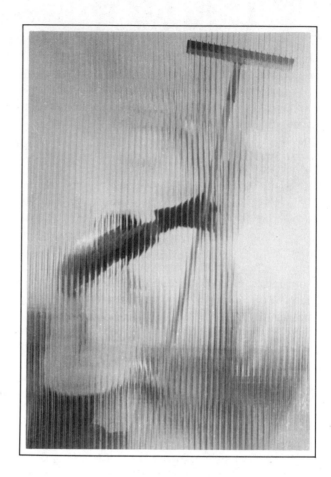

Photo III-9: Double-wall extruded acrylic glazing has a unique look to it that can enhance the beauty of some kinds of glazing applications, as well as providing some privacy by its lack of total clarity. Because it's extruded, it can be made in long 16- to 20-foot lengths, which can help to simplify large-area installations.

tance and strength. As an exterior glazing layer on greenhouses, polyethylene is normally replaced after one to three years of service. Mechanical failure, accelerated by buffeting winds, is usually the last straw. But when used as an inner glazing, its life expectancy is greatly increased. It is not, however, a totally clear glazing, but about halfway between clear and translucent. Polyethylene is selected by commercial growers for new greenhouse construction more often than any other glazing material. Even with annual replacement growers find polyethylene economically advantageous. For residential applications this may or may not be an advantage, depending on how you work your own time-money equation.

The Du Pont Company makes two *fluoropolymer* films, Teflon and Tedlar, which are specifically for solar applications. Teflon is an exceptionally stable plastic with very high transmittance (93 to 95 percent). It's used primarily as a second (or third) *very* thin glazing in a multiple glazing scheme (1 mil or one-thousandth of an inch thick). It withstands higher temperatures than most plastics and is suitable for use in flat plate collectors. It does, however, have a high coefficient of thermal expansion, which can result in sagging at high temperatures. Among films, Teflon is fragile and difficult to handle, and many users have commented on static buildup and dust as problems encountered during installation. But as long as it is protected by another glazing layer and its stress limitations are not exceeded, Teflon film can last practically forever. Tedlar is cheaper than Teflon and has similar ultraviolet stability and transmittance. It is normally used in a 4-mil thickness and is thus easier to handle. It has a lower thermal expansion coefficient, hence less tendency to sag. In fact it is possible to shrink Tedlar after it's been fastened to the frame by giving it a hot air treatment with a heat gun or other source of hot forced air. The film will appear to sag when it's being heat-treated, but when it cools, it will shrink and pull tight to the frame. After this shrink treatment the film will have little or no sag at normal operating temperatures. These advantages combined with its lower cost make Tedlar the film of choice among many solar designers.

Several companies make *polyester* films, *acrylic/polyester laminates* and other plastic film products that have good transmittance and other properties useful for low-temperature solar designs. Consult table III-1 for information about these other types. Shop around, and remember when evaluating a film to find out about its aging characteristics under solar exposure. Look for such adjectives as ''weatherable'' and ''ultraviolet-stabilized,'' and ask for test data regarding transmission loss and other degradation over time.

Fiberglass-reinforced plastic glazings (FRP for short) are very popular in solar applications, and there are several brand names on the market (Kalwall, Filon, Lascolite, Glasteel). The fiberglass materials are lightweight, strong and have high transmittance. Although their transmittance drops slightly over time, they are durable in exterior use, unlike plastic films. FRP glazings present a good cost/performance compromise between films and rigid sheet products. They are also available in corrugated form and prefabricated into multiple-layer panels (Kalwall).

Light is diffused through FRP glazing primarily because of the reinforcing fibers embedded in the plastic resin. Thus it's not a glazing ''with a view,'' which may be useful when, for example, more privacy is desired. Overall, their light weight, high impact strength, ease of handling and modest cost make them an excellent choice for many solar glazing systems.

Structures for Glazing

Any glazing application requires careful attention to structural support, not only so that the glazing will be supported, but also that it can respond to a variety of changing loads and stresses. Structural designers typically distinguish between ''dead,'' constant or unchanging loads, and ''live'' or dynamic loads. In most structural analysis, dead load consists of the weight of the structure itself. Your glazing material has weight, as do the mullions and studs or rafters that carry it. These are dead loads. In most situations the design live loads represent the worst possible conditions and are much larger than dead loads. Live loads include snow accumulation, wind and impact loads. In the case of rigid

Photo III-10: The New Alchemy Institute in Wood's Hole, Massachusetts, has built solar green-houses with curved FRP glazing sections. The upper greenhouse's convex panels give it an interesting appearance and also increase the glazed surface area of the greenhouse by about 10 to 15 percent. The concave glazing in the lower greenhouse also increases surface area, but it also directs water away from the seams, which helps to simplify the glazing seals.

TABLE III-2 HOW MUCH GLAZING?

Even though the area of a retrofit glazing installation is usually limited by the available surface area of existing wall or roof, you can still make use of these sizing guidelines to get an idea of the optimum ratio between glazing area and heated floor area. These numbers apply to vertical and steeply pitched glazings.[1]

Average Winter Outdoor Temperature (°F) (degree-days/mo)[2]	Square Feet of Window Needed for Each One Square Foot of Floor Area[3]
Cold Climates	
15° (1,500)	0.27–0.42 (with night insulation over glass)
20° (1,350)	0.24–0.38 (with night insulation over glass)
25° (1,200)	0.21–0.33
30° (1,050)	0.19–0.29
Temperate Climates	
35° (900)	0.16–0.25
40° (750)	0.13–0.21
45° (600)	0.11–0.17

SOURCE: Edward Mazria, *The Passive Solar Energy Book,* 1979, with permission of Rodale Press, Emmaus, Pa.
NOTES:
1. These ratios apply to a residence with a Home Heating Index of 8 to 10 (Btu/day/sq ft/°F). If space heat loss is less, lower values can be used. These ratios can also be used for other building types having similar heating requirements. Adjustments should be made for additional heat gains from lights, people and appliances.
2. Temperatures and degree-days are listed for December and January, usually the coldest months. Consult table I-8, column B, for the average daily temperature for your location.
3. Within each range, choose a ratio according to your latitude. For southern latitudes, i.e., 35° NL, use the lower window-to-floor area ratios; for northern latitudes, i.e., 48° NL, use the higher ratios.

sheet materials, dead loads are described in manufacturer's specifications as the material's "bending" or "flexural" strength, and in their product literature they often call out the maximum spans or pane area possible for various thicknesses and types of materials.

A corresponding strength evaluation for plastic films normally involves "tensile" rather than flexural strength. Because of the negligible thickness of films, bending strength is virtually nil. Instead, loads applied over a plastic film glazing are "stretching" or tension loads. With films the critical areas of concern are the edges, where tensile stress is likely to be highest. Thus it's important to make edge supports as smooth and even as possible to avoid stress concentrations. For this reason mere stapling is inferior to groove-and-spline, tape, adhesive and other fastening techniques.

Impact Loads

Not all loads, of course, are created equal; that is, evenly distributed. An impact load is applied over a very short duration and over a limited glazing area. This concentrated load results in stresses that can far exceed a material's strength. Many manufacturers provide in their product literature some index of impact resistance, particularly for rigid sheet plastic and FRP glazings, since these are often used instead of glass specifically to protect against vandalism or the hazards ensuing from accidental breakage. Be careful, though, in making direct comparisons between manufacturers' claims. There are several standard tests for impact, and only results achieved by the same method are comparable. Table III-1 summarizes glazing impact strength.

If the potential for breakage is an especially important issue, your strength evaluation should involve consideration of failure modes: *If* it breaks, how will it happen; what will be the likely consequences, and how will repair or replacement be accomplished? Glass, of course, shatters, or crumbles in the case of tempered glass. Shattering can also occur with acrylic and some other plastic materials, but with the polycarbonate and FRP glazings you can forget about shattering. Small stress cracks, however, can occur in both glass and plastics, as the result of either impact, thermal or distributed loads, especially when they are repetitive. *Stress cracking,* also called *crazing,* usually doesn't result in immediate or dramatic structural failure. In plastic glazing materials, crazing appears as minute, scarcely visible cracks, which may produce leaks, but not necessarily. The important thing is that stress cracks reduce the capability of a glazing pane to withstand its normal rated loads. Once a crack is produced, it is likely to propagate. Another less severe load will increase its size or produce similar cracks elsewhere. The most common cause of stress cracking is not impact, but improper glazing fastening and support.

Impact failure of plastic films is again quite different from that of rigid glazing materials. The flexibility and high tensile strength of films are advantageous in resisting impact loads, and when failure does occur, it is usually a repairable puncture or a tear. From a safety standpoint this kind of failure is relatively benign: There is no shattering and the glazing is still in place.

Thermal Loading

The other major structural load is caused by temperature changes. Solar applications can subject glazings to wider temperature variations than they normally encounter in traditional building applications. Minimizing temperature changes in the immediate glazing environment is one way to defuse potential problems. South-facing glazing, for example, should not be obstructed in such a way that heat builds up around it.

There are three ways temperature change affects glazing materials. The first, *thermal shock,* occurs when the glazing temperature is greatly raised or lowered. You may have observed the effects of thermal shock if you ever poured boiling water into ordinary glassware. In most solar applications, temperature variations can be large but gradual, and thermal shock usually results in cracks that seem to have appeared for no reason. Compared with most types of glass, plastic glazings are nearly immune to damage from thermal shock. If thermal shock does occur in plastics, it is more likely to result in buckling or crazing.

The second temperature-related structural effect is *thermal expansion*. As mentioned previously, this property is usually expressed in terms of a unit measurement of expansion per unit measurement of glazing dimension (length and width, it doesn't matter) per degree Fahrenheit change in temperature. From table III-1 we see that glass has an expansion coefficient of 0.49 times 10^{-5} inches per inch per °F temperature change. If, for example, a 34-inch-wide piece of glass experienced a temperature increase of 80°F, it would expand $0.49 \times 10^{-5} \times 34 \times 80 = 0.013$ inches. That's not a great deal of movement, but if, for example, a glass edge came up against the metal shaft of a screw or bolt, that much expansion could cause a crack. Thermal expansion is more pronounced with plastics than with glass. The movement must be dealt with at the edges or fastening points by giving the glazing enough room or "edge clearance" to expand freely.

The edge of a glazing support must provide enough overlap or "bite" with the glazing to carry the glazing when it contracts. If proper edge clearance isn't provided, or if the glazing is restrained at the edges, expansion will cause material stress that can cause bowing, crazing or breakage. If proper bite is not provided, contraction can cause gaps between the glazing and its supports or seals, producing leaks and air infiltration. At worst, the glazing may fall out of its frame.

The final type of thermal effect applies to plastics, mainly plastic films: It's called *sag* and it's essentially the result of thermal expansion. With plastic film the problem can be reduced by prestressing it (installing it with a slight stretch). With rigid and semirigid plastic glazings more intermediate supports should be used between the edge supports than would be used with glass.

We've presented you with all these various factors that affect glazing performance mainly to make you aware of them, not to put you into the role of materials analyst. In summary, the structural "behavior" of installed glazing materials directly affects their appearance, performance and longevity, and all these factors are important in a solar heating system. Good attachment, adequate support and appropriate sealing details and techniques are critical to the quality of the whole project, and fortunately there is an abundance of experience in right methods of handling glazing. The product literature of some glazing manufacturers provides thorough descriptions of glazing attachment details. If you're hiring an architect or a professional glazier, you can be assured of good detailing. And as you've seen from the projects in the previous domestic water heating section, glazing details are always included in the construction illustrations.

Glazing design is the heart of many a solar application, and if you don't get it right, you'll most certainly have to keep coming back until it is right. Please take heed of the details that are presented; know thy glazing; treat it right, and it will faithfully return years and decades of useful service.

Lawrence Lindsey

⌂PROJECT
A South Glass Addition

The glazed area on a south wall is increased for direct gain space heating.   A small window is removed, and a larger opening is created with standard carpentry techniques to make room for the new double-pane glazing units. Provision is made for venting the hot air to the outside to keep the room cooler in summer.

When they first started thinking about a solar retrofit, Karen and Dave Parker were thinking "greenhouse," and in the end they got a greenhouse of sorts, but they did it with just a simple addition of vertical south glass. Like many homes, the Parkers' had a special mudroom or "junkspace" to which was relegated everything that wasn't being used but that wasn't quite ready to be thrown away. This room happened to be on the south end of the building, a 256-square-foot no-man's-land that was uninsulated, unoccupied (except by domestic flotsam) and unheated. When they finally cleared away all the stuff and took it to the dump, the Parkers got their first glimpse of how a south glass addition could totally transform the junkspace into a sunspace.

Once the choice was made, it seemed like such an obvious one. Options for solar retrofits are often more easily defined than design alternatives for new construction. The existing building and site can be welcome limiting factors in the planning and decision-making stages of a project. Imperfections, on the other hand, are another matter. Renovation can be tricky with sagging beams, wavy walls and, worst of all, rotting frame members. In such cases it may be necessary to make general improvements during the construction of solar components, especially where new loads (framing, thermal mass) are concentrated. Because of the vagaries of old house conditions, the design and construction phase is a prime arena for the Battle of the Trade-offs, and it's at this point that the design should remain flexible so it can change

with unforeseen changes in construction.

Even with the apparent simplicity of a south glass addition, the work at the Parkers ran into these sorts of problems, but nevertheless, they got what they wanted. A south glass addition in a frame wall is essentially a simple undertaking. The basic task was simply that of removing one 10-square-foot double-hung window and replacing it with 54 square feet of fixed double-pane glass. The new window wall was made up of three 34 by 76-inch sliding glass door replacement units, which cost about $2.75 to $3.25 a square foot. (These units are relatively low in cost because they're mass-produced and usually available from local glass outlets.)

First, the room's interior wall finish and its 2×4 frame were removed from the entire south wall to allow for an enlarged rough opening framed with 2×6's. Removing the studs presented no problems since the house frame was post and beam, and the studs were not load bearing. (Editor's note: With regular stud wall construction the studs are usually bearing the load of second-story framing and/or roof rafters, which means that new studs should be put in place as soon as an old stud is removed. If you're removing all the old studs at once, put in some diagonal shoring braces running from wall to ground. It may also be necessary to strengthen the horizontal span of the window opening with a new header.)

Sizing the openings for the glass required adding ½ inch to both the length and width of the glazing module to create a ¼-inch perimeter clear-

Photo III-11: It wasn't a big change, but when the Parkers added more south glass to a once unused room, shown here on the right, they suddenly had themselves a solar room, complete with a little growing bed for plants and vegetables.

ance between the glass and the framing members. Without this clearance around the glass there is a chance that any settling or racking of the building could result in breakage. Framing materials must be very straight, and needless to say, calculations and cuts must be accurate, and all openings square and equal.

The trickiest operation was cutting from the inside to remove the exterior siding. This was done with a reciprocating saw using the rough window opening as a guide. You can also transfer the dimensions of the opening to the outside face of the wall (where the cutting is easier) by drilling holes through the siding at each corner of the inside cutout. Then go outside and snap chalk lines between each of the holes to make your cutting lines.

Next, two studs were accurately spaced and securely nailed to form the openings for the three glass units. Inside glazing stops were installed around each opening and standard butyl glazing tape was stuck on them. In this case the glass was installed from the outside, but placement from the inside works just as well when outside stops are nailed on first. In any case, you must decide just where the glass will go on the face of the studs: if it will be closer to either the inside or outside edges or placed right in the middle. These decisions involve how you intend to finish and trim the new opening and how

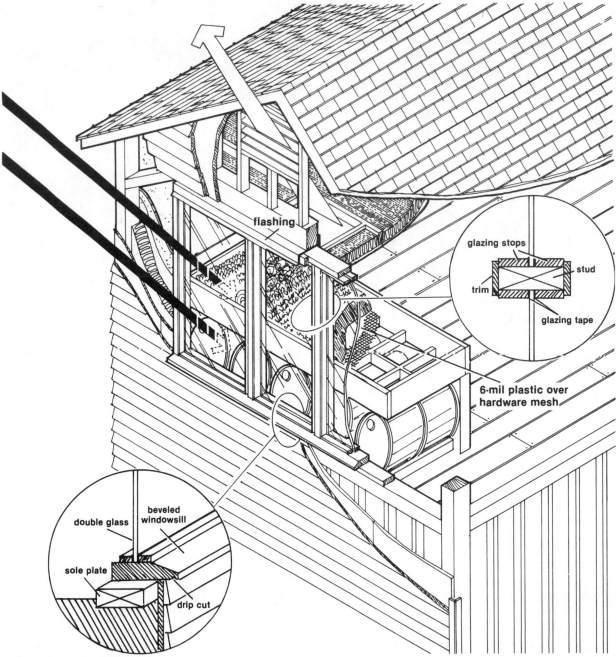

flashing

glazing stops

stud

trim

glazing tape

**6-mil plastic over
hardware mesh**

double glass

**beveled
windowsill**

sole plate

drip cut

Figure III-4: Only a minimum of structural work was needed to install the new windows. In any job like this careful attention should be paid to sealing out all possible air and water leaks. The detail at the left shows how the design of the bottom sill will shed water, and the glazing detail at right shows the proper technique for ensuring stability and airtightness. In summer hot air is exhausted through ceiling and attic vents. Direct gain is absorbed by the water-filled barrels and the growing bed.

rainwater will be drained off the bottom sill.

Just before laying up the glass, neoprene spacer/cushion blocks were placed on the sill near the corners of each opening. The glass units were carefully placed on these blocks and then tilted up into the opening. Final side-to-side positioning was done before the glass came in contact with the sticky butyl. With the unit centered in the opening, it was pressed firmly against the inside stop and glazing tape to produce a total perimeter seal. More tape was applied around the edges of the glass before the outside stops were nailed on. To further improve the seal and ensure against infiltration of cold air or moisture through the stops, a final bead of silicone was run around the seam between the exterior stop and the glass. It's also important to follow standard conservation techniques for sealing off possible air infiltration points around the perimeter of the window wall framing.

Photo III-12: With this simple growing bed the Parkers have realized an additional benefit from their retrofit. The sun heats the room and grows plants and vegetables.

All exposed framing materials were treated with a preservative to prevent damage from moisture. Green Cuprinol #14 was used, because the manufacturer assures that it is compatible with plants and safe for greenhouse use. To finish the opening, the clapboard siding was cut back to accommodate 1×4 trim; then standard drip-cap flashing was installed over the outside head casing (top trim piece). Inside, the new 2×6 wall was insulated with R-19 fiberglass batts, covered with a 6-mil polyethylene vapor barrier and finished with drywall. Then trim was installed around the window, and all was primed and painted.

The Parkers' greenhouse took the form of a 3 by 10-foot plant bed that was built just inside the window wall. The sides and the ends were made of 2×12's. The floor of the bed was built with 3-foot 2×4's, placed 16 inches on center. One end of each 2×4 was notched and rested on a 1×2 ledger nailed flush along the bottom edge of the 2×12 nearest the glass. The other end was simply face-nailed to the other 2×12 that formed the front of the bed. Blocking was placed between these joists to prevent them from twisting, and then all the lumber was treated with Cuprinol #14 preservative. The framed-in floor of the bed was covered with galvanized hardware mesh and a polyethylene liner to contain the soil and irrigation water. One side of the completed bed was nailed to the window wall framing, and the other side was supported off the floor with 4×4 legs.

The height of the plant bed was designed to allow a thermal mass of five 55-gallon water-filled drums to be placed underneath with one end facing south. The drums rest on two 2×4's placed on edge, which creates a toespace and allows air circulation. In the floor under the drums, there are two operable registers that draw cool air from the house's foundation crawl space into the room for summer ventilation. There also happened to be a ceiling trap door right over the growing bed which could be opened in summer to allow heat to rise up and be exhausted through a gable vent in the attic. For cold weather a roll-up type of window insulation (Window Quilt) was a final important detail in the plan. The face width of the 2×6 studs allowed enough room to install the edge-sealing running tracks for the shade. With a heat resistance of at least R-5, the energy-saving value of some kind of movable insulation system is significant and should always be included in a window wall plan if the budget allows. If it doesn't at the time of the retrofit, at least plan the work so that window insulation can be added later. The shade can also be used to limit solar gain in warm weather.

Another improvement related to the retrofit was the removal of the wall separating the kitchen from the new solar room, which made for better heat and light distribution. One of the initial problems was that the south end of the house always seemed colder than other parts, a condition that's been turned around with insulation and south glass.

Nowadays winter daylight reaches far back into the kitchen, whose view includes an apple tree, a woodpile and a stream along with heat-producing sunlight. What was once an empty room is now full of flourishing greenery, with two dogs and a kitten basking in sunny spots on the floor. The Parkers like it too.

Materials Checklist

Window Wall

new framing: 2×4 or 2×6 studs; 4-by or doubled 2-by material for a new header, if required
fixed window units: 34×76-inch sliding glass door replacement
1×2 inside and outside stops

neoprene setting blocks
butyl glazing tape
silicone caulk
movable window insulation (see Section III-B)

Planting Bed

framing: 2×12's, 2×4's, 1×2, 4×4 lumber
hardware mesh
6-mil polyethylene

wood preservative (Cuprinol #14 is compatible with plants)

Rick Schwolsky

☀PROJECT
Build Your Own
Multi-Pane Windows

This is a simple procedure for making double- or triple-pane windows to reduce glazing heat loss. It involves modifying the existing window sash and replacing the single glazing with a multi-paned unit that is built with the same materials used by commercial manufacturers. Adding south glass means increasing the potential for building heat loss, and since multi-pane windows are a must for most climates, you can save a bundle by making your own.

There are at least half a billion energy-wasting windows in houses around the country. Summer and winter, these windows leak heat in the wrong direction: outward when you're trying to stay warm and inward when you want to stay cool. In fact, heat conducted through windows can account for as much as 70 percent of a home's heating load and up to 46 percent of the cooling load.

Chances are, your windows aren't up to snuff, and one way or another that's going to cost you money. You can leave your present windows alone, and watch your energy costs spiral upward forever, or you can pay a little extra to upgrade your windows now, and watch your lowered energy costs pay for the improvements in just a few years. For starters, see the box, "Tightening Up Your Window Frames" in this project.

If you're not sure it's worth the bother, consider this: In an average American house, approximately 45,000 Btu are lost through each square foot of single glazing over the heating season. That's the energy equivalent of about 13 KWH, nearly a half a therm of natural gas and about a third of a gallon of fuel oil. In summer, about 86,000 Btu enter the house through the same square foot (25 KWH, 0.86 therms, 0.62 gallons). Of course, the actual figures depend on the climate in your area, but these averages indi-

cate just how serious a problem inefficient windows can be.

In most cases, you can cut the heat loss figures nearly in half by installing double glazing, which is simply two sheets of glass assembled into a single unit with an airspace between them. Summertime heat gain can be reduced by only about 15 percent due to reduced light transmission. Thus the main benefits of multiple glazing are with heat *loss* reduction. Triple glazing, which is three sheets of glass arranged in a glass-air-glass-air-glass sandwich, cuts conductive heat loss through the window by about 65 percent and heat gain by about 30 percent over single-pane windows.

Another form of double glazing is with the use of storm windows over single-glazed windows. The important factor with storms is that they be very well sealed to the window frame, and that the existing window also be tightened up with weather stripping. Under these conditions the window-and-storm combination is as efficient as double glazing installed in a tight window.

Which type of glazing is most cost-effective? Proponents of triple glazing point to its higher R-value, but triple glazing tends to be costly, and the finished glazing units are heavy and thick and possibly difficult to retrofit. Although double glazing

(R-1.8) does not have as high an R-value as triple (R-2.7), its insulating ability is still twice that of single glazing, while the costs, weight, and thickness are much less. In fact, recent studies indicate that a combination of double glazing and movable insulating curtains, shutters, or drapes constitutes the most cost-effective window treatment now available (see Section III-B). The combination produces high R-values and significant energy savings with a reasonably short payback time. Because of these benefits and the drawbacks of triple glazing, we'll focus primarily on making double-pane units.

Buying double glazing can be expensive. If you were to remove your home's existing windows and replace them with new high quality double-glazed units, you could easily spend $100 to $200 per window—even doing the installation yourself.

But if your present windows are in reasonably good condition and have wooden sashes (as most do), you can upgrade just the glass at a much lower cost. There are two ways to go about this. One is to remove the glass from your present windows, enlarge the glass-holding rabbet, and insert a double-glazed, "welded glass" unit. With this method, the cost will run around $5.50 per square foot for a unit with a ¼-inch airspace. You'll be increasing your window's R-value from 0.88 for single glazing to 1.64 for double glazing, which just about cuts window heat loss in half.

Welded glass is made from a sheet of glass that has, in effect, been folded over on itself. The edges are melted together so that the glass becomes a continuous sheet, open only at one corner. Air is removed from between the panes, dry nitrogen is pumped in, and the little hole is sealed with molten lead. This creates a thin, light, and easy-to-handle glazing system that's great for new windows. But retrofitting welded glass to an existing window can be a problem: It comes in a limited range of standard sizes (custom sizes are prohibitively expensive); costs are fairly high, and the window's existing glass must be scrapped.

A second way to recycle your windows is

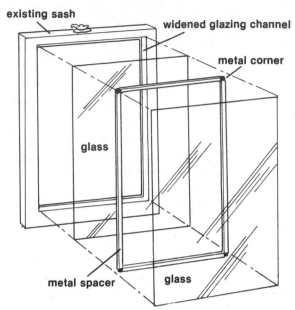

Figure III-5: Making multi-pane glazings is a simple job: Metal spacers and corners form the frame to which the glass panes are joined with an adhesive sealant, creating an insulating airspace. Then the unit is held into widened channels in the window sash with standard glazing points and finally caulked with glazing putty.

to build your own multi-pane units using special metal spacers fitted between separate sheets of glass (figure III-5). Because at least one of these sheets can be recycled from the glass that's already in your window, you can build your own hermetically sealed windows for much less than the cost of welded-glass or factory-built, double-glass units. Furthermore, you can build your multi-pane windows to virtually any size and thickness.

For example, you can build an R-1.82 double-pane window (using glass from your present window for one of the panes) for $2.50 to $4.00 per square foot. If you can obtain all the glass you'll need free from salvage, the cost drops to just what you spend for the hardware and sealant, which will be

$1.00 to $2.50 per square foot. In both cases, as window area increases, cost per square foot drops, but in any case you come out with better insulation at less cost than for a welded-glass unit.

Clearly, home-built double glazing is the least expensive way to increase your windows' insulating value. All you need are windows with wooden sashes thick enough to hold the new units, or fixed windows with frames that can be modified (usually by moving the stops).

The total thickness of a glazing system is determined by the thickness of the panes of glass plus the airspace. For example, a double-glazed unit with two ⅛-inch-thick panes and a ¾-inch airspace has a thickness of 1 inch overall. You'll need to check your present windows to make sure there's enough room for double glazing. Measure the thickness of the sash and subtract ½ inch to allow for a minimum ¼-inch lip on either side of the glass. The result is the maximum glazing thickness you can insert into the sash without having to perform major surgery, i.e., building up the thickness of the sash. Usually, though, you can get away with using the existing sash by widening the glass-holding channel with a router and chisel.

You should select as thick a unit as your sash will allow, to maximize the airspace. A ¾-inch airspace is about optimal for insulating. Larger airspaces do little more to hold in the heat, but cost somewhat more. If your sash won't carry a 1-inch-thick unit, you can use a narrower spacer between the panes of glass with no great sacrifice in insulating value. Spacers are also available in ¼-, ⅜-, ½- and ⅝-inch widths.

Spacers

Spacers are long, hollow rectangles made of thin, roll-formed aluminum. Because they are hollow, they also provide a place where *desiccant* can be stored to protect the window's inner surfaces from fogging with condensed water vapor. (Desiccant absorbs water that might find its way into the airspace.) Although some spacers are precise rectangles, most have beveled or chamfered edges to create a superior bonding surface for the sealants that hold the two panes together. Spacers are made in hundreds of sizes and shapes and typically cost 10¢ to 15¢ per lineal foot.

Corners

Since spacers come only in straight sections, you'll need special right-angle inserts to form corners. Naturally, the insert must fit snugly into the spacer, so it's best to buy spacers and corners from the same supplier. Corner inserts are available in three materials: zinc, aluminum and plastic. Metal corners are best for do-it-yourself applications because they form a tight friction fit with the spacer, and because sealant adheres better to metal than to plastic. They cost about 13¢ apiece.

Sealants

Sealants provide the mechanical bond that holds the spacers and glass sheets together and also protects the airspace from water vapor. A failure in the sealant's bond will allow water to enter and condense inside the unit and fog the glass once the desiccant is saturated. Because a good seal is essential, buy only high quality sealants that are specifically recommended for use with multi-pane windows. At $3 to $5 per caulking-gun tube, these sealants cost more than standard caulks, but they're really the only ones that can do the job properly.

Polysulfide sealant cures to a hard, dense, rubberlike substance, and although it resists liquid water well, it doesn't absolutely halt the transmission of water vapor. It is also vulnerable to the destructive action of the ultraviolet radiation in sunlight. Polysulfide (sold in standard caulking-gun cartridges) is usually used for "single-seal" construction. Single-seal windows rely on the desiccant inside the metal spacers to absorb whatever water vapor passes through the seal. Typically, polysulfide is applied in

Tightening Up Your Window Frames

Adding multi-pane windows or insulating shades won't solve your heat loss problems if cold air can seep into the house through cracks in and around your window frame or sash. While you're adding layers of glazing or window insulation, take the opportunity to weather-strip all around the sash of an operable window and caulk any gaps where infiltration may occur. A spring-metal weather strip can be used if the sash is loose in its channels (double-hung window) or if it doesn't make a tight seal when closed (hinged window). This strip is tacked to the outside edge of the sash where it presses against the inside of the jamb, thus sealing the gap between them. A lower-cost alternative is a vinyl tube weather strip that can be tacked directly to the jamb so that it presses against the inside of the sash. These two types stand up well to the friction caused as the sash is opened or closed. Where the upper and lower sashes of the double-hung window meet in the middle of the frame, again use either the spring-metal or vinyl-tube materials. Adhesive-backed foam would eventually wear down and should only be used where parts of the window press together rather than slide. Where the top and bottom sashes meet the window frame (double-hung), use a thin (⅛-inch-thick) strip of foam attached to the horizontal edges of the sash. You don't need a thick foam strip to make the proper seal, and an overly thick piece might prevent the window from closing completely.

Another big source of infiltration is the gap between the windowsill and jamb and the rough framing around the window. Often this space is filled with nothing more than the leveling wedges used to position the window. If you are rebuilding your windows and can gain access to the gap by removing the inside or outside trim, stuff it with hunks of fiberglass insulation. For a really tight seal, lay a bead of caulk under the interior window trim and nail it back into position. For filling large gaps pressure can foam caulk makes the job easier. A single can of foam (urethane) is said to equal several tubes of standard caulk. If you aren't rebuilding the windows, caulk outside the window between the exterior window trim and the siding. If you do that, don't seal the interior casing as well because moisture might get trapped inside the frame. If you have some fixed window units, you can also caulk around the seam between the window and the frame to seal out any nasty little drafts.

Figure III-6: The best weather stripping materials for any operable window are the folded metal or plastic tube types, shown in the right-hand details of this double-hung window. Where a double-hung sash meets the windowsill and the top of window, use the thinnest (1/8" thick) self-adhesive foam or felt weather stripping you can find.

fairly thick coats to a thickness of ⅛ to ¼ inch.

Silicone sealant forms a strong mechanical bond that resists liquid water and ultraviolet radiation but, like polysulfide, it can pass some water vapor to the airspace. Because of this, some manufacturers use silicone as the outer seal of a double-seal window —one that uses two kinds of sealant.

Butyl (short for *polyisobutylene*) comes in tubes, and unlike polysulfide and silicone, it resists water vapor very well, but forms a weak mechanical bond and degrades if exposed to liquid water. Because of this, it is used only as the inner seal of a double-seal window, with polysulfide or silicone forming the outer seal. The result is an edge seal that

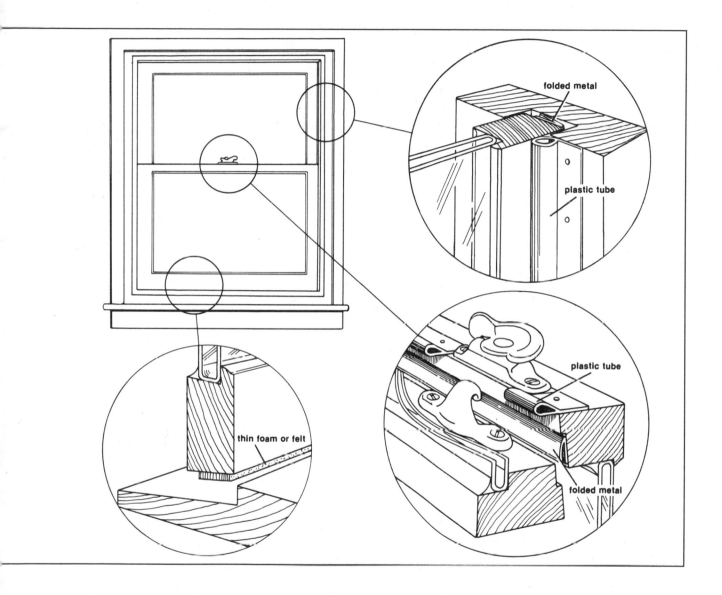

stops liquid and water vapor and is mechanically strong.

In summary, use polysulfide or silicone in a single-seal window; use butyl only as the inner seal of a double-seal window with either silicone or polysulfide as the outer seal. If you want to go the full route, you can also tape the sealed edges of your window (after the sealant has cured) with aluminum tape made by the 3-M Company.

Desiccant

As was mentioned previously, a desiccant is a substance that absorbs water. *Silica gel,* which resembles coarse white sand, is the desiccant most commonly used in window manufacture. It is poured inside the spacers where it sits for years absorbing any moisture that finds its way into the airspace. In time, if excess vapor penetrates the seal, the desiccant can become saturated and unable to absorb any more water until condensation finally begins to fog the window. This is why the proper application of the right sealant is of utmost importance. The desiccant in a well-made, single-seal window can function effectively for 10 to 20 years and beyond, but a bad seal can admit enough moisture to saturate the desiccant in a matter of months. A 5-pound can costs about $14 and holds enough desiccant for dozens of windows.

Cushioning Blocks

Double- and triple-glazed windows should be "floated" in their sashes to prevent mechanical or thermal expansion stress from rupturing the sealant. Pieces of neoprene, called "setting blocks," are commonly used for this purpose, and they're available from almost any glazier or glazing supplier for about 15¢ each, or they can be cut from larger pieces of neoprene.

Sources and Suppliers

Local glass dealers or window manufacturers listed in the commercial pages of your telephone directory may be able to sell you the components you'll need, or inform you of another local source. If there is no supplier nearby, there is an excellent mail-order source that has several outlets around the country: CR Laurence Co., Inc., Box 21345, Los Angeles, California 90021. Telephone: (800) 421-6144; in California, (800) 372-6361; in Alaska and Hawaii, (800) 421-6088. Call them for locations of branch offices in Illinois, New Jersey, Texas, and Georgia. This company requires minimum orders of $25, so ask for all the details of billing and shipping before you place an order. For example, corner inserts come in lots of 100. If you're building fewer than 25 windows (each with four corners), you'll have some

left over. Perhaps you can combine orders with a friend, or you can simply accept a small amount of waste as inevitable. At 13¢ apiece, buying a few extra corners isn't going to break you, and you'll almost always find that the cost of do-it-yourself windows—even counting waste—is still only a fraction of the cost of store-bought units.

If you're not able to make the right hardware connections, here's a trio of manufacturers that might be able to help you out.

Metal spacers and corners:
Higrade Metal Moulding Manufacturing
   Corporation
540 Smith Street
Farmingdale, New York 11735

Metal spacers:
Custom Rollforming Company, Ltd.
500 Barmac Drive
Weston, Ontario
Canada M9L 2X8

Metal spacers and corners:
All Metal, Inc.
636 Thomas Drive
Bensenville, Illinois 60106

Glass

Double-strength glass (about ⅛ inch thick) is best for do-it-yourself double glazing because it is less likely to break than single strength, which is about 1/16 inch thick. If your windows are small, you can probably work safely with single strength, but be careful! You can also save a few dollars by cutting your own panes from a large sheet, but a novice should practice on a throwaway piece before turning the cutter onto the real thing. Having glass cut by a glazier adds to the cost but ensures that the edges will be straight and the corners square, the two prerequisites for effective insulating windows. Uneven edges may mean an uneven seal so if you're not confident about your ability, have it done.

If you're interested in recycling glass, keep

Figure III-7: There are two kinds of spacers, single seal and double seal. The double-seal spacer has an indentation on each side where the butyl sealant is placed, and both types are set in from the glass edge to create a channel for the edge sealant. Both kinds can be used to form double-pane units that rest in a widened glazing channel as shown at left.

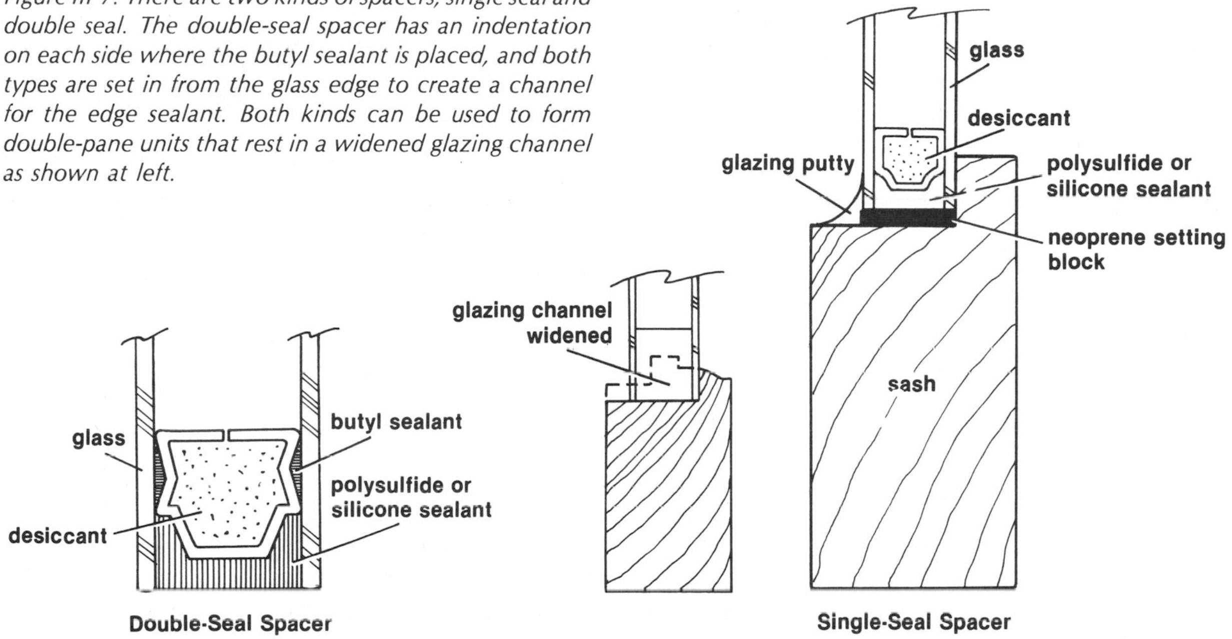

glass

desiccant

glazing putty

polysulfide or silicone sealant

neoprene setting block

glazing channel widened

sash

glass

butyl sealant

polysulfide or silicone sealant

desiccant

Double-Seal Spacer

Single-Seal Spacer

your eyes open for renovation or razing of local buildings. Ask glass dealers or contractors if they sometimes scrap old windows during a renovation. "Salvage plate," as it's called in the trade, is ¼-inch-thick glass taken from store windows, etc., and it's always cheaper than new.

Work Place and Tools

It's important to work in an area that's clean, protected from the elements and free from large amounts of airborne dust. Don't try to assemble your window at temperatures below 40°F (4°C) because the sealants may not bond properly. Also, don't assemble the units when humidity is high. If you must work when it's humid, work indoors, preferably in a heated or dehumidified room. Keep your work space uncluttered. You'll be working with a lot of glass, and an accident could be costly and dangerous.

You'll also need a clean, rigid work surface

large enough to hold at least two panes of glass side by side. If it can hold four or six panes, that's even better for mass production. The work surface should be about waist high; lumber or plywood placed across sawhorses will work. Cover your work surface with several thicknesses of newspaper to avoid marring the glass. Lean hard on your work surface to make sure it's not flexible. If it sags, add extra support, another sawhorse perhaps. Don't risk cracking the glass.

Construction Steps

1. Remove the glass from the window frame. With double-hung windows, it's usually best to remove the sash first. Remove the stops that hold the sash in place by prying them up carefully with a stiff-blade putty knife or small crowbar. Then disengage the counterweights. Working on a firm, flat surface, loosen the glazing compound, remove any glazing points, and lift the glass free of the frame. In

Materials Checklist

glass	supply of newspapers (Save those Sunday editions!)
hollow metal spacers	
metal corner inserts	several rolls of paper towels or lots of clean, absorbent rags
silica gel desiccant	
sealant	a supply of fresh, clean water
caulking gun	brand new, soft-bristle paint brush
neoprene setting blocks	stiff-bristle scrub brush or pad
heavy-duty knife	bottle of household, nonsudsing ammonia solution
fine-grained file	clean, cotton work gloves
hacksaw with a fine-toothed blade	breathing mask
putty knives or other scraping tools	framing square, ruler and pencil
glass cutter	funnel and small cup for handling desiccant

fixed windows, the job is even simpler. Simply remove the stops that hold the glass, cut through any sealants that might be present and lift out the panes.

Carefully stack the glass in a clean location near your workbench. If you lay the glass flat, insert newspaper or cardboard to keep the panes from sticking together.

Remove only one window at first, and perform all the following steps of double glazing and reinstallation to discover any problems arising from your home's particular construction or window type. After you've satisfactorily completed one unit, you can set up a production-line operation if you like and work on several windows at once.

2. Thoroughly clean all glass, even brand new panes, on their inside surface. Adhesion of the sealant is of prime importance and determines how long the units last. Dirt, sawdust or old putty will obviously weaken the sealant's bond, but so will an invisible film of grease. The outer surfaces can be cleaned in a conventional manner after the finished windows are remounted.

Caution: Some household glass cleaners contain chemicals that form a dirt-repelling film on glass. This film may interfere with proper bonding of the sealant. Avoid this possibility by cleaning the window's inner surfaces with a simple solution of 1 ounce of ammonia to a half gallon of water.

Once you've cleaned the glass, polish the surface with crumpled sheets of newspaper. Most newsprint is a coarse, fibrous paper with ink made of carbon particles suspended in mineral oil. This combination happens to make an excellent glass-buffing material.

When you finish polishing, put the two panes side by side and cover them with fresh sheets of newspaper to protect them from dust.

3. Next cut the hollow metal spacers to length. The installed spacers will be inset from the perimeter of the glass by a small amount to create a groove for the sealant.

Although manufacturers recommend that the spacers be recessed from the edges of the glass by about $\frac{1}{8}$ inch, the life of the windows is related to the thickness and effectiveness of the seal, so increase the indentation of the spacers to about $\frac{1}{4}$ inch from the edges. To calculate the proper length of your spacers, start with the dimension of the windowpane and then subtract the indent from each end. For example, if your window length is 48 inches and you want a $\frac{1}{4}$-inch indent on each end, you'd need a spacer length of $47\frac{1}{2}$ inches.

Remember to allow for the corner pieces, which also take up some room. Because of the range of sizes and styles, there are no infallible sizing guidelines. Your best bet is to use a full-sized piece of

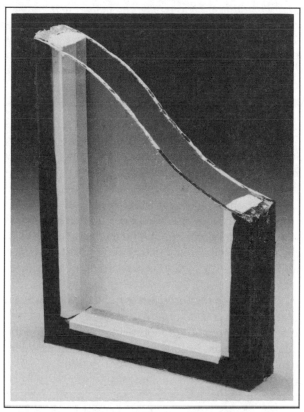

Photo III-13: When it's all done, your double glazing looks like this: glass, spacer, desiccant, sealant, glass. If you're interested in triple glazing for some of your windows, you simply add another spacer and another layer of glass.

cardboard or an extra pane, and experiment until you have found the exact spacer length that produces an even 1/8- to 1/4-inch indentation all around the edge of the glass. After you've got the first spacer frame just right, you'll know just how much to subtract from each dimension because it'll be the same no matter what the length or width of the glass is.

Important: Most spacers are thoroughly degreased at the factory and shipped clean. Prevent skin oils from weakening the sealant bond by washing your hands or slipping on cotton work gloves before you handle the spacers.

Once you've determined the proper lengths, mark the cutting lines and make the cuts. If you feel like getting fancy with mitered corners, don't. The corner inserts are designed to be used with square cut-offs. The spacers are lightweight and easily deformed, so use a hacksaw with a fresh, very fine-toothed blade. Work carefully, and you'll have no problems.

After you've cut the spacers to length, use a fine-grained file to remove any burrs or roughness that could prevent the glass from lying flat on them. Finally, inspect the spacers inside and out to make sure they're still clean, straight and square.

4. Now you're ready to assemble the spacers and fill them with desiccant. Start with the spacer that will be at the window's bottom edge and insert a corner piece. The seamed or perforated side of the spacer always faces toward the center of the glass. In some cases, the corner insert may not slide into the spacer easily, but don't be tempted to spread the spacer. That tight fit will help keep your window sealed.

Open the can of desiccant and insert a funnel into the open end of the spacer. Using a cup or scoop, pour the desiccant into the spacer until it is almost filled. The desiccant flows easily and has the feel of coarse sand or salt. Leave only enough room in the spacer to insert the second corner piece. *Caution:* Some brands of desiccant produce silica dust when poured, so it's a good idea to protect yourself with a filtered respirator.

Insert the second corner, and place the desiccant-filled spacer on a clean floor or a sheet of paper. Add two vertical side spacers to form a U shape. Again using the funnel, fill both side spacers with desiccant, and add the two top corner spacers. Finally, fill and fasten the top spacer to form a complete, desiccant-filled rectangle. The assembled spacer frame may be a bit floppy, so be careful not to twist or bend it.

When desiccant is exposed to air, it immediately begins absorbing moisture, so don't delay finishing the job and keep the desiccant can tightly closed when you're not actually using it.

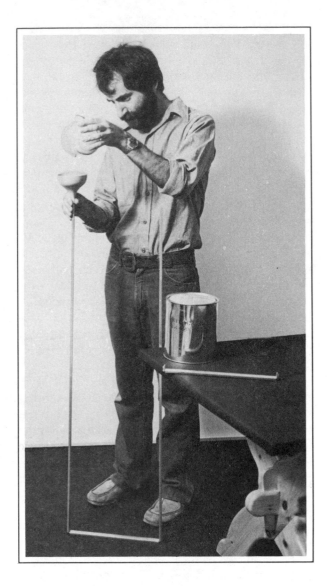

Photo III-14: When assembling the spacers with the corners and filling them with desiccant, care must be taken. The assembly is wobbly, and it wouldn't take too much of the wrong movement to distort the spacer. The fellow in this picture should be wearing a breather mask to fend off silica particles.

If you intend to make a double-seal unit, do it immediately after assembly of the spacers. Apply butyl sealant to both sides of the spacer frame completely filing the grooves made for that purpose, and lay the frame onto one of the panes.

Next, square the spacer frame to the glass to make sure that it's set in equally from all edges of the glass. Carefully pick up the second sheet of glass and place the clean side down on top of the spacer frame.

Once the top sheet is in place, check again that the spacer assembly is completely square and centered between the sheets of glass and that the top and bottom glass exactly overlap one another. Work carefully and get it right while corrections are easy. Once you apply the sealant, the assembly will be locked in place.

When the window is ready for sealing, place a weight on the top pane to hold everything firmly in place. Use anything that's reasonably heavy and large enough to spread its weight over a wide area.

6. Now you're ready to "butter" the windows with sealant. Don't try to fill the space at the edge of the window in a single pass. With ¾-inch spacers, three passes are ideal. On the first pass, seal the seam where the spacer meets the bottom glass. On the second pass, fill the seam between the spacer and top glass, and on the third pass fill the rest of the indentation out to the edge of the glass. Any voids in the sealant will shorten the life of the window, so work carefully. Push the bead of sealant ahead of the caulking gun nozzle, and work it firmly into the edge space. The sealant should completely fill the space, out to the edge of the glass. Be especially sure to fill the corners completely, right out to the edges of the

5. Remove the newspapers covering the cleaned, polished glass, and lay the desiccant-filled spacer assembly on one of the panes. Check for dust or other debris that may have fallen onto the surface of the glass, because anything that's on the glass when the window is sealed will be there forever. Use a clean, soft-bristle paint brush to remove any final traces of dust or dirt.

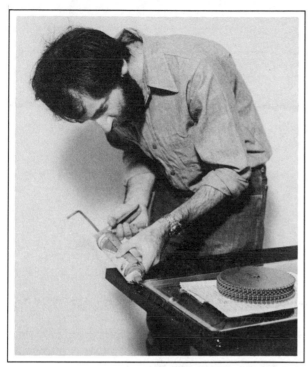

Photo III-15: A heavy but well-distributed weight will stabilize the glazing sandwich while you're making the edge seal. Be very generous with whatever kind of sealant you use. If you're into a mass production thing, you can stack several sandwiches, one upon the other, and "butter" them all at once.

glass. You may find it useful to slightly overfill the indentation and later go back with a knife to trim away the excess.

Curing times for different sealants range from one hour to one day. Check the label. Once you've sealed your units, let them lie undisturbed for the full cure time to ensure that shifting the units won't break the seal.

When the sealant has cured completely, check along the entire perimeter to make sure it hasn't sagged or settled or shrunk, and fill any voids

that may have developed. Using a sharp knife, cut off any bulges or blobs that extend past the edges of the glass.

7. While you're waiting for the sealant to cure, you can measure the exact thickness of the assembled unit and begin modifying the window frames or sashes to accept the new, thicker glazing.

With operable windows, you can deepen the glazing channel with a router and chisel or attach a new lip made of corner molding to support the new glazing. With fixed windows your modifications can usually be done by simply repositioning the stops that hold the glass. But because of the great diversity in window styles, there are no steadfast rules, so you'll have to improvise as best you can if you've got a unique situation.

8. With the sealant fully cured and the sashes ready, you can remount the windows. This procedure is the reverse of Step 1 with one small addition: Insert four neoprene setting blocks into the bottom of the glazing, one at each corner and two more spaced equally between the corners. Stand the window units on these blocks. (They'll compress out of the way.) Then slip in two more setting blocks at the top of the window, spaced equally between the corners. With fixed windows you need only use blocks along the bottom edge. On both types fix the glass into place with standard glazing points.

Replace the beads, stops, or trim pieces that support the glass unit, then thoroughly seal the glass to the wood with glazing compound (fixed and operable windows). You may find that you have to increase the mass of the counterweights on double-hung windows to compensate for the weight of the extra glazing.

After cleaning off the outer surfaces of your new double-glazed windows, go ahead and settle back for some truly energy-efficient window gazing.

Frederic S. Langa

PROJECT
Skylights

In recent years skylighting has become a popular retrofit home improvement, with the availability of dozens of ready-to-install products. Skylights will transform the dimmest of rooms into the brightest, and with proper placement they'll add an extra bonus of solar heat. You can buy them, you can build them, and you can install them without too much trouble—a good project for the budding solarizer.

I used to live next door to a building contractor who couldn't stand skylights. "Nothing but a hole in the roof," he'd say. "Rain in, heat out." Mick's friends wouldn't even mention the word around him to avoid triggering a tirade, but they never told me about it. I found out one evening when I was over for dinner after I casually mentioned that I might install one in a cabin I had just built. After his brief outburst we finished the meal in silence, and I only learned why a few days later. Mick had once agreed against his better judgment to help a friend repair a home-built, operable skylight. The design had proved too complex for the friend's skill: The thing leaked like a sieve, and Mick had to go back several times to work on it, twice repairing rain-damaged drywall. He finally stopped the leaks by sealing the unit so it couldn't be opened, but by then he wanted nothing more to do with skylights.

A few years later Mick finally came around, and like a true convert he's installing skylights all over the place. He insists on adding some kind of movable insulation to minimize heat loss, and he prefers to use skylights that can be opened for ventilation, though he only uses store-bought units for that. He's even designed some shading devices that can be mounted on the roof to prevent excessive heat gain in summer. The lesson is simply that one bad experience can sour you on a good thing, so it's always worth your while to think through exactly what you're doing to

make sure the job is done right.

Skylights Are for Daylight

Skylights are coming into their own as a practical way to brighten up the interiors of old and new houses. A well-placed skylight can transform the dark recesses of a room into an inviting living space, complete with hanging plants and a view of the sky.

A modest 2 by 2-foot skylight, even on a bright, cloudy day, will flood the room below it with the light equivalent of roughly three 100-watt light bulbs. This diffuse, evenly distributed daylight has a very pleasing quality. The shifting play of light as clouds drift by, as the sun travels its daily arc, and as the seasons change is uniquely beautiful. A skylight can be located so that it will frame a view of nearby hills or trees, or it can be set for day and night sky views.

A south-facing skylight can also contribute direct-gain heat to your home, and if you fit it out with some kind of movable insulation, the net heat gain will be significantly increased. In summer, an operable skylight can help cool your house and prevent heat buildup at the ceiling. If hot air escapes through a ventilating skylight, convective or fan-assisted currents can pull cooler air into the house, possibly from a basement or crawl space. Some units

Photo III-16: For houses that are short on windows, skylights will turn those dark spaces into bright ones.

keep out rain even when they're open, and you can leave them open all summer to let the house breathe.

Sizing and Locating a Skylight

You may already know where you want to put your skylights, but if you're still thinking, a few guidelines will help you locate it in the best possible spot. Unless you are building a new home, you've got to consider just how much of your house you want to tear up to put in a skylight, especially if you are paying someone else to do the work.

For general lighting, the unit can face in any direction, although a south roof pitch will provide the most light (and the most heat). In summer, you may want to block out the direct light, so find out which trees or buildings cast a shadow on the opening. When heating is a concern, you naturally don't want a spot where the skylight is shaded in winter and in direct sun during the summer, so a little site analysis (à la "Your Place in the Sun") is in order.

Be careful when considering a high attic space for a skylight. The light received at the floor level is greatly dissipated if the opening is too high. In a single-story house, a skylight can usually be located up to 14 feet away from the floor without too much loss of light. If the house has a ceiling, an opening obviously must be cut to let in the light. You can separate the space under the skylight from the surrounding attic by making a *lightwell,* which consists of four insulated walls that couple the roof pitch to the ceiling. An optional design for reflecting the light downward is to have the sides of the lightwell flare out from the skylight in all directions so that the ceiling opening is somewhat larger than the skylight opening. The daylighting effect is more widely distributed with a flared lightwell. Building straight walls creates much more of a shaft effect that results in greater variations in light intensity. The sides of the lightwell should be painted a light color, preferably gloss white or, if you like the look, they can be lined with reflective foil.

The amount of "light gain" that is created by a skylight is of course a function of its size and

Photo III-17: There's always going to be some distance between the roof plane and the ceiling that has to be boxed-in and finished off, and for a softer, less "boxy" look, the skylight opening can be flared as it is above. This operable skylight (Velux) has a blind for limiting direct gain in summer, while at the same time it can be opened for ventilation and some diffuse light gain.

orientation, and if you can't always control orientation, you can certainly control size. A rule of thumb for sizing a skylight that's located at a height of 10 to 14 feet above the floor is that it have an area around 5 to 7 percent of the floor area *to be illuminated.* In a large room, for example, you may be interested in skylighting just a part of it, so make a rough measurement of that particular area and size accordingly. If you want skylighting throughout a large room, it may be a better plan to install two, three or more smaller units rather than a single gigantic one. The overall materials' cost may be higher, but you'll end up with better light distribution. Also, it's easier to create a

movable insulation system for a smaller skylight.

If you live in or near a large city, smog may cut down your light appreciably. Skylight manufacturers thus suggest sizing the opening to 10 percent of the floor area to compensate for frequent smog or clouds. Of course, skylights are like solar collectors: They aren't made in a close range of sizes, so your sizing will always be approximate.

The glazing material and number of layers used also have some effect on the percentage of available light that reaches into a room. The reflectivity of a lightwell and even the walls of your room noticeably alter the extent of light gain and distribution. Chocolate brown walls, for example, absorb 90 percent of the light a skylight brings in.

If you also want a skylight to be a significant heat source, it will naturally have to be sized larger than one intended to be primarily a light source. The ratio of the skylight size to the net heat gained changes according to the same factors that affect any solar installation: the azimuth and tilt of the glazing, the weather-tightness of the installation, the quality of the movable glazing insulation, the amount of thermal mass (added and/or existing), the level of wall and ceiling insulation and so forth.

Since skylights are normally installed parallel to the roof axis, the azimuth usually isn't adjustable. Tilt, however, is adjustable if you're willing to build the proper structure to support the skylight. But if your plans take you upward past a 45-degree tilt, maybe you're really planning a clerestory, which is discussed next in this section.

Most skylights mounted flat to roof pitches won't be optimally tilted for direct gain. While an azimuth 30 degrees east or west of true south is acceptable for a skylight to gain significant thermal energy, a tilt of 0 to 30 degrees is not (in latitudes above 20 degrees), and that's a common range for roof pitches. Low winter sun angles of course deliver more energy to steep pitches; for example, at 40 degrees north latitude the best glazing tilt is in the range of 55 to 65 degrees, depending on the local climate. Glazing that is tilted 35 to 45 degrees below the optimum tilt will collect 25 to 40 percent less

A Lightwell for All Seasons

When Rich Kline bought an old farmhouse in the hills of eastern Pennsylvania, he wasn't too concerned about how much sunlight came into the house. He just wanted a sturdy dwelling in a secluded spot. The person who had built the place wasn't too concerned about letting in sunlight either, and when Rich moved in he found the kitchen to be a very dark place. That didn't seem right for a kitchen.

Rich was also wondering what to do about a big upstairs room that served no real purpose other than allowing access to the other rooms. Then he was seized by a flash of architectural inspiration that sometimes strikes in the most timely fashion. The total solution evolved into this: Cut a hole in the roof, another larger hole in the second-floor ceiling, and an even larger one in the floor of that useless room and, Kline reasoned, the light would come showering in. He was right. The south-facing roof pitch means an optimum light source for the skylight, and the lightwell delivers the abundant light of day to the once-dim kitchen area. The insulated (R-19) walls of the lightwell keep heat from leaking into the unheated attic. The opening in the upstairs floor is surrounded by a railing festooned with elegant hanging plants, and the space has become a kind of year-round atrium. For three holes, some framing and some finish

Photo III-18

work, the entire house has been brightened and transformed.

---

energy per square foot than glazing that is tilted to correspond with winter sun angles. It's fair enough to say that a shallow-pitched south-facing skylight won't be a boon to your solar space heating plans. By the time you plan an opening that's large enough to warm the interior, you may well have exceeded practical limits for expense and complexity, and you'll certainly have more heat loss to deal with in a large skylight, plus the likelihood of the greater problem of summer heat gain.

So most skylight placements will probably be for light first with heat gain being a secondary

consideration, and there are other planning considerations that may also supersede the gallons of heating fuel that might be saved. If you are concerned about your privacy, a skylight can provide a way to let in sunlight without opening your living room to the world if, for example, it faces a busy street or is close to another house. If the best location for a skylight does allow an unwanted view *in,* you can consider using translucent or diffusing glazing (FRP, double-wall extruded acrylic, milk-white plastics). Security may also be a concern because an operable skylight can be a way for someone to break in. The minimum guard is a locking operator mechanism; the maximum, à la New York City, is steel bars.

Locating a skylight above humid areas will encourage condensation. Condensation occurs in poorly designed skylights where two glazing layers touch at the perimeter, but it can also happen in humid rooms. You can expect some if you locate the unit in a bathroom or greenhouse, though some of the better units have a gutter around the perimeter to collect condensation. Greenhouse plants won't mind a few drops from above, but in the bathroom you might find dripping water a bit too invigorating. The kitchen is another typically humid room, but there a ventilating (operable) skylight might be worth the slight bother of condensation because it can provide a path for heat and odors to escape.

A final consideration is that your sky window be accessible from the inside and the outside. Who knows? You might have to repair the flashing some day. Even if you don't, the thing will need occasional cleaning. If summer heat gain is a problem, you may also need access to mount a seasonal shading device out on the roof. You should be able to reach the inside to operate it, to handle movable insulation and to clean it. We make no bones about movable insulation: Don't add glazing (in climates with more than 2000 heating degree-days) without adding glazing insulation. Otherwise the glazing is simply not a completed system but rather a continual drain on your heating system.

What Kind to Get

The first thing to look for in a quality skylight is double glazing with properly assembled glazing layers. A single layer of glass or plastic hardly slows down the heat rushing to escape, yet there are many single-glazed skylights on the market. Also, because single glazing doesn't separate the warm interior air of your home from the cold outside air, you are sure to get condensation no matter what the location. Double glazing minimizes this effect because the surface of the inner glazing is warmer than the dew point of the room air. But not all double-

Photo III-19: A spic-and-span skylight maintains daylighting and keeps your view clear. When cleaning plastic skylights, be sure to use the right kind of nonabrasive cleaner.

glazed skylights are created equal. In some, the layers actually come together where they attach to the skylight frame (see figure III-8). Condensation occurs there, and heat leaks out because two glazing layers in contact are really single glazing.

In the better skylights, the layers are separated by spacers (thermal breaks). The airspace between the glazing is vented to the outside so the humidity in the space is similar to the outdoors, which helps to eliminate condensation between the layers. Since the vent holes are small, heat loss is not significantly increased. But check carefully. Some cheaper units have holes that allow air to escape through both glazing layers to the outside, a detail that really wastes heat.

The ideal distance between the layers of glazing is about ¾ inch. If the distance is much greater, miniature convective air currents may form which can transfer heat from the inner to the outer glazing. Before you buy, check the cross-sectional diagram, which should come with the unit, to be sure the design is adequate. Be cautious with one that doesn't have a diagram.

Skylights are made with either a domelike "bubble" plastic glazing or with flat glass or plastic panes assembled much like a double-glazed window. The bubble kind may provide a slightly broader view of the outside if it's located where you can actually poke your head into the opening. Bubbles do distort the view a little, if crystal clarity is a concern. Any decent plastic-glazed model is made with material that will not degrade appreciably in sunlight.

The main consideration with glazing is safety and durability. Plastic scratches easily and could lead to problems where rough treatment or an abrasive environment is likely. Regular glass, while it resists scratches, does shatter and shouldn't be used as an overhead glazing. Even tempered glass breaks into potentially harmful glass beads, and some local building codes don't allow it to be used in skylights for that reason. Some codes will only allow wired and/or laminated safety glass in overhead applications. Avoid locating a glass skylight under branches or anything else that could fall and break it. In particularly risky locations plastic glazing or wire-laminated glass should be used.

Another thing to examine carefully is the flashing package offered with a skylight, especially if you plan to install your own unit. Most good quality skylights have some kind of integral flashing system, but the better ones may have long-life copper flashing and a built-in curb to help shed water. The best flashing system doesn't depend on roof cement to produce a weather-tight seal, although it doesn't hurt to add some anyway.

When you shop for features, give some consideration to an operable unit for the benefit of natural ventilation. A skylight that can open is going to cost more, but it can be worth the expense to be able to have a ventilator that's well placed for exhausting heat. Make sure that you can somehow operate the unit remotely if you can't actually reach the unit itself. You can get motorized units that respond to the touch of a switch, but these involve more work (electrical) and considerably more expense.

From the many dozens of commercially available skylight systems, we've culled in the "Hardware Focus" section following this essay what we believe are the brands that fulfill the minimum quality standards described above. Undoubtedly there are others of adequate quality, and you'll be able to select a good unit from the Hardware Focus or by using the guidelines. Some units have attractive features that go beyond what's been discussed, and though they are naturally more expensive, the added

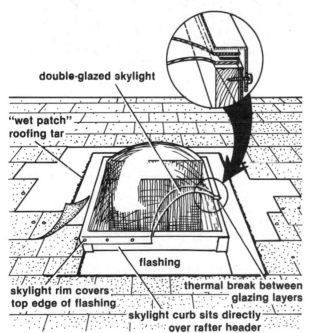

double-glazed skylight

"wet patch" roofing tar

flashing

skylight rim covers top edge of flashing

thermal break between glazing layers

skylight curb sits directly over rafter header

Figure III-8: When choosing a skylight, it's important to know how the edges of the double glazing are sealed. The detail shows the proper gasketing, which seals and separates the glazing layers, reducing conductive heat loss. You'll find that just about all commercially made fixed and operable units come with perimeter flashing that lies over the curb and roof flashing you install. On home-built jobs using flat plastic glazing or safety glass you can make your own perimeter flashing using standard metal corner flashing. Apply a bead of caulk to the glazing perimeter and nail or screw the perimeter flashing to the sides of the curb.

Skylights for Space Heating

Jim and Maureen Greisemers' passive solar home isn't a retrofit, but it's certainly an ode to the value of skylights for space heating. Over half the area of the south roof pitch is covered with skylight glazing, which lets in enough solar energy to provide about half of their heating needs. Because the roof is pitched so steeply (45 degrees), there was no need to build in extra tilt for the skylights, which kept the installation from becoming too complicated. The big skylight is really a "skywall," a window wall turned toward the sun, the clouds and the stars. It was made with 160 square feet of double-wall extruded acrylic glazing that was set right over 4 $\times$ 12 rafters placed on 4-foot centers. The direct gain through this glazing is for the Greisemers' greenhouse, with excess heat going into the rest of the house. Movable insulation for the skywall is the Skylid insulated louver system. The R-5 insulation of these louvers greatly reduces night heat loss through the glazing, and the louvers also provide summer shading.

The six other skylights are the operable Velux-brand units, which add another 90 square feet of glazing with all the convenience features for which Velux is famous. The units are double glazed with glass, and not only are they operable for summer ventilation, but the sash can also swing into the house for easy cleaning of the outside glass. The exterior awning shades direct sunlight, while admitting diffuse and reflected light, and the unit can be open all the while for

Photo III-20

ventilation. Velux also has an interior roller shade and an interior insect screen, making this brand one of the most "fully featured" of all commercially available skylight systems.

convenience may justify the extra cost. Additional features include a built-in insect screen, a built-in interior curtain-shade, an operable exterior awning or louvers for shading; there may now even be some that include movable insulation. One brand that has several of these features is the Velux Roof Window. It's expensive, but it's got convenience written all over it. At least give it a look; it may give you some ideas that you can build into your own skylight system. A good quality, bubble-type unit is the Ventarama. Some models are operable; they come with copper flashing, and they can be fitted with screens and shades supplied by the manufacturer. Again, quality always costs.

Photo III-21

Home-built Skylights

If you would consider making a screen or a shade or movable insulation for your skylight, you've probably got the wherewithal to make the skylight itself. You can certainly save big by making and installing your own, but you can also lose big if it's not made or installed properly. A fixed skylight certainly invites fewer problems for the home builder. A flat, double- or triple-pane glazing unit can be home-fabricated and installed as shown in figure III-9. A home-built operable skylight is likely to be a lot trickier at all points in the project, but nothing's impossible when experience is brought to bear on a

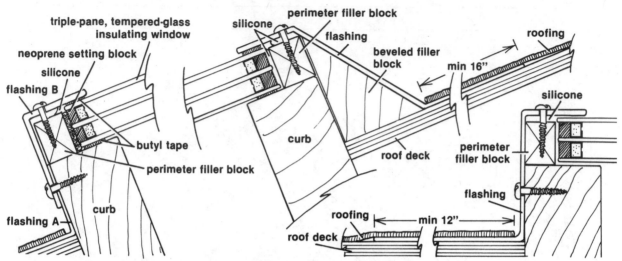

Figure III-9: Here's a plan for installing a triple-pane, tempered glass insulating window (as described in "Build Your Own Multi-Pane Windows," found earlier in this section) into a skylight opening in a roof pitch. The cross section at left shows the down side detail in which flashing A is run between the glass and the curb, and flashing B runs over the window and over flashing A. A neoprene setting block cushions the window against the perimeter filler block on the down side only. The middle cross section shows the up side detailing in which a beveled filler block is used to carry the flashing back to the roofing. This reduces the amount of water that can stand above the skylight and possibly work its way up under the roofing. In the side cross section at right the flashing is run at least 12 inches out from the curb, and the outer edge is finished with a hook bend. Note that even though butyl tape seals all the flashing to the glass, a thin bead of silicone is also run between the glass edge and the perimeter filler. When using aluminum flashing, it's a good idea to use aluminum screws for fastening.

problem. Look at commercially made units to see how they get it right; talk with a professional glazier about making flashing and sealing an operable skylight. A local glazing shop may be able to supply you with the right hardware.

Insulation

It's common to think of skylights sitting in the roof as being static—like a chimney. But the totally "unattended" skylight will be an energy liability, just like any uncontrolled glazing system. All skylights are windows, and windows lose a lot more heat than the neighboring insulated wall or roof section. When a window is mounted in the roof, heat loss is even greater because the warmest air collects at the ceil-

ing. Unless there's good insulation, a square foot of skylight will lose somewhat more heat than the same area of vertical glass. Double or triple glazing provides some degree of insulation, but it's not enough on cold nights. More protection can be had by using insulating louvers, quilts, curtains or foam panels.

Louvers consist of slats in a frame that admit light easily when open. Closed, the overlapping, insulating slats are a barrier to radiant and convective heat loss. The only commercially available insulating louver system is made by Zomeworks (P. O. Box 712, Albuquerque, New Mexico 87103); it's called Skylid, and it's made for large openings that are 4 to 6 feet wide and 4 to 10 feet long. A roller-type shade can be used as both a shade against summer heat gain and as insulation against winter heat

Photo III-22: The Zomeworks Skylid system is a unique insulating louver that opens and closes automatically, depending on the indoor air temperature and the availability of sunlight.

loss. Some commercially available insulating shades run on edge-tracks, and these can be modified to work with an overhead skylight. In Section III-B you'll find a project for an insulating shade that might be modifiable for skylights.

Rigid foam panels are probably the simplest way to insulate a skylight. They can be cut to fit precisely into an opening and be held in place by friction, Velcro, magnetic strips or some sort of mechanical latch. The panel can be stored next to the opening when it's in regular use and perhaps go to a more out-of-the-way storage area when the weather turns warm.

It's tempting to call these friction-fit panels energy corks. Imagine a Styrofoam block shaped to fit snugly into a lightwell. Pop it out for star viewing or solar gain; pop it in for cold weather protection. These panels insulate best when they are covered with a reflective foil to reflect radiant heat back into the room. When either naked polyurethane, Styrofoam or polystyrene is used, it must be protected from ultraviolet radiation. Foil-covered foams such as Thermax (isocyanurate, an excellent insulator) are protected from sunlight but still need edge protection. All these foams have the disadvantage of producing poisonous fumes when they're exposed to flame or very high temperatures, but their light weight and good insulating value make them hard to beat.

If handling foam panels is a problem, especially with large skylights, consider permanently mounting the panels on tracks as shown in figure III-10. Two half-panels can be "tracked" to rest on each side of a large opening. A half-panel system would be useful when there isn't enough ceiling area to accommodate one large panel, and in the open position two half-panels might present a more balanced look.

An alternative to sliding panels is panels mounted, with some ingenuity, with hinges and a rope-and-pulley system. The advantage is in having remote operation and a relatively unobtrusive installation. This system would be appropriate for skylights that have long lightwells.

It's also possible to add a third and fourth layer of glazing in the form of a movable glazing panel. North-facing skylights in particular are definitely net energy losers in cold weather, even when the sun is shining. A double layer of a thin film, for example, would maintain daylighting without greatly reducing overall light transmission, and heat loss would be somewhat reduced relative to the heat loss from a double-glazed skylight.

Also protect the skylight against wind-induced heat loss. Strong winds can ferret heat out

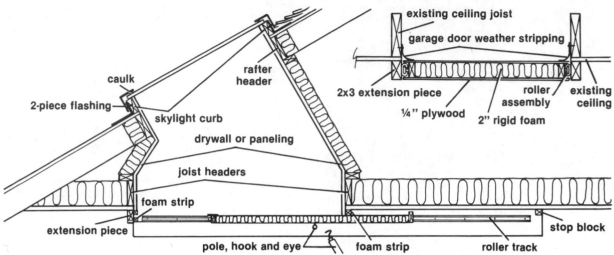

Figure III-10: With the insulation supported by a roller-and-track mechanism, thicker and therefore heavier foam can be used to get increased insulating value. Two inches of high quality foam (isocyanurate) is rated at R-14 to R-16, and 3 inches gives an R-21 to R-24 rating, both of which come close to standard levels of insulation for ceilings. (Design adapted from William K. Langdon, Movable Insulation, *1980, with permission of Rodale Press, Emmaus, Pa.)*

of the tiniest of cracks, increasing general glazing heat loss. If your planned skylight location is likely to be in the face of prevailing winter winds, you can protect it by building a low barrier around three sides, leaving the south side open so as not to block direct gain. If you want to get fancy, you can fasten reflective material to the inside of this barrier to bounce more sunlight at the skylight.

Shading

When solar altitudes start to reach high into the midday sky, hot weather can't be too far off. The home energy manager's thoughts and actions turn toward keeping cool rather than keeping warm, and shading windows is a high priority item on the list of natural cooling options. External shades for skylights are generally the most effective, although an internal shade combined with an open operable unit is also a good way to resist heat gain. The possible conflict with operable units is that internal shading might block ventilation. This is where external shading

again has an advantage.

An external shade can be designed to exclude direct sunlight while admitting some diffuse and reflected light. A deciduous tree is about as good a shade as any; maybe tree locations around your house should be considered when planning skylights. For want of a tree, an exterior panel or awning can be rigged for seasonal placement. An arrangement of fixed louvers would be an interesting shading device. Fixed louvers have been used on roof overhangs to admit low-angle sunlight to south-facing windows and exclude high-angle beams while admitting some reflected and diffuse light. For large, steeply pitched skylights that contribute to direct gain heating, an adjustable louver system could be a way to exercise a great deal of shading control. Realistically speaking, the making of such a device is best left to the craftier home builder.

Internal shading has simplicity in its favor. Movable insulation can be a shade, but that blocks all light. A plain white roller shade can be rigged for remote, horizontal operation, and some light would still get through. It's not necessarily true that every skylight, no matter what the size or orientation, should absolutely be shaded. As with any glazing system, the bigger the aperture, the more important controls become. If you have a house with several small, operable skylights, they could conceivably remain unshaded in summer if you instituted other natural cooling procedures. But if a big skylight is beaming in the summer sun, it'll naturally be an enemy of your efforts at keeping cool. Take a wait-and-see approach to shading: Wait 'til you can see a real problem, and then, if you're so moved, take remedial action.

Installation

The actual carpentry required to install a skylight is exacting, but not terribly difficult. Everything points toward making the installation waterproof and free of air leaks by following standard construction procedures. Skylights are nowadays a regular construction event, and there is no mystery about what it takes to get it right. You've also got to be prepared for working on the roof, and there's a big difference between horizontal or shallow-pitched roofs and steep ones. The steeper the roof, the more physically demanding the job will be. If you have reservations about your ability to pull it off, you may be better off hiring someone with experience in skylights.

Once you've decided on the actual location of your skylight, find the rafters under that particular roof section. A small skylight (up to 24 to 36 inches wide) may fit between two rafters without requiring any to be cut, but often you will have to cut out and head-off one or two to make room for the unit. If you cut away no more than two rafters, there's probably no need to double the ones on each side of the opening, but check a carpentry text to find the recommended spans for the size and type of framing in your house to be certain these members aren't already undersized. It's not that the skylight adds a big load, but that the loads of cut rafters must be borne by the two rafters at either side of the skylight. If you end up cutting away three or more rafters for the skylight opening, the two side rafters will probably have to be doubled between the roof ridge and the top plate of the exterior wall.

If the inside of the roof pitch is finished, it's going to be harder to install a large skylight that will require doubled rafters. The interior wall finish will have to be stripped away to allow installation of the rafter doublers, but if you must, you must. If the inside of the roof is unfinished, you can locate the skylight on the inside and drill a hole through the roof at each corner of the planned rough opening. The rest of the installation can be done up on the roof.

Find the four holes you've just made in the roof and start stripping away the roofing in and around the rough opening. Be neat when stripping away shingles; it's better to take out whole shingles rather than snapping four chalk lines and just cutting through them. When you pull whole shingles, it's easier to roof back to the skylight, and the reroofing will blend better with the old roofing.

With the shingles removed, cut away the

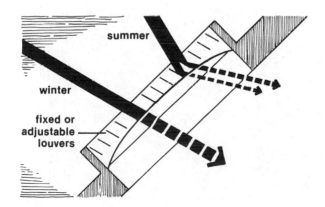

Figure III-11: Shading for large skylights can be achieved with fixed louvers set at the appropriate angle or with movable louvers that are perhaps adjustable from the inside. In summer, even though the direct beam is blocked, there will still be daylighting from diffuse and reflected light.

A Home-built Skylight

Bob Ayers is the sort of craftsman who will rip out a whole drywall taping job if he can see the seam. He was no less meticulous when he installed a large south-facing skylight above the living room of his smallish home in the foothills of Pennsylvania's Lehigh Valley.

Since there was an attic above the living room, Bob cut a big hole in the ceiling and made a three-wall lightwell that tapered to the south-facing exterior wall. (This gave him the added opportunity to further perfect his drywall technique.) The skylight itself is 6 feet wide by 5½ feet high; the center section is glazed with ½-inch Plexiglas. Two narrow side sections will carry a pair of stained-glass windows, also Bob's creations. The glazing is mounted on a 6-inch-high curb that is weatherproofed with aluminum step flashing. Counter flashing is fastened at the top edge of the curb to overlap the step flashing. The glazing slightly overhangs the bottom edge of the curb to direct water over the flashing. It's sealed to the curb with a bead of durable silicone caulk, and a glazing cap of ⅛-inch-thick aluminum angle stock is screwed through the Plexiglas. The screw holes for the Plexiglas were predrilled, filled with caulk and sealed between the screw head and the cap with a neoprene washer.

Photo III-23

Photo III-24

Inside, a thin sheet of Plexiglas was fastened to the bottom of the curb as a second layer of glazing. For added insulation, Bob is building a roll-type insulating shade that will slide on runners along the bottom edge of the curb.

To prevent warm weather heat buildup in the lightwell, Bob will add an exhaust fan at the highest part of the well, venting through the attic to the outside. He may run another duct from this vent to a back room to allow the winter option of redistributing heat collected in the lightwell. A semitransparent shade will also be used to limit summer heat gain.

The house may be small, but the big skylight has made a little dining nook into an emporium of light where people and plants alike seem to flourish.

Photo III-25

roof decking or sheathing to expose the rafters, again using the four holes as end points for your cutting lines. Now the framing begins. If rafters must be cut, mark the rough opening dimensions and cut them away. Then cut and install the headers for the top and bottom of the opening. If there's no rafter cutting involved, you just install cross-blocks between the two rafters, and if the width (between rafters) of the rough opening is narrower than the rafter spacing, you'll have to nail filler blocks between the cross-blocks to get the right rough opening.

For skylights that come with a flashing package, you usually specify the pitch of your roof when you order, and the flashing you get will be suitable for that pitch. For pitches under 20 degrees, the flashing is designed to come up higher from the plane of the roof and is used with a curb. A curb is a good idea in any case, because it gets the critical seams above the roof, reducing the chance that wind can drive water past the flashing or that ice can back up and get inside. Simply build a 2×4 or 2×6 box directly over the sheathing, rafters and headers of the opening, and toenail it to the roof. On steep roof pitches, fixed skylights can usually lie flush with the roof plane.

You can also build the rough opening and the curb in one operation by using lumber that's wider than the existing roof framing. For example, if the roof is framed with 2×6 rafters, and the skylight requires a 3- to 4-inch-high curb above the roof deck, you could use 2×10's or 2×12's for rafter headers, cross-blocks and filler blocks to get instant curbing.

Most store-bought skylights have a built-in continuous flashing shield for fast installation. The perimeter of the rough opening is generously daubed with roofing cement, then the unit is put in place and tacked down. The lower edge of the shield must lie *over* the existing roofing, while the other edges are roofed over with the replacement shingles (which can be recycled from the shingles that were previously removed). The replacement roofing goes to within about ¾ inch of the metal skylight curb.

Other skylights come with a package containing step flashing and counter flashing. The step

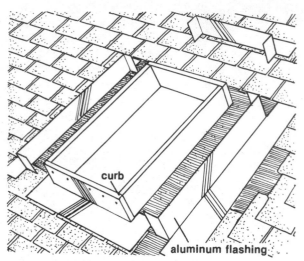

Figure III-12: Flashing can make or break the success of a skylight installation. Use common sense: Make sure that uphill flashing sections lap over downhill pieces; before nailing down a piece trowel on a fat bead of black goop ("wet patch" roof tar) under all sections except the downhill section, which simply laps over the shingles. Shiny aluminum flashing can be painted to match your roof.

flashing is interleaved with the replacement shingles, starting from the bottom, after a larger piece of flashing has been laid over the bottom course of shingles and against the lower face of the curb. Step flashing has a 90-degree bend that allows it to rest against both the roof and the curb, and it is nailed into the side of the curb. A cap is placed at the top of the curb and the counter flashing (which may be part of the skylight) is set in place and fastened to the top edge of the curb. The skylight is then set in place, sometimes with additional counter flashing attached that covers the screw or nail heads in the first counter flashing. The skylight is fastened on the inside to the rafters and headers and finished with drywall and trim.

Standard flashing systems like these work well because they present a series of interlocking barriers to the weather, and they minimize the number of seals that are made only with roofing cement, which can eventually fail. If you install your own home-built skylight, imitate these methods by cutting step and counter flashing from a roll of aluminum coil or roll stock and by building a curb to hold the glazing. To flash the top of a wide skylight with a high

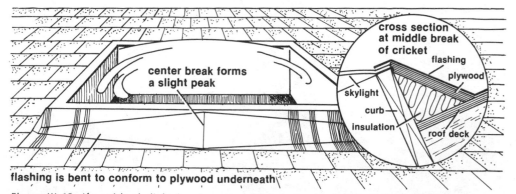

Figure III-13: If a wide skylight is set up on a high curb, the flashing along the top edge must be formed so as to direct water to the sides and prevent pooling. A cricket can be made using two triangular sections of 3/4-inch plywood nailed to the curb and the roof deck. The flashing is bent to the same diamond shape. Two pieces of aluminum flashing can be used, with an overlapping seam at the middle break point. The space enclosed by the plywood is insulated to reduce heat loss through the curb.

Photo III-26: This is the inside view of the skylight pictured at the beginning of this article. This rather gigantic skylight made a dark attic into a beautiful plant place, with lots of greenery on both sides of the window.

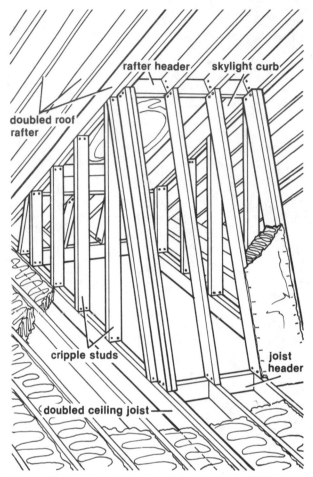

curb, a cricket can be built up from the roof to shed water to either side of the unit (see figure III-13).

If the skylight is above an unoccupied attic, you'll have to build a lightwell to help direct the light downward and seal heat out of the attic (see figure III-14). It can be framed with 2×3's, 2×4's, or 2×6's, the latter giving a bigger cavity for insulation. The ceiling opening is framed the same way as the skylight rough opening, with headers and blocks as necessary. It should be at least as large as the skylight opening, but to really benefit from it, you should build the lightwell somewhat larger than the skylight to spread more light around. The lightwell framing connects the bottom of the roof rafters to the top of the ceiling joists. Fiberglass batts are stapled between the studs and a vapor barrier tacked over the inside edges of the frame. Drywall or some other wall finish is fastened over the vapor barrier. As discussed earlier, a white-painted wall finish bounces the most light down to the living space.

A good skylight, meaning one that is properly located, installed and controlled, is much more than just a hole in the roof. It's a view to the sky, the stars, the treetops; it's a special view that houses usually don't give you. Properly controlled, a skylight is thermally efficient and responsive to changes in weather and season. It brings a gracious daylighting and perhaps a bit of space heating: changes for the better for your home.

John Blackford

Figure III-14: A careful scan of the framing of this lightwell shows that connecting a hole in a roof with a hole in a ceiling isn't too terribly complicated. The rafter and joist headers tie into the doubled roof rafters and ceiling joists, and the cripple studs serve to carry the finish wall and insulation. In doubling either a joist or a rafter it's important to run the ends of the doubler over vertical walls that can bear the extra load.

Hardware Focus

Skylights

APC Corp.
44 Utter Ave.
Hawthorne, NJ 07506
(201) 423-2900

Sky-Vue Skylight and Kleer-Vue Skylight
Double acrylic dome; fixed or operable units; numerous sizes; smallest 14¼" × 14¼"; largest 46¼" × 46¼".

Bristol Fiberlite Industries
401 E. Goetz Ave.
Santa Ana, CA 92705
(714) 540-8950

Bristolite Insulated Flat Skylights
Double-walled acrylic; fixed unit; 15 sizes; smallest 14¼″ × 14¼″; largest 37″ × 75″.

Circle Redmont
P. O. Box 853
Darien, CT 06820
(203) 323-2103

Outlook Skylighting Systems
Double-glazed glass; 10 standard sizes; smallest 21¾″ × 53¼″; largest 53¾″ × 53¾″; fixed unit.

Dur-Red Products
4900 Cecelia St.
Cudahy, CA 90201
(213) 771-9000

Dur-Red Products Skylights
Double acrylic dome; fixed or operable units; numerous sizes; smallest 14½″ × 14½″; largest 92½″ × 92½″.

Faulkner Plastics
P. O. Box 11266
4504 E. Hillsboro Ave.
Tampa, FL 33610
(813) 621-4703 or toll-free in Florida:
    (800) 282-2780

Solar Lite Skylight
Outer dome of Lexan with inner twin wall of polycarbonate sheet; fixed or operable units; 15 sizes; smallest 14″ × 14″; largest 95″ × 95″.

Kalwall Corp.
P. O. Box 237
Manchester, NH 03105
(603) 668-8186

Sun-Lite Skylight
Sun-Lite glazing sheets bonded to both sides of sealed aluminum I-beam core; 9 sizes; smallest 14¼″ × 14¼″; largest 30¼″ × 46¼″.

Kennedy Sky-Lites, Inc.
3647 All American Blvd.
Orlando, FL 32810
(305) 293-3880

Kennedy Sky-Lite
Double-, triple- or quadruple-glazed Lexan domes; fixed or operable; numerous sizes; smallest 14½″ × 14½″; largest 46½″ × 46½″; optional screen for ventilating unit; sunshade under development.

O'Keeffe's, Inc.
75 Williams Ave.
San Francisco, CA 94124
(415) 822-4222 or toll-free number:
    (800) 227-3305 (except California);
    in California (800) 622-0721

Double acrylic dome; fixed or operable units; 4 models; 26 standard sizes; smallest 16″ × 48″; largest 120″ × 144″.

Roto Frank of America, Inc.
Spencer Plains Rd.
Old Saybrook, CT 06475
or
P. O. Box 157
Centerbrook, CT 06409
(203) 399-7158 or toll-free number:
    (800) 243-0893 (except Connecticut)

Roto-Roof Window/Skylight
Double glass; operable units; 14 sizes; smallest 21¼″ × 33½″; largest 52¾″ × 57″; screens, blinds or shades optional.

Solartron Corp.
100 S. Ellsworth Ave.
San Mateo, CA 94401
(415) 342-8142

Solartron Insulating Skylights
Double-walled fiberglass-reinforced acrylic-polyester; translucent; fixed unit; 11 standard sizes; smallest 24″ × 24″; largest 48″ × 96″.

Tub-Master Corp.
413 Virginia Dr.
Orlando, FL 32803
(305) 898-2881

Skymaster Skylight
Double-walled plastic; fixed or operable units; sizes range from 16″ × 16″ to 48″ × 72″.

Velux-America, Inc.
74 Cummings Park
Woburn, MA 01801
(617) 935-7390 or 935-7848

Velux Roof Window
Double insulating glass; operable unit; accessories include blinds, awnings and remote controls; 9 sizes; smallest 21⅝″ × 27½″; largest 52¾″ × 55″.

Ventarama Skylight Corp.
75 Channel Dr.
Port Washington, NY 11050
(516) 883-5000

Ventarama Skylight
Double acrylic dome; fixed or operable units; sunshades, triple panels or electrical motorization available; 6 sizes, smallest 22″ × 30″; largest 45½″ × 45½″.

Wasco Products, Inc.
P. O. Box 351
Sanford, ME 04073
(207) 324-8060

Wasco Skywindow
Double acrylic dome; fixed or operable units; numerous sizes; smallest 22¼″ × 22¼″, largest 92½″ × 92½″.

Movable Insulation for Skylights

Sunflake
625 Goddard Ave.
P. O. Box 676
Ignacio, CO 81137
(303) 563-4597

Sunflake Skyshield
Power-driven, pulley or manually operated; slices in frame from storage section to cover viewing section; storage section above, below or to side of skylight aperture; rated R-14; Sunflake custom designs to fit any application.

Zomeworks
P. O. Box 712
Albuquerque, NM 87103
(505) 242-5354

Skylid Insulating Louvers
Two types: (1) louvers built with 2 sheets of aluminum curved over wood ribs, filled with fiberglass insulation; (2) flat honey-comb core louver has Kraft-paper core between aluminum sheets. Requires no external power source; system opens and closes in response to sun by means of self-contained gravity balance system. Rated R-3 to R-5. In standard 2 or 3 louver models to cover approximately 4′ × 4′ to 6′ × 10′ skylight openings. Custom sizes available.

PROJECT
Clerestories

Clerestories (and dormers) are really just big, propped-up skylights. If you want a large roof window for solar gain, consider the clerestory. With vertical or slightly tilted glazing, a clerestory will give you more control over when and where sunlight will enter. It will also give you more headroom if you're expanding a cramped attic or top floor space. To do all this you've got to know some basic carpentry and roofing techniques; you'll end up with what amounts to a mini-addition and a line of solar windows that will give you direct gain space heating and daylighting like never before. Go ahead, raise high those roof beams.

There are many ways to capture the sun's warmth in a house, and most of them share one thing —glazing on the south wall. But a south wall may be ineffective if it is shaded in winter, oriented too far from true south, or otherwise unsuited to the addition of large areas of glazing. In such cases clerestories may be better for retrofitting because they are located up on the roof where they can be clear of shading obstructions that fall on the wall below. They can sometimes be oriented at a more favorable azimuth than the roof itself, and, perhaps their greatest advantage, they can admit heat and light to rooms too far from the south wall to receive much benefit from it.

A clerestory is a structure that is built up from a roof to support vertical or tilted glazing. When properly designed, built and controlled, a south-facing clerestory is an effective direct gain solar heating system. The vertical glazing admits a greater percentage of winter's low-angle sunlight than a skylight, which is usually built with its glazing parallel to the plane of the roof. In summer the roof overhang and vertical glazing of a clerestory shade and reflect away much of the high-angle sunlight, preventing excessive heat gain.

For homes that weren't designed to benefit from solar gain, a clerestory may offer the only practical way to admit enough of the sun's energy to heat the interior, especially to heat rooms on the north side. For example, many houses have a narrow wall facing south that is not large enough to warm more than a fraction of the living space. If the roof in such cases is of an amenable design and the appropriate azimuth, extra solar heat can be gained through clerestories.

Simply collecting heat from the sun is not enough, though. Some of that warmth must be stored or distributed, or the living space under the clerestory may overheat during the day and become too chilly at night. A well-designed clerestory should be aided by enough thermal mass to minimize temperature swings by absorbing excess heat during the day and releasing it at night. Where thermal mass cannot be added, improved heat distribution throughout the rest of the house is also an option. If neither is possible, the clerestory window area must be sized down to provide some light but only a limited amount of heat.

Planning a Clerestory

A well-designed clerestory may be a significant source of heat, but not all houses can be retrofitted with one. In some cases the structural changes required in the roof to accommodate a clerestory

Photo III-27: Like their cousins the skylights, clerestories can be selectively placed on a roof to bring in light and heat, but because of their design they can carry more glazing area.

may be so complex as to make the project infeasible. On a steeply pitched roof it may not be possible to locate the clerestory close enough to the living space to warm it effectively. Also, a steep pitch would require a long and possibly unattractive run of vertical glass to enclose the clerestory, in which case a skylight or a dormer with a down-sloping roof might be more appropriate. Roofs with shallow pitches are certainly good candidates for clerestories, and even a shallow north-facing pitch can take one on if conditions require. Where the roof ridge is not aligned close enough to the east-west axis (within 30 degrees), it may not be structurally or aesthetically feasible to orient the clerestory toward the south. Clerestories do combine well with flat roofs where they can be oriented to receive the maximum solar gain no matter which way the house faces. This makes them useful in the Southwest, where flat roofs are common.

As with any solar remodeling, shading must be carefully analyzed. Clerestories will, of course, not work as heat collectors if they are shaded in winter by another building, evergreen trees or other obstructions to the south. Because clerestories are on the roof, however, shading is less likely to be a problem than with south-wall glazing. You may find that a clerestory is the only possible solar option if the south wall is shaded by a neighbor's house or trees.

Consider the structural changes that will be required if you add a clerestory. The project may involve alterations to the rafters, ceiling joists or other structural components, and it always involves cutting a sizable hole in the roof. Roof work must be done with great care, perhaps with assistance from an experienced roofer. A poorly weatherproofed clerestory will haunt the homeowner during every rainstorm until the problem is fixed. A good time to add a clerestory is when it's time to reroof, especially when old roofing must be stripped away. Perhaps you are planning to add insulation to your ceiling; this might offer a good excuse to look into the practicability of adding a clerestory. It is the sort of venture to keep in the back of your mind until some other roof work becomes necessary, but someone who is ready to do it now shouldn't be deterred. Just as with skylights, clerestories don't have to wait for anything but a good plan and good weather.

Keep in mind that a clerestory may change the appearance of your house dramatically. You should make a drawing of your house and sketch in possible clerestory locations and shapes. An architect or designer would of course be glad to make professional-quality drawings, and it may well be better to spend money on some useful drawings than to risk creating a permanent, expensive eyesore.

Consider, too, how a clerestory will affect the interior of your house. How will the extra light fit into the house layout? Will light and solar heat get to

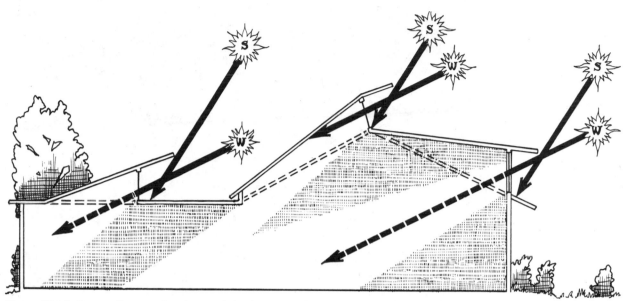

Figure III-15: Depending on the shape of the house, a clerestory can be located anywhere it has a clear aspect: on the south side, in the middle, on the north side. The north side clerestory is a likely candidate for a flat roof, while the middle one could be built up from a roof ridge. The south side clerestory is actually more like a dormer in that it has a down-sloping roof pitch. The dotted lines show the original roof lines of this imaginary building.

areas where they are needed? Where will heat be stored? Will modifications to the interior be necessary to improve heat or light distribution? Measure up and draw a floor plan of the affected rooms, or use original blueprints (if they are available) to make your analysis easier. If there is an unheated upstairs crawl space or attic between the ceiling and roof, you'll have to build a lightwell through the space to connect the clerestory with the ceiling. As well as helping reflect sunlight down to the living area and preventing warm air from escaping into the unused space, the shaft will present a good finished appearance.

If your attic has been remodeled and finished as a heated living area, your clerestory could be used to heat just that space. In such a case you should be careful not to build too large a clerestory, since there is not likely to be much thermal mass available to store the solar heat unless you add some.

Another design variation to consider is the use of a "sawtooth" or multiple clerestory arrangement to admit more heat and light into the house than would be possible with a single unit (photo III-28). Sawtooth constructions are particularly useful where the axis of the building runs north to south, because the units can be placed one behind another to utilize most of the roof area. Designing for proper drainage is especially important with a sawtooth array because a series of clerestories creates several valleys where water can stand.

If you want a clerestory primarily as a light source, a modest-sized opening of around 30 square feet will be fine for lighting a 250-square-foot room. Even a north-facing clerestory can do the job of providing diffuse, but never direct sunlight. If you want a significant level of direct gain heating from your south-facing clerestory, you'll need a larger opening, with a glazing area that's at least 15 percent of the

Photo III-28: When a building offers zero opportunity for adding south glass to the vertical walls, clerestories can come to the rescue. Because the roof is flat, this sawtooth arrangement of multiple clerestories could be perfectly oriented for direct gain into south, middle and north rooms.

immediate floor area. To repeat, there must also be some kind of heat storage medium or heat distribution system to prevent overheating during the day. The clerestory should be provided with an overhang that is sufficiently extended to block out direct sunlight in summer, and it should have some ventilating capability as well, through the glazing and/or through side wall vents. The flow of air throughout the living space should allow warm air circulation in winter and ventilation in summer.

Collecting solar heat is the primary function of a clerestory. The glazing should be positioned (azimuth and tilt) so that, in winter, sunlight arrives at a low angle of incidence to the glazing surface (for maximum gain), while in summer it arrives at a high incident angle and is shaded by the overhang (for minimum gain). Collection of sunlight is also affected by obstructions to direct gain, by the surface area of the glazing, the type of glazing material and the num-

ber of layers of glazing. Thus, as with any glazing addition, the proper glazing area must be determined according to the size and heat loss rate of the living space and by the amount of solar gain actually available per square foot. Use the essay "Finding Your Heat Gain" in Section I to determine the appropriate glazing area for your project, while considering some of the following design guidelines.

Site and Orientation

Although clerestories have the advantage of often being clear of ground-level obstructions, it's still important to survey the site as part of your planning. A building or tree that is well below the high summer sun may completely block the sun in winter, and except for deciduous trees, any obstruction that blocks summer or spring and fall sunlight will definitely block winter sunlight. As has been discussed,

Photo III-29: An interior detail of the retrofit shown in Photo III-27 shows that a clerestory doesn't have to be built in line with the existing roof framing.

warm weather shading is a must for south glazing, but the problem for clerestories is that trees must be growing practically on top of the house to be effective. For homes lucky enough to be shaded by tall, close-in deciduous trees, all the better, but in most cases it's likely that you won't be able to count on flora to cool your clerestory, which again brings up the need for overhangs or built-in shading techniques.

In most cases glazing can face up to 28 degrees east or 40 degrees west of true south with less than 10 to 15 percent reduction in overall performance. If you have a choice, lean west because southwestern orientations are usually better than southeastern ones for heating, since the higher outside air temperature in the afternoon reduces heat loss through the glazing. You may still prefer a more eastern orientation if your clerestory is to be used for heating a breakfast room or some other space that needs morning light.

Tilt

The optimum tilt for winter sunlight collection varies with latitude, and as a rule of thumb, figure the optimum as latitude plus 15 degrees. For south-facing clerestories, winter energy gain using the optimum tilt can be more than 15 percent higher than with vertical glazing. However, overheating in the summer is more of a problem with tilted glazing because even in summer much of the sun's energy can still penetrate the tilted glazing and heat up the interior. To understand why, consider the angle of incidence, the angle between the path of the sun's rays and a line perpendicular to the glazing surface. Assume that the glazing faces due south in a clerestory located at 40 degrees north latitude. If the glazing is at an angle of 55 degrees for optimum winter gain, the angle of incidence at noon on December 21 is only 10 degrees. At a 10-degree angle of incidence the glazing will transmit better than 98 percent of the available solar energy, a truly optimum figure. However, on June 21 the noon angle of incidence for the glazing will be only 35 degrees, yet it will still transmit over 80 percent of that sweltering solar energy, which makes overhang or curtain shading of utmost importance for tilted glazing.

Vertical glazing, on the other hand, reflects away more summer sunlight because it presents a sharply angled surface to the sun's rays. Vertical glazing mounted in the clerestory in the previous example has a noon angle of incidence on December 21 of roughly 35 degrees and will transmit better than 80 percent of the available solar energy. On June 21

the solar altitude will be 70 degrees, which equals the angle of incidence for vertical glazing. At this angle the glazing will transmit only 34 percent of the incident solar energy, almost half as much as the glazing set for maximum winter gain. Still, however, overhang shading is a low-cost must even for vertical glazing. See "Air Conditioning without Air Conditioners" in Section V for a simple way to calculate the overhang needed.

Unfortunately a shading overhang for tilted glazing must extend well out from the clerestory to block summer sunlight, and that may look unattractive. An alternative to a large overhang is to set exte-

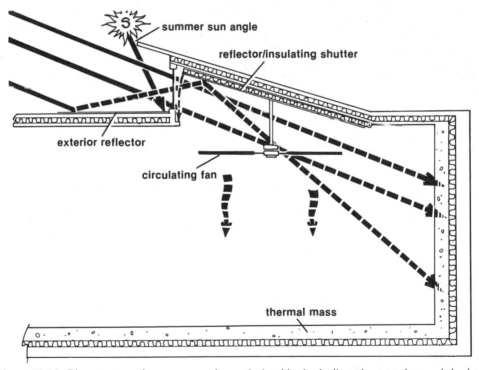

Figure III-16: Clerestory performance can be optimized by including elements beyond the basic framing, glazing and insulation. An exterior reflector might bounce 10 to 15 percent more sunlight at the glazing. The interior reflector improves interior light (and heat) distribution, and in its second function as a movable insulating shutter it greatly reduces night heat loss. The ceiling fan blows warm air from the clerestory pocket down into the living space. If thermal mass is not already present, it does represent a major addition, but one that can be considered for large areas of glazing.

rior movable shading over the glazing in warm weather—if you don't mind getting up on the roof twice a year to install and remove it. All things considered it may be best for the do-it-yourselfer to design for vertical glazing; construction, weatherproofing and shading are a bit easier than with tilted glazing. But if you decide that tilted glazing better suits your overall needs, by all means, go for it.

Glazing

Planning for the proper glazing area is just as important as planning for tilt and orientation. If there is too much glazing, the space will overheat, and if there is too little, the space will not receive enough heat to make the project worthwhile. The required glazing area is determined by several factors: the floor area and volume of the space beneath the clerestory, the quality of insulation and weatherproofing of the heated space, the amount, surface area and location of thermal mass, the orientation and tilt of the glazing, and the effectiveness of the insulation in the clerestory itself.

If the clerestory glazing is improperly designed (a single-glazed clerestory with no nighttime insulation), it could be a net loser of heat rather than a provider, in which case any size would be too large for the energy-conscious builder. Thus the more heat the clerestory itself retains, by having double glazing and night insulation, the more heat loss from the living space is minimized. Of course no matter how much glazing is best for optimizing heat gain, you must work within the structural and space limitations of your house. The roof might not support extensive additions without costly modifications. Also, oversized additions could look unsightly. Always let your heating and lighting goals be tempered by the changes in structure and appearance that a clerestory will cause.

Glazing materials are either transparent (glass) or diffusing (FRP), and for clerestories diffusing glazings are useful since they can distribute the solar gain more evenly throughout the living area. Glass will of course preserve a view of the sky with its cloud shows and star patterns, which may be no less important than heating and lighting. If glass is used, care must be taken when working with it, especially on a roof, and you should be wary of installations near tree branches, especially where the glazing is tilted. In overhead applications where there is a potential for projectiles, tempered glass should be used.

Thermal Mass

The goal of all passive solar heating systems is to maintain nearly constant temperatures in the living area. Rather than letting air temperature rise above comfortable levels during the day, the floor, walls and even furniture should absorb some of the excess heat. As the interior temperature begins to drop, this stored heat is naturally released (reradiation) to keep the room warm. Thus, the function of thermal mass is to minimize temperature swings, and with sufficient "sun-charged" mass, the room can even have some carry-over of stored heat that will continue providing warmth to the room during a cold, sunless period.

If you're considering adding mass, it's important to know that simply having volume is less important than having adequate surface area. Concentrating thermal mass in a very thick wall with small surface area would be far less effective for heat storage than a spacious wall and floor area composed of only a few inches of storage material such as a brick veneer wall and a ceramic or masonry-tiled floor.

Before actually adding mass, make sure that your floor joists are sturdy enough to support the extra weight (check with an experienced builder or building inspector). (The essay "Moved-In Mass," in Section III-D, discusses options for thermal mass additions.) You might find that it's more practical to add mass in the form of water-filled containers, which can provide the same heat storage capability with less surface area. Keep in mind that adding mass can significantly change the appearance of the living area.

Another option for thermal mass is to use phase-change materials. Certain materials such as salt hydrates or eutectic salts (e.g., Glauber's salt) can store a large amount of heat in a small volume. Although many problems have existed in the past with salt-hydrate storage, improved packaging techniques have improved performance, and nowadays reliable materials are available from several companies (again, see "Moved-In Mass").

With all types of thermal mass (masonry, water, and phase-change materials), heat storage is most effective when sunlight shines directly on the mass. It is therefore important to determine in the planning stage where light from the clerestory will fall throughout the day. Water and phase-change materials work well for heat storage even if sunlight shines directly on the containers in a small area, because both materials readily absorb heat from the surface. With masonry materials it is more effective to have diffuse light fall on a larger area. For primary heat storage in the floor, the walls should be a light color to reflect sunlight down. For storage in walls, dark colored walls are more effective in absorbing heat where the sun shines on them.

The amount of available or added thermal mass is a major factor for determining how large the glazing aperture of a clerestory should be. With too little glass for the amount of glazing present, daytime overheating will be a problem. As a rule of thumb, provide at least 3 square feet of mass area for each square foot of glazing in a direct gain system. If the thermal mass is not in direct sunlight and is heated instead by air movement, a much larger surface area may be needed to provide adequate storage. This is important for clerestories if they will be over rooms that can be opened up to other parts of the house; natural convection air flow helps distribute heat, and this reduces the amount of thermal mass required under the clerestory.

Some Design Details

All exposed wood and wall materials (e.g., drywall) used to frame in a clerestory should be painted white (or at least a very light color) so that the sunlight striking the sides or ceiling of the clerestory will be reflected down toward the floor instead of being absorbed. Reflection is especially important if an unused attic space must be bridged with a lightwell.

Warm air will collect by convection in the clerestory "pocket," which can lead to the loss of a lot of valuable solar heat back through the glazing, even during the day. For this reason, it's a good idea to install some type of fan or blower to push the ceiling-level heat down to the floor. Large, slow-moving "Casablanca" fans serve well for this purpose. They are quiet and available in styles both simple and elegant.

In summer it is also important to prevent too much heat from collecting up in the clerestory, especially if slanted glazing is used, and along with

Photo III-30: A combination of side vents and operable windows can guarantee positive ventilation in warm weather. Note how the sill of this clerestory is built perpendicular to the tilted glass so as not to block sunlight.

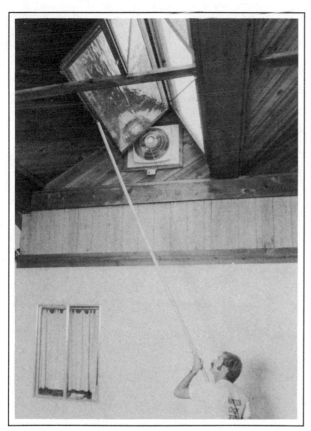

Photo III-31: A simple, pole-operated movable insulation system can maximize the net heat gain from a clerestory. This rigid shutter is also a daytime reflector for bouncing more light into the living space.

shading, vents should be provided. Ideally vents should be built in above the glazing because the largest amount of heat will collect there, but casement or awning windows will provide adequate ventilation. Vents can also be installed in the sides of the clerestory, near the top where a side overhang (if one is built) can protect the openings from rain. Side vents on both ends of the clerestory will provide effective cross ventilation. A good idea is to use a reversible fan or blower that will be useful for both hot air distribution and ventilation.

Less Loss, More Gain

To ensure that overall heat gain is greater than heat loss, clerestories in most parts of the country should be provided with movable insulation. The suggestion in photo III-31 shows a hinged, rigid insulation panel that also serves as a reflector by day. You can also consider insulating shades or roll-up curtains with remote manual or automatic controls. There are several varieties available commercially, and they are eminently home-buildable. The best arrangement from an energy standpoint is to use double glazing with movable insulation that has at least an R-5 rating (see Section III-B). Aside from its energy liability, single glazing—even when it's aided with movable insulation—can collect a great deal of condensation, liquid and frozen, on its cold inner surface, which can eventually run off the glazing and stain or damage

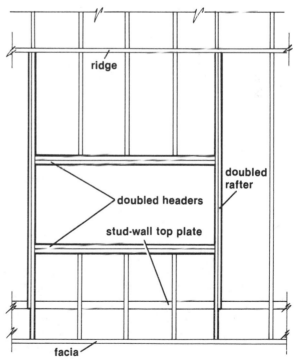

Figure III-17: This overhead view shows how to modify the existing roof framing for a clerestory opening. Doubled headers and rafters are required to support the rafters that have been cut.

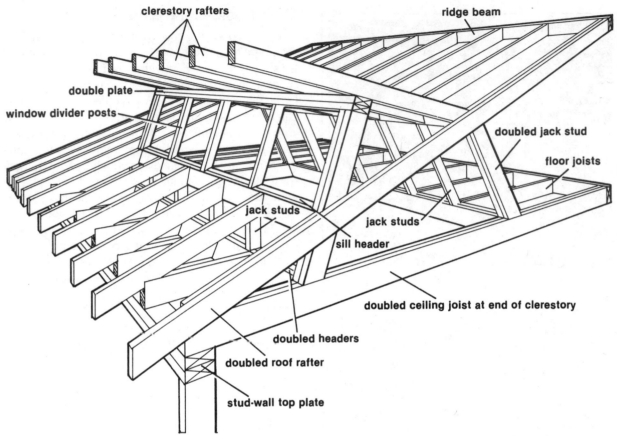

Figure III-18: This framing skeleton of a clerestory addition shows where the existing roof rafters and ceiling joists have to be doubled up to carry the added load at both ends of the clerestory. In most climates the end walls should be studded out with 2×6's to contain at least 6 inches of fiberglass insulation (R-19). The rest of the clerestory framing can be done with 2×6's and 2×4's, except perhaps for the clerestory rafters, which could be 2×8's to carry more insulation.

wood framing or drip onto furniture or the floor below. Double glazing stays drier.

You can make your clerestory collect more energy by adding exterior reflectors near the glazing. On a flat roof, white gravel spread out in front of the clerestory can increase heat gain up to 10 percent. Snow is even more effective. On shingled roofs, white or light-colored shingles will increase reflection into the clerestory. In very cold climates, where it is important to maximize the collection of every bit of sunlight, it may be a good idea to build a more effective reflector with polished metal or mirrorlike film,

or to use a sheet of plywood painted with a highly reflective paint.

Construction

For all your thinking and planning and designing, you've made a pretty big job for yourself. It's not prohibitively difficult by any means, but it's no piece of cake either, and it's assumed that someone tackling a clerestory has had some experience in framing, roofing and finish work.

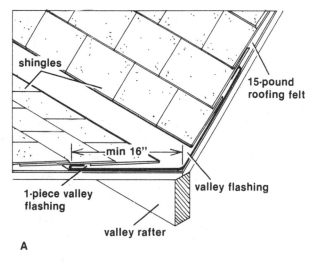

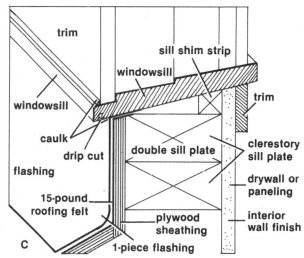

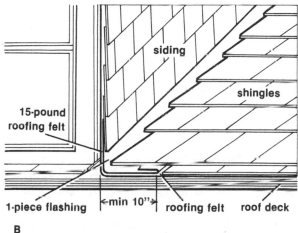

Figure III-19: As shown in A, it's important to extend the valley flashing way up under the shingles to eliminate the possibility of water getting around it, especially when built-up snow begins to melt. With plank and plywood siding, a continuous run of flashing is adequate (B). When wood shingles are used as siding, step flashing must be used; consult your local roofer. Both the flashing and the roofing felt can be run up under the finished clerestory windowsill with a bead of caulk to make a good water and air seal (C).

Locate the rafters in the section of roof you've chosen, and with a chalk line lay out the perimeter lines of the hole to be cut. It makes sense to put each end of the clerestory right over a rafter. Cut through and remove the roofing and decking to expose the rafters. You may decide to leave the rafters exposed for good looks. If you're going to cut them away, you'll have to provide more support for the load that they carried: Double up the two end rafters; run doubled headers between them at the top and bottom of the opening, and hang all the cut rafters on these headers.

Once you know the location of the lightwell, if there is to be one, cut through the ceiling. (Again you can reduce structural complexities and gain a nice look by leaving the ceiling joists in place. Painted white, they won't appreciably reduce solar gain. But if you do cut them away, follow the same procedures as described for the rafters.) As with skylights, it's good to make the ceiling opening bigger than the clerestory opening for better light distribution. A lightwell for a clerestory is similar to that for a skylight: Each side is a small insulated stud wall, and you have a great deal of latitude in designing the way it opens into the space below. A flared design helps to bounce more light down and exposes more of the clerestory to view from below. The result is nice, but the work requires carpentry with experience, or with a lot of free time.

Figure III-20: The finished clerestory gets the same treatment that any construction finish work does. Insulation of all newly created exterior surfaces is essential. You'll have a great deal of freedom in the design of your windows. It's best to have at least two operable units in place for adequate ventilation. Siding and trimwork can be used effectively to blend the addition with the look of the rest of the house.

The clerestory itself can be framed with 2×4's, 2×6's, even 4×4's, all depending on your loads and spans. One general framing system is shown in figure III-18, but it is only an example; there are many possibilities. Of course, to keep the building inspector happy, your clerestory should comply with local building codes.

Besides proper structural detailing, the most critical element in clerestory construction is weatherproofing. Each seam between the clerestory and the roof and between the glazing and the framing is a potential leak. The joint between the clerestory sidewalls and the roof should be sealed with metal flashing that extends at least 12 inches in both directions from the cover. This is exacting work that should probably be left to a professional roofer or contractor if you're not sure you can pull it off. The seal between vertical glazing and the clerestory frame presents fewer problems for the home builder.

After framing in the clerestory, add the glazing you've chosen and rough in your vents and any wiring and switches needed for fans or lights. Finish the exterior with whatever sort of siding and roofing is planned, and seal all joints, edges, and nail holes with caulk. After this you can finish the interior work without risk of getting wet or ruining your drywall. Add insulation between the clerestory rafters and jack studs, and line the inside of the framing with a plastic sheet vapor barrier prior to putting on the finish wall. Then you can finish with drywall and white paint, paneling, boards and batts, etc. Make sure in all this work that you've properly planned for your movable insulation system. Since it's likely to be remotely operated, you might want to give the system a test run before finishing everything off, in case you've got to make any changes.

Now that you've got this direct-gain solar collector sitting over your head, it has to be operated properly. Outside of any fans you might have installed for air handling, this is a passive system that will pretty much operate itself with a little help from its owner. The winter routine will simply consist of operating the movable insulation system in the morning and evening. Fans and blowers should be controlled thermostatically with a manual override. In summer the clerestory pocket can collect excess heat, and you can use vents, operable windows or reversible fans to exhaust the warm air from the house. If you've built the proper overhang or other shading device, the clerestory won't contribute to heat gain. But year-round it will be filling rooms with a fine daylight, a window to the sky.

Alex Wilson

Photo III-32: Solar cats have a natural affinity for clerestories.

WINDOW
B. INSULATION

Photo III-33: Quilts on the bed, quilts on the windows—they both do a good job of increasing comfort by reducing body heat loss.

The Cover-Up

We're going to be very hard-nosed about this: You shouldn't put in solar glazing without putting in glazing insulation. Period. Well, you can go without the insulation if you want, but you know that glazing is a two-way street. A day of sunshine will fill your solar room with light and heat, but a night's worth of hard cold will induce mass quantities of glazing heat loss. Glazing is a friend when it's controlled; when it's uncontrolled, you're once again at the mercy of the climate. You need a cover-up.

Window insulation has some history in old wooden shutters and things like that, but it's really an offspring of the new shelter. High energy costs have been the mother of invention to an enormous family of options for solar and energy-efficient design for buildings big and small, new and old. And it's not just a solar-related improvement. As a group, the windows scattered over all your four walls represent a big thermal hole for conductive and infiltration heat loss. Relative to an insulated roof or wall section (R-11 to R-40) a window with a storm (R-1.8) can have a conductive heat loss rate that is 6 to 16 or more times faster than that of the surrounding wall. When a house is weather-stripped and well-insulated, the windows become the biggest energy losers.

The good news is that thermally effective window insulation systems can be easily home-built, and they can look good in what they do. The various designs range in do-it-yourself difficulty from zero to middling, but with all of them you're the decorator. You get to use your favorite fabrics, natural woods and paint colors to make your window insulation a lively contrast or a quiet complement to your interior decor. You can also enjoy the fact that some designs will pay for themselves in energy savings in as little as one or two heating seasons. With your solar glazing, window insulation will increase the net heat gain by 25 to 50 percent. It will also increase your comfort by raising the surface temperature of that once naked glass area. Neither your body nor your budget has to suffer the chills and costs of exposed glazing. The rules of the new shelter game require a cover-up. It's the right thing to do.

Photo III-34: When the pop-in goes up in the night, heat loss goes down by 60 to 80 percent.

PROJECT
Pop-In Panels
Things that go up in the night

Presenting the simplest, lowest-cost way to warm up your windows. Pop-in panels can be nothing else but featherweight slabs of foam insulation, a perfect example of the KISS philosophy (Keep it simple, stupid). With a little wood and fabric they can acquire the status of wall art with only a small compromise of their KISS roots. The right kinds of foam can squeeze in a lot of insulation value per inch, and when used faithfully, they'll quickly pay for themselves. The author of this project can testify to that: Lately his neighbors have been averaging $1000 oil-fired heating seasons, while Jack and his family are stuck at around $300. Their pop-ins aren't completely responsible for the difference, but they're definitely worth at least a couple of hundred gallons.

Odds are high that pop-in panels are the best system for you, if you intend to make your own window insulation. Pop-ins are sometimes regarded, mistakenly, as the least elegant and most bothersome of all window insulation options: They're too simple to have good looks, and, in the "popped-out" position, they somehow get in the way of domestic activity. Don't be misled by either myth. I've known a good number of people who have built very handsome panels, and without exception the ones that *always* get built *and* used are the pop-ins. Their simplicity makes them attractive to any level of skill, and there always seems to be a place for them. Except for insulating very large or inaccessible windows, pop ins score high on convenience and low on obtrusiveness, either in your rooms or in your life-style.

The look of the window is scarcely changed when you use pop-in panels. The curtains stay right where they are when you slide the panels in and out of the window frame. In place, a white pop-in looks about like an ordinary window shade. You can draw curtains over them, or decorate them in any paint, fabric or paper you like. In my experience few people ever notice they're there unless I tell them.

My first set of shutters was conceived and built fast. One of our bedrooms was unheated, and the exterior walls were uninsulated brick. Needless to say, in winter it got very cold. I was intending to build some hinged shutters for all the bedroom windows, but I hadn't settled on the details of the design or the materials nor found the time to put them together. The cold room needed a quick solution, so I decided to make some kind of inexpensive shutter for it right away and replace them when the permanent shutters were ready. Enter the pop-ins.

I built ¾-inch-thick panels using styrene beadboard (R-3 to R-4 per inch) inside a light wooden frame. The edges are weather-stripped with felt weather stripping, and aluminum building paper covers the outward side. The inside is bare, white foam. I didn't use a rigid plywood skin on the panels because I was keeping costs, work and weight down. The wooden frames take all the wear and handling well. The aluminum building paper is tough; it increases the rigidity of the foam, and it has lasted several winters (and summers, when I used the panels for shading). When they're not in use, they hide behind the bedroom door, out of the way. Though I now see some minor design improvements, four years later I'm still using the originals, and I don't anticipate building new ones for that room for many years.

After that room, I did all the downstairs windows and doors. On most of these I used Nightwall clips, available from Zomeworks (see the "Hardware Focus" section at the end of this project), which hold foam to glass by magnetism. A magnetic strip is glued to the perimeter of the foam, and a metal strip goes on the glass. Nightwall is perfect for windows with frame edges that are too shallow to hold a panel by friction alone, such as the window in a door. If your window sashes are glazed with a single piece of glass instead of several little panes (colonial style), then Nightwall can be used.

With no friction or stress on the "magnetized" foam, the panel needs no frame. All you need do is cover it with fabric, latex paint, paper or foil-faced building paper for dirt and sun protection. When the frame isn't used, the difficulty and the expense drop even further. I tried Nightwall on a few windows, and I soon got more. When warm weather comes around, we store these panels in racks in the pantry.

Nightwall

I'll talk about Nightwall construction first, because it involves little more than cutting foam. First, measure the perimeter of the glass very carefully, because you want the panel to fit the glass precisely. Lay the foam on a clean, solid work surface. Use a straight-edge to guide your (very sharp) cutting knife, and work the knife along slowly. Beadboard will require a sawing stroke, while other foams can be neatly sliced in one or two passes. Smooth rough edges with sandpaper or a Surform abrader, but be careful not to trim off too much or you'll lose the tight, measured fit.

As you probably know, there are a few different types of foam that can be used. I used white styrene beadboard for my first shutters and Du Pont Styrofoam blueboard for the Nightwall panels. I much prefer the blueboard. It is easier to keep clean, and it resists wear better than the beadboard. In general, you should go for the highest R-value material that you can find. You want panels that are both light and compact, while getting the most insulation value over your windows and doors. Consider using a polyisocyanurate foam (trade names: Thermax, R Max, High R, and others). It is rated at R-8 per inch, compared to R-4 for beadboard and R-5.4 for blueboard. It is usually covered on both sides with shiny foil, which increases its insulating value.

(Editor's note: It must be noted that all these foams will produce toxic, sometimes deadly gases when exposed to flame or high temperatures (above 200°F, 93°C). Because of this some will insist that foam panels should never be used uncovered, but rather protected with some kind of fire-retarding or fireproofing material. Drywall comes to mind, but even just an inside layer would mean a ten-fold increase in weight. We did, however, manage to track down a fire-resistant fabric, Du Pont's Nomex, which may prevent flame, but not necessarily heat, from reaching the foam. Nomex is described in the "Hardware Focus" section at the end of this project. The question of whether foam window insulation presents a serious hazard in the event of fire is in some ways moot. Other plastic furnishings also emit toxic gases when they burn. How far gone is a house whose windows are afire? Perhaps by the time the fire was that extensive there would be several other deadly conditions. That is not to say that adding one more possible hazard is okay; the real hazard is fire itself, and *any* house should be equipped with smoke detectors, fire extinguishers and practice a plan for a family fire drill. You may also be able to request safety inspection from your local fire department.)

The Nightwall fasteners go onto the four corners of the window glass and the foam panel. With large window panes the fasteners can be spaced up to 24 inches without surpassing the strength of the magnets. Each side of the Nightwall fastener, the magnetic and the steel strip, is self-adhe-

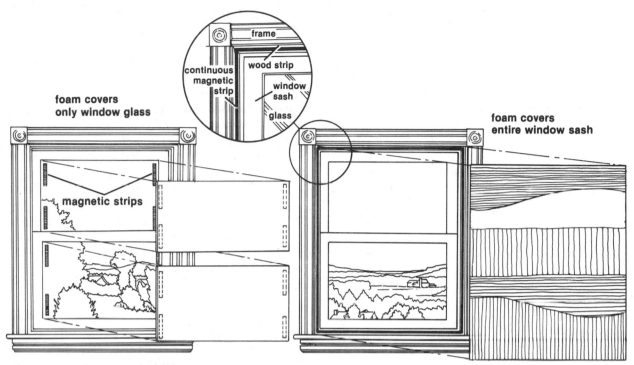

Figure III-21: If your window frame has a suitable wood mounting strip, or if you can add one yourself, run a continuous magnetic strip around the sash to hold an insulating panel, as shown here on the right. But if there isn't room for a continuous strip, mount shorter magnetic strips directly to the glass. You'll get a little more infiltration, but still a very effective window insulation.

sive and comes with a paper backing that is peeled off for installation. You can also make your own version of Nightwall. Magnetic "tape" is a common hardware store item. Mounting can be done with silicone or some other bonding agent. You may want to finish the inside of the panels for looks, but it's also important to cover the outside faces with paint, paper or fabric to protect them from the sun. (Cover the magnetic strips with adhesive tape to keep the paint off.)

Another possible fastener for these panels is Velcro. Strips could be glued or stapled to the panel and the window frame, and the tenacity of Velcro ought to make it a reliable fastener without the need for using more than a couple of inches per strip,

depending on the weight of the panel. Velcro also comes in a few different colors.

My first panels insulated a somewhat larger area than Nightwall does because they covered both the glass and all framing right to the edges of the window opening. Nightwall only covers the glass. I tried to improve that situation by attaching the clips to the wooden window frame, but the paint soon flaked away and the glued clips came off. The idea would work if the metal strips were fastened to the frame with small, flathead brads. The advantage of total coverage is of course better insulation. Also, a tight-fitting, full-coverage panel helps to all but eliminate whatever infiltration there might be after the window was caulked and weather-stripped. Most types of rigid foam (Styrofoam, beadboard and isocyanurate) are available in 4×8-foot sheets, which makes it possible to cover most windows with a single sheet. If you end up with pieces you don't want to waste, they can be held together inside a wooden frame or bonded to a sheet of cardboard.

The thickness of the Nightwall fastener creates a 1/16-inch gap between the foam and the surface it's mounted to. The gap can provide a small amount of additional insulation if it is closed around the perimeter. This is the key to total success with window insulation: having a tight seal between the panel and the window. A wide gap (over 3 to 4 inches) between the panel and the window allows convection, which short-circuits the insulating effect to some degree, but more important is the seal, because a poor seal will admit relatively humid room air to the gap. Some have said that the gap between Nightwall and the glass is so narrow that it isn't worth worrying about. The gap *is* tiny, and it certainly does greatly reduce air flow, but moisture from the room air does form ice patterns on the glass. That indicates some air flows, when you remove the panel, and the ice melts and water runs down the glass onto the sill. I don't like it, and the problem is very easy to correct by making a gasket on the foam panel with strips of cardboard or felt of the proper thickness glued to the foam with white glue. This edge seal is meant primarily to eliminate air exchange between the room and the panel-window air gap. It is *not* intended to stop infiltration through a leaky window or storm/window combination. You can also carefully recess the magnetic strip into the foam so that the foam actually comes into contact with the glass, thereby eliminating the air gap.

The first three orders of business with windows are caulking, weather stripping and adding a second glazing layer. Window insulation is at its best when it can capitalize on these improvements. When cold winds rage, you want to stop them at the point of entry, not the rear guard. Window insulation is for stopping conductive, convective and radiative heat loss and only secondarily for stopping infiltration heat loss.

A Pop-In with a Frame

I had only the basic hand tools—hammer, screwdriver, drill, saw, plane and chisel—when I set to building these shutters. Since I couldn't very easily rip down the frame pieces to any old size I wanted, I matched standard lumber to the available insulation sizes. That meant buying trim, which is usually a pretty good quality wood. I used 1×1's, which actually measured ¾ × ¾ inch to use with ¾-inch beadboard foam. Note that insulation comes in the exact thickness advertised, while lumber standards dictate that wood is always a little smaller than the dimensions it goes by. If you don't have a table saw, you will have to match things up or have the lumber yard rip down the size you need. A 1×6, for example, will yield five 1-inch by ¾-inch sticks for use with 1-inch foam, and it shouldn't cost that much to have the cuts made.

The first step is again to measure the window opening to get the length of all four sides and its true shape. (Is the opening square? A parallelogram? A [shudder] trapezoid??) If you're lucky, it will be perfectly square. Often it won't be, especially in old buildings. Make the frame ¼ inch undersized in height and width so that there is a ⅛-inch gap all around. The gap is for the felt weather strip that is tacked around the frame to make the friction fit work.

The frame is joined with simple lap joints, glue and two 4d finish nails at each corner. If the lap joint is cut carefully, the union will be strong. If any joint turns out loose, use a flathead wood screw to make it tight. Now check to see that the frame fits into the opening properly. If it's a little oversized, you can plane down the frame. If it's undersized, it will need extra felt on the edge.

After the foam is cut to size, press it into the frame. There are several possible ways to secure the foam, and it boils down to the tools you have on hand and what kind of appearance you want. I have used thin wooden battens to secure the foam edge on the inside face of the panel. The battens (1½ × ¼ inch) are secured to the frame with small brads and glue, and the glue is also used on the foam. To

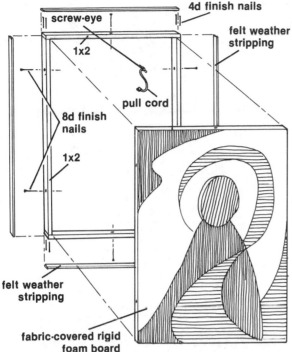

Figure III-22: A pop-in panel is made of fabric, glued or stapled to rigid foam insulating board, inserted in a 1×2 frame and secured with glue and 1/4-inch wooden dowels. Felt weather stripping around the outside face of the 1×2's assures a tight seal with the window frame, and a short length of cord tied to a screw eye makes removing the panel easier.

minimize waste of the insulation, I sometimes used two pieces of foam, with the vertical seam right in the center of the panel. The seam is also covered with a batten, and the natural wood in contrast with the white foam has a nice appearance. If you're going to wrap the panels with fabric, that should be done before the foam is pressed into the frame. The edges of the fabric can be glued to the foam.

Another method used 8d finish nails or 1/8-inch dowel pieces inserted through the frame edge into the foam. First drill holes into the edges of the frame at 8- to 10-inch intervals with a bit that is just slightly smaller than the shaft size of the nail or the dowel. Then the finish nails can be lightly hammered through the frame and into the foam. This will elimi-

nate "foam flop," and if they're not treated roughly, the panels will last a long, long time.

If you have a table saw, or access to one, you can make some wooden channels to simplify the framing and edge-securing procedure. As figure III-23 shows, the channels for 1-inch foam are made from 1¼ × ¾-inch pieces of wood using a dado blade or several passes with a single blade. Mitering the corners makes a nice finished appearance.

It's very important to protect the foam from ultraviolet degradation in direct sunlight. In summer you may also want to use the panels to keep direct sunlight out of south, east and especially the west windows. Used this way, the panels do a great job of keeping rooms cool. I recommend covering the outside face with foil-faced building paper. It's very easy to apply to the frame with glue or staples and it's an excellent material for rejecting sunlight. You could use white latex paint to protect the foam, but it isn't nearly as good as the foil.

The last steps are to paint or varnish any exposed wood surface so it will be easier to clean. Then fasten the felt weather stripping all around the perimeter of the frame with staples spaced every 2

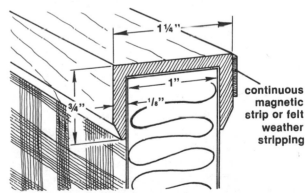

Figure III-23: A wood channel can be milled on a table or radial-arm saw, or even with a router, to make a good-looking frame for a pop-in. The drawing above is shown actual size with the channel surrounding both the insulation and the fabric, if any is used. The continuous perimeter weather stripping is tacked onto the outside face of the channel.

to 3 inches. The panels will also need a knob or a tab of fabric at the top or bottom of the frame so they can be easily pulled, instead of pried, out of the frame.

Pop-In Problems, Solutions and Performance

No other window insulation system could hardly be cheaper or easier to build than these designs. Their humble looks don't really matter when they're hidden behind curtains or drapes. During the day they stack behind doors and other furniture. Dressing them up with paint, wallpaper or fabric is a simple improvement and though it can double the cost, the panels would still be inexpensive. My own Spartan framed versions cost me about 40¢ per square foot in 1981 prices.

The only real problem with the framed panels was that the frame wood warped a little. Even though trim is supposed to be high quality, kiln-dried wood, the stuff I got was not stable. The twisted trim has not affected the performance of my panels. They still fit tight, but a worse twist could create a gap between shutter and window frame, breaking the edge seal. Warpage may be a bigger problem with large panels, so if you have big windows, consider making two smaller panels. That may make handling easier, too. In any event, be very selective with the wood you use, and be sure to seal it with paint or varnish.

Another idea for assuring a flat panel is to cover one face with ¼-inch plywood, which is very stable and will give the panel much more strength. A good grade of plywood (mahogany, birch, etc.) or paneling could become the inside finish of the panel. An air gap between the plywood and the foam would also slightly increase the overall insulation value.

Corrugated cardboard is another underused material that would work well to strengthen and perhaps flatten pop-ins. It is worth about R-1.5 in insulating value and affords another opportunity for building in a dead airspace.

If I were forced to choose between the

Photo III-35: Night windows are dark anyway. You can give them a little style with a pop-in that's wrapped in nice fabric. With super-nice fabric the pop-in could double as wall art by day. Just hang it next to the window.

framed, friction-fit panel design and Nightwall, I would stick with the Nightwall idea. Handling is easy and it's strong, durable and simple to repair. (I've only made two repairs in three years; the metal strips came off the glass and were rebonded with silicone.) Its cost-effectiveness is undeniable, and the unframed panels are so simple to build that they barely qualify as a construction project.

A simple calculation can be used to model the performance of these panels and show just how cost-effective they are. In this example we'll use a panel made of the highest quality foam, isocyanurate, in a 1-inch thickness (R-8, the reciprocal of which is U, which defines heat loss in Btu's per

TABLE III-3 PERCENTAGE OF HEATING LOAD SAVED WITH MOVABLE INSULATION

R-Value of Movable Insulation in Addition to Window	Percentage Saved* (with single glazing)	(with double glazing)
1	32	22
2	42	31
3	46	38
4	49	41
5	51	44
6	52	46
7	53	48
8	54	48
9	55	50
10	55	50
11	56	51
12	56	52

SOURCE: William K. Langdon, *Movable Insulation,* 1980, with permission of Rodale Press, Emmaus, Pa.
*Based on average indoor-outdoor temperature difference of 35°F with movable insulation over the window 14 hours per day.

square per hour; our R-8 means a U of 0.125). We'll assume that the window in question is north-facing and is either double-glazed or single-glazed with a storm (R-1.8; U-0.56). The window frame has been caulked, and the sash has been weather-stripped. The panel covers both the window glass and the sash frame, and in this case the window has an area of 12 square feet. First we'll look at costs. At about $15 per 4 × 8 sheet the foam has a cost of about 45¢ a square foot. With a Nightwall-type system (unframed panel) the rest of the materials add about 10¢ to the square-foot cost, bringing the total cost to 55¢ per square foot, or $6.60 for a 12-square-foot unit.

The window is in a climate that has 6000 heating degree-days (HDD). The total heating season heat loss from the window is found by multiplying U times HDD times 24 (hours in a day) times window area. In this example the multiplication becomes:

$0.56 \times 6000 \times 24 \times 12 = 967,680$ Btu per heating season. If electricity at 5¢ per KWH were used to supply that heat loss, then $967,680 \div 3413$ (Btu/KWH) = 284 KWH would be used, worth about $14.20. Now we can recalculate the window in place. Table III-3 shows the percentage of the heating load saved with movable insulation, depending on the insulation value of the panel itself. The table shows that R-8 insulation saves about 48 percent of the heating load when it's used over a double-glazed window. The calculated heat loss is then multiplied by this percentage to show how many Btu's are saved: $967,680 \times 0.48 = 464,486$, which means a savings of about 136 KWH, worth about $6.80. In this case the window insulation pays for itself in less than one heating season. Other window insulations can cost somewhat more (up to $5 per square foot and higher), especially commercially made systems, which would mean payback periods of several years. Clearly the home-built pop-ins have a strong cost advantage.

As to the choice between a pop-in panel and the much more elaborate permanently mounted systems, I feel that my initial instincts were right. Outside of cost, the important thing is to get something in place so you can start saving energy and feeling more comfortable. And for that pop-ins are tops. The seconds it takes to move the panels back and forth every day are time well-spent, and it's hardly any more time than would be spent handling permanently mounted, higher cost systems. Personally speaking, it's been easy to make the daily pop-in and pop-out a regular routine. At night the windows aren't big black holes anymore. If I'm late putting them in, I feel sort of naked, especially knowing how fast my heat is leaving. When they're in place, I feel more secure and the place has a definitely cozier feeling. Saving energy this way is more than a dollars-and-cents deal; it's peace of mind.

Jack Ruttle

Hardware Focus

Fire-Resistant Fabric

Du Pont Co.
Nomex Marketing
Centre Road Building
Wilmington, DE 19898
(302) 774-1000

Nomex
One possible alternative to the asbestos cloth recommended as a covering for foam insulation is Du Pont's Nomex, an aramid fiber used in fabric or paper form. Du Pont manufactures the paper, but several other manufacturers fabricate cloth from the fiber. It is widely used in such applications as industrial work clothes and suits for race car drivers in fabric form and in electric motors and aircraft in paper form. In both forms Nomex is UL-rated at 428°F (220°C) and can withstand short-term exposure to 500°F (260°C). The paper comes in thicknesses from 3 to 30 mil, and the cloth is available in a variety of widths and thicknesses. Contact the company for additional technical information and a listing of distributors and manufacturers of Nomex products.

Magnetic Seals for Movable Insulation

3M Co.
Industrial Electrical Products Division
3M Center
Saint Paul, MN 55101
(612) 733-4977

Plastiform
Self-adhesive, flexible magnetic strip. Thicknesses: 0.030" and 0.060"; widths: ½" to 24".

Zomeworks Corp.
P. O. Box 712
Albuquerque, NM 87103
(505) 242-5354

Nightwall clips
Flexible magnetic strips that adhere to insulation board by adhesive backing. Clips are ½" wide and 3" or 6" in length.

⬛PROJECT
The Hinged Shutter

Beyond the pop-in panel is a wide array of options for permanently mounting window insulation. In the realm of the do-it-yourself project, the hinged shutter is a relatively simple one whose toughest requirement is precision measuring and careful mounting, and that's not so tough. These shutters are like quarter-size versions of the mass-produced, hollow-core door that seems so light and insubstantial, as a door that is. As an insulating shutter the lightweight construction is perfect for the job at hand. Once again, you're the decorator; the plans call for plywood with nice veneers or you can go cheap on the plywood and use your "perfect" fabric. If maximum convenience is your game, a hinged shutter will always be there, winter and summer, ready and waiting to serve. And since it is always there, making it look good is part and parcel of making it right.

The better your home is insulated, the more you should worry about windows. That's where the big opportunities for energy savings lie. Windows can account for as much as one-half of the heat loss in a well-insulated home even when it's equipped with storm windows. In an uninsulated house windows may be the source of only one-quarter of the heat loss, because much more heat escapes through walls and roofs. But even then it's often quicker, cheaper and easier to do a good job of shuttering windows against the cold than it is to insulate the rest of the building.

Windows affect comfort even more than these figures suggest. A room with an air temperature of 70°F (21°C), but with large expanses of glass, may seem cold because your body radiates its heat to the cold windows, causing a noticeable chill. Insulating shutters can make a room feel warmer by blocking some of this body heat loss at the same time that they're doing their main job of slowing heat loss from the room.

The hinged shutter consists of a light wooden frame, built around fiberglass or foam insulation and covered with fabric or thin paneling. The finished assembly is hinged to the interior window frame, where it is conveniently operated. As long as there is room for the shutters to swing open, they are unobtrusive and won't take up any living space. If the ceiling is low, a single shutter can be hinged at the top to swing up and be held against the ceiling with

Photo III-36: Take a pop-in, cut it in half, put on some hinges, and voilà, the hinged shutter.

latches or counterweights. The same shutter can also be hinged at the bottom, assuming there is enough room under the window for the shutter to swing down and lie flat to the wall. Hinged on the side, a single shutter may not be as successful unless the window is very narrow, because it might look unbalanced when open, or you might have to move furniture away from the path of its wide swing.

Double shutters (hinged on both sides of the window frame) are less likely to interfere with your living space, although they are a little more exacting and time-consuming to make and install. This style looks symmetrical when the shutters are open, and the two halves don't sweep as far into the room as does a single shutter.

Where the window is next to a perpendicu-lar wall or two windows are side by side, a bifold shutter may be the best choice. The bifold is a single shutter that is split vertically and hinged in the middle and at one side of the window frame. This type folds together as it opens so that only half the shutter is visible against the wall. One bifold unit can open away from an adjacent wall, and two can be mounted on opposite ends of side-by-side windows. A variation of this type is the double bifold: the two halves open away from each other, and each half is hinged in the middle. The double bifold is symmetrical when open and requires even less clearance around the window—only one-quarter of the window width on each side—although with four shutter panels per window you must cut and hang the shutters very carefully if they are to make the all-impor-

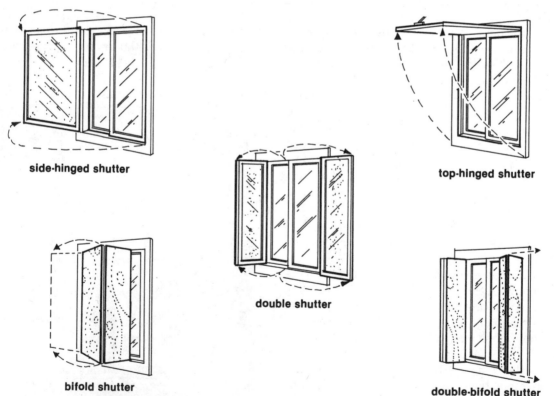

side-hinged shutter

top-hinged shutter

double shutter

bifold shutter

double-bifold shutter

Figure III-24: You can choose from among several styles of hinged shutters. Select the one that will work the best for each window: the single-hinged types are easiest if space is not a problem; the bifolds and the double shutters are less likely to get in the way, but must be hung carefully. If you use more than one style in the same room, using the same fabric or other finish covering will help to unify slight appearance differences.

tant snug fit in the window opening. The double-bifold style is likely to be a good choice for wide windows and, on a large scale, for glass sliding doors.

Measurements

Once you've decided which style you want, measure the window frame very carefully. That's the first step to ensuring that the completed shutter will operate smoothly, even in old houses that may have settled out of square or in new ones where green lumber may have twisted the window frame.

To begin, take three measurements for both the height and width of the window at different points in the window frame. Check for squareness by measuring both diagonals of the window and by using a carpenter's square in each of the corners. Check the flatness of the opening with your eye to see if the wall section around the window is excessively warped, and use a level to see if the frame is plumb and level. Record your measurements and findings on a working sketch of the window.

Now you are ready to design the shutters to fit the opening. For clearance reduce the shutter dimensions by 3/16 inch on all four sides of the perimeter. The main difference among the various types of hinged shutters is in the placement of the hinges and the number of sections that make up the whole. Allow 1/8 inch between each section of double or bifold shutters (in addition to the clearance around the perimeter) by reducing the width of each section by 1/16 inch. This is where the squareness of the opening is important. If the opening isn't square, you can build the shutters to the largest dimension and plane them down to fit the opening.

Before you start cutting materials, check all your measurements again and decide where hinges will be placed. They should be evenly spaced from the top and bottom and allow the open shutter to lie flat against the wall. Collect everything you'll need: butt hinges, latches, screws, nails, glue, framing, paneling and insulation. The hinges should be 1 inch long for each 12 inches of shutter length. Thus a 3-foot-long shutter should have 3-inch-long hinges. Get 4d finish nails and 5/8-inch brads. You'll also need 120-grit sandpaper and something to finish the wood: paint or stain, or fabric if you're taking that route.

Assembly

Shutter construction is not hard. Use 1 × 2's for the frame (actually about ¾ inch by 1½ inch) and ⅛- to ¼-inch panels of plywood or composition board for the skin. You may be able to find yourself some "door skins," which are made with usually

Photo III-37: A long tall bifold like this one would need extra internal bracing to prevent warping. Usually it's difficult to shade tall windows in summer with an exterior overhang, and this is where shutters like these can also be useful.

good quality veneers of mahogany, birch and other woods. The frame can be pine, spruce, fir or redwood, but pine and redwood seem to shrink the least. The most important criteria are that the wood be free of warp-producing knots and that it be old, and therefore dry, or kiln-dried. The frame pieces can be simply butted together and held with glue and nails (figure III-25). This is the simplest joint to make, and by itself it's not very strong, but the shutter gets its real strength from the rigid skin panels.

For those working without a table or radial arm saw, a few tips on cutting the panels: A chalk line is an easy way to make the long cut lines accurately. A *straight* board clamped to the panel as a saw guide will guarantee an undeviating cut if you feel you're a little wobbly with your handsaw or circular saw. You can buy a special plywood blade for a power saw for $3 to $4 to give a smooth, clean cut. When cutting the panels with a circular saw, keep the finish side down so that any splintering occurs on the inside

face. Masking tape is another way to reduce splintering. Set the saw blade only a little deeper than the wood so many teeth are in simultaneous contact with the wood—that also reduces splintering. And remember, cut only to your chalk or pencil line, not on it, or the finished measurement will be off by the width of the saw blade.

Now put the frame together. Set two 4d finish nails in both ends of the long edge pieces. To prevent splitting blunt the nails by lightly tapping them against a hard surface with your hammer. Then glue the ends of the shorter edge pieces and nail the frame together.

With a bifold shutter you will have four skin panels to set on two frames. To start, line up all the panels in proper order (perhaps to get a nice grain pattern) with the good sides facing out. Lay a bead of glue on each frame and set on the panel. Nail ⅝-inch finish brads first at two adjacent corners to square the frame to the panel. That may require a

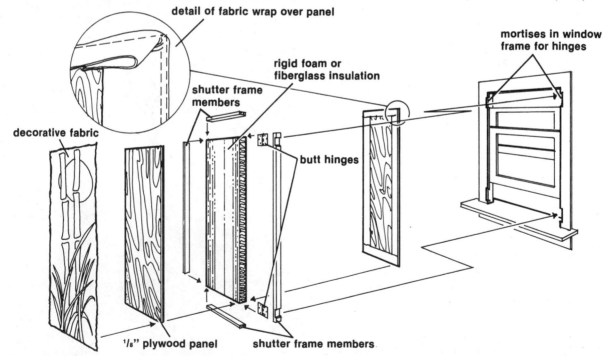

Figure III-25: If you get your measurements exact, it's easy to see that making and installing hinged shutters is a fairly simple undertaking. The finished appearance is up to you, be it natural wood or some nice-looking fabric. Don't forget perimeter weather stripping!

little bending of the frame. Then secure the other two corners, and drive more nails along all the edges, about 6 inches apart.

After you secure one panel to the frame, turn it over and place the insulation in the cavity. If the discussion about foam insulation in the previous project makes you not interested in using it, you can fill the panel with fiberglass and mineral wool insulation, which do not pose the same risks, though they have lower R-values (about R-3 per inch). You can make a decent shutter without foam insulation by making the frame with 1 × 3's or 1 × 4's and inserting 2½-inch (R-7) or 3½-inch (R-11) fiberglass batts covered with a plastic sheet as a vapor barrier (on the interior side of the batts).

When the insulation is in place, nail on the second skin panel. Then set the shutter aside for the day or night. After the glue has dried, set the nails and brads with a nailset, and fill the holes with wood putty that matches the color of the wood. When that's dry, finish-sand the edges and the face to smooth all roughness. Then put the shutters into their respective window openings to check their fit. Oversizing errors can be sanded or planed or trimmed on the table saw. (Don't forget to set the frame nails well below the surface.)

The first hinges to install are those that will attach the shutter to the window frame. Mark their actual location on the shutter by holding a hinge in place and outlining it with a sharp pencil. Make sure the hinges are aligned exactly by lining up the center edge of the hinge (the pin side) with the edge of the panel.

Mortise the spot for each hinge with a chisel. Cut the outline of the mortise first, with the chisel held vertically and the beveled side facing the wood to be removed. Tap the chisel firmly, so it penetrates no deeper than the thickness of the hinge plate. Then, using the chisel with the bevel down, very carefully carve out enough wood so the hinge will rest flush with the edge. Then mark the screw holes, make a starter hole, and drive in the screws. Use the same procedure to mount the hinges that will join the two halves of bifolding shutters.

The last step before mounting the shutter to the window is to finish the wood. Use an oil stain to emphasize the wood, or paint the shutters to match the room. Fabric glued to the panels will work, but you have to account for the thickness of the fabric if it wraps completely around the shutters or covers the edges. Another way to cover with fabric is to glue it to the panels (wrap about ½ inch over the edges) before you attach them to the frame. This eliminates any clearance problems and leaves the wood edge of the frame visible.

Mounting and Weather Stripping

After you've decided how to hinge the shutters and have built them to fit, set each section

Photo III-38: The simple, natural look of a wood-faced insulating shutter can be accented with nice operator hardware. In this installation brass fittings add a little sparkle; the slide bolts at top and bottom to seal the shutters to the weather stripping.

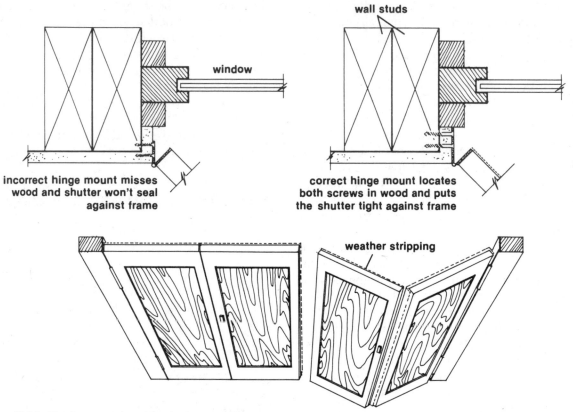

Figure III-26: The importance of having a virtually airtight seal between the shutters and the window frame cannot be stressed enough, especially in colder climates where ice buildup on window glass can be a big problem. The first step is proper positioning or secure mounting of the shutter hinge. Make sure that all hinge screws are sunk into wood, either into the window frame or into wall studs, depending on window style. The hinges should be positioned so that when the shutter is closed, the perimeter weather stripping makes total contact with the window frame.

in the window frame as precisely as possible, using wood shims to achieve equal spacing all around. Then mark the top and bottom of each hinge on the window frame with the attached shutter hinges as a guide. Next, outline the mortise and cut it out. Have a friend hold the shutter in place when the time comes to drive the final screws. If the window frame is boxed in with drywall, make sure the hinge screws actually go into wood, not into the butt end of the drywall.

The final step is to weather-strip the shutter's edges. It's relatively easy, but if it isn't done right, you won't get an airtight fit. Use a compress-ible, adhesive-backed weather stripping (soft foam or felt) on all the vertical edges and sweep-type or bulb-type weather stripping for the top and bottom.

It's going to take quite a while to get the whole house shuttered unless you've got plenty of free time to give to the task. So start making shutters for the windows that lose the most heat (the biggest ones or those in the warmest rooms, or north-facing windows). Besides saving you a good bit of money, these shutters will make you feel more comfortable right away, and that's the main reason you're building them: cost-effective comfort.

David Bainbridge and Denny Long

☼PROJECT
An Insulating Roller Shade

Solar design succeeds most when components are integrated into the building structure or finish-work to save space, improve appearance, and increase user convenience. To that end, even the common window shade has been redesigned for energy efficiency. What better place to store window insulation? It's over the window in a hooded valance where it's out of sight, out of the way and easy to roll up and down with a unique cord-and-pulley mechanism. In the line of window insulation options, this is indeed the most complex design, and it requires careful attention to detail. But the people who developed it have built it time and time again to get it just right, and if you stick closely to the instructions and the tables, you'll get it right too.

Roller-shade window insulation systems are convenient to use and good looking, but the commercially made brands tend to be quite expensive for the R-value they provide. With that in mind, the Rodale Plans Department developed a build-it-yourself design that costs about half as much as store-bought units, but is just as effective.[1] The finished shade will reduce conductive heat loss through a single-glazed window by 81 percent and by 63 percent through a double-glazed window.

We actually found that owner-made shades offer advantages not available in commercial models. Perhaps the biggest problem with commercial shades is fitting them to the many varied types of windows, but the way this shade is designed, that's no problem as each shade is custom-made. By using a data sheet and taking just two measurements on your window, you can calculate the size of each and every piece in the shade assembly. Thus every shade fits exactly, making it an effective seal against air infiltration and giving it an attractive custom-built appearance.

But the main advantage of making your own shade is simply economics. In cold-room tests we found our shades to have an R-value of 4.8, using the standard recommended fabric. This equals or exceeds the R-value of most commercially available

roller-shade systems and at a cost of only $2.00 to $2.50 per square foot, instead of $4.00 per square foot and up for commercial versions. With their higher cost most commercial systems have at least a four-year payback period while our build-it-yourself design can pay for itself in two years.

Table III-4 shows the calculated savings for our shade in many regions of the country. In some areas the shade will pay for itself in less than two years, based on a fuel oil price of $1.00 per gallon and a furnace efficiency of 60 percent. With more expensive energy sources (electricity) and as the price of fuel goes up, the payback period naturally gets shorter. If you put an insulating shade over a standard 6 by 7-foot sliding glass door, your savings from that one shade could run anywhere from $63 a year in Sault Ste. Marie, Michigan (9048 heating degree-days) to $18.90 a year in Fresno, California (2611 heating degree-days). For costlier electric heat each square foot of shade will save an average of about $1.25 per year, while savings per square foot of shade with wood or coal heat will be about 65¢.

Somewhat surprisingly, your savings will not differ much as the window orientation changes. Table III-5 shows the savings from a shade on windows facing the four points of the compass. Note that while south-facing windows gain substantial energy

Photo III-39: In older masonry buildings with thick walls, the insulating shade can be mounted inside the frame to minimize the gap between the glass and the insulation. In either position, inside or outside the frame, operation is simple: The shade is pulled down from its lower edge and raised by the pull cord.

Table III-4 What's Window Insulation Worth?
(savings per year per square foot of shade with R-4.83)

Location	Single-Glazed Window
Sault Ste. Marie, Michigan	$1.50
Madison, Wisconsin	$1.40
Portland, Maine	$1.30
Hartford, Connecticut	$1.10
Denver, Colorado	$1.05
Salt Lake City, Utah	$1.05
Indianapolis, Indiana	$1.00
New York, New York	$0.80
Seattle, Washington	$0.70
Nashville, Tennessee	$0.60
Fresno, California	$0.45

SOURCE: Calculations based on methods given in *Movable Insulation,* William K. Langdon (Emmaus, Pa.: Rodale Press, 1980).
NOTE: Figures are based on the use of oil at $1.00 per gallon at 60 percent efficiency = $11.70/MBtu.

from the sun, the overall savings in energy is roughly the same regardless of what direction the window faces.

The larger the window is, the more cost-effective the shade will be. Larger shades cost less per square foot, and they're saving proportionately more energy. Smaller windows require more material per square foot of shade installed, which makes them good candidates for some other form of movable insulation such as pop-ins or shutters. In a well-used part of your house, however, the convenience of roll-ups may be worth the extra cost. Small or large windows in out-of-the-way places, on the other hand, may not warrant the convenience of a roll-up shade, especially if the insulation is likely to be on more than it's off. Sliding glass doors are normally the biggest single energy loser in a home, and they're the first place you should install a shade.

Design

The basic shade consists of two layers of quilted fabric and one layer of reflective, aluminized Mylar that rolls down from a valance roller box. Closing devices on each side of the window seal the edges of the shade and a draw cord raises or lowers it. The shade is designed to mount outside the exist-

TABLE III-5 WINDOW ORIENTATION AND THE VALUE OF INSULATION

The following numbers represent the dollar value of the net energy gain or loss during the heating season for a 10-square-foot window in New York City. Because of the solar gain through a south-facing window, both the single- and double-glazed categories show a "plus" dollar value without the insulating shade. But that value is increased three-fold in double- and single-glazed windows when the insulation is added. The east/west- and north-facing windows also move into the black with movable insulation. This calculation is based on a fuel oil energy source at $1.00 per gallon with the furnace operating at an efficiency of 60 percent.

Window Type	Window Orientation		
	North	East/West	South
Single-glazed	−$8.77	−$3.33	+$ 4.63
Double-glazed	−$2.29	+$2.43	+$ 9.21
Single-glazed with shade	−$0.62	+$4.68	+$12.79
Double-glazed with shade	+$1.30	+$5.92	+$12.71

SOURCES:
1. Ray Wolf, *Insulating Window Shade* (Emmaus, Pa.: Rodale Press, 1980).
2. Robert G. Flower, "An R-5 Insulating Window Shade Designed for Construction by the Homeowner," In *The 5th National Passive Solar Conference,* eds. John Hayes and Rachel Snyder (Newark, Del.: American Section of the International Solar Energy Society, 1980).

ing window trim to avoid the problem of adapting the mounting system to the limitless trim styles. This design also allows the shade to be installed without making any alterations to the existing trim. Another advantage with this mounting is that when the shade totally covers the window opening, infiltration heat loss is virtually eliminated. Various studies have shown that 60 percent of the heat loss through windows occurs as air infiltration through the trim and other parts of the windows. Caulking and weather stripping does away with most of that air leakage,

and by sealing outside the trim this shade blocks nearly all the remaining infiltration.

To live up to our design goals, the heart of the shade system, the roller mechanism, had to be easy and inexpensive to build and smooth in operation. We developed a roller box that holds two dowel rods. Small plywood pulleys are fastened to the ends of the dowels, and the fabric is fastened to the dowels. An endless loop of drapery cord is wound around the two pulleys and runs out the bottom of the roller box. Placing rubber bands around the pulleys gives them enough traction to pull the weight of the largest shade. The cord hangs down to the windowsill, where it passes through a drapery cord tensioner, which keeps the cord taut. When installed, the entire system looks and operates exactly like a standard set of drapes.

The valance or roller box holds the two dowels and stores the fabric out of view above the window. The box is simple to make since every piece but one is a rectangle, and there are no complicated joints or other intricate cuts. The box is built in two sections, a front and back unit that are hinged together. The hinging allows easy access to the inside for cleaning the fabric or adjusting the shade.

We found that the best material to use for the box is plywood. If you wish to stain your box, a good quality veneer plywood is best. From a single 4×8-foot sheet of plywood you can get all the pieces needed to build three roller boxes for 3 by 5-foot windows. For the utmost in economy we use standard ¾-inch plywood for parts of the box that do not show and veneer plywood for the few pieces that do. Although thinner grades of plywood could be used for the roller box, we recommend ¾ inch because it makes a sturdy box that nails together easily for little extra cost. Dimensional lumber is more expensive than plywood and can warp and bow, making the roller box more difficult to build and operate. A good compromise is to use quality kiln-dried lumber for the pieces that show and plywood for those that do not. For assembly, flathead wood screws or 6d or 8d finish nails are countersunk, with their holes filled.

The sides of the shades are sealed with two hinged, wooden closer assemblies. One piece of the assembly, the mounting board, is fastened to the wall next to the window trim. In some cases, fastening the mounting boards to the wall may be difficult, depending on what type of wall construction you have. Consult your hardware store for the proper wall fastener; it may turn out that bonding the mounting board with construction adhesive (Max Bond) is the easiest solution.

The second piece, the closer, is fastened to the mounting board with spring-loaded (self-closing) hinges. The closer overhangs the window trim and presses the fabric firmly against the trim when the shade is down. To make the side closers more effective, a bead of caulk is run between the mounting board and the window trim as part of the overall window caulking and weather stripping.

Perhaps the most important design feature of the shade is the type of fabric used. After testing a number of quilted fabrics, we found that material made for table placemats works best; it's sold in most fabric stores and comes in a wide variety of patterns and colors. Normally one side is plain fabric, and one side is patterned with a ⅛- to ¼-inch layer of foam between. We have found that the thicker foam is best, giving an R-value of 4.8 and no difficulty in rolling up into the valance. We also tested ½-inch-thick ski jacket material, which at R-8 reduced heat loss by 88 percent, but it was too bulky for the roller system.

The shade is designed so the fabric needs little sewing. The fabric is cut to size, and edging material is sewn on all four edges. We have found double-fold bias tape is the best and easiest way to finish the edges. First approximate the length of fabric you need, sew the bias tape on three edges, install the shade, then cut off the excess material and finish the last edge when installing the fabric. Attach fabric layers to the dowels with standard fabric snaps attached to the fabric and a *stud fastener* on the dowel. Stud fasteners—nothing more than snaps with wood screws on them—screw into the dowel, and the fabric snaps to them. This makes for easy

installation and removal of the fabric. Stud fasteners are sometimes hard to find (try boating suppliers), so be sure to get the stud fasteners first, then the fabric snaps to fit them. Snaps about ⅜ to ½ inch wide work best.

The second layer of the shade, the aluminized Mylar barrier, is very important because it reflects radiant heat back into the room, heat that would otherwise escape through the window. Glass, although nearly opaque to infrared radiation, does absorb and conduct some to the outside, but the Mylar almost eliminates this loss. It also serves as a vapor barrier, preventing moist room air from reaching and condensing on the window. We designed the shade so the quilted fabric does not touch the Mylar across most of the shade. This air gap adds to the shade's insulating value, and it also preserves the reflective properties of the Mylar.

As an added feature, the Mylar layer can be reversed in summer by unsnapping it from the inside of the shade and reattaching it outside the exterior layer of fabric. This enables the Mylar to bounce sunshine back out the window, greatly cutting down on heat gain to the room. With double-hung windows, you can cool the room by opening the top sash and pulling the shade down to within about 6 inches of the windowsill. A "thermal chimney" will be formed as heat builds up between the shade and the window: The heated air rises and exits at the top, pulling warm room air into the opening. This helps to set up a ventilating air flow if north-facing windows or a basement door are opened to allow sources of cooler air to enter.

Measuring Up

Accurately measuring the window and calculating the size of the pieces to be used for the shade are essential to a successful project. Everything depends on the accuracy of these measurements. First, measure the width of the window from the outside edge of the trim on one side of the window to the outside edge of the trim on the other side, and record this as measurement A on worksheet III-1. Take the measurement in several places on the window to be sure that the window is square and record it as measurement B. If the window is out of square, use the larger of your measurements, and later trim the shade back to fit the actual opening. Now calculate the finished fabric width by subtracting ¼ inch from measurement A, and record the result on the worksheet. Then calculate the rough fabric length by adding 12 inches to measurement B; enter this figure on the worksheet.

Now determine how much space there must be in the roller box between the two dowel rods, measured center to center. This space must be large enough to allow the shade to roll up completely. If the space is too small, the two rolls of fabric will press against each other, and rolling up the shade will be difficult, if not impossible. The distances given in table III-6 are for place mat material with ¼ inch of foam. If you use that type of material, correlate the

TABLE III-6 CENTER-TO-CENTER DISTANCES

Rough Fabric Length	Center-To-Center Distance
Up to 30 inches	3 inches
31–36 inches	3¼ inches
37–42 inches	3½ inches
43–48 inches	3¾ inches
49–54 inches	4 inches
55–60 inches	4 inches
61–66 inches	4¼ inches
67–72 inches	4½ inches
73–78 inches	4¾ inches
79–84 inches	4¾ inches
85–90 inches	5 inches

rough length of your fabric to the required center-to-center distance between the two rollers; record this on your worksheet.

If you buy material with a different thickness, cut it to the rough length noted on your data sheet, then tack one end of the fabric to a 1-inch diameter dowel, roll up the fabric, and measure the diameter of the roll. The center-to-center distance should be the same as the diameter of the roll of fabric. If you roll the fabric tightly, add an extra ½ inch to the size you calculate, and record this measurement as the center-to-center distance. Continue filling in the worksheet by completing the calculations based on measurements A and B.

Assembly

Once the worksheet is filled in, you are ready to start cutting out the pieces of the roller box (see figure III-28). The measurements given for the two rod holders are approximate outside dimensions

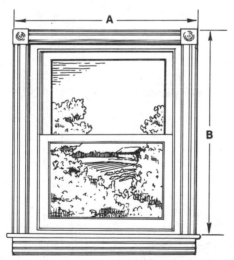

Figure III-27: Accurate measurement of the window width (A) and height (B) are critical to gaining smooth operation from your shade. The two measurements are shown at just one part of the window, but if you think your window is out of square, take these measurements in a couple of places and always use the widest measurement. The shade fabric can be trimmed later to account for out-of-squareness.

of the boards that the holders are cut from. And likewise, the sizes given for the pulleys are the rough sizes of the pieces from which you'll be cutting the round pulleys.

You'll need a saber saw or jigsaw to cut out the rod holders. Each holder is comprised of two pieces, making a total of four pieces per roller box. First mark a center line along the length of one of the rod-holder pieces. Make two marks on this line along the length of one of the rod-holder pieces. Make two marks on this line at one-fourth and three-fourths of its length. Using these marks as the center point, drill

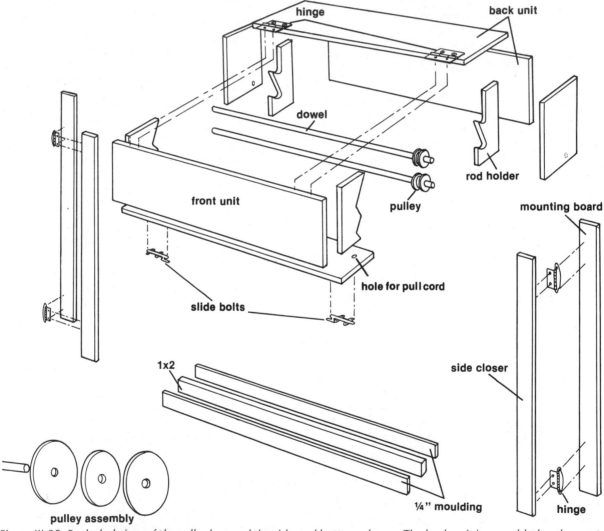

Figure III-28: Exploded view of the roller box and the side and bottom closers. The back unit is assembled and mounted as one piece. The front unit remains separate, attached to the back with hinges that allow it to be swung out of the way for maintenance or seasonal repositioning of the Mylar sheet.

Worksheet III-1 Wall-Mounted Shade

Window Location _____

        Measurement A _____   Measurement B _____

Finished Fabric Width: _____

        Measurement A MINUS ¼″ = _____ Finished Fabric Width

Rough Fabric Length: _____

        Measurement B PLUS 12″ = _____ Rough Fabric Length

Fabric Classification Number = _____

Center-to-Center Distance (from table III-6) = _____

Name	Quantity Needed	Calculation	Dimension	
Back	1	Length = A plus 3″ Width = 2 times center-to-center distance	Length Width	_____ _____
Top	1	Length = A plus 3″ Width = center-to-center distance plus 1½″	Length Width	_____ _____
Sides	2	Height = 2 times center-to-center distance plus 2¼″ Width = center-to-center distance plus 1½″	Height Width	_____ _____
Rod holders	2	Height = 2 times center-to-center distance Width = center-to-center distance plus ¾″	Height Width	_____ _____
Front	1	Length = A plus 4½″ Height = 2 times center-to-center distance plus 2¼″	Length Height	_____ _____
Bottom	1	Length = A plus 3″ Width = center-to-center distance	Length Width	_____ _____
Dowel rods	2	Length = A plus 2⅞″	Length	_____
Pulleys	2	Diameter of sides = center-to-center distance minus 1¼″ (not to be smaller than 2″) Diameter of center = diameter of sides minus ½″	Side Diameter Center Diameter	_____ _____

a hole 1/16 inch larger than the diameter of the dowels you are using for the roller (1-inch-diameter dowels work well). Refer to figure III-29 for the layout of the cuts to be made. If you're using solid wood, you're less likely to have problems if the grain runs perpendicular to the sides of the pieces. If you use plywood, the grain direction doesn't matter. Use care in marking the lines and cutting the rod holder, because if the cuts are not exact, the finished roller box will not close smoothly.

The pulley parts (three parts for each pulley) are made from plywood. The two side pieces are

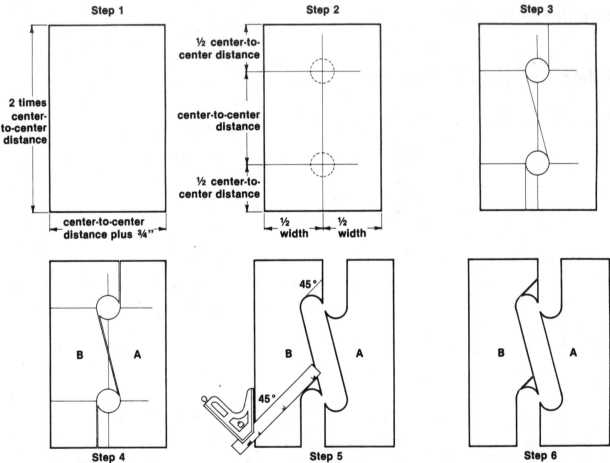

Figure III-29: The two-piece rod holder is the most "complicated" of all the wood cuts in this design, but it's not really that difficult. In Step 1 the raw block has been sized according to the calculation done in the worksheet. In Step 2 two holes are laid out as shown and drilled 1/16 inch larger than the diameter of the dowel rod. Step 3 shows the cutting lines which separate the B side from the A side (Step 4). In Step 5 the final cutting lines are laid out using the 45-degree or miter side of a combination square. After the cuts are made (Step 6), all cut edges should be sanded until they're very smooth. The B side then attaches to the front section of the roller box, where it helps to lock in the rods that sit on the A side, which is attached to the back section.

made from ⅛-inch plywood, and the center piece is cut from ⅜-inch plywood. Altogether you'll need four side pieces and two center pieces to make the two pulleys required for a single shade. Cut the plywood circles to the diameters noted on the worksheet, then drill a hole in the center of each one the same diameter as the dowel. Place the three pieces over a piece of scrap dowel, and glue and tack them together. After the glue has dried, remove the scrap dowel, and glue the assembled pulley onto a full-length dowel. The outside face of the pulley should

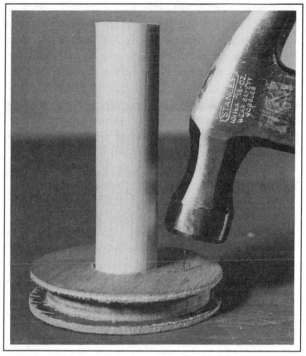

Photo III-41: The scrap piece of dowel used in the assembly of the pulleys ensures that each part of the sandwich is centered over the other.

be 13/16 inch in from the end of the dowel. This allows ¾ inch of dowel to rest in the rod holder with 1/16 inch of clearance.

To assemble the roller box start with the back unit. Glue and nail the back to the top, with the top overlapping the back; then nail on the two side pieces, and lastly nail the A pieces of the rod holders using small nails and plenty of glue. To assemble the front of the roller box, attach the two B pieces of the rod holders to the bottom, then attach the front. Note that the front should overhang the bottom rod holder assembly by ¾ inch. This overhang is very important, since the rod holders won't fit without it.

When the two units are assembled, slip them together to be sure they fit. If they do, all that remains is to hinge them together. Standard rectangular hinges are screwed into the top of the back and front units. Two hinges should be enough for most

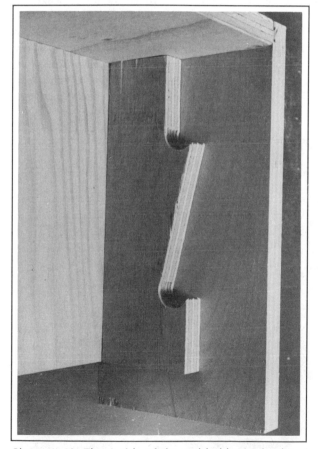

Photo III-40: The A side of the rod holder is glued and nailed inside the back of the roller box. Excess friction in the cupped sections, where the dowels rest, can be reduced with candle wax.

roller boxes, but wide units will need three or four.

Drill a ½-inch hole in the bottom piece of the back unit to allow the pull cord to pass through. The location of the hole is shown in figure III-28. The metal splicer that holds the two ends of draw cord together must pass through this hole, so bevel it well so that it doesn't get hung up. The hole should be directly below the center of the pulleys and 1⅞ inches in from the outside face of the side piece. The draw cord can be placed on either side of the box (wherever you mount the pulleys), and the pulleys are easily reversed to suit your needs.

You'll need a small latch to keep the roller box fastened shut. The simplest way is to mount slide-bolt latches on each of the bottom pieces, and drill a ¼-inch hole in the side piece to receive the end of the slide bolt. When the roller box is shut, just push the slide bolt into the hole, and it will hold the box shut. Without some type of closing latch the

pressure of the fabric on the rod holders would push open the box.

When all is assembled, you're ready to hang the box. The roller box should be centered exactly over the window. Be sure to fasten the roller box securely to the wall studs or the masonry because pulling the shade up and down puts a lot of stress on the box. Again, the combination of bonding plus fastening will ensure a sturdy mount. The bottom of the back piece should sit directly on the top of the window trim. If your window is out of square, use a level to scribe a horizontal mounting line. It's very important to get the box level so the fabric will hang straight.

With the box mounted you are ready to cut the mounting boards and side closers. Each mounting board should be 1½ inches wide and should run from the windowsill to the bottom of the roller box. The mounting boards should be the same thickness

Photo III-42: Use a level to make sure that the roller box is horizontal.

as the window trim to allow the closers to hold the fabric firmly against the trim. Measure and cut the mounting boards and then the side closers. The closers should have about ¾ inch of clearance from the bottom of the roller box and the windowsill. Thus they should be 1½ inches shorter than the mounting boards, but the same width and thickness. Attach the mounting boards and fasten the closers to them with the spring-loaded hinges. The closer board should overhang the window trim by about 1 inch when it's installed. Be sure that the two mounting boards are parallel to each other and are perpendicular to the roller box.

After the hardware goes on, the "software" goes in. All that remains is to sew the fabric and attach the Mylar. The worksheet gives you the finished width of the fabric; cut it to that size. Then cut one end square with the two vertical sides and sew on the edging material around the three edges of the fabric.

To attach the fabric to the dowel rods, prop the front of the roller box open, and put the bottom dowel rod in place. Open the side closers, and hold the finished edge of the fabric against the dowel rod. Be sure one side of the fabric is right next to the side of the pulley. Using thumbtacks, adjust the position until the fabric is parallel to the closers for the entire length of the window. Remove the dowel and fabric, and punch a series of holes through the fabric into the dowel. Remove the fabric from the dowel, attach stud fasteners to the dowel and snaps to the fabric; then snap the fabric to the dowel, and return the dowel to the roller box to check the fit. If everything fits, go on and repeat the process with the second fabric layer on the other (upper) dowel. The snaps in the dowels can be moved to make adjustments in the hang of the fabric.

With both pieces of fabric attached to the dowels, lay the bottom dowel on a flat work surface and cut the Mylar sheet about ¼ inch less in width than the fabric. Center the Mylar on the fabric with the top edge about 2 inches below the dowel. Make a series of marks through both the Mylar and the fabric for snap placement, as shown in figure III-30.

Photo III-43: The first step in hanging the quilted fabric is to lay it up to the dowel and tack it in place after it's been adjusted to hang parallel with the side closers.

When installed in the roller, the Mylar should hang between the two pieces of fabric. Make sure you have the snaps installed on the Mylar and the fabric so that the Mylar will hang on the proper side of the fabric. The fabric on the bottom roller will face the room.

When you've got the Mylar attached to the front piece of fabric, place it in the roller box to check for a good fit. Then attach snaps to the outside fabric layer (the top roller) in the same manner. For summer use, the Mylar will face out the window when it's snapped to this layer.

With the Mylar and the fabric snapped in place, return the dowel rods to the roller box. The fabric should overhang the bottom of the windowsill. Remove the dowels from the roller box, and cut each piece of fabric to length, which is about 4 inches longer than the point where each layer reaches the sill. The two fabric layers will end up with different lengths because one dowel is above the other. *Don't cut the fabric too short.*

The bottom molding of the shade is a sandwich of three pieces of wood. The center piece should be ¾-inch plywood, while the two outer pieces should be ¼-inch-thick solid wood. Turn the fabric layers up into a double hem with each hem taking up 1 inch of fabric, and sew the hems. To reduce the thickness of these hems, remove a 2-inch-wide strip of insulation from between the quilting material at the hem. The plywood piece of the molding goes in between the two fabric layers, and the two outer pieces are screwed to it, with the screws passing through both fabric layers. Leave about ¼ inch of the hemmed fabric exposed below the bottom molding to serve as an air seal when the shade rests on the sill. Note that the Mylar isn't part of this sandwich so that it can be moved for the summer mode. Cut the Mylar off about 1 inch above the bottom molding.

The final steps involve setting up the pull cord system. Before you put the dowels in the roller box, stretch two wide rubber bands over the face of each pulley to give the cord added traction. The shade may not roll up without these rubber bands. Figure III-31 shows the cord path. Run the cord over the top pulley, and cross the two ends before going around the bottom pulley, forming a figure eight. Then pass both pieces of cord through the passage hole in the bottom, and close the roller box.

Find a place to mount the cord tensioner (usually the windowsill) and then cut the cord to twice the approximate length of the distance from the roller box to the tensioner. An electrical butt splice suitable for 10- to 12-gauge wire will fit standard drapery cord perfectly and can be used to fasten the two ends of the cord together. Put one end of the cord in the splice and crimp it; then insert the

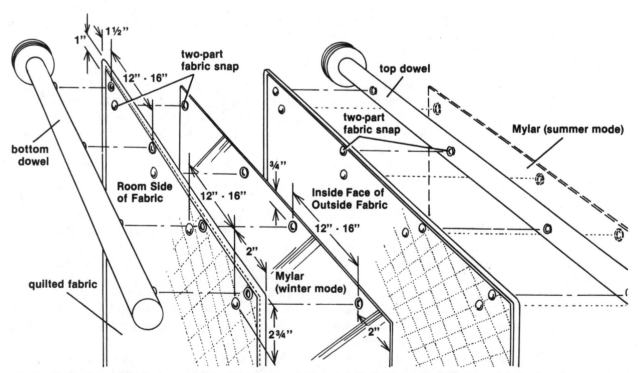

Figure III-30: Accurate placement of the fabric snaps is the most important part of getting the shade to hang straight and run true. The dimensions given for the spacing of the snaps are closely approximated.

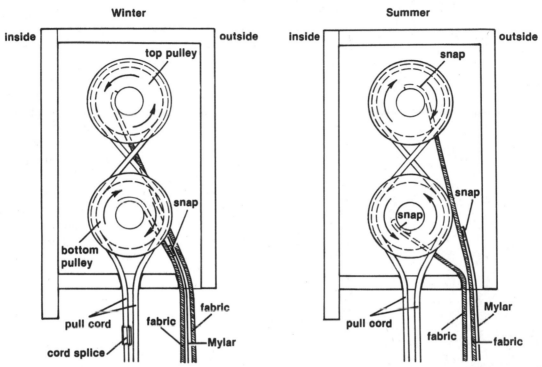

Figure III-31: In the winter mode the reflective Mylar is fastened in between the two layers of quilted fabric; the shade is lowered by pulling down on the pull cord closest to the window. With the Mylar snapped to the outside of the quilted fabric in the summer mode, the shade is lowered by pulling down on the opposite side of the pull cord.

other end, crimp the splice again, and presto! You have an endless loop of cord. Attach the cord tensioner to the wall, or the sill, and follow the manufacturer's directions for establishing tension on the cord. In most cases the butt splice will not pass through the cord tensioner, so you will have to adjust the position of the splice to assure that it does not go through the tensioner when you operate the shade. Over time the splice may work its way around until it gets in the way of the tensioner. If this happens, you simply have to release the tension, open the roller box and readjust the position of the splice.

That should finish your shade. To operate it, open the side closers and pull the bottom molding to get the shade into place. Then close the side closers to seal the edges. To raise the shade, open the side closers and use the pull cord. (When the Mylar

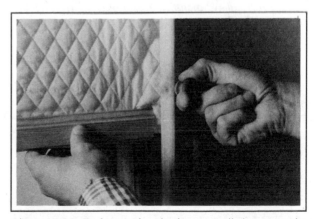

Photo III-44: To lower the shade you pull down on the bottom molding with the side closers open. When the shade is down, the side closers clamp the fabric to the window trim.

Photo III-45: To raise the shade you use the pulley cord. The cord tensioner and the rubber bands wrapped around the pulleys maintain the proper friction for the cord.

is in the middle of the shade for winter use, pull the front cord. When the Mylar is on the outside for summer use, pull the back one.) The Mylar must always be on the outside of the fabric layer it's attached to. In winter this means the fabric rolls in one direction and in summer in the other direction. This is why different cords must be pulled in the two modes of operation. If you pull the wrong cord, the fabric will bunch up and not roll properly.

If your shade operates stiffly, apply some wax to the ends of the dowels and the rod holders to reduce friction. If you have any other problems, check the cord path and the way the fabric rolls up. Otherwise the shade will operate for many years, saving energy long after it has paid for itself and making your home more comfortable from day one.

Ray Wolf

Note

1. Ray Wolf, *Insulating Window Shade.* (Emmaus, Pa.: Rodale Press, 1980). This book includes step-by-step instructions for building and sewing a shade which has a rated R-value from 4.3 to 8.5, depending on the fabric used. Eight pages of blueprints are included.

References

American Society of Heating, Refrigerating and Airconditioning Engineers. 1979. ASHRAE symposium, window management as it affects conservation in buildings. In *ASHRAE Transactions,* vol. 85, part 2, New York.

Hastings, Robert S., and Crenshaw, Richard W. 1977. *Window design strategies to conserve energy.* Washington, D.C.: U.S. Government Printing Office. no. 003-003-01794-9.

Langdon, William K. 1980. *Movable insulation.* Emmaus, Pa.: Rodale Press.

Lawrence Berkeley Laboratory. 1980. *Windows for energy efficient buildings.* Berkeley, Calif.: University of California.

Shurcliff, William A. 1980. *Thermal shutters and shades: over 100 schemes for reducing heat loss through windows.* Andover, Mass.: Brick House Publishing Co.

Hardware Focus

Commercial Movable Insulation Systems

We don't expect everybody everywhere to be raring to build window insulation, and for those who would rather buy than build we've compiled this list of current (as of late 1981) manufacturers of a variety of different styles. The R-values listed below are those claimed by the manufacturers on their product literature. All products are for interior application unless otherwise indicated.

Aardvark & Sun Solar, Inc.
167 Webbers Path
West Yarmouth, MA 02673
(617) 394-6391

Thermal Window Shutter
Double-hung, double-glazed window unit with sliding wooden shutter with Thermax insulation; R-8; numerous sizes.

Abox Corp.
629-3 Terminal Way
Costa Mesa, CA 92627
(714) 645-0623

Roll-Awn
Exterior roll-down shutter; polyvinyl; operates from inside; R-2.5 over single glazing; numerous sizes.

American German Industries
14611 N. Scottsdale Rd.
Scottsdale, AZ 85260
(602) 991-2345

Rolladen
Exterior rolling shutter on track; made of interlocking extrusions of PVC or aluminum; motorized operation; PVC with dead airspace rated about R-2 with single glazing; custom sizes available.

Appropriate Technology Corp.
P. O. Box 975
Brattleboro, VT 05301
(802) 257-4501

Window Quilt
Aluminized polyester plastic between 2 layers of fiberfill; 2 outer surfaces of polyester-rayon fabric; unrolls along edge-sealing side tracks; R-4.25 (single glazing); R-5.15 (double glazing); UV resistant; many sizes.

Ark-tic Seal Systems, Inc.
P. O. Box 428
Butler, WI 53007
(414) 276-0711

Insealshaid
3 plastic roll-up shades (transparent, reflective, heat absorbing) mounted on same frame. Appropriate deployment of 1 or more shades provides night insulation, summer sun control or winter passive solar gain. Reduces air infiltration. Models available for windows and patio doors.

Conservation Concepts Ltd.
Box 376
Stratton Mountain, VT 05155
(617) 547-0495 or (802) 297-1816

Warm In Sealing Drapery Liners
Kit of partially assembled components or custom-made drape/curtain liner; 1 layer of bubble polyethylene and 1 of cotton; R-2 to R-2.5 (double glazing).

Creative Energy Products
1053 Williamson St.
Madison, WI 53703
(608) 256-7696

Window Warmers Do-It-Yourself Kits
Quilted Roman shade; spring-loaded clamps seal shade against window frame; shade of 4 layers: drapery fabric, vapor barrier, Thinsulate insulation, lining fabric; R-4 includes airspace between window and shade; kits or plans.

Energy Industries
Solar Shutter Division
2010 N. Redwood Dr., Rte. 1
Independence, MO 64050
(816) 257-1919

Homesworth Corp.
18 Main St.
Yarmouth, ME 04096
(207) 846-9934

Independent Systems Corp.
P. O. Box 329
Durham, CT 06422
(203) 349-1078

Insul Shutter, Inc.
Box 338
Silt, CO 81652
(303) 876-2743

Kipling's
110 4th St., N.E.
P. O. Box 1087
Charlottesville, VA 22902
(804) 295-6030

Minute Man Anchor Co.
305 W. Walker St.
East Flat Rock, NC 28726
(704) 692-0256

National Metallizing
P. O. Box 5202
Princeton, NJ 08540
(609) 443-5000 or toll-free number:
       (800) 257-5116 (except New Jersey)

Pease Co.
Ever-Strait Division
7100 Dixie Highway
Fairfield, OH 45023
(513) 867-3333

Shutters, Inc.
110 E. 5th St.
Hastings, MN 55033
(612) 437-2566

In-Sol Sliders
Double-walled, rigid polypropylene plastic clip in panels; R-2.43; 7 sizes or do-it-yourself panels.

Sun Saver
Interior shutter in kit form; pine frame sandwiched between reflective board stock with ¾" dead airspace; shutter alone R-4.4; 3 standard sizes that can be cut down.

Independence 10
Four layers of plastic film on guide frame; rolls down; maximum width 72"; maximum length 120"; UV resistant; fire retardant; R-10.71 (over double glazing).

Insul Shutter
Rigid polyurethane core between wooden panels; double foil radiation barrier; foam strip around frame; panel widths 8" to 16"; panel heights 24" to 80"; R-9.1 over single glazing.

Decorator Thermal Panel
Pop-in rigid styrene foam panel encased in removable fabric (choice of 6 colors); secured with magnetic clips; R-4; custom sizes.

Minute Man Storm Window Shade
Roller shade of transparent plastic film; reusable tape seals edges and bottom; sizes: 37¼", 46¼" or 55¼" × 72".

Nunsun Shade
Roller shade; 2 layers of polyester film with a layer of aluminum; bronze, silver or gray/silver; rated R-values range from 5 to 10 (double glazing); UV resistant.

Pease Rolling Shutter V or VM
Exterior shutter of hollow PVC slats; motor operated; rolls down on track; R-values when installed not more than 1¾" outside single glazing: Model VM—R-1.76, Model V—R-2.47; custom sizes; UV resistant.

Thermafold Shutters
Exterior shutters for sliding glass doors; 1" urethane core between wooden panels; R-9.9 over double glazing; open dimensions: 122" (W) × 81½" to 86½" (H).

Solar Energy Components, Inc.
212 Welsh Pool Rd.
Lionville, PA 19353
(215) 644-9017

Sunflake
625 Goddard Ave.
P. O. Box 676
Ignacio, CO 81137
(303) 563-4597

Sun Quilt Corp.
Box 374
Newport, NH 03773
(603) 863-2243

Sunway Heating Systems, Inc.
409 Lawrence St.
Petoskey, MI 49770
(616) 347-6213

Thermal Technology Corp.
P. O. Box 130
Snowmass, CO 81654
(303) 963-3185

Wind-N-Sun Shield, Inc.
P. O. Box 2504
131 Tomahawk Dr.
Indian Harbor Beach, FL 32937
(305) 777-3558

windowBlanket
Rt. 7
Box 322A
Lenoir City, TN 37771
(615) 986-2115 or toll-free number:
    (800) 257-7850 (except New Jersey);
    in New Jersey (800) 322-8650

Zomeworks
P. O. Box 712
Albuquerque, NM 87103
(505) 242-5354

Thermo-Shade Thermal Barriers
Segments of hollow, rigid PVC on roll-down track; suitable for standard or very large windows; R-5 (single-pane window); R-7 (double-pane window); UV resistant.

Sunflake Bi-Pass
Insulating panels covered by white textured steel for large areas of glass; panels interlock and move in series across glass; R-14; sizes up to 8' in height.

Sun Quilt Thermal Gate
Washable polyester/cotton or nylon tricot and waterproof nylon; manual or automatic operation; size and R-factor to customer's specifications; most effective for large window areas.

Movable insulation window shutters built to the customer's specifications.

Insulating Curtain Wall
Several layers of metallized reflective films; automated and thermally controlled; suitable for very large windows; R-9 to R-12 with ICW alone; UV resistant.

Wind-N-Sun Shield
Drapery liner panels or roller shades; aluminized polyester on one side; white vinyl on other; R-4; UV resistant.

windowBlanket
Curtain; cotton with polyester fiberfill filling; backing of Roc-lon, insulated cotton lining; R-2 to R-2.4.

Beadwall panels
Double-glazed windows which turn into insulated walls when white foam beads are sprayed into airspace; panels become windows again when beads are vacuumed from the bottom and go back into the concealed storage unit; R-8 when full; standard sizes: 34" × 76", 46" × 76", 80¾" × 41¾", 80¾" × 81", 80¾" × 53¾", 80¾" × 105"; plans also available.

C. PORCHES

Photo III-46: Instead of building a greenhouse onto her porch, Sheila Darr of Berkeley, California, turned her porch into a greenhouse, complete with all the necessities, especially the hammock.

Solarizing the Great American Porch

Ah yes, porches . . . sultry evenings, hammocks and chair swings, relaxation, romance. In all of its varied styles—the porch, the patio, the deck, the balcony, the veranda—it's an ever-present fixture of American architecture, and thank goodness we have them. Before houses were crammed together in the manner of sardines, a porch was a front-row seat to Nature's play. People sat to take in views, sounds and airs; they were outside, but still at home, free from the confines, yet still secure. Of course we still sit out, though the views may include more houses than trees, and the sounds are as much neighbors and traffic as Nature, and the air may have distinctly unnatural flavorings. But when I walk through town on a hot summer night, the porches are filled with people, conversation and meditation.

It's a pity that winter shoos us and our deck chairs indoors, that porches are just a warm weather event, but for those that have even a little solar aspect it doesn't have to be that way. With thorough sealing and insulating and either seasonal or permanent glazing you can have a year-round affair with your porch. Solarizing the Great American Porch may just be one of the best solar deals going because the structure is already built, and that's where most of the cost goes for a new enclosure. And being what they are, porches are open in design and usually don't require major remodeling in the solarizing plan. Glaze the walls, glaze the roof, make sure the space is tight and that hot air can enter the house in winter and leave the porch in summer (a vote in favor of seasonal glazing); put in thermal mass if you want, back-up heat if you're highly committed; put in greenery, planting beds, some comfy seating and voilà: a building addition on a budget, a special sunroom that will get you back on your porch in the coldest of times, where you can resume your summer's meditations and relaxations and have a romance with the sun.

Photo III-47: There weren't any major visual changes in adding a single layer of glass to this porch. The real change was in big energy savings.

PROJECT
A No-Frills Porch Glazing Retrofit

If you've got some old houses in your town, you might find some with glassed-in porches. Very fancy. Nowadays they don't get built that way very much, but it's not hard to get there. This project starts with a porch that's all but ready for solarizing, and it finishes with a little carpentering and some glass work.

When I decided to solarize my house, I quickly found out that there wasn't enough space on my lot for the greenhouse I wanted. There was, however, a south-facing porch running across the front of the house, so I decided to glass it in. The major design restriction was clearly described by my wife, "I don't want to look out the picture window at a bunch of window frames"—a sensible request indeed. Armed with that directive, I put together a stretch of glazing that was interrupted only by a storm door and the existing porch posts.

The retrofit consists of seven large panes of plate glass (about 260 square feet total) butted together and sealed into a simple framing system. It's by no means complex and certainly not expensive. You might call it your basic, no-frills retrofit since it represents the most elementary change that can be made to a porch: just glazing—no movable insulation, fans or reflectors, just glazing—and it works.

The porch is about 36 feet long and faces almost due south. Twenty-five feet of that length is a narrow, 42-inch-wide corridor, and the remaining 11 feet is about 8 feet deep and forms the main entrance to the house. All in all the enclosed floor area of the porch slab is about 125 square feet. The ends of the porch are enclosed by the house wall, and a wide roof overhang extends over the floor slab, supported by the porch posts. The two 6×6 porch posts carry a 6×6 header at the edge of the overhang, so there was no need for any structural work prior to the glazing work.

Materials Checklist

1-by wood for glass stops, spacer and inside fascia/trim; use cedar or redwood or pressure-treated wood 2-by or 4-by wood for sill material; again use rot-resistant wood	lag or through-bolts for fastening sill closed-cell foam strips Hypalon and silicone caulk tempered glass, single or double pane

To begin the glazing, a 1×8 was fastened to the inside face of the 6×6 header (see figure III-33). Then a 1×2 spacer and a 1×4 inside fascia were fastened through the 1×8 into the header with long, fat wood screws. The continuous sill was a 4×6 cedar beam, chosen for its natural rot resistance.

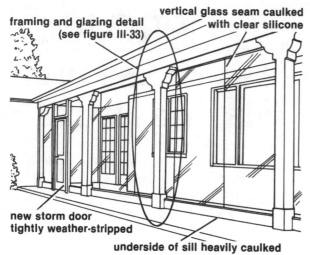

Figure III-32: The key to success with this and any porch glazing retrofit is in making tight seals between new framing members and existing walls and floors.

Because the porch slab was slightly raised above grade, a through-bolt was used to fasten the sill. The tiny crawl space was just high enough to allow me to crawl in and set the nut. A slab-on-grade floor would require a standard lag bolt and a lead shield or other kind of masonry fastener. With a wood floor the sill can simply be nailed or lag-bolted to the floor joists. Be sure to run a couple of fat beads of caulk between the floor and the sill before cinching everything down to seal out air infiltration. The stops for the inside and outside were 1×2 redwood, again chosen for its natural rot resistance.

Since the retrofit was to be used primarily for heat gain, single-pane glass was used. I recycled some 5 by 7½-foot tempered plate glass that had been used in store fronts. They were delivered to my door for 80¢ a square foot, a great price. (Editor's note: If heat retention were more of a concern, double-pane insulating glass would be used, possibly along with some movable insulation.) The actual glazing installation was about as simple as the framing. At the top and the bottom the glass was sealed first with closed-cell foam strips inserted on both sides of the glass. These strips also served to center

the glass between the stops and thereby form a backing for the Hypalon caulk that was used for the final seal along all horizontal seams. Hypalon is a black rubberlike material that is not degraded by sunlight, a good solar caulk. The vertical edges of the glass were separated by ¼ inch, and this gap was very neatly filled with clear silicone. The job was completed with some framing at the west end of the porch for a storm door. In the end the total materials' cost was $550, a rock-bottom price for a glass-glazed porch retrofit.

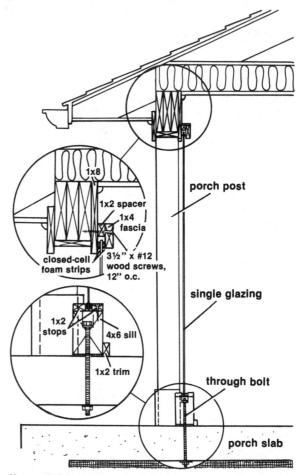

Figure III-33: With such simple framing for the glass, this kind of a retrofit would probably take all of one weekend. It's very important to seal all new framing as well as other parts of the porch to eliminate air infiltration.

Performance

The glazing addition has a few thermal benefits to its credit. Because the temperature of the porch space is somewhat higher in winter, heat loss from the house's exterior wall is somewhat reduced. I calculate my savings to be about 2.4 million Btu per year. Heat loss from infiltration was also greatly reduced by an estimated 15 million Btu per year. The best number of all was the estimated 28 million Btu solar heat gain for the heating season. These estimates are nicely verified by our lower gas bills, which show savings of about 450 therms per heating season (45 million Btu). This represents a 36 percent reduction in our heating load. So far, my total savings for four heating seasons has been about $400. By taking the state of Kansas' solar income tax credit, the net cost of the system is about $400, so the cost of the improvement has already been paid for. And now that once-cheap natural gas is being deregulated, the savings are going to pile up to enormous amounts over the years.

Cost of Materials		Savings in Heating Costs	
glass and caulking	$355.50	1976–1977	$ 90
wood framing and stops	$ 79.01	1977–1978	$ 93
glass storm door	$ 94.99	1978–1979	$106
	$529.50	1979–1980	$116
state tax credit	−$128.28	total savings	$405
net cost	$401.22		

Operation

During sunny winter days, the temperature in the porch space usually rises up to 60°F (33°C) above the outside temperature. When the porch temperature reaches 70 to 75°F (21–24°C), the doors to the house are opened, and the warm air is allowed to circulate. With the doors open, the porch temperature generally does not exceed 80°F (27°C), while the house temperature may get as high as 75°F. Since this retrofit is single glazed and the doors are closed at night, no attempt is made to store heat in the porch or to keep the porch heated, but the nighttime temperature on the porch usually remains 10°F above outside temperature. Thus, conductive heat loss from the house wall is still reduced through the night, compared to what it would be if the porch were unglazed.

During the summer, the roof and gutter overhang and the deciduous trees reduce solar gain, and the outside porch door is left open to prevent heat from building up. A screen was added to the storm in the second summer, after we found that the open porch was a perfect trap for flies and their friends.

Photo III-48: The space enclosed by the glass gets quite warm on a sunny winter's day. Imagine going "outside" to find the warmest place to sit and soak it all in.

A Seasonal Glazing System for Porches and Other Sunny Spaces

The Japanese do it with their *shoji* panels. We do it with storm windows. Why not do it with certain kinds of solar glazing? A seasonal glazing system would make your solar porch into a not-so-solar porch when warm weather returned, because by removing the glazing you remove the greenhouse effect, and the space becomes one with the ambient temperatures. It also becomes open to "the world," as it were, by not being hemmed in by the boundary, albeit transparent or translucent, of glazing. Your sunspace becomes a summer place.

I decided that I wanted my porch to be this way because it's such a nice place to sit on a summer's eve. I can listen to the street and the trees, and talk with passersby. When the chill returns, I can reglaze in about 30 minutes and enjoy the warm weather of my solarized porch. It's a simple system, and for ease in handling it's meant for use with lighter weight plastic glazings. I'm using a double-wall extruded acrylic glazing, but you could also make your own frame and double-glaze it with plastic sheet stock or film. (Since the double-wall acrylic and other plastic glazings like FRP aren't totally transparent, I do

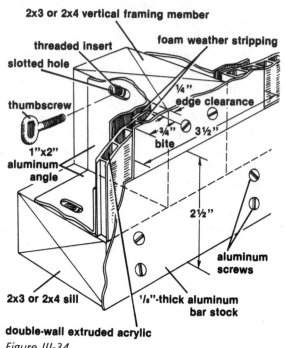

Figure III-34

The only maintenance thus far has been glass cleaning. After a very cold night the inside of the glass becomes covered with frost, but the sun usually melts the frost the next morning. Before it dries, all the glass can be cleaned in about five minutes with a squeegee. So the condensation is not considered a problem but rather a convenient cleaning aid, a real time-saver.

Still another benefit of this retrofit is a significant reduction in street noise, which is especially beneficial in the warm months. The retrofit is simple and for a modest investment it has turned a little-used porch into a solar collector that has reduced our heating load by better than one-third.

Ronald Wantoch

have one large piece (34 by 76 inches) of fixed, double-pane glass so I can see out when I'm glazed in.) As the illustration shows, the vertical and horizontal wood framing members are capped with aluminum flat stock, which becomes the receiving flange for the glazing. To ensure an airtight seal, strips of foam are bonded to the glazing; they will compress when the glazing is clamped in place. The clamping is done with 1×2-inch aluminum angle stock. The 1-inch face is pressed onto the edge of the glazing, and then it's secured with a thumbscrew that screws into a threaded insert in the framing. The hole in the angle stock for the thumbscrew is slotted to allow the angle to be pushed into the glazing and compress the foam. It's simple. When spring rolls around, I back out the thumbscrews, which are spaced about every 2 feet, pull off the angle and the glazing, and strap the glazing flat against a wall of the porch with a "seatbelt" out of the way 'til fall.

Joe Carter

Hardware Focus

Threaded Inserts

Albert Constantine and Son, Inc.
2050 Eastchester Rd.
Bronx, NY 10461
(212) 792-1600

Constantine Wood Insert
Inside of insert threaded to receive ¼", ⁵/₁₆" or ⅜" common stove bolt.

Mutual Hardware Corp.
5-45 49th Ave.
Long Island City, NY 11101
(212) 361-2480

Knife Thread Insert
Sizes: ³/₁₆", ¼", ⅜". Threads per inch: 24, 20, 16.

Improved Stage Screw and Screw Plug
A ½-inch hole accommodates plug which can easily be removed. Screw designed to be used with plug.

Glazing for a Two-Story Porch

Old houses often have porches and balconies that stick in rather than out, perfect for a solar retrofit. With three walls, the floor and the ceiling already in place, all that's needed is a glazing wall to transform it into a heat collector and a place to bask on a sunny winter's day. The retrofit shown here was done to a brick building in Pennsylvania by the Lancaster County Solar Project, which has for several years been building inexpensive porch enclosures as a way to reduce heating costs for low- and fixed-income people. The house's brick construction gave the enclosure the added feature of having some built-in thermal mass for heat storage. Because there are several tons of this mass within the enclosure, and because it receives heat loss from the house, a sunspace like this could also double as a greenhouse. Plants would be protected from freezing in all but the most frigid climates. Some kind of roller window insulation setup would guarantee absolute freeze protection.

The first steps of the retrofit involved a little reconstruction to do away with some rotten framing members and make both floors structurally sound. Any porch that's been weathering the elements for several decades is bound to have some weak spots. When the balcony floor was replaced, the new boards were spaced about ⅜ inch apart to allow warm air to rise up from the lower porch. Also, a lot of attention was given to sealing out air leaks so that cold air infiltration wouldn't deplete the solar gains.

The glazing is supported by a conventional stud wall frame. As figure III-35 and photo III-49 show, the top half of the framing was slightly tilted to get it in under the balcony roof. An alternative to this would be to extend the roof. Again, the perimeter of this framing was carefully sealed to the existing wall to seal out infiltration. When gaps were very wide, regular fiberglass insulation was stuffed in as a filler and a backing for a surface bead of caulk.

Two layers of plastic glazing are separated by 2×2's. The inner layer, a thin film mate-

Photo III-49

rial, is stapled to the stud wall, followed by the 2×2's, which are nailed to the studs. The outer FRP layer is tacked to the 2×2's, and the whole assembly is finished with an outer 1×2 batten. All the seams made by the 3-foot-wide FRP are overlapped and sealed with caulk. The stud wall framing included openings for two vents and a storm door. Frames for the vents were made with 1×4's and glazed with two layers of clear Plexiglas to preserve a view to the outside. In winter the vents are tightly closed, with weather stripping providing a positive seal, and in summer they are permanently propped open to exhaust heat. The site-built storm door was made to carry removable glazing panels that are replaced by screens during the warm months.

The heat collected in this sunspace en-

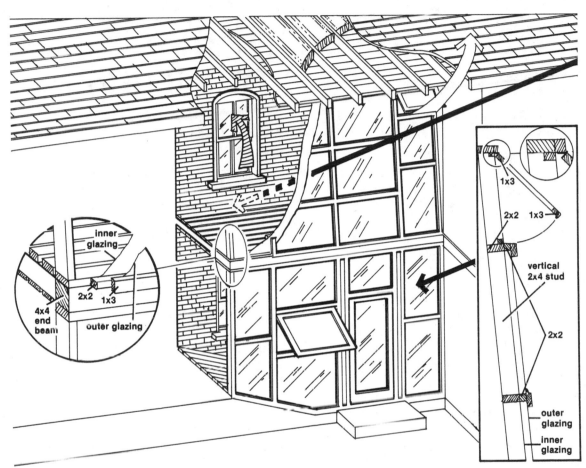

Figure III-35: The key to the success of a retrofit like this is weather-tightness. First of all, make sure the roof is leakproof. It's important to have a good seal (with caulk and/or foam gasketing) used wherever the retrofit meets the brick side walls. Doors and windows should be meticulously weather-stripped to keep out wind and rain. If you're concerned about losing the use of your favorite sitting porch in summer, modify the glazing design so that some sections can be removed and, if need be, replaced with screens.

ters the house through the open top sashes of double-hung windows, while cooler air returns through the raised bottom sashes. Fortunately, several rooms of the house open onto the upper and lower levels via these windows, which makes heat distribution a lot easier. Certain windows can be opened and others can be left closed depending on where the owner wants the heat to go. This enclosure had a materials' cost of just $600, a small outlay considering the benefits of solar heat and increased living space that a simple glazing wall created.

M.R. Carey

☼ PROJECT

A Glassed-In Porch with Fan-Powered Heat Distribution and Summer Ventilation

Winter in northern New Mexico typically adds up to over 6000 heating degree-days, but fortunately that combines with an exceptionally rich solar resource. It can get very cold up there, and in summer a sunspace can get pretty hot, so the solar design has to account for both seasons. Limiting heat gain is the first ounce of prevention in summer, and ventilation can do a lot to cure overheating ills. When blessed winter heat gain overwhelms, heat distribution "cools" the passive sunspace, raising efficiency and reducing heat loss. This porch retrofit does it all.

Sometimes the best solar addition is the one that makes good use of a house's existing potential, such as an underused room with southern exposure. When the Sterns began thinking about solar heating for their 2600-square-foot home, they considered a large active flat plate collector system that could be connected to their hot water heating system, but a solar designer suggested a passive solar improvement as being cheaper and much less complex. It turned out that their back porch was almost custom-made to be a direct-gain collector, with over 400 square feet of vertical southwestern exposure, a brick floor and concrete walls for heat storage, and roof overhangs for summer shading. In the end, the Sterns enclosed the porch with double-pane glass, and later they added two ventilation towers to exhaust hot air from the porch, either to another room in the house (winter) or to the outside (summer).

The Site

When the Sterns built their house in 1965, they were naturally more concerned with the view from their lot than with solar heating, and by good fortune their Santa Fe property sits on a gently sloping, south-facing hill, 7000 feet above sea level. This cool, mountainous region is dotted with prominent peaks, and by orienting their house to the southwest, the Sterns managed to obtain impressive views from nearly every window in the house.

The heating season in Santa Fe is long and demanding. State weather records show over 6000 heating degree-days as average for this location, with heating required periodically even in summer. Despite cold mountain air temperatures, the region is exceptionally sunny and is usually clear for about 70 percent of the winter daylight hours.

The porch itself is 60 feet long and 8 feet wide with a brick floor and a full-width overhanging roof supported by posts. The roof effectively shaded the sunny side of the house from the summer sun. The porch also had a good exposure to the cooling southwesterly summer breezes. The house has never needed summer air conditioning, and it has relied entirely on natural cross ventilation during the not-too-hot summer. The prevailing winter winds come from the northwest, and the house is well-protected on that side by many small piñon trees.

The original house was insulated to the prevailing 1965 standards. All windows are double-pane glass, and the cement slab floor has 1 inch of rigid foam perimeter insulation, 2 feet deep. The exterior walls are built of 8-inch hollow-cell masonry blocks, and the cells are filled with pumice, a light, porous volcanic stone, which provides about half of the insulating value now required by current building codes. The flat roof originally carried 1 inch of rigid polystyrene insulation over which 6 inches of pumice (R-2 per inch) were laid and sloped to a thickness of 2 inches near the edge of the roof to provide drainage. A standard built-up tar and gravel roof sealed the pumice. This type of roof construction provided about one-third of the presently accepted insulating value for roofs.

Yet even with this level of insulation the Sterns' boiler burned no fewer than 2651 gallons of butane gas in the winter of 1976 to keep their house between 68 and 70°F (20–21°C). When butane started getting expensive, the Sterns began to look seriously into energy-conserving alternatives.

Design

The most inadequately insulated part of the building was the roof, and this seemed to call for an extensive reroofing and insulating job. The Sterns called in Bill Lamoreux, who had designed and built the house ten years earlier, to get his opinion. Mr. Lamoreux concurred on the need for reinsulating and refinishing the roof, and he also related his recent experience of installing an active solar space heating system in a local Santa Fe bank building. The same kind of active liquid heating system could be added to the Sterns' existing hydronic, radiant-floor circulating system with little trouble, technically speaking. All that was missing was a big bank of collectors and a large hot water storage tank.

The problem with that plan was in the high cost of materials and installation for such a system, and so the Sterns decided to look into other solar heating alternatives that might be less expensive. In a conversation with Steve Baer (long-time solar in-

Photo III-50: The Stern porch isn't much different in appearance from what it used to be, but the glazing addition has made a dramatic improvement in the way the house works to keep itself warm and cool.

ventor and designer from Albuquerque), the Sterns learned that their southwest-facing porch was already ideally located to be a passive heat collector. All it needed was the glass! If the perimeter of the entire porch were enclosed with double glass, it could not only heat most of the house, but the new sunspace would also add nearly 500 square feet of living space. This sounded far simpler and more practical, and it was certainly much less expensive than the active collector system. The Sterns decided to go for it.

The final design of the porch enclosure grew somewhat organically from three fundamental goals: conserving heat and fuel, collecting and distributing solar heat and cooling the porch during warm months without an air conditioner. In order to conserve heat it was decided to insulate the previously uninsulated porch roof just as well as the rest of the roof. The fact that the porch faced southwest, rather than south, was a minor drawback in the ultimate thermal performance of the system. Glass facing southwest will gain about 80 percent as much of the available solar energy as glass that faces true south. It was thought that the large aperture of the porch would make up for its less-than-perfect orientation, and it was decided that the entire porch should be glassed-in to maximize the collector area. This resulted in 412 square feet of vertical glass with a view of the sun that is equivalent to about 330 square feet of true south-facing glass. This effective aperture equals about one-fifth of the floor area of the main living space. By rule of thumb, direct-gain windows in this climate are optimally sized to equal from one-fifth to one-third of the floor area of the living space. Since this retrofit is at the low end of the sizing scale, it could be expected to provide a significant fraction of the space heating load, though not 100 percent.

To save money when buying tempered glass, the framing was designed to accept standard patio-door glass sizes. The benefit of using standard sizes is that you get two layers of sealed, tempered glass for about the price of two unassembled sheets of ordinary glass. Standard sizes come in a variety of widths and lengths, and the double-pane units are as easy to install in a simple frame as are single sheets. Prehung sliding-glass door units were also designed into the enclosure for use as passageways and as ventilators.

The question arose during the design stage whether or not there was enough thermal mass in the brick floor of the porch and in the adjoining house wall to prevent excessive daytime overheating. It was decided that 55-gallon drums full of water could be lined up against the house wall to increase the thermal mass. The Sterns adopted a "mañana" policy toward further additions or alterations to the house: If the house needed more heat storage or more ventilation after the glass was installed, it could be added later. As it turned out, extra mass was not required, but due to the western orientation of the glass, the porch did need more ventilation during hot summer afternoons.

A tight, well-insulated structure is an essential ingredient in the successful, efficient operation of any heating system, and with a passive solar heating system, this is a top-priority requirement. In the Sterns' house the 1965 insulation was no match for 1977 fuel prices, so that was the first thing to be improved. The existing roof was stripped down to its tongue-and-groove decking, and 1×12's were framed in on edge to provide a 1-foot-deep attic cavity above the roof deck. This cavity was filled with two layers of 6-inch fiberglass batt insulation, thereby increasing the roof insulation from R-10 to R-40. The cavity was decked with plywood and finished with a standard built-up roof. This improvement was made over the entire roof, including the porch roof. The only other conservation measure taken was the addition of three insulated storm doors to entrances on the northeast side of the building.

Structural Details

The construction of the solar aspects of the Stern residence was very straightforward and can be duplicated by anyone with basic carpentry skills. The framing for the glass was attached to the outer face of the existing porch posts. Finish trim and flashing were installed around the frame to tidy up the appearance, and all new wood was painted a rich brown. Prehung, sliding double-glass doors were installed where exits were needed and the rest of the frame was filled in with fixed, double-glass panes. The finished frame enclosed about 412 square feet of southwestern exposure and also about 108 square feet of southeastern and northwestern exposure.

After the porch was enclosed, it was discovered that, due to the southwestern orientation, the porch overheated late in the day in summer. Even though the porch roof shades the sunspace during most of the day, it can't keep out the low afternoon sun. To remedy this problem the Sterns added the two cooling towers, with 24 by 20-inch louvered openings that face away from the prevailing summer breeze, a placement that helps to pull air out the vents. Two 20-inch, ¼-hp, manually controlled fans are mounted in the porch ceiling, and a sliding insulating cover is mounted beneath the fan to seal the vent when the tower is not in use. An automatic backdraft damper installed inside the tower opens when the fan or the wind forces air up the tower but closes to prevent air from flowing back down. (See Section III-E for more information on backdraft dampers.)

Before the cooling towers were installed, the Sterns were surprised to discover that they also had an overheating problem in the middle of winter! As the sun dipped lower in the western sky, it beamed directly at the southwest-facing glass and often drove up the temperature in the porch to over 100°F (38°C).

Following their "mañana" policy, the solution was improved heat distribution. One of the cooling towers was also fitted with an insulated 10-inch duct and a smaller fan that forces hot air to two northeast bedrooms that are 40 feet away. The duct enters an insulated plenum box on the roof, where two 6-inch hot-air ducts branch off to the bedrooms.

Photo III-51: On a sunny winter's day the porch is a most inviting, cozy spot.

Operation

The Sterns like the simplicity and effectiveness of their passive solar retrofit. "It's ideal!" Mr. Stern remarks. "By 10 o'clock in the morning the sun is pouring into the porch and the temperature out there jumps up to 70°." This is what happens every sunny winter day, regardless of the temperature outdoors. Even on bright, overcast days, the porch often provides warm air to the rest of the house. It's essentially a large, inhabited solar collector, controlled by the people who live in it. On a typical winter day at around 10 o'clock, Mr. Stern opens all the house windows and doors that lead to the porch, and he is met by a mass of warm air, at 70 to 85°F (21–29°C), which gently rises into the house. Cooler house air returns to the porch through the doors and windows, establishing a convective loop.

As the sunshine beams in on the brick floors, they heat up rapidly, and so does the back wall of the porch. They both act as a thermal heat storage mass, and the wall in particular provides a great deal of nighttime heating to adjacent rooms. As the wall surface heats up, a heat wave begins to travel through the wall by conduction. It takes hours for this heat to travel through the 8 inches of concrete and pumice, so by early evening the first of the stored afternoon heat begins to radiate to the indoor living space. Stored heat will continue to flow out of the wall until the wall temperature drops below the interior room air temperature. Then the hot water baseboard heaters take over late in the evening.

At night the Sterns close all the windows and doors to the porch to minimize heat loss. With this routine the porch provides space heat all day long, and because of the heat stored in the bricks and the insulating properties of double-pane glass, it somewhat reduces the amount of heat lost from the southwest wall at night and during sunless days. The system relies mostly on convection of warm air and conduction of heat through the southwest wall to transfer heat into the house, but these passive methods weren't enough for moving warm air to the two northeast bedrooms. But the fan-powered duct system was. The fan and the dampers in the outlet vents are manually operated and used only when these rooms are occupied.

The backup heat from the gas-fired boiler circulates hot water through the cement floors and baseboard convection heaters throughout the house. The generous 187,600 Btu per hour output of the boiler is large enough to heat the house easily. On a typical winter day Mr. Stern runs the radiant floor system for about an hour in the early morning, and after about 10:00 A.M. the solar-heated air shuts off the room thermostats. No additional heat is required until late in the evening. If back-up heating is required out on the porch, the Sterns fire up a stove to chase the chill of a cold night or a dark cloudy day.

The Sterns have discovered that the masonry structure of their home can also be used to "store" coolness in summer, just as it stores heat in winter. Just after sunset in summer the north windows and doors are opened and the cooling tower fans are turned on. The fans draw cool evening air through the whole house from north to south, flushing away heat from the house. The concrete and stone structure of the house stays cool well into the middle of the next day. The Sterns' home is not air-conditioned, but the storage mass, the shading and fan-powered venting all combine to meet their cooling needs.

Performance

The thermal performance of a passively solar-heated space is characterized by indoor temperature fluctuations that follow the outdoor temperature changes. In the Sterns' porch these fluctuations are quite pronounced in the winter. Mr. Stern has recorded both indoor and outdoor daily temperature records for two years with a "max-min" thermometer, which registers the highest and lowest temperature of the day as well as the current temperature. In the dead of winter a solar porch or greenhouse space will show a big temperature difference between indoors and outdoors if it gains heat well. The average

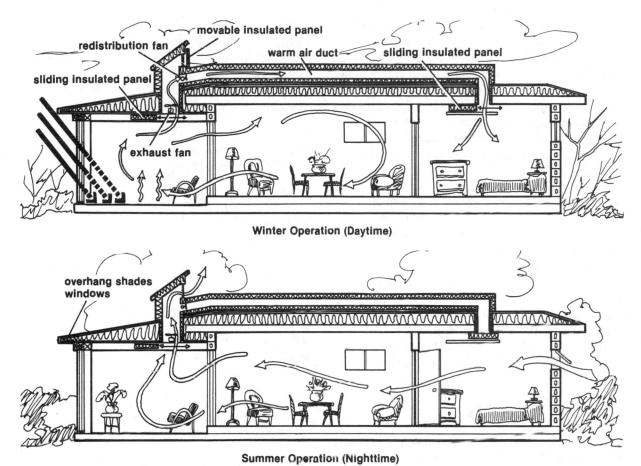

Winter Operation (Daytime)

Summer Operation (Nighttime)

Figure III-36: The ventilation tower and the air distribution system both have winter and summer operation modes. In winter, during the day, the redistribution fan directs warm air that collects in the tower to cooler rooms on the north side of the house. When the fan is not on, sliding insulation panels are placed over the ceiling vents to prevent cold air from leaking in. In summer, at night, the large exhaust fan pulls cool air in from the north side of the house and exhausts hot air out the tower.

porch temperature should be 20 to 40°F (11–22°C) warmer than the average outdoor temperature, and it should always be above freezing. At the same time the temperature in the porch should not fluctuate more than the outdoor temperature does in terms of net degrees of temperature change or "swing." Generally, the temperature difference at night indicates how well your nighttime insulation and heat storage are working. The range of the temperature swing

indicates how effective the storage mass is at absorbing and later releasing heat. Thermal mass helps to make the interior swing less than that of the ambient temperature swing.

The Sterns' January temperature records are quite revealing. While the average outdoor temperature that month was about 21°F (−6°C), sometimes dipping below 0°F, the average porch temperature was about 61°F (16°C). Twenty-five out of 31 days were sunny enough to supply heat to the house at 70°F (21°C) or hotter. It is interesting to note that on cloudy days when the porch temperature stayed below 70°F, the outdoor temperature was warmer than usual because of the cloud cover. This reduced the overall demand for space heat. This is a fairly typical weather pattern in many parts of the country. Sunny, cold weather and cloudy, warmer weather in winter enhance the conservation effect of passive solar heating. The records show that on the average it was 40°F (22°C) warmer in the porch than it was outdoors. The indoor temperature swings were somewhat more pronounced than the outdoor fluctuations, which indicated a lack of thermal effective storage mass in the porch. Window insulation has not been used either. In the Sterns' case this condition is only a minor drawback, since nobody lives in the porch full-time, and temperature extremes can be avoided by moving indoors. If it becomes uncomfortable during the day, the heat can be dumped through the cooling towers, either back outside (summer) or into the north bedrooms (winter).

Cost and Benefits

The success of the solar heating system is shown in the Sterns' annual heating bills for the past few years. Before the retrofit, in the winter of 1976–1977, the boiler gobbled up 2651 gallons of bottled gas. In the winter of 1977–1978 it only used 700 gallons, and in 1978–1979, 910 gallons. The solar retrofit and added roof insulation are saving the Sterns 70 percent of their annual heating bill. The temperature records in summer show that the porch temperature rarely rises above the outdoor air temperature. The shading and venting successfully prevent heat buildup on the porch. Since the fans are also used at night to cool down the house, it is usually at least ten degrees cooler during the hottest part of the day than it was before the towers were added. In this climate and at this altitude, that is all the cooling required to maintain summer comfort.

The state of New Mexico's tax credit covers up to 25 percent of the Sterns' solar equipment costs, including the passive elements (the double glass and framing). The federal tax credit is not yet very strong on passive solar but covers part of the cost of the added roof insulation. The approximate costs and savings of the alterations are as follows:

solar retrofit: (double glass, framing lumber, vents, ducts, fans, no labor)	$ 4,000
conservation retrofit: (insulation, roofing, storm doors and all labor for solar and conservation improvements)	$17,000
total cost:	$21,000
federal tax credit:	− $ 2,500
state tax credit	− $ 900
net cost	$17,600

The Sterns save an average of 1846 gallons of bottled gas each year, which costs about 80¢ per gallon at current (early 1981) prices. Using these prices and assuming no inflation rate of money or fuel, the simple payback is:

payback period of total job with tax credits: 15.8 years
payback period of solar equipment cost,
    based on calculated total solar heat gain: 6 years

Of course, when the inflation rate of fuel is included in the analysis, the payback period becomes shorter. However, it can be seen that the solar equipment cost was about one-fifth of the total retrofit cost, and it would pay for itself somewhat sooner if the job had no added labor cost. South-facing double glass is definitely a good investment for the do-it-yourselfer.

Evaluation

Everyone concerned with this retrofit is satisfied with the results. The only change the Sterns plan to make is to seal the east cooling tower in winter so that cold air does not get into the heat

Photo III-52: Part of Mr. Stern's standard operating procedure is opening and closing the sliding panel under the exhaust vent. It's designed to seal tight to the ceiling.

distribution duct and then to the back bedrooms. Mr. Lamoreux, the builder, remarked that the only change he would make would be to use belt-drive, squirrel-cage vent fans if he had to do it again. He feels that the present 20-inch axial fans are too noisy, although the Sterns do not mind them. The long payback period for this retrofit does not bother the Sterns, because they were not just installing a heating system. They were also buying a much-needed new roof, 500 square feet of additional living space and a warm, pleasant atmosphere for their home.

The passive system fits right in with the Sterns' home life. Since they are home most of the time, they don't mind opening and closing vents during the day. In fact, they enjoy having control over thier comfort level, and they installed manual fan controls for this reason.

In the future the porch space could be used as a greenhouse, or with the addition of window insulation and more thermal mass, its thermal efficiency could be raised somewhat. For the moment, though, it remains a pleasant daytime and evening room, and its simplicity and effectiveness outweigh the need for any additional improvements.

Variations

The solarized porch naturally has a multitude of possible applications and variations. The Sterns' retrofit was exceptionally large and contractor-built, but the same simple principles and construction techniques can be applied to simpler, smaller additions. The one essential ingredient is of course the glazing. Tempered double-glass door panels in standard sizes can be obtained at present prices for between $2.50 and $3.00 per square foot. Thus a 200-square-foot opening could be framed and enclosed by a homeowner for perhaps $600 to $1200, depending on the features included and the amount of new framing required. As a direct-gain system it could supply significant heat to a home. Plastic glazings could be used to keep costs lower. A double-layer extruded acrylic glazing (Exolite) costs about $2.60 per square foot and has the added feature, if it's desired, of being a nearly no-see-through glazing.

It is best to use vertical rather than tilted glass in south windows. Tilted glass gains more heat in the winter, but it also gains more in summer and is more difficult to seal against water leakage. Vertical windows are also easier to shade in summer and easier to build. The exterior glazing should be mounted as close to the outside edge of the window frame as possible to avoid the possibility of the window being shaded by the frame in the morning and afternoon. In summer, shading is desirable and awnings or overhangs can be used to reduce heat gain in either vertical or tilted windows.

In a well-insulated, well-sealed house, each square foot of south glass can be expected to supply heat to 3 to 5 square feet of heated living space. The solar heat should be stored for nighttime use if possible. If your solarized porch has a brick or cement floor, it will help to store heat, but if the house is wood-frame construction, the wood can't store much heat, and you will need to add heat storage mass in some form, probably water in containers. (See "Moved-In Mass" in Section III-D.)

Ventilation must always be provided to the

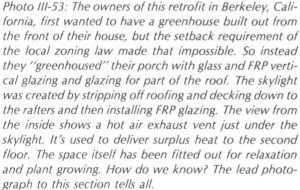

Photo III-53: The owners of this retrofit in Berkeley, California, first wanted to have a greenhouse built out from the front of their house, but the setback requirement of the local zoning law made that impossible. So instead they "greenhoused" their porch with glass and FRP vertical glazing and glazing for part of the roof. The skylight was created by stripping off roofing and decking down to the rafters and then installing FRP glazing. The view from the inside shows a hot air exhaust vent just under the skylight. It's used to deliver surplus heat to the second floor. The space itself has been fitted out for relaxation and plant growing. How do we know? The lead photograph to this section tells all.

outdoors in summer. This is most easily accomplished by providing a door and some operable windows in the porch glazing. Some vents should be near the ceiling (the downwind vent), and others (upwind) near the ground. They should be about the same size, and together they should equal about one-sixth of the porch floor area. If your openings can't be that big, place fans in the windows or add vents to the porch roof. Other kinds of passive roof vents such as turbine vents and boat-hatch vents can also be used. A boat-hatch vent is simply a hinged, flat door that covers a hole in your ceiling. A turbine vent is a wind-powered device that mounts on top of a short metal chimney. The former can be built, and the latter can be bought from a heating and ventilating supplier. All exterior vents must be sealed and insulated during the winter.

A fan and duct system can be used to draw hot air from the sun porch and deliver it to any remote room. The duct should be well-insulated (R-6), as big in diameter as the fan and as short and straight as possible. The fan should be sized to deliver at least one cubic foot per minute (cfm) for every square foot of floor area in the room to be heated. This should give satisfactory heat delivery in most cases. A variable-speed fan controller is recommended for varying heating demands. The heated air should be drawn from the peak of the porch roof and, if possible, dumped into the room at the floor level. Insulated ducts should be placed indoors in attics, crawl spaces and hollow walls whenever possible. If this kind of heat distribution isn't needed, the warm air is better circulated inside the porch with a Casablanca-style circulating fan or a fan-powered, drawdown duct.

The use of window insulation over your collector windows can greatly increase heat retention. It's a must if the porch is to be used as a living space at night or if it will be used as a greenhouse in a cold climate.

Construction Steps

The following is a general review of the steps taken to enclose and solarize the Sterns' south porch, steps that can be applied to other buildings of different construction. These steps begin with the assumption that a level porch floor exists, that an insulated roof supported by porch posts extends over the porch floor and that fixed, double-pane glass will be set into wood frames.

Before you build, call local building suppliers to find out what sizes of standard double-glass panes are available and also what sizes of prehung sliding-glass doors are available. Make a sketch and measure-up and record the height and width of all openings, posts, roofing and flooring members. Then sit down with your calculator and figure out whether or not the standard glass sizes will fit into your openings. If they won't fit between your existing posts, you can build your frame outside or inside them. Small gaps at the ends of your framing can either be filled in with nonstandard glass sizes or simply filled with insulation and covered with siding and trim.

Materials Checklist

Window Framing

2×6's, 2×4's, 2×3's or 4×4's for posts between windows; lumber should be straight and kiln-dried	6d finish nails
	2-inch flathead wood screws
1×6's from which battens and stops can be ripped; again, use high quality lumber	8- to 10-inch gutter spikes or long lag bolts
	flashing material
wood shims	lag bolts and lead anchors (for cement floors)
10d common nails	neoprene setting blocks
16d common nails	butyl glazing tape
	silicone caulk

Air Handling System (Some of the Basic Parts)

framing lumber	duct fan or blower
exterior plywood or siding	exhaust fan
rigid foam insulation	operable register
metal ducting	

To begin the framing, cut the pieces for the frame as laid out in your sketch so that the glass units will fit into the finished openings with ¼ inch to spare on all sides. Fasten the sill to the floor with caulk in between. When framing members are attached to the floor, the porch posts or the porch header, you can use the wood shims to make the frames straight, plumb and level. To attach the vertical 2×4's, 2×3's, 2×6's, etc., on edge to the existing structure, use the long gutter spikes or lag bolts. With the spikes, always predrill a ¼-inch hole to avoid splitting the wood. These primary fasteners should be spaced about every 16 inches.

Sections of the frame that will receive prehung sliding-glass doors should be measured to the specified "rough cut" dimensions for the unit you have chosen. Next, install the inside window stops ripped from 1-by stock to hold the edge of the glass against the inside of the frame. They should be glued and nailed to the frame using 6d finish nails (see figure III-37). To ensure straightness with the stops, which is essential, snap chalk lines down each vertical and horizontal member that will receive a stop. When framing, keep checking things with a framing square and level to be sure all is plumb and square. Even though the stops may be straight, the window opening may be twisted a little or a lot, which could put excessive stress on the glass. Before the glass goes in, paint the framing with a primer or stain that is compatible with your house's color scheme.

When the framing is complete, fill large gaps (that won't be glazed) with insulation and siding. At all points around the perimeter of the new framing, caulk all seams that might be sources of air infiltration. This is a good time to do an "energy audit" on your porch to find and seal air leaks.

Any prehung (operable) windows and doors should be installed before the fixed glass panes are mounted. Rip the outer stops for the fixed glass. They are screwed to the outside of the frame and will overhang the face of the glass by about ¾ inch to form an outside stop and trim strip. The edges of the inside window stops receive the butyl glazing tape, all the way around the frame. Then lay in three pieces of neoprene setting block on the bottom of the frame. Two people are needed to handle the double-glass pane, which weighs 60 to 90 pounds. Care must be taken not to upset the edge seal. The bottom edge of the glass is set onto the neoprene setting blocks, and when the glass is centered in the opening, tilt the pane into the frame, pressing it against the sticky glazing tape. While one person holds the panel in place, the other can run another strip of tape around the perimeter of the glass. Then the exterior stop is tacked in place after being pushed up tight to the window.

Once all the windows and stops are in place, go back and screw the stops into the frames, with flathead screws spaced every 6 to 8 inches. The bottom flashing is installed before the bottom horizontal stop goes on (figure III-37). Note that the bottom stop is beveled to shed water. Any top or side flashing that may be used will generally go over the outside stops. The final sealing is done with silicone run neatly around all the glass stop seams. After the silicone cures, finish painting or staining can be done.

Making an Air Handling System

This component is an option to consider if improved heat distribution is required and/or if summertime overheating necessitates additional fan-powered ventilation. The vent is installed between a pair of roof rafters, so pick a place that is close to the peak of your porch roof, away from obstructions to air flow. If a cool air intake is located at one end of the porch, the vent should be located at the other end to ensure that ventilation air flows through the entire space. If the main cool or return air inlet to the porch is through, say, a door, try not to place the vent in that vicinity, or the ventilation air path will be short-circuited between the doorway and the vent. The rest of the porch space may not get sufficiently ventilated.

Once you've located the vent (or vents), draw an outline of the opening on the ceiling, then cut away the ceiling material. Pull insulation out of

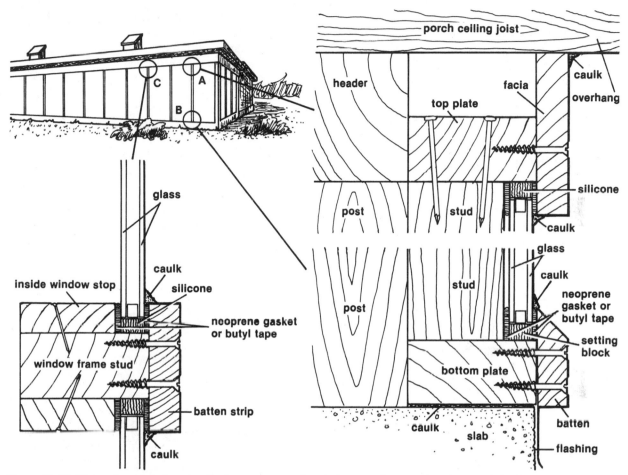

Figure III-37: These three glazing details show cross-sectional views at the top (detail A) and the bottom (detail B) of the glass; detail C is an overhead view showing how to handle the glazing where two separate panes meet. The primary sealing agent is on butyl tape or neoprene gasketing, which can be fortified with a bead of silicone.

the ceiling cavity and draw an outline on the under-side of the roof sheathing, the same size as the ceiling opening and directly above it.

Drive long nails through the roof deck at each corner of the outline. Climb up on the roof, and scribe a line between the nails. Then cut through the shingles, roofing paper and built-up roof along the chalk mark. Strip away the roofing, then cut away the roof decking with the saber saw.

Build a curb for the hole by nailing together a box frame out of 2-by material, at whatever face width is needed for the curb to rise above the roof plane by about 4 inches. The outside dimensions of the box should be just slightly less than the inside dimensions of the hole. Slip the curb frame into the hole and nail it to the rafters on either side. Wrap metal flashing around the box, fasten it with nails and seal it to the roof with roofing cement. If only the exhaust fan is to be installed for summer ventilation, it can be fastened to the top of the curb. Most attic exhaust fans have their own covers for keeping out rain and snow. The interior cover for the vent can be a sliding insulated panel, as at the Sterns', or a pop-in or a hinged shutter type of movable insulation, in the styles discussed previously (see Section III-B).

If you want to go the full route and build a cooling tower/heat distribution combination, take a close look at figure III-38, which shows the system in cross section. All framing members can be 2×4's or 2×2's. The base is a simple frame whose outside dimensions are ¼ inch larger than the outside dimensions of the vent curb. The corners of the base frame should be mitered and nailed together, and the risers (studs) are nailed to the base frame, standing vertically with their tops cut to the angle of the tower roof. This roof angle has no exact requirement; you can match it to other pitches on your house.

Frame an opening for the louvered vents near the top of the downwind side of the tower. If you are going to run a heating duct from the tower to another room, make the duct frame at the same time. Cover the outside of the tower frame with ½-inch exterior plywood, leaving holes for the louver and duct. The plywood sides should be long enough to extend several inches below the frame to overlap the curb. The tower roof should overhang the tower walls by about 3 inches all around. Three-and-a-half inches of fiberglass insulation should be packed into the tower and then covered with a plastic vapor barrier to keep the fiberglass dry. Where batt insulation won't fit, use rigid foam. Comprehensive insulating of this component is important for minimizing

Photo III-54: The warm air distribution ducts snake across the roof to their appointed destinations. In climates that experience a lot of snowfall, the duct covering must be virtually like a roof, one reason why it's likely to be easier to run ductwork indoors.

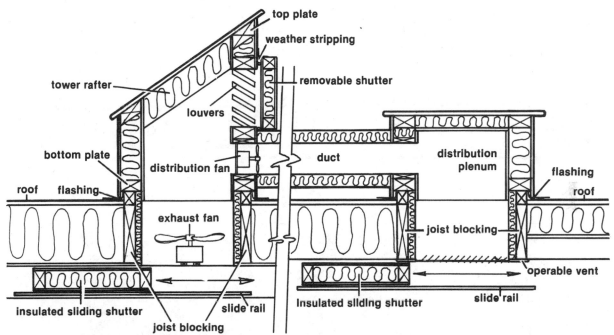

Figure III-38: The exhaust tower vents hot air in warm weather. In winter, a removable insulating shutter is placed over the exhaust louvers, allowing hot air to be directed through the duct and distribution plenum to other rooms.

heat loss in the heat distribution mode. The open lower section must also have a removable insulation shutter, which is put in place for heat distribution.

The main ventilating fan should be an axial (propeller blade) type, sized to fit the vent opening. The fan for the duct system is sized to fit the 10- to 12-inch diameter of the duct; they're called duct or duct booster fans. They can be controlled manually and/or thermostatically, with separate adjustable controls for each fan. If the duct is run outdoors, it should be insulated with at least 2 inches of fiberglass and covered with plastic. It can be protected from the weather with roofing material or with a sheathing of larger-diameter ductwork. Naturally, it's a lot easier to run the duct system indoors.

The outlet end of the heat distribution duct is essentially a carbon copy of the cooling tower, but without the exhaust louvers. Again, thorough insula-

tion is important if the duct is run over a roof to this outlet, and an insulating panel should be fitted to the opening in the ceiling.

This vent tower and hot-air plenum can be adapted to any roof pitch with minor changes in framing. What's presented here is at least an idea of how to do it, but there are certainly many possible variations on this theme. The construction is not complex, although particular care in sealing is required anytime a roof is penetrated. The most important concern, after the tower and duct are fully insulated and weather-tight, is correctly sizing the ventilating fans to the porch volume and the heating fans to the required run of duct. But for a modest investment, a ventilating and distribution system can be a useful adjunct to a passively heated living space.

Bristol Stickney

D. THERMAL MASS

Photo III-55: Water thermal mass is often the easiest to retrofit, and when used in this way it can be an attractive, artful addition to a room.

On Mass

We've been hinting at this almost from the beginning: thermal mass. It's a good thing, but it's rather different from the other components of solar heating systems. Basically, it's solar energy's solution to sundown, to cloudy days and to the fact that a lot of south glazing can lead to overheating.

Thermal mass is heat storage, and actually you've already got some. Your walls, floors and furniture, everything in your house has a certain *heat capacity,* and thus can store some heat. But those things can't store as much as the materials that are commonly used in solar systems, namely masonry and water. Of the two, water is a higher quality thermal mass because it can store about twice as much heat as can the same volume of masonry (brick, stone, concrete). When barrels filled with water are put into a solar greenhouse, the mass of the water absorbs some of the surplus solar gain that isn't needed to keep the greenhouse warm or that isn't transferred to the house. This stops overheating, which can hurt plants and which wastes solar energy. In solar retrofits for masonry buildings, the south wall is glazed to create a wall-full of thermal mass. The sun heats the darkened masonry surface and the air between the glazing and the masonry. Air flow into the house satisfies daytime heating needs, and at the same time heat migrates through the masonry to the inside. By the end of the day it begins to radiate heat into the living space. The sun is gone, but its effects persist until sunrise when the "recharge" cycle begins again. In both the greenhouse and the mass wall, if the recharge were followed by cloudiness, the mass would have some heat to contribute to the space. This is called *thermal flywheel,* another way of saying that heat is stored. The flywheel helps to limit both high and low temperature extremes that normally coincide with sunrise, sunset and changes in weather. By flattening those temperature swings, less auxiliary energy is needed to maintain comfort. New homes that are passively solar-heated use tons upon tons of this stuff, but it's also possible to put a few tons into a retrofit house and get the same good effects.

☀PROJECT
Mass under Glass
for Passive Space Heating

Hard-to-insulate masonry walls are a significant source of heat loss in many buildings. Here, a section of south-facing stone wall at the second-floor level was converted to a Trombe-type wall with the addition of double glazing. Air flow is by natural convection with inlet air coming up from the first floor between the wall and glazing and back into the house along several possible circulation paths. There is also a fan-powered draw-down duct that can distribute solar-heated air back to the first floor. This renter-built retrofit, combined with careful weatherization, reduced the heating load by 56 percent, and it's yet another example of the versatility of glazing for solar heating.

When I moved into Star Route Farm in 1973, the house was in such poor shape that I've been renovating, weatherizing and retrofitting ever since. It's a plastered fieldstone structure built around 1850 on a ridge overlooking the Delaware River—made to last but without much thought for thermal efficiency. My efforts have reduced the house's heating load by 56 percent. The solar retrofit is a 28-foot-wide by 5-foot-high, double-glazed Trombe wall built on the second story of a south wall. The collector system is small (140 square feet) relative to the size of the house (1000 square feet of heated floor area), but I calculate that it handles more than 20 percent of the heating needs of the house, which has no exterior wall insulation.

The solarized part of the wall contains about 375 cubic feet of stone weighing about 27 tons. It receives direct sunlight and reflected light from the silver-painted shed roof that comes out from the original stone wall. A large portion of the solar heat conducts through the wall and radiates to the second floor. Also, a natural convection air flow draws air from the kitchen up between the wall and glazing and through second-floor window openings and ductwork. When the second floor begins to overheat, or when I want to direct more heat to the

first floor, I can activate a forced-air circulation system. My back-up heat comes from a woodstove. There is also an oil furnace in the basement, which isn't needed anymore.

Conservation

My first tasks involved stopping the huge air leaks through the open chimney, through open floor boards and through the poorly sealed attic with its slate roof and disintegrating windows. I closed the chimney by stuffing fiberglass at both ends of the flue and by sealing the fireplace with a sheet of plastic.

To isolate the drafty attic from the house, I stapled more polyethylene plastic sheeting to the floor and used multiple layers of old carpet to create a respectable level of insulation. I also added 8 inches of fiberglass insulation to some of the attic's kneewall, floor and wall sections. My simple window insulation consists of 2½-inch-thick foam rubber mattress material that is wedged tightly into the window openings. I also use recycled plastic bubble packing material as another low-cost window insulation. The attic door seals tightly and is usually closed, but when it's opened, heated air rising up from the second floor makes the bedroom comfortable in a short time.

Since the plastered stone exterior of the house has been long neglected, I had to chip out and replaster many cracks. Then I applied two coats of cream-colored Thoroseal-brand masonry paint to improve the looks of the old place and also to seal out moisture penetration from the outside. This helped to reduce some of the dampness that is common to stone houses. I also caulked the perimeters of all the window and door frames before repainting them.

The basement held a big energy waster in the form of an electric water heater. Wrapping the heater and pipes in a few dollars worth of 6-inch (R-19) fiberglass insulation made it possible to lower the thermostat 20 degrees without significantly reducing the delivered water temperature. This simple effort cut my electric bill in half.

Keeping the stone walls warm in winter is an old problem for which there is an old but effective solution: wall hangings. Hanging fabric over the walls increases comfort by raising the mean radiant temperature of the room. Even though the R-value of ¼ inch of fabric is minimal, this layer still helps to keep the flow of heated air away from direct contact with the cold stone. The carpeting on the floor also helps to raise the mean radiant temperature.

In the kitchen addition I insulated the ceiling and added a new south-facing picture window for direct-gain heating. Now the room warms up quickly on sunny mornings, and it holds the heat into the evening much better than before. Even though the picture window is double glazed, it was still uncomfortable to sit next to on a cold evening, so I made an insulating window shade for it. This has made the "cold seat" a lot cozier.

Solar Design

The addition of an 18-square-foot picture window does not a solar retrofit make, but I wanted to harness the sun as a heat source, and a Trombe wall seemed the best way to do it. The slate roof, though still serviceable, was very old and well worn, and it couldn't have supported a roof-mounted collector system. The best southern exposure was the 28-foot-long by 5-foot-high section of stone wall above the sloped kitchen roof. About one-third of this space was taken up by double-glazed windows, which left about 100 square feet of solid masonry wall. I wasn't about to convert the 18-inch, solid stone walls into a direct-gain system, so the Trombe wall option was the natural choice, and I felt confident that it could be attractive and consistent with the traditional look of the house. It also had to be low maintenance and durable for years of service.

I decided to use clear glass as the glazing material because I didn't want to block the pleasant view of fields and orchards through the existing windows. As luck would have it, a salesman offered me a case of slightly flawed glass for a mere 10¢ per square foot. Accepting the offer gladly, I began framing out the Trombe wall.

Construction

With so much glass on hand, I was able to design the system to accept double glazing. I bought standard construction grade 2×4's and ripped them down to 2×2's. Then I cut double rabbets in the 2×2's to receive the two layers of glass (see figure III-39). I wanted to ensure that there would be a free flow of air throughout the solar wall, and I determined that an airspace of about 3 inches between the stone and the inner glazing layer would be sufficient without encouraging internal convective currents (heated air flowing back down against the cooler glazed surface). I mounted the framing members with pipe spacers to hold the framework away from the wall. To fasten the framework, I drilled into the wall with a hammer drill and a masonry drill bit and set in lead shield anchors. Then I countersunk lag bolts into the framing members through short sections of plastic pipe (the spacers) and bolted them into the lead shields. Since the stone wall was fairly uneven,

The Meaning of Mass

Thermal mass is probably the subtlest of all solar system components. It doesn't go on and off; it doesn't have water or air running through it; it certainly doesn't rise and set like the sun. It just sits there, quietly absorbing and radiating Btu's. Its effects are usually barely perceptible, which is ironic because thermal mass can have such a significant effect on interior comfort levels. But like solar input, the performance of thermal mass is predictable in the sense that it is known how much heat a given weight or volume of a certain material either stores or releases with a given change in temperature. *Specific heat* measures the energy flux (in Btu's) through 1 pound of a material for every degree Fahrenheit of temperature change. Water, for example, has a specific heat of 1, which means that 1 pound of water absorbs 1 Btu when its temperature is raised 1°F, and conversely it releases 1 Btu when its temperature is lowered 1°F. With this information you can calculate how much heat a water wall stores as its temperature is raised by direct or indirect gain. Suppose that 500 gallons of water were exposed to the sun for a day, and its average temperature was raised by 20°F by the end of the day. One gallon of water weighs 8.34 pounds, so there are 500 × 8.34 = 4170 pounds of thermal mass. To calculate the Btu's stored, the temperature rise is multiplied times the pounds, and the result is multiplied times the specific heat of the mass. For the 500 gallons the calculation is: 4170 (pounds) × 20 (temperature rise) × 1 Btu/pound/°F (specific heat) = 83,400 Btu stored. *Heat capacity* deals with mass on the basis of volume, specifically how much energy is stored in 1 cubic foot of a given material for every degree Fahrenheit of temperature rise. As table

TABLE III-7 THERMAL MASS HEAT STORAGE CHARACTERISTICS

Material	Specific Heat (Btu/lb) X	Density (lb/ft³) =	Heat capacity (Btu/ft³/°F, no voids)
Water	1	62.4	62.4
Concrete	0.22	140	30.8
Stone	0.21	90 to 170	18 to 36
Brick	0.21	130	27.3
Stucco	0.22	116	25.5
Drywall	0.26	50	13
Dry soil	0.21	105	22
Wet soil	0.44	125	55
Wood, hardwood	0.30	45	13.5

III-7 shows, water has a heat capacity of 62.4. One cubic foot of water weighs 62.5 pounds, and heat capacity equals the weight per cubic foot times the specific heat. Brick has a specific heat of 0.19, only about one-fifth that of water, but since brick weighs 120 pounds per cubic foot, its heat capacity is: 120 × 0.19 = 22.8 Btu/cubic foot/°F. So in terms of volume, which may be a more accurate comparison of different kinds of thermal mass, brick is really about one-third as effective as water at storing energy. Use table III-7 to calculate the energy storage potential of mass that you already have or plan to add. In the project "Moved-In Mass," found later in this section, David Bainbridge presents another way to use this information to figure out just how many cubic feet of mass should be added in conjunction with a solar heating system.

Photo III-56: Durability is a key element in any solar heating system, and with this glazing retrofit the basic materials, glass and wood, were chosen for maximum life with minimum upkeep.

I used black plastic pipe for the spacers, as they could be easily cut and/or compressed with the lag bolts to allow me to true-up the framing surface despite the irregularities of the wall. The goal here is to end up with a flat glazing wall, for good appearance, no matter how wavy the existing wall is. To minimize any ultraviolet degradation of the plastic pipe, each spacer was painted white.

  I attached the top framing member (the plate) to the roof soffit and the bottom one (the sill) to the kitchen roof, and then added the intervening vertical members (muntins). Having been assured that the still-undelivered glass would come in 32 by 32-inch panes, I carefully laid out the muntins so they wouldn't obstruct the window openings. This was a big mistake. After all of the framework was in place,

my glass shipment arrived: a full case of panes that measured 30 by 30 inches. My heart sank, but at the price I couldn't refuse delivery. Instead, I had to rework almost all the framing to accommodate the smaller size. Take my advice: Don't frame out your system until you know *absolutely* the dimensions of your glazing material. Plastic you can cut, but even then you should be certain of the stock dimension to minimize waste. With a vertical run of nearly 5 feet, it was necessary to add a horizontal muntin halfway up each run, which nicely coincided with the meeting rails of the existing double-hung windows. The horizontal muntins were made from the same stock as the vertical pieces and were notched at both ends to lap over the rabbets in the vertical members. as shown in figure III-40.

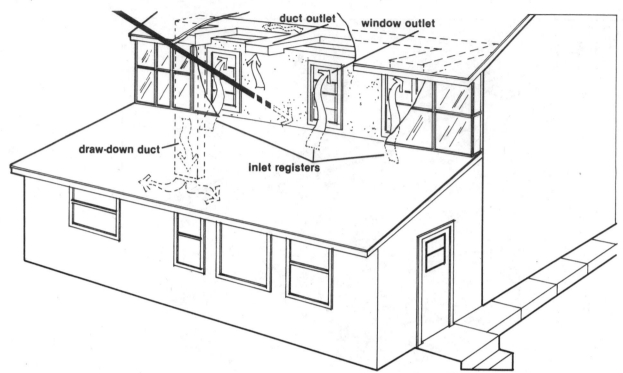

duct outlet   window outlet

draw-down duct

inlet registers

Figure III-39: To move heated air to different parts of the house, there are two sets of outlet vents in this particular system. The windows direct air to the second floor, while the duct outlets send air to a main draw-down duct that is fan-powered. The point is that warm air distribution systems can be varied to meet a house's exact needs.

The wall surface had already been plastered and painted with the Thoroseal that was used on the rest of the house, and to darken it I used two coats of flat black exterior latex paint. I primed the raw framing members and added a coat of alkyd enamel in gloss white to keep the wood from getting too hot and thereby degrading. To avoid condensation buildup, I drilled two 3/16-inch holes in the horizontal muntins to create an exit path for water vapor.

Before installing the glass, I sized and cut in all the necessary vents. The rule of thumb is that both top (outlet) and bottom (inlet) vents should each be equal to one-half the area of the horizontal cross section of the collector. Since the collector measured about 28 feet long by 3 inches (or 0.25 feet) deep, I needed to distribute about 4 square feet each of inlet and outlet vents along the solar wall ($\frac{1}{2} \times 28 \times 0.25 = 3.5$).

At the same time that I insulated the kitchen roof, I also installed six 3 by 12-inch inlet air registers in the ceiling where it met the stone wall. These registers were connected with corresponding inlet vents in the roof at the base of the Trombe wall, in the gap between the glazing and the masonry. This made it possible to draw air into the collector from downstairs, which was generally cooler than the second story. I made two outlet vents by cutting two holes into the roof soffit and ran short sections of air duct into an existing hot-air duct left over from the unused furnace. Fortunately this old duct ran the full length of the southern eave, making the connections easy. This duct can deliver heated air either to the second or first floor with the use of a fan, or to the

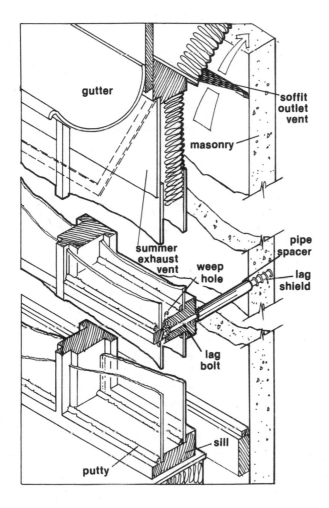

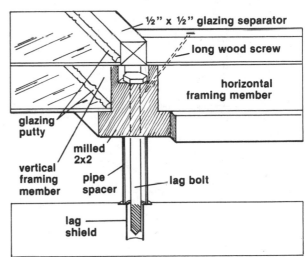

Figure III-40: The above illustration shows a typical cross section of the framing, glazing and wall attachment. At right the solar wall is shown from top to bottom, starting with the soffit detail, which shows the summer exhaust vent. The middle detail shows again how horizontal framing runs into vertical framing. Note the weep hole between the two glazing layers; this is a vent for possible condensation. The bottom detail shows how the sill of the solar wall is fastened solid to the masonry.

attic space above by natural convection (see figure III-39). Because of the existing windows in the solar wall, there were three more outlet vents already in place. I can open the top window sashes about half-way and, voilà, three 10 by 28-inch outlet vents. With ample air circulation between the two floors, I've had little overheating on the second floor. Poly-ethylene antibackflow flaps cover the tops of the windows, and these are critical for reducing night-time heat loss. If the outlet vents were unprotected, the collector could draw warm air from the house when it cooled down after sundown. Cool air in the collector airspace would fall and pull house air out

through the outlets, but the flaps close over the outlets to eliminate reverse air flow. When the collector is supplying heat, the feather-weight flaps simply flutter up out of the way by the force of the natural convection air flow. These vent controls are used in several forthcoming mass wall and air heating collector systems. For more on how they're made, see figure III-41.

Because the framing was arranged so as not to block windows, it became necessary to cut each piece of glass to a custom fit. It was easiest just to take a whole stack of panes up to the roof, hold each one in place, mark the cut lines with a felt-tip pen, and cut them right there. Before a pane was put in place, the opening was backputtied (a thin layer of putty applied to the rabbet). With the pane set into the opening, standard glazing points and linseed oil putty were used to secure and seal. It is important to clean

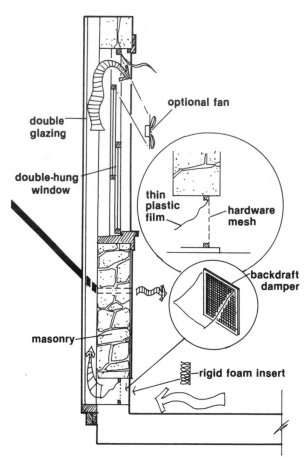

double glazing

double-hung window

masonry

optional fan

thin plastic film

hardware mesh

backdraft damper

rigid foam insert

Figure III-41: In a typical system built over a wall with double-hung windows, an inlet vent must be cut through the masonry wall, while the open top sash of the window serves as the outlet vent, which can be optionally fan-powered. The backdraft damper at the inlet vent stops cold air from falling back into the house after sundown. Heat loss through the vent and the window can be further reduced with movable insulation.

each piece carefully because only the exterior surface will ever again be accessible for cleaning. Polishing the glass with a soft rag and whiting (powdered chalk, available in paint stores) is most successful. The whiting picks up the excess oils and also helps

to set the putty.

With the fixed size of the glass panes, there remained a gap of about 4 inches between the top of the glazing and the roof overhang. Since this top 4 inches is shadowed by the overhanging gutters, there was no loss of solar gain. Instead of filling the gap with solid wood, which would have been a solution, I let the top edge of the inner glazing "float" with no contact with the framework (it experiences little wind loading or other physical stress, and three of the four sides are secured to the frame). The space between the two glazing layers was packed with fiberglass ceiling tile insulation, and the outer glazing was shingle-lapped with a 4-inch-wide strip of Masonite and sealed with silicone caulk. This allowed the space between the glazing layers to breathe, eliminating any problems with condensation. This alternative to sealed, desiccant-filled double-glazed windows has proved to be trouble-free, with no concern about a broken seal or saturated desiccant. For summer venting of the Trombe wall, I simply hinge up a few of these Masonite panels and dump the heated air to the outside. To finish the solar wall the framework was carefully caulked around the perimeter with butyl caulk and given another coat of enamel paint. That same season I painted the kitchen roof with aluminized roof coating. This roof slopes away from the wall at a 15-degree pitch, and with reflection the amount of sunlight striking the collector was increased by at least 10 to 20 percent.

Later on I installed a 1/3-hp squirrel-cage fan in the old furnace duct system. The duct in the attic that is tied into two of the collector outlets ultimately runs through the living room and thence to the furnace. By dampering this duct just before it runs into the basement and installing the fan near the floor in the living room, I can pull surplus heat from either the collector, the second-floor bathroom or the bedroom ceiling and exhaust it into the living room. This heats the living room and the kitchen with air from the collector and augments the whole-house circulation loop between the collector and the return (inlet) vents in the kitchen ceiling. Since I don't usually want this forced-air system to go on until the solar wall is well charged and almost overheating, the fan is controlled manually or in conjunction with a standard household timer. It could also be thermostatically controlled. Actually, since I'm rarely home in the afternoon when the second floor is most likely to overheat, I have little use for this fan.

Operation and Performance

In normal operation the solar wall draws air from the kitchen ceiling and delivers heated air to the second floor. At the same time the blackened stone wall is charged with heat for nighttime radiation. When there is insufficient insolation, the air floor system stops automatically with the closing of the polyethylene flap valves.

If I want to put more energy into charging the thermal mass (perhaps after a long period of very cold, overcast weather), I close the registers on the kitchen ceiling to create a stagnant condition in the solar wall. In this mode the exterior surface of the wall gets much hotter, and while it causes some loss of efficiency (by increased reradiative heat loss), the charging time for the thermal mass is shorter.

It's very important in a system like this to be able to dump heat to the outside in summer. The whole house is cooled at night by venting through the attic door and first-floor windows. To hold in the cool air, the house is closed up during the day. Since the kitchen is of a light frame construction, and more likely to overheat, it is vented through the inlet registers of the solar wall, which ultimately vents to the outside via the collector outlets. Neither air conditioning nor fan-powered ventilation has been necessary in this house.

As was expected, the most impressive gains have come from my weatherization efforts. I conservatively estimated the initial infiltration rate in the house at three air changes per hour, and the original heating load to have been in excess of 167 million Btu per heating season, the equivalent of about 1200 gallons of heating oil. But at the present rate of heat loss (with higher R-values and an infiltration rate re-

duced to one air change per hour, but not including any solar gain), the heating load has been reduced to about 91 million Btu per year, the equivalent of about 650 gallons of oil. That's a 46 percent reduction just with conservation.

To determine the solar contribution, I estimated that the reflective roof section in front of my collector wall boosts the overall gain by 20 percent and determined that the total solar contribution of the Trombe wall is about 21 percent of the present heating load, or over 19 million Btu per year of usable energy input. This is the equivalent of 140 gallons of oil, giving me a materials' cost payback period of less than one heating season. The materials for the solar wall cost about $125, an unusually low figure because of the low cost of the glazing.

The total energy savings for the entire project (weatherization and solar retrofit) is about 94 million Btu per year, which would represent a savings of about 680 gallons of oil a year, if I burned oil. Since I burn wood for back-up heat, the calculations indicate that I would need about 2 2/3 cords of red oak to maintain 65°F (18°C) throughout the heating season. In fact I use somewhat less that 2 cords of firewood per season, since I allow the temperature to drop below 65 at night and when I'm not at home.

This is about as far as I will go with my energy improvements, since I don't own the house. Much more is possible though. Its overall rating of 13.28 Btu per degree-day per square foot is still not good by current standards. If I owned the place, I would eliminate three-quarters of the ducting in the basement and maintain the oil burner as a minimal-use back-up system. I would insulate under the entire first floor. All windows and doors and the basement entrance would be reworked with new storms, weather stripping and caulk, with added insulation for the windows. I would also insulate the stone walls on the outside to greatly increase the building's thermal mass. (See the box "Exterior Insulation for Masonry Buildings" in this section.) Some of these efforts would be fairly expensive, but would save about 250 gallons of oil a year; they would surely be worth considering.

Evaluation

I'm certainly pleased with the results. Since I knew that 140 square feet of collector area was a little less than half of the recommended area of one-third to one-half of the 1000 square feet of living space, its performance has exceeded my expectations. The glass wall hasn't greatly changed the traditional appearance of the house. All too often people see restoration and retrofitting as being two conflicting goals, but this certainly need not be the case.

I find only two negative aspects of living in a solar house. The first comes on cold, sunny mornings when I tend to bear chilly temperatures rather than fire up the woodstove. Knowing the house will soon warm up with solar heat and that it will be quite warm by the afternoon, I convince myself that burning the stove is just wasting wood, and I go sit in the sun by the picture window. The second problem, also peculiar to cold and clear days, is that because it is so glorious to be indoors with the sun streaming through the windows and the collector gently delivering 90°F (32°C) air, I just don't care to go out!

Your Place

One of the nicest aspects of the Trombe-type wall system is its wide applicability. While the idea works in the Northeast on the thousands of old stone houses like mine, it also works equally well on the brick houses of the South and the adobe houses of the Southwest. In metropolitan areas, where buildings are frequently built with masonry materials, I see almost any unshaded, south-facing masonry wall as a potential Trombe-type wall.

Most newly constructed walls intended for this kind of application are 10 to 14 inches thick, and made of either poured concrete or concrete-filled block. Obviously, for a retrofit you have to use what you've got. In my case, the 18 inches of fieldstone were probably excessive, in that there was too much mass to charge. This limits the migration of heat to the inside and lowers the radiant temperature of the

wall. Most brick walls are of two- or three-course construction (9 to 15 inches thick), and they're fine for this kind of retrofit. But some old stone houses have walls that are 3 feet thick, way too fat for this kind of solarizing. In this case it would be a better idea to add exterior insulation to the wall and then build an air heating collector system over it, as described in the next section (III-E).

The opposite extreme would be a south wall built of unfilled, 8-inch hollow-core concrete block. In this case, the available mass would be insufficient. It might be possible to have concrete pumped into the core cavities, but this could be expensive. Thickening the wall with a brick or filled-block facing is another alternative. But if 8 hollow inches are all you've got, you'll have to make sufficiently large inlet and outlet vents to keep the mass from overheating. In operation most of the day's solar gain would be directly transmitted to the living space, with a small amount being retained in the wall. Think of a low-mass Trombe wall as more of an air heating collector than a heat storage system. Double glazing in this case is not as critical as with a truly massive wall, since night heat loss is less significant when a collector is used more for air heating. But to retain more of the heat stored in a Trombe-type wall, double glazing is essential. Some new-construction Trombe walls also use movable insulation to further minimize night losses. This is usually more difficult to do with a retrofit if the glazing structure is fastened to the wall. Thus double glazing is *the* primary means of controlling heat loss from the masonry. In the future we can look forward to the availability of new, low-emittance films, like Heat Mirror, or selective surface coatings, both of which may significantly increase the efficiency of retrofit Trombe-type walls.

In actual construction you should repair any plaster or stucco or defects in pointing before you darken the surface, because once the glazing is in place you won't be able to make these repairs. Select a paint suitable for masonry surfaces that won't fade over time. A good quality to look for is paint or masonry stain that is made with inorganic pigments. Flat black paint is the best choice for absorption, but not necessarily the best for looks. You could, for example, mix a dark brown or a red with the black to get a slightly warmer tone without significantly sacrificing energy collection. An inexpensive solution is to mix black masonry pigment with mortar cement. Make a thin slurry and apply the mixture with a wallpaper brush.

As a true believer in the value of this kind of system, I would advise glazing as large an area of south masonry wall as is available. If you buy new materials and do most all of the work yourself, you can expect to spend $5 to $8 per square foot of collector, and the cost per square foot usually drops

Photo III-57: This retrofit was also designed and built by the author, and it shows how the glazing was arranged to cover a large wall area. This is an important consideration in system design: Just how will the lines of the glazing enhance overall appearance? Note the use of operable windows in the upper part of the solar wall for summer ventilation.

as collector area increases. If you can, build at least 8 feet of vertical height into the collector to improve the natural convection air flow. Should your Trombe wall cover two or more stories, you would do well to zone the collector with baffles and an array of inlets and outlets to balance the heat delivered to different levels. Otherwise all the heated air tends to rise up to the uppermost story and stratify there instead of circulating evenly throughout the house.

Some studies have indicated that there is not a great deal of difference in efficiency between a standard air-circulating Trombe wall and a stagnating Trombe wall (one where there are no air vents at all and all the heat transfer is by conduction through the wall). But my own experience suggests that a static Trombe wall causes the exterior surface of the masonry wall and the air within the cavity to rise to very high temperatures, which leads to greater reradiative heat loss to the outside and therefore decreased efficiency. I feel that air flow should be part of the design so that you have both options. Double-hung windows can work as either inlet or outlet vents or both. And it isn't too difficult to chop through brick or block construction with a small jackhammer, which can be rented.

A gap of about 3 inches (plus or minus ½ inch) between the glazing and the masonry is usually recommended for Trombe walls. With a very narrow gap (less than 2 inches) natural convection air flow would be restricted and fans would be needed to deliver warm air. With a gap that was overly wide the potential for convective loops between the wall and the glazing is greater, which would reduce overall efficiency. As was noted, if you use vents, they should be equal in area to no less than one-half of the area of the horizontal cross section of the airspace (air gap × length of wall ÷ 2). The rule of thumb for fan-powered air flow in air heating collectors is to have at least 1½ to 2 cfm of air for each square foot of collector, but this rate isn't strictly necessary for a Trombe-type wall. They usually have enough thermal mass to prevent rapid overheating, but with a

new installation you should keep an eye on the situation. You may need to boost the air flow if temperatures in the collector reach above 115°F (46°C).

In general, I always recommend making air circulation inside the house as flexible as possible to maintain the ability to shunt heat to different rooms. If you have several inlet and outlet vents to the collector, they can also be controlled to balance heat delivery to different rooms.

Air flow inside the solar wall (between the masonry and the glazing) can be controlled with baffles as well as with the opening and closing of different vents. Say for example that in a particular wall section there were no windows, and the inlet vent was in the same line as the outlet vent, though several feet below it. The air flow, be it natural or forced convection, will have a tendency to follow the path of least resistance and make a short circuit between the two vents. Thus parts of this wall section will have little or no air flow, and they will be hotter than parts that do experience heat transfer by convection. This condition reduces overall efficiency because with "hot spots" reradiative heat loss is increased. Horizontal or slanted baffles can be used to spread the air flow throughout the wall section so that the air will pick up more of the collected heat. Another solution is to increase the number of vents. Instead of having one big inlet and one big outlet vent per a given wall section, you can make four half-sized vents and space them apart to gain a more evenly distributed air flow (see figure III-42).

Any standard glazing material will work with this kind of retrofit, and your choice will depend on the appearance you want and the budget you've allowed for the project. Plastic glazing such as FRP can be used in a double layer or as an exterior layer over an inner glazing of thin film. Double-wall acrylic glazing (Exolite) is a good choice for looks, relative economy and ease of handling. My method for using single panes of glass could be repeated using large module sizes to reduce the amount of framing required. When you get into standard sizes such as the

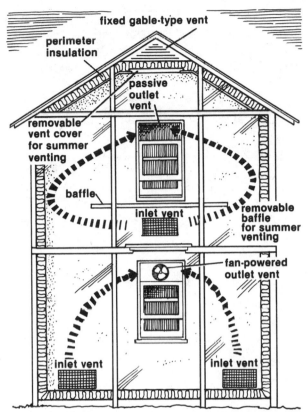

Figure III-42: Many inlet/outlet combinations are possible in a glazing over masonry retrofit. On the lower floor of this model system two inlets were cut through the wall, while upstairs one vent is used with a long baffle. Also shown is the summer ventilation system, which involves removing the baffle at the midpoint of the wall and opening the louvered vent at the peak.

start playing the wholesale game for all it's worth. Spend some time on the phone; identify yourself as a contractor, and you stand to save a lot. The other glass alternative is to buy prefabricated tempered double-pane insulating glass units, which are available in a few standard sizes that will determine your framing layout. Local double-glass fabricators can make custom sizes for places where standard sizes don't fit. In figures III-43 and III-44 we've detailed several possible glazing attachment details that use the above-mentioned materials.

Whatever kind of glazing system you use, the framing members should be good quality wood, preferably a variety that is naturally rot-resistant such as redwood or cedar. Any wood should be sealed with a stain-sealer or a light coat of paint. To protect

34 by 76-inch module, square-foot prices can be lower, even with tempered glass. If you have a large wall (over 600 square feet) that you want to double-glaze with glass, do some research and find yourself a glass factory. A case lot of glass (about 1500 square feet) can sell for less than half the square-foot price of a smaller order bought at retail. Remember, when you're buying large quantities of anything, you can

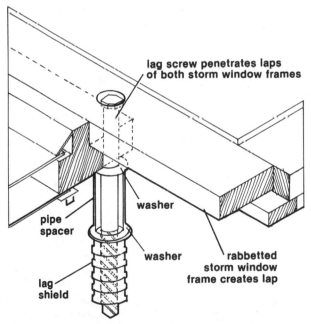

Figure III-43: This fastening detail shows how storm windows with wooden frames might be used for this kind of retrofit. The edges of the frames are rabbetted on opposite sides so that the storms can lap over each other.

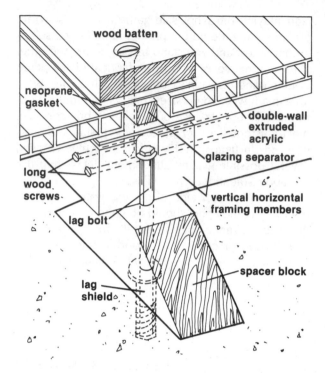

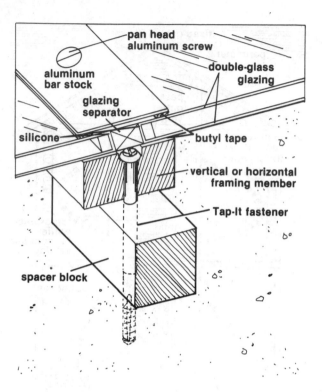

Figure III-44: This detail shows how sheets of FRP glazing are fastened. The Stud Anchor (U.S. Expansion Bolt Corp., York, Pa.) is another kind of masonry fastener that can speed up the job at hand because it requires that only one size hole be drilled through the wood framing and into the masonry. At right is shown a glazing system using prefabricated double-glass panels (typically 34 by 76 inches). A gap of 1/2 to 1 inch is left between the panels so the aluminum batten can be fastened. The Tap-It fastener is also made by the above company, and it too can simplify and speed up the job of fastening the framing.

exterior wooden battens that lie over the glazing, I prefer a dark oil stain. For a puttied glass arrangement, as in my system, enamel paint is best for protecting the glazing seals from deteriorating.

As was mentioned earlier, it's entirely possible, especially with old buildings, that the walls are out of true, either in flatness or squareness. This adds to the difficulty of the framing, which must in any case be very precise, particularly when glass is used.

But if the wall isn't flat, a large (34 by 76-inch) glazing opening could actually be twisted. This could lead to a breakage of glass or a less-than-clean appearance with more flexible plastic glazings. If there's a great deal of waviness in the wall, uncorrected framing will simply duplicate that irregularity, which at the least will be a compromise of appearance. But it doesn't have to be that way. The width of the spacers can be controlled to eliminate irregularities, and figure III-45 shows the standard carpentry method for trueing-up long runs of framing. It takes more time, of course, but doing it right always does.

Clearly this is the kind of job that should be taken on by experienced hands. Every wall will present its own problems that will require lots of planning and an awareness of what things can be done with what kinds of materials. When you're in that planning and design stage, when all the questions come up, don't hesitate to ask about for advice from

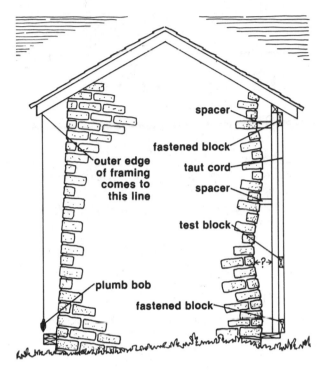

Figure III-45: Masonry walls, especially old ones, can tend to be out of plumb or wavy, and it's a good idea not to repeat such imperfections in the solar wall. This exaggerated view shows how to use a plumb bob to find where the framing should come to at the top of the wall. If the wall appears wavy, as it does at left, the spacers must be custom-sized to ensure that the framing runs up, down and across on straight lines. A string is run from two fastened blocks at both ends of the framing. Then a test block of the exact same thickness is run down the framing to the points where spacer blocks will be placed. At those points measure for the length of the spacer when, by pushing or pulling the framing in or out, the string just touches the best block. The same method is used to true-up framing that is bowed in or out.

local professionals. A solar retrofit like this combines the talents of the painter, the carpenter, the glazier, the mason, perhaps the electrician, and a lot can be learned by sharing your ideas with their experience. You may even find that you'll be better off hiring out parts of the job instead of struggling through them yourself and perhaps making costly errors as well. This is smart do-it-yourself-ism: knowing when to do it and when to have it done.

Tom Wilson

Reference

Wilson, Alex. 1979. *Thermal storage wall design manual.* Sante Fe: New Mexico Solar Energy Association.

⌂PROJECT
Solarizing a Basement Wall

This variation on Trombe-type walls makes use of what would otherwise be just an ordinary basement wall, but because it faced to the south the owner had the foresight to plan for solar heating when he built his frame addition. Any old or new foundation wall can get this treatment if it has the right orientation. Heat can be directed into the room behind with air handling and solar charging of the masonry thermal mass, or it can be used to heat rooms above the wall, as it is in the project. Is one of your basement walls only realizing part of its potential? Give it a chance for total fulfillment with a simple solar retrofit.

Sometimes an array of solar collectors can have a jarring visual effect when installed on older homes. To get around that problem I developed a low-cost, unobtrusive Trombe-type collector system that makes use of a south-facing basement wall to heat not the room behind it but the living space above. This system is part of a passive solar addition to my house in Norwich, Vermont. The solarized wall section measures about 18 feet long by 8 feet high. The masonry component was built primarily as a foundation for a large building addition; it was made of poured concrete with the inlet vents cast in at the base. The above-grade walls sit on a frost wall that extends fully 5 feet below grade because of the deep freeze that the ground gets in this 7700-degree-day climate. Adding the air vents was a simple matter of placing small wooden forms on top of the frost wall before the basement wall was poured. The forms were made of 1×4 and 1×6 stock, 5 inches high and 7 inches wide and placed at 16-inch intervals. The arrangement of reinforcing steel (rebar) within the wall had to account for the presence of the vents, but that's not difficult for any competent foundation subcontractor.

Soon after the wall was poured, holes were drilled in the exterior face for inserting lead expansion anchors. Since concrete continues to harden for months, in fact years after pouring, the sooner this

drilling is done the easier it will be. A hammer drill makes fast work of it. The 2×6 support sill for the glazing was screwed to a 2×4 ledger that was fastened to the wall. Care was taken to leave a space of about 3 inches between the ledger and the earth

Photo III-58: Along with a lot of south glass for the upstairs part of his addition, the author made plans in advance for converting his basement/foundation wall into a solar heating system. Be it ever true: A little advance planning can save a lot of effort, time and money later on.

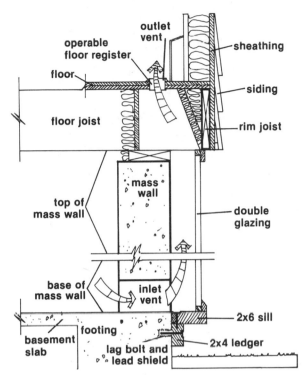

Figure III-46: The top of this illustration shows how the author's building addition was slightly cantilevered over the foundation wall so that the warm air outlet could be located right on the finish floor between the floor joists.

beneath it to prevent rot. The wall was painted with black masonry paint, and the concrete was allowed to cure for several weeks. Then 2×6 uprights were nailed to the support sill and to the overhanging joists of the first floor of the addition. To conform with the glazing materials we used, the uprights were placed on 3-foot centers, which didn't always match with the 16-inch on-center placement of the floor joists. When an upright came up between two joists, a sturdy metal strap was run between them and nailed to the upright. All support structures were then stained with a dark preservative stain.

The 2×6 uprights perform another function besides that of supporting the glazing. They break up the glazing-masonry airspace into vertical channels, which promote a comprehensive, uniform

flow distribution that utilizes the complete area of the wall with fewer "dead zones" where heat can build up and be wasted. To enhance the channeling effect each upright was sealed to the wall with an adhesive caulk. The uprights do cast a shadow on the wall, except when the sun's rays are perpendicular to the wall, but some of the sunlight striking the sides of the uprights can be reflected back onto the wall by bonding a reflective surface to both faces of the uprights or by painting them white.

The glazing panels were made with an outer glazing of FRP and an inner glazing of Tedlar film, fixed to 3 by 8-foot frames made of 1×2's. The 4-mil Tedlar film was stretched over the frame and stapled to the edges. A bead of adhesive was applied to the outside face of the frame, and the FRP was laid in place. An aluminum angle strip was attached to the perimeter of each frame to distribute pressure uniformly and also to help stiffen the entire assembly. The angle (¾ inch on each side and 1/16 inch thick) fits over the stretched film and makes a neat package.

As each glazing panel was completed, it was fastened to 1×1 stops that were screwed to faces of the uprights. A bead of caulking between the stops and the uprights was the first step toward making a good air seal. Caulk or butyl tape or foam weather stripping is applied between the stops and the panels to complete the "weatherization" of the solar wall.

Structurally, the most unusual feature of this system is a cantilevered living room floor that extends 8 inches beyond the foundation. This allows heated air to flow directly from the collector into the living room via standard floor registers. A 2-inch-thick Styrofoam panel was used as a baffle to direct the air to the floor vent, which was a little offset from the actual collector airspace. This baffle also reduces heat loss through the rim or header joist. It's also important to insulate the sides and the sill of the collector to limit edge heat loss. Styrofoam has proven adequate but it should be covered with a protective layer of aluminum foil or thin plywood to prevent deterioration. Black paint alone is *not* sufficient because heat absorption still produces a break-

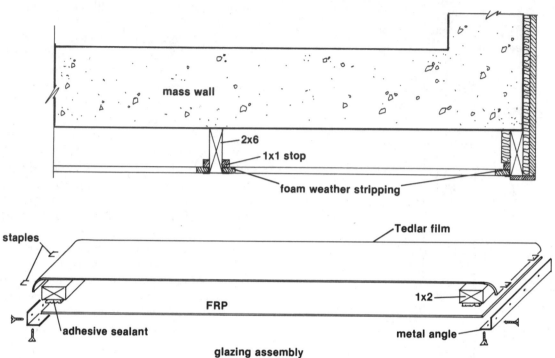

Figure III-47: The lower detail shows how the FRP and the film glazing layers are attached to the glazing frame and then, above, how the glazing panels fit onto the framing. Care must be taken when handling the film to avoid starting little tears that could lengthen over time. It's a good idea to double over the edge of the film before stapling it.

down of the insulation, and the paint falls off along with the powdery remains of "solarized" Styrofoam.

Since most basement walls aren't built under a cantilevered first floor, in other retrofits there will be a need to develop some kind of vent coupling to connect the collector with the living space. If the top of the collector has to have some kind of mini-roof arrangement to cover the tops of the uprights, perhaps it could be placed a few inches above the line where the wall framing and the finish floor come together; the collector uprights could be extended to support the mini-roof. With this arrangement outlet vents could be cut into the face of the frame wall, which would eliminate having to cut holes through the header joist. No matter how the outlet vents are rigged, it's important that they be operable so they can be closed at night to prevent heat-wasting re-

verse thermosiphoning. It is also important to install the plastic film flapper or one-way valve described in the previous project.

Cost and Performance

Since any house must have a foundation, the "solar" cost of this collector system consists only of the cost of framing, glazing, fastening and vent materials. This particular wall is 10 inches thick, though 8 inches would be sufficient for thermal mass requirements. The glazing materials at current prices cost about $270. Structural lumber, paints, floor vents, Styrofoam and miscellaneous hardware amount to about $150. It is thus possible to build six panels for about $400 or about $2.90 per square foot of collector area.

The system was gradually completed over two years while other energy-related improvements were also being made, so before and after figures on heating costs are not available. However, a computer simulation indicated that the 140-square-foot collector would contribute about 10 percent of the heat normally required by the 2000 square feet of the entire house or 20 percent of the 1000-square-foot addition. Actual performance was monitored for the month of January 1978, by installing a group of thermocouples in the wall. The data for a day of full sunshine illustrate the thermal behavior of the system. The outside low temperature was 5°F (−15°C) just before sunrise. Thermocouples embedded in the concrete began to show a temperature rise about an hour after sunrise, and by 2:00 P.M. the temperature

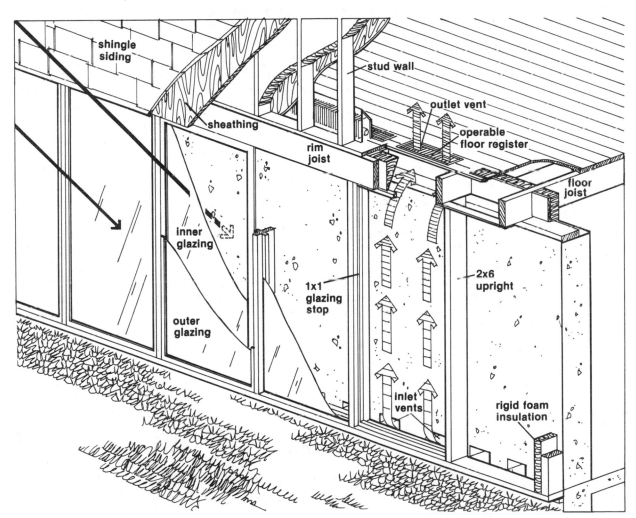

Figure III-48: If you have a basement or foundation wall that faces in the vicinity of south, you can give some thought to a system like this one. In planning you can decide whether to heat the room behind the wall or the one above it or both. It doesn't have to be a masonry wall either. A frame wall can be solarized as an air heating collector, the likes of which are described in the next section (III-E).

of the masonry 1 inch in from the sunlit surface reached 98°F (37°C). At the same time the temperature of the airspace just beneath a closed floor vent was 93°F (34°C). Heat was conducted slowly through the thermal mass of the concrete, and the inner surface of the wall did not reach its maximum temperature of about 64°F (18°C) until 6:00 P.M., more than an hour after sunset. The air in the unheated basement reached 50°F (10°C) at the same time and was still 45°F (7°C) at sunrise the next day, with the inner surface of the concrete at about 50°F.

Improvements and Variations

Review of the temperature record suggests that this system has not too little but too much heat storage capacity, which means that the temperature doesn't get as high as would be necessary to provide more radiant space heating after sunset. My colleague, Al Converse, evaluated a similar retrofit, whose wall consisted of 8-inch hollow-core concrete blocks, that reached 140°F (60°C) during the day. Although it ultimately cools to below freezing at night because of its reduced thermal mass, it does continue to heat the space later into the evening.

There are other steps that would have improved the performance of this system. If the basement were actually a heated living space, foam insulation around the perimeter of the floor slab would reduce basement heat loss. Significant heat loss also occurs from an unglazed area of concrete next to the collector. My plan is to glaze this section, though another option would be to insulate it on the outside. Of course as in many Trombe-type walls a substantial amount of heat is also lost as the outer surface of the wall radiates back out through the glazing when heat collection is over for the day. One way to get around that is with some sort of movable insulation either in the collector airspace or over the glazing. Another way to reduce reradiation is to use one of the selective absorber coatings that have both a higher absorptance and a lower emittance than regular flat black paint. Either improvement adds to the cost of the system but would substantially reduce heat loss.

If the basement is to receive no solar heat at all, insulating the inside face of the masonry wall would minimize the amount of heat it could give to the basement and increase daytime temperature buildup in the collector. Also, ductwork could be run from the living space to the collector inlet, supplying warmer air to the collector to promote higher temperature buildup during the day.

As it is now the system contributes space heat to the living room only during the day, but when some of these modifications are made, it should warm the living space further into the night by making better use of the heat-storage capability of the masonry.

Irving Thomae

Reference

Converse, Alvin O. 1979. *Performance of the Norwich School retrofit Trombe wall.* Hanover, N.H.: Thayer School of Engineering, Dartmouth College.

PROJECT
A Water-Wall Retrofit

Frame houses can have thermal mass too. Adding a water wall is a way to gain about twice the heat storage capacity of a masonry wall of the same volume. Water walls can be dealt with a few gallons at a time, which makes it easier to put them together. This project used 5-gallon cans that are rectangular in order to tightly "stack" more water into a limited area. The wall section is taken down to the bare studs and brought into the solar age with a simple glazing system.

What has 64 honey cans, a ton-and-a-half of water, and stays in a garage in New Jersey? Don't rack your brain over this because the answer—an innovative water-wall retrofit—would probably be about the last thing you'd guess. Gary Settles is the owner of this retrofit, which is a blend of down-home improvisation with a smattering of high-tech solar sophistication. The basic statistics about the retrofit are pretty simple: A $600 investment created a 96-square-foot glazing system, 85 square feet of which is clear aperture. Twenty-six square feet are used as direct-gain window area, leaving about 59 square feet of glazing that is coupled directly with the water wall. It also operates simply: When the water thermal mass has been heated by the sun, it gives up its heat to the adjoining room by radiation and natural convection. This design theme is often repeated in solar construction, but it's the wall's actual components and the way they were put together that make Gary's wall a unique and practical variation.

For one thing, the water storage containers are those 5-gallon honey cans. At $2 apiece the total cost for storing 2700 pounds of water came to $144. Another unique feature is that each can's sunward side is faced with a state-of-the-art selective absorber surface, a special high absorptance, low-emittance solar foil that was applied with spray-on adhesive. Flat black paints and the selective surface can both absorb up to 96 percent of incoming solar radiation, but the low-emissivity selective surface radiates away only about 10 percent of its thermal energy while the flat black radiates away 96 percent. This means that the sun's heat is more efficiently conducted into the thermal mass by the selective surface.

Gary's entire wall was designed with a clear ideal in mind: "I wanted to create a simple, straightforward type of do-it-yourself retrofit," he said. "It had to be duplicatable in other locations; it had to be economically justifiable for middle-income families, which meant, of course, that it had to work." The idea was to keep structural changes simple and to use inexpensive materials.

To reduce convective heat losses at night and during sunless periods, the inlet vents were equipped with the standard flapper or check valves made of wire mesh and plastic food wrap. With the flapper valves to minimize convective heat losses, and with the selective surface to reduce radiant heat loss, Gary's wall was planned to get by without the expense and bother of a movable glazing insulation system. Gary explains: "The cheapest movable insulation system we could use in the wall costs around $3 or $4 per square foot, but our combination of inexpensive air valves and the selective surface cost about $1.50 per square foot, so it's a pretty attractive alternative." Actually, tests have shown that movable insulation would upgrade performance substantially, but even without it the system keeps the room at a comfortable temperature throughout the winter.

Photo III-59: This retrofit is essentially the same as adding south glass for direct-gain space heating, with the simple addition of thermal mass for heat storage. With frame houses especially there are usually several solarizing options that can be pursued singly or in combination.

The Site and the Plan

About three years ago some people at Princeton University's Center for Energy and Environmental Studies were looking for a representative suburban home in which to experiment with their novel solar ideas. Gary (working in another part of the university) heard of the search and offered his 1400-square-foot ranch-style home as a test site. The researchers readily accepted for it was just the sort of home they wanted. The house was built with 2×4 stud walls and 8-foot ceilings, so it represented "typical" construction. In addition, Gary's home had already been partially equipped with the necessary prerequisites for energy efficiency: weather stripping and extra insulation.

The room selected for the water wall was an electrically heated study, originally a garage. It was a major drain on the home's wintertime energy budget, but on the plus side, the largest exterior wall faced just 10 degrees east of true south, well within the limits for acceptable solar orientation. No trees or bushes shaded the wall, and the only impediment to direct gain was a ham radio tower placed in front of

the wall. The base was sheathed with sheet metal to stop the neighborhood children from climbing up, and as a result a fairly large shadow swept across the wall each day. The solution was simple enough: Remove the sheet metal and replace it with clear FRP panels, leaving the tower as unclimbable as before, but transparent instead of opaque.

The Princeton researchers chose components that had low cost, reliability and ease of assembly. The glazing they used was Exolite (double-wall extruded acrylic) because it was easy to handle and has a long service life. It transmits 83 percent of available sunlight, better than two layers of glass; it has an R-value of 1.75, and now costs about $2.80 per square foot, cheaper than double-pane glass. Water was selected for the thermal mass because it's inexpensive and can store about two times more heat than the same volume of masonry. Honey cans were chosen because of their low cost and because their shape and size (9½ by 9½ by 14 inches) allowed them to fit snugly between the existing studs (16 inches on center) to save space, reduce framing costs and simplify installation. And because the cans are metal, they transfer heat much more efficiently than the plastic storage containers that are sometimes used in water walls.

To get optimum performance, water or masonry-mass walls require some sort of movable insulation to keep in the heat at night or on cloudy days, but in the case of this project, the design objectives of simplicity and low cost didn't include movable insulation, though it would be a desirable component. The way they were used in this project, the cans would block direct access to the inner face of the glazing, and any insulation next to the glazing would require some sort of remote control operation. Insulating shades could be rigged to track between each pair of studs. Another approach would have been to replace the studs with a load-bearing header to create an unblocked gap between the cans and the glazing. This would have permitted use of a single roller-type insulating shade. Studies have shown that net heat gain can be increased by 25 to 50 percent when movable insulation is used with double glazing. Movable *exterior* insulation in the form of rigid re-

flector panels would be another option, although in this project the radio tower was in the way. In the end the researchers decided to stick with the basics and see how well the system would perform with the selective-surface foil and double glazing.

With the selective surface, they hoped the tins would make the most out of the available solar energy and also be to some degree self-insulating, because compared to flat black paint the selective surface would prevent more of the captured solar energy from reradiating back out through the glazing. The material used was "Berry's Solar Strip," which is made in three layers. The top layer is black chrome, a highly efficient absorber. The middle layer slows down reradiation of heat, and the copper backing allows the captured heat energy to be rapidly conducted away from the surface, into the storage medium. This material can be purchased in bulk lots that can be cut to the appropriate size and bonded directly to the thermal mass. (See the "Hardware Focus" section at the end of this project.)

Materials Checklist

water containers	extruded aluminum batten, or flat bar stock
rust-proofing paint	aluminum screws
framing materials	silicone caulk
glazing	flashing
neoprene gasket	wire mesh and plastic food wrap (flapper vents)

Installation

Prior to the work done on the water wall Gary's home received a visit from some local "house doctors," who inspected the building and tracked down the major heat leaks. "We made the house a lot tighter than it was," said Gary. "It's not perfectly sealed, but now it's pretty well buttoned up."

After figuring out how much of the wall could be "massed" with honey cans, the researchers found they could use nine bays (spaces between the wall studs) and make stacks that were eight cans high, for a total of 72 5-gallon containers. The total

mass addition was about 360 gallons, weighing about 3000 pounds. Thus for every degree Fahrenheit the water wall rose in temperature, about 3000 Btu would be stored. This didn't represent the total solar energy gain, however, because the system also included some natural convection air flow for daytime space heating. Important: Before any cans are put in place over a wood floor, be sure to check the underpinnings; the joists may need to be strengthened. Slab floors should be able to support the added weight as is.

When the cans were stacked, they were placed to create a 2-inch air gap between them and the glazing. The cans were supported by shelves attached to the existing studs and to a series of studs added along the interior face of the can stacks (see figure III-49). To save space these room-side studs were laid face-on to the cans rather than edge-on (making the total protrusion of the cans into the room about 10 inches).

The bottom sill for the new room-side framework was bolted to the slab floor, and the stud wall was raised in the normal fashion. Using heavy-

duty angle steel (L-shaped brackets) and ½-inch plywood, the first shelf was raised about 6 inches off the floor to create a free airspace beneath the tins for air circulation (see figure III-50).

Since stacking the cans more than four high might deform or crush the one on the bottom, a second shelf was put in about 38 inches above the first so that courses five through eight would be independently supported. Finally, a third shelf was built across the top of the framework, creating another

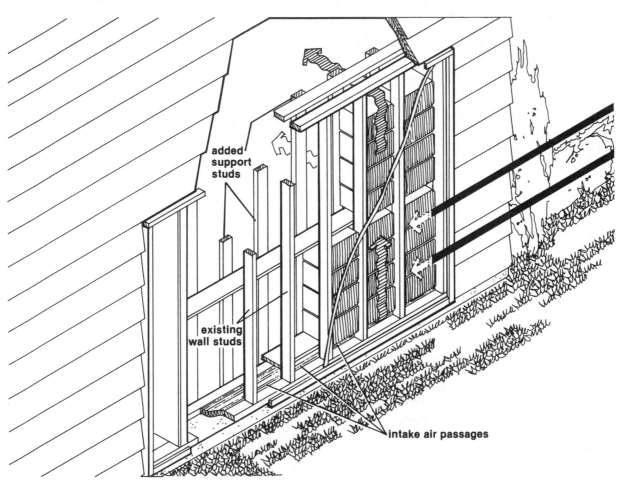

added support studs

existing wall studs

intake air passages

Figure III-49: This cutaway reveals the basic framing used for the water wall. The wall studs are left in place after the siding is removed, and they become the dividers between the honey cans as well as retaining their structural function. When trimming and finishing the exterior of the wall, careful sealing is of the utmost importance in eliminating infiltration. Be sure to caulk all seams between the siding and the solar trim work.

6-inch space for air circulation.

Once the shelves were in place the entire assembly (old studs, new studs) formed a single strong unit, ready for installation of the honey tins.

Stopping Corrosion

The use of metal containers for storing the water posed an obvious problem: An inexpensive container that rusted out and sprang leaks in a year or two would be no bargain at all. The researchers wanted a container with a minimum life expectancy of 10 to 15 years. To get that the honey cans were painted inside and out with two coats of rustproofing paint. The brand doesn't matter: Any commercially available paint containing the antirust chemical zinc chromate will work. Coating a can on the inside was a simple matter of pouring in some paint, screwing the cap on and sloshing the paint around until the entire inner surface was smothered. The excess was poured off, the paint allowed to dry and then the procedure was repeated for the second coat.

The sun side of each can was then covered with the selective surface foil. The bonding was done with an ordinary aerosol adhesive (such as Scotch Spray Mount). When the glue had set, all the cans were lined up on the lawn and filled with tap water. They were left uncapped in the sun for several hours so that rust-promoting dissolved air would have time to "rebubble" and then escape from the slowly heating water. While the tins vented, the workers added a rounded tablespoon of technical-grade sodium sulfite (available through drugstores or chemical supply houses) to each can. The sodium sulfite immediately combines with any remaining dissolved oxygen in the water to produce a noncorrosive solution of

Figure III-50: This side view of the water wall's construction shows how the existing studs are used to carry support shelves for the honey cans. The upper and lower glazing details are yet another variation of a compression-type gasket seal. If you're working over a wood floor, check to make sure that there's enough strength in the joists to carry the load.

sodium sulfate that helps to ensure long life for the can. Finally, a sealing compound was applied to the spout of each tin, and the cap was screwed down tightly. Then the tins were stacked in the wall.

When the solar wall was just a wall, it depended on the siding and perhaps some wooden bracing for its shear strength, but when it was stripped, some kind of bracing was needed that wouldn't shade the cans. After the cans were put in place, a diagonal tension brace of ⅛-inch steel cable

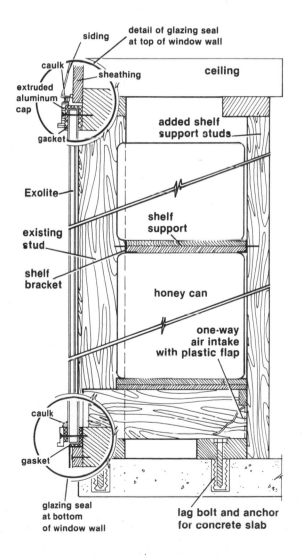

was run across the wall to resist any side-to-side (shear) forces that might harm the wall, the cans or the glazing. With a standard turnbuckle the cable could be pulled as tightly as needed.

The Glazing

The mounting system for the glazing was made up with a standard aluminum extrusion, neoprene gaskets and caulking. These materials proved to be just as effective as the mounting system sold by the glazing manufacturer, but much less expensive.

To begin, the Exolite was held in place (after being cut to size) over the inner neoprene gasket perimeter seal, which was fastened to the framing. Then the outer gasket strips were bonded to extruded battens, and this assembly was then screwed into place with aluminum screws. As the screws were tightened, the gasket strips were compressed into a water- and air-tight barrier. The top and bottom edges

of the glazing panels were fitted with aluminum flashing and caulked to seal out water. If it all sounds terribly simple, it is, but it's a standard mounting system for this kind of glazing. Once the glazing was in place, standard one-way flapper (antibackdraft) vents were installed at the bottom of the water wall in the space beneath the lowest shelf, and this passive space heater was ready for action.

Operation and Performance

With the presence of the water thermal mass the wall doesn't rely solely on natural convection through the inlet and outlet vents. Because the room-ward sides of the cans aren't coated with a selective surface, they are 8 to 10°F (4–6°C) warmer than the room, and can freely radiate heat into the room. "It's like having a giant woodstove," Gary exclaims. "You walk in here, and you can just feel

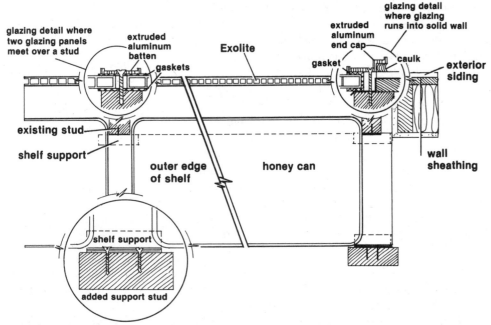

Figure III-51: This is the view from the top. As in the preceding illustrations the same compression sealing system is used at all glazing seams. It's important to create a totally airtight seal where the glazing meets the existing wall; otherwise you'll be losing heat via infiltration.

Photo III-60: The outside surface of the thermal mass is the absorptive side, while the inside surface is the radiant side, which can reach temperatures up into the 80s (°F). The way the framing is arranged, it would be tempting to use the wall space for object storage as well as heat storage, but any objects in front of the honey cans would block heat radiation.

the heat beaming out of the wall." Like the other mass wall retrofits that have been discussed, this system also has the benefit of a thermal "flywheel" by providing a low-temperature source of radiant heat after the sun goes down.

The system is effective in itself, but because of its small size it provides only about 19 percent of the Settles' space heating needs or approximately 8.4 million Btu per heating season. The area of this collector equals only about 4 percent of the house's total floor area, and a larger system would certainly take care of a correspondingly larger share of the space heating load. On a clear midwinter day the water in the cans starts out in the morning between 67 and 74°F (19–23°C) and climbs to 75 to 90°F (24–32°C) by late afternoon. If the average temperature rise were 15°F (8°C), the water mass would have stored over 40,000 Btu. Early afternoon air temperatures in the space between the glazing and the cans have reached up to 130°F (54°C), which makes for effective daytime space heating.

Gary says everyone in his family likes the system, and that includes the pets. "They come to sleep in the sunlight that hits the floor beside the wall, and I'm sure they'd try to get into the wall itself if the vents weren't screened."

But the wall isn't perfect: Remember, it was built to test new ideas, and some of the basic premises turned out to be at least partially wrong. The selective surface, for one thing, did not work as well as had been hoped. Although it did significantly increase the amount of heat gained by the wall and did help to limit reradiative losses back through the glazing, there were still fairly large heat losses through the glazing.

To find out more about how the system worked, the researchers ran tests to generate some hard data: One row of cans was removed and replaced with identical cans that had ordinary flat black surfaces rather than the selective surfaces. At night, a 2-inch-thick layer of R-11 Styrofoam insulation was placed over the flat black cans, and a side-by-side comparison was made between these cans and uninsulated cans with the selective surface. During the daylight hours, the selective surface absorbed more thermal energy, until by noon water in the selective-surfaced cans was an average 1°F warmer than water in the flat black ones.

Next, tests were run between just the flat black and the selective surfaces with no night insulation, and in this case the selective surface was the clear winner. An order of efficiencies began to become apparent: At the bottom of the totem pole (least efficient) was the plain flat black surface with no insulation. The selective surface with no insulation was the next best with 10 to 15 percent higher efficiency. Better still was the flat black surface with night insulation, which was about 30 percent better than the uninsulated flat black alone. The best of all possible combinations would of course be an array that had both the selective surface and night insulation. It would also be the most expensive at about $6 to $7 per square foot for commercially made insulation and the selective surface (the cost of movable insulation could be reduced somewhat if it were home-built).

The study done on this system has affirmed that costs and complexity naturally increase with increases in efficiency, but the cost increases aren't prohibitive, and the rewards of higher efficiency are generally worth the investment. Thus a homeowner can select the combination of components that would give the best thermal results for the money available for a given project.

Continued study revealed more about the glazing that was used, information that is significant for all types of glazing. This double-wall extruded acrylic is advertised as having a transmittance of 83 percent, but the average measured value at the Settles' wall was lower at about 72 percent. Generally a manufacturer's numbers are based on a reading taken when the solar angle of incidence is zero, which in the real world occurs only once during the day, no matter what the orientation of the collector. As the angle of incidence increases, transmission decreases, so the average transmittance over the day will tend to be less than the optimum. There's nothing that can be done about this; it's a fact of life that could only be corrected with a glazing system that tracked the sun—not too practical for a passive solar retrofit.

Summer overheating is another area of concern, because the mass absorbs and stores heat in August and in this system there are no vents connecting the airspace to the outside to exhaust unwanted heat. Gary presently uses supermarket aluminum foil draped in sheets between the glazing and the cans to reflect the summer sun back out, but, he admits, this is where the aim of simplicity and low cost ran against the needs of proper design. Exterior vents should be built into a wall like this, and it's something that can be done without a great deal of difficulty. Another improvement that could go with vents is overhang shading with the horizontal overhang extended just enough to block high-angle sunlight. A movable insulation system can also be used effectively to limit heat gain in summer, another point in favor of including it in a retrofit like this. Clearly, you should consider using movable insulation if it is feasible for your project. It will greatly increase heat retention (by 25 to 50 percent) and reduce summer heat gain by up to 50 percent.

Finally, the wall had one totally unexpected problem with, of all things, spiders. This double-wall glazing is stiffened with internal vertical ribs that divide the space between the inner and outer layers into channels that measure about ½ inch by ½ inch. It turned out that little spiders could find their way through inconspicuous voids in the caulk and gaskets and set up a sort of invulnerable cozy, web-spinning nirvana *inside* the glazing. To my knowledge, no one's ever actually studied the solar transmissivity of spider webs, but it's a safe bet that they're slowly reducing the amount of sunlight striking the honey tins. CY/RO, the manufacturer of Exolite, now sells a sealing strip intended to close off the open ends, and events like Gary's spiders are a good reason for using it.

The Bottom Line

As a test wall and as a demonstration of an option for a low-cost, do-it-yourself water wall, Gary's installation is pretty much a success. Its price is reasonable, and the basic plan can be scaled up or down to meet the requirements of different homes. The idea of using a wall's existing studs to provide a supporting frame for rectangular metal water containers is an excellent one. The careful rustproofing done at Gary's site guarantees long life for a low-cost container.

With a total outlay of $600, the cost per

square foot of glazing, including the direct-gain area, was about $7. With a net gain per heating season of about 8.4 million Btu, the gain per square foot of glazing was about 100,000 Btu. Adding movable insulation (R-11) would result in an increase of 15,000 to 25,000 Btu per square foot for a cost of $2 to $4 (per square foot). Gary is considering adding an exterior roller insulation system to realize this increase.

If you do get into a project like this, use imagination in selecting and placing the thermal mass. The rewards of cleverness can include lower costs, ease of assembly and less wasted space. The range of design options discussed here can help you choose the most effective combination you can afford. Whatever design you come up with, this is certainly a job you can do yourself without great cost or difficulty and with the certainty that it will yield positive benefits.

When I asked Gary if he was satisfied with his water-wall retrofit, Gary looked surprised that the question even came up. ''Satisfied?'' he said. ''Why, yes. Totally.''

Frederic S. Langa

Hardware Focus

Selective Surfaces

Berry Solar Products
Woodbridge at Main
P. O. Box 327
Edison, NJ 08817
(201) 549-3800

Solar Strip

Strips of 0.0028", 0.005" or 0.010" copper with black chrome selective surface and nickel underplayment. Widths up to 25". 0.0028" Solar Strip can be bonded to water walls, mass and Trombe walls to act as nighttime insulator.

Water Wall

Crimsco, Inc.
5001 E. 59th St.
Kansas City, MO 64130
(816) 333-2100

Heat Wall

Crimsco, Inc. and Communico, Inc. of Sante Fe, New Mexico, are developing Heat Wall, a water wall for passive solar applications. The Heat Wall tank, made of 16-gauge steel, is 46 inches wide and 76 inches high and holds 178 gallons. When full it weighs about 1700 pounds. The module has 5 water chambers each lined with a 6-mil polyvinyl bag. The interior side of the tank is sheathed with ½-inch Sheetrock. Crimsco and Communico have also designed a 46 × 46-inch water tank to allow for direct-gain heating in the glazed upper portion of the opening. Additional information is available from Crimsco, Inc.

Water Tubes and Stud Space Modules

Kalwall Corporation and One Design, Inc. offer two modular alternatives to the 55-gallon drum and other "found" containers that are often used to create water walls. Kalwall manufactures freestanding, translucent cylinders made with their FRP glazing material, and One Design fabricates polyethylene or polycarbonate containers that can be freestanding or installed between wall studs.

Kalwall's Solar Storage Tubes, available through their mail-order services, are suitable for direct gain heat collection or remote heat storage. The tubes are available in standard sizes of 12- and 18-inch diameters in 4- and 8-foot (12-inch) and 5- and 10-foot (18-inch) lengths. The FRP cylinder wall has a thickness of 0.040 inch. The smallest tube (12 inches by 4 feet) holds 23½ gallons and weighs 204 pounds when full. The largest tube (18 inches by 10 feet) holds 132 gallons and weighs 1122 pounds when it's full. The service temperature ranges from 34 to 170°F (1–77°C), though the tubes can withstand intermittent temperatures up to 200°F (93°C), which are unlikely in a passive solar application.

Since the tubes are installed vertically, a safe installation requires a level floor that can bear the weight of the water. Kalwall also supplies nontoxic dyes in black, yellow, bronze or blue to increase the absorption efficiency of the water, if that's desired.

For remote storage applications, Kalwall offers assistance in designing the Kal-therm Storage Module which is a clustering of the tubes in an insulated enclosure with south glazing. An-

Photo III-61: The water in these Kalwall tubes was dyed black to increase solar absorption. When the water is left clear, the tubes have a very subtle, translucent appearance. Either way they have function and beauty. Thermal performance can be somewhat enhanced with night insulation between the mass and the glazing.

other option is in using the tubes to store electric heat. Individual tubes can be charged with an immersion heater. Storing electric heat in thermal mass is more economical if your utility has time-of-day pricing for the power it sells, a policy

Photo III-62: One Design's water mass module fits right into a stud wall, which makes them suitable for retrofitting if the wall can be opened up and glazing can be added to the outside. As with the Kalwall tubes the water can be dyed to a desired color that might match or contrast with the room color scheme. (Photograph courtesy of One Design.)

that will become increasingly common in the near future. If the tubes aren't sufficiently charged by the sun on a given day, they can be "topped off" with electricity during the off-peak or discount period, usually during nondaylight hours. More information about the tubes is available from Kalwall Corporation, Solar Components Division, P. O. Box 237, Manchester, New Hampshire 03105. Telephone: (603) 668-8186.

Like the Kalwall tubes, One Design's Stud Space Modules are suitable for placement behind south-facing glazing or on the north wall of a greenhouse. The modules can also be installed in the floor of a greenhouse and then covered with clear Plexiglas to provide not only thermal mass, but a walkway, too.

Each module is 46½ inches long, 45 inches high and 7 inches thick and holds 53 gallons. When it's full it weighs 442 pounds. If the modules are installed two or more high, the manufacturer recommends building a framing system with 2×4's or 2×6's to support individual modules. Otherwise, a water wall one module high can be secured with a couple of steel L-brackets. These containers are designed to fit between studs on 48-inch centers. A possible application is to use them to form the wall between a greenhouse and the living area. Or, they can be placed in an existing wall and double glazed as a water-wall retrofit.

For more information contact One Design, Inc., Mountain Falls Route, Winchester, Virginia 22601. Telephone: (703) 877-2172.

Margaret J. Balitas

◈PROJECT
Moved-In Mass

In the previous project water thermal mass was added to create a solar water wall for a frame house. Moved-in mass takes things a step further: Frame or masonry houses can be retrofitted with water or masonry thermal mass that's distributed around the house. Some of it can be exposed to direct gain; some of it can be standing in the shadows and still perform useful service. All of it can be engaged in double-duty: thermal mass disguised as a shelf support system, a table or counter support, a planter, even freestanding sculpture.

Once you have weatherized, insulated and solarized, it may be worth your while to add thermal mass to further boost your home's energy efficiency. Before making those improvements, your heating system generally delivered just about as much heat as the house was losing in order to maintain comfortable temperatures, and whenever sunlight came in through the windows, the added heat gain didn't really amount to much. But with a tight house and a solar heating system, it's quite possible to overheat a room, a floor or even an entire building if the rate of solar gain is much greater than the rate of building heat loss. One solution is to use natural or forced convection to improve heat distribution through the house and thereby cool off the hot spots and warm up the not-so-hot spots. But this doesn't give you the benefit of thermal mass which, as has been shown in the previous projects, is heat storage. Masonry or water walls absorb heat and then deliver it to the living space as the space cools off after sundown. You can get the same kind of benefit by moving in thermal mass and integrating it into different parts of the house.

Energy always radiates from a hot surface to a cooler surface. During the day thermal mass will usually have the coolest surface and will readily absorb radiant energy from the sun and from other warmer surfaces that are exposed to it. At night it's the mass that has the warmest surface, and it

becomes the radiator. The amount of heat that can be stored will of course vary somewhat with the type

Photo III-63: Moved-in mass can take many forms and functions. The very basic and by now classic 55-gallon water drum comes in several "styles" that allow for vertical or horizontal use. In this bedroom they stand as quiet solar sentinels, protecting the living space from becoming either too hot or too cold.

and the volume of the material used (refer back to the box, "The Meaning of Mass"). Wood has a relatively high specific heat, but it's not a good material for thermal mass because its heat capacity is low. Masonry, because of its density, has a higher heat capacity than wood even though it has a lower specific heat. It's useful as a "moved-in" thermal mass, but it's not nearly as effective as water, which has about two to three times the heat capacity of masonry. Water that's being solar-heated in a container is also a "moving mass" in that a convection loop is created that causes the heated water to rise and be replaced by cooler water at the surface being heated. This gentle mixing within the container improves overall heat collection. For these reasons our emphasis is on using contained water for several moved-in mass options. Masonry, however, does have uses in situations where water wouldn't be practical, such as in adding thermal mass to a floor.

Before you can make thermal mass an effective part of your energy system, you should be certain there's enough energy available to heat the mass along with the living space. If your solar-heated spaces seem to be chronically overheating, then more mass will probably be a worthwhile addition. But if the solar system isn't being terribly overproductive, then it may be better to improve heat distribution rather than increase heat storage. Too much mass will rob heat from the room if it isn't adequately energized by the sun, so while there is quite a bit of latitude in sizing mass additions, it's important to size correctly. Generally it will be true that it's difficult to add too much mass in a retrofit situation, given the limits of space in an existing house.

In terms of location thermal mass naturally works best for heat storage when it is fully exposed to direct or reflected sunlight. Mass is also reasonably effective in spaces that receive sunlight even though it doesn't shine directly on the mass. Mass that is totally removed from solar rooms is beneficial, but not nearly so much as are the previous two situations. As a general rule of thumb, mass that receives direct gain is "worth" three times as much as mass that is in a sunny room but doesn't receive direct

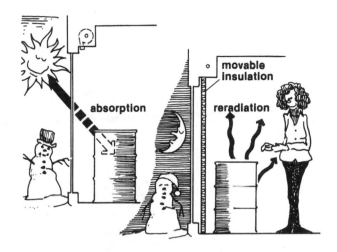

Figure III-52: In winter thermal mass is used to store direct and indirect solar gain, which is reradiated into the room if it starts to cool off. At night window insulation is important for minimizing heat loss from both the room and the warm mass.

radiation, and it's about ten times more effective for heat storage than mass that doesn't "see" any sunlight at all. Clearly, we're talking about adding mass in direct-gain solar heating systems as yielding the optimum benefit.

Thermal mass can also help to cool a house. In warm weather the mass soaks up heat during the day and helps to moderate temperature increases. Then at night cool air is brought in from outside and blown over the mass and then exhausted back outside. Since the air is cooler than the mass, it will extract the heat from it, cooling it down so that it will be able to absorb heat again during the next day. On a very warm day you will find it very comforting to sit near to thermal mass so your body can radiate heat to it. You're cooling off; it's heating up.

Thermal mass used for cooling can be effective in any room of the house that is well ventilated at night and can be closed off in the daytime. Thus in a warm or hot climate it is desirable to spread the mass throughout the house rather than concen-

Photo III-64: If metal water drums aren't quite in line with the interior decor, they can be artfully concealed with some simple woodworking, as with this slatted shelf system.

trate it on the sunny, south side as one would do in a cold climate. The key to using mass for cooling is not only the volume and quality of the mass, but also the rate of air movement. There must be enough cool air moving around to lower the temperature of the mass. Where nighttime temperatures don't usually fall below 70°F (21°C), as is often the case in regions with high humidity, the mass will not cool off enough to act as an effective heat sink during the day.

Thermal mass is a relatively subtle energy improvement. It does not, of course, go on and off like a switch; nor does it bathe you in radiant heat the way a beam of sunlight does. It's quiet, but it is effective. The table on page 439 shows the calculated benefits of a masonry and water thermal

mass retrofit in a house in southern California, which received a number of other energy improvements. The building that was modeled was a standard tract house with a good southern exposure, and the improvements in energy efficiency are listed as increases in the percentage of solar heating and natural cooling (without an air conditioner).

How Much Mass?

You can use a simple calculation to estimate the amount of thermal mass that could be useful in your situation. Multiply the area of south glass you have by the average daily winter energy gain per

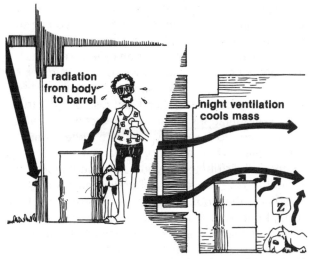

Figure III-53: During a summer day the goal is to keep thermal mass shaded so it stays cool. Then it can absorb more room and body heat, which can be "dumped" outdoors at night with ventilation and reradiation.

square foot of glazing (from the calculations that were done in "Finding Your Heat Gain" in Section I). About one-half to three-quarters of this energy can be absorbed by the mass; the balance goes into daytime space heating and building heat loss. If you have a house with an open plan, multiply the result of your calculation by ½, and if the direct-gain space is partitioned off from the rest of the house, multiply by ¾. Now divide this figure by the heat capacity of the material you're going to use (see table III-7). Then multiply the result times the temperature rise you can expect with sun-exposed mass; usually exposed mass undergoes a 10 to 20°F (6–11°C) temperature rise over a day. This calculation will give you the number of cubic feet of mass that you can use, ideally speaking that is. Don't be surprised or despairing if this number is more cubic feet of masonry or water than you could ever fit into a given space. Know, however, that every pound of mass you put in will be better than no mass at all.

For example, if a south wall has 200 square feet of south glass that collects 750 Btu per square foot per day, the calculation would be: 750×200

$\div 2 = 75{,}000$ Btu of storage capacity that could be added, assuming the direct-gain space is well coupled to the rest of the house. If water (with a heat capacity of 62.5 Btu per cubic foot per °F temperature rise) is to be used as the thermal mass, and a 15° temperature rise is assumed, the calculation continues: $75{,}000 \div (62.5 \times 15) = 80$ cubic feet of water that could be moved in. This would require 12 55-gallon drums or a water tank that measured 2 by 4 by 10 feet. If masonry were to be used, then about 192 cubic feet would be needed, or nearly two-and-a-half times the volume of the water mass.

A rough calculation for the mass requirements for cooling is done using these factors: the maximum allowable indoor temperature, the daily heat gain and the average nighttime low temperature.

	Percentage of Solar Heating	Percentage of Natural Cooling
house as is, some direct-gain solar heating	35	10
window shades added	35	26
perimeter of floor slab insulated	38	27
double-pane windows installed	39	28
increased summer shading, some floor tile added, 2500 pounds (40 cubic feet) of water mass	60	64
another 2500 pounds of water added	78	68
5000 pounds of water and 40 square feet of south glass added	92	86
insulating drapes added (R-5)	98	96

Note: The energy savings from these improvements was calculated to be $320 a year. (Source: M. Hunt et al., *A Solar House for California's Mass Market,* staff draft [Sacramento: California Energy Commission, 1976]).

The daily heat gain is divided by the difference between the indoor temperature and the nighttime low. For example, if the heat gain is 150,000 Btu, the maximum indoor temperature is 80°F (27°C) and the nighttime low is 60°F (16°C), this is the calculation: 150,000 ÷ (80 − 60) = 75,000. The result is divided by the heat capacity of the material to be used. Say, for example, that two floors are to be covered with brick or tile and mortar, which has a heat capacity of about 26. The mass requirement would be 7500 ÷ 26 = 288 cubic feet of masonry. In a 1600-square-foot house the thickness of the masonry covering would be about 2 inches (288 ÷ 1200 = 0.24 feet = 2.2 inches).

Although adding mass is usually desirable, there are some situations where it wouldn't be beneficial. For example, a recreational cabin or house that is used only occasionally isn't a likely candidate. When the house is shut up, it will get cooler and cooler (or hotter and hotter in summer), and when you arrive even a hot fire may take hours to warm up a lot of cold mass. The house will seem cool because the radiant surface temperatures remain low. By the end of the weekend the mass will be recharged, but after you leave the house will slowly cool off again. Houses that are used intermittently are often better off with little thermal mass but lots of insulation.

Thermal mass can also work against you if your house is thermally inefficient. If temperatures get too high in the house during summer, then thermal mass may keep it hot long into the night after it would otherwise have cooled off. And conversely if the house gets too cold in winter, added thermal mass may keep it cold longer. Before adding mass you should always optimize your house's thermal efficiency. Adding mass is really an option for fine-tuning a house's thermal behavior.

Structure and Space

Calculations are one thing, but the structural and spatial limitations of your house are an-other. The numbers may show that your house can use tons upon tons of thermal mass, but you've still got to figure out if your place has the strength and the space to carry it. Space can be analyzed pretty easily: The previous calculations give you the number of cubic feet, and you can do some planning to see how that volume can be arranged and distributed. Structural considerations must be researched carefully. Water mass, for example, can be a rather concentrated load, while masonry spread over a floor is pretty well distributed. If you have raised floors, you should look at the structural underpinnings of your house and then look into some engineering literature to find out what kinds of maximum loads the floor joists can bear. If major reinforcement is required, it may be too costly to make the addition worthwhile. Or, you can reduce the amount of weight to be added, or distribute it so it's not so concentrated. This all has to do with not only the safety of the house, but the safety of the people in it as well. If you live in earthquake country, this includes planning for safety when things start to move around. Recent studies of ground movement indicate that lateral acceleration is greater than was assumed when building codes were written, so it's a good idea to look into earthquake-related structural safety if you live in an area that's visited by quakes. In the moved-in mass ideas and illustrations that follow, some suggestions about structural reinforcement are included.

Adding Thermal Mass

The easiest house to add mass to is one made of brick, stone, or concrete block because you can insulate it on the outside by adding sprayed or rigid foam insulation and covering it with stucco or siding. You aren't adding mass, just improving what you've got. This insulation is usually added to the west, east, and north sides of the building, with double glazing added to the south wall for space heating. With these improvements the building will go from being a very inefficient building to being an excellent passive solar building. The mass in this type of build-

ing is excellent for cooling because it is widely distributed with a great deal of surface area that can be
exposed to ventilation. The large tonnage helps to
stabilize interior temperature swings in cold and hot
weather. In a sense, adding glass over a masonry
wall, as was done in the earlier project, "Mass under
Glass for Passive Space Heating," is also another
form of moved-in mass. (See the box, "Exterior Insu-

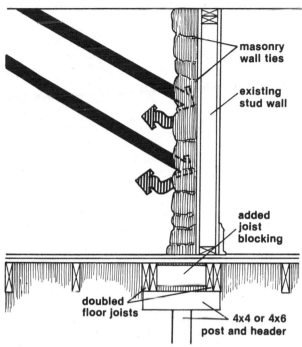

Figure III-55: An interior partition may or may not be a
bearing wall, but if you stack up a pile of masonry against
it, it's definitely going to need reinforcement underneath.
In this illustration there are two reinforcement methods
shown. By calculating the added load you may find that
you only need to double up the floor joists and run connecting blocks between them. If more support is required, a post with a tee head can be used to carry the
load directly to the basement floor.

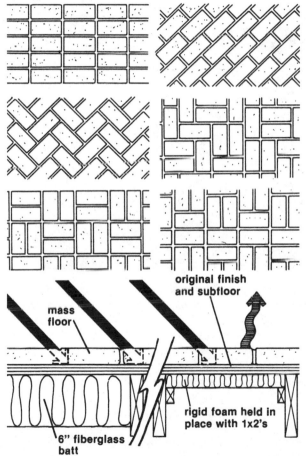

*Figure III-54: Added-mass floors should be insulated un-
derneath whenever possible to limit the conduction of
heat away from the living space. When adding masonry
wall toppers, try out different layout patterns that might
give the illusion of widening a narrow room, shortening
a long one or enlivening a dull (!) one.*

lation for Masonry Buildings.")

If you live in a standard frame house, you
may also have considerable mass if you have a concrete slab floor. To expose this mass to the sun or
cooling night breezes you should replace the carpet
with paint, tile, parquet, linoleum, vinyl flooring or
any other material with little insulating value. Ceramic tile or paint is best but the other coverings are
suitable. For heating you needn't do this throughout
the house, but only where the sun reaches the floor
in winter. If you don't have a slab floor or a masonry
(continued on page 444)

Exterior Insulation for Masonry Buildings

In terms of today's energy standards, your standard old, new or in-between brick, block or stone house is essentially unfinished. It may be plumbed, wired, furnished and even inhabited, but it's not finished; it needs insulation, badly. Masonry is virtually a noninsulator with its rather lowly R-value of 0.08 to 0.25 per inch. A brick house with walls that are 8 to 10 inches thick rates a shivery R-2.7 to R-3 for the exterior wall, and an old stone house might have walls that are 2 to 3 feet thick, and all of about R-3 to R-4 to keep the heat in. What that means is progressively colder walls in winter and progressively warmer walls in summer, when the reverse would make for a better comfort level. But with a new overcoat of exterior insulation, masonry walls can in fact more closely approach the warmer-in-winter, cooler-in-summer condition. Exterior insulation is a way to move tons of thermal mass into a house without lifting so much as a single stone or brick and without making any structural reinforcements.

There is certainly more than one way to skin a building with insulation, but here is a method that worked for me and that seemed to present a fairly efficient way to go about it. First, the right insulation had to be selected. Rigid foam is the natural choice because it generally has a higher R-value per inch than fiberglass. This reduces the extent to which windowsills and doorstops have to be built out. Fewer inches of foam also simplify the job of fastening to the wall. But all foams are not created equal. Polystyrenes are rated at R-3 to R-5 per inch, while urethane comes in higher at R-6 to R-7. Another variety has them both beat: Isocyanurate foam is rated at R-7 to R-8 per inch. It's sold in 4 × 8-foot sheets under several brand names (Thermax, R Max, High-R). Two inches of this stuff would add R-16 to my own R-2.7 walls, and a thick coat of stucco would add another R-1, bringing the total near to R-20. When oil is a dollar a gallon, you've got to have at least R-20 in your walls, or

Photo III-65: No matter what the house size, adding exterior insulation is a big job. Scaffolding is essential both for fastening the foam and applying the stucco. Two and better yet three people should be on the job so that work will flow smoothly. Note that on this house one wall isn't being insulated. It's the south wall, which awaits a glass-over-masonry solar retrofit.

winter will be more than just a season; it'll be an enemy.

The heart of the fastening system was a 4-inch-long expansion anchor called Tap-It (made by the U.S. Expansion Bolt Corporation, Box 1589, York, Pennsylvania 17405). It con-

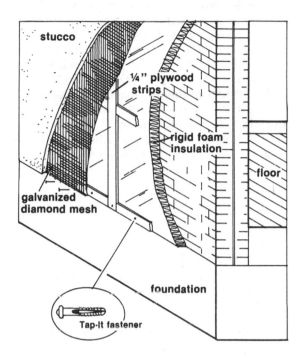

Figure III-56: An exterior insulation job is essentially simple in terms of how the foam-mesh-stucco sandwich goes together. It's really the scale of the job that causes difficulty. Ideally at least half the house should be scaffolded; the crew should consist of at least three people. Detailing should be carefully considered: How will the foundation line be handled? How will window frames be treated? With good planning even the biggest of jobs is cut down to a more manageable scale.

the foam and the stucco, given that seven to nine fasteners are used per sheet of foam. But it's not just foam and stucco. The traditional stucco application requires a galvanized metal mesh as a base, something for the mortar to hold onto. *Diamond mesh,* as it's called, also increases the strength of the stucco.

The Tap-It was used to fasten the foam and the mesh at the same time. Between the two there was a 2-inch-wide strip of ⅜-inch plywood, which primarily served as a spacer to allow the mortar to completely engulf the mesh. The strip also served as seat for the head of the Tap-It and kept it from burying itself in the foam. (Now Tap-Its are available with 2-inch diameter plastic washers that serve the same purpose. They also now come in a 6-inch length that would allow a 3- to 4-inch-thick foam retrofit.) The strips were spaced about 24 inches apart, and the edges of the mesh (which comes in strips that are about 27 inches wide) were overlapped about 3 inches. The fasteners go through both layers of mesh, through the plywood, the foam and into the brick, all in one operation. The beauty of this system is that the edges of the foam are butted right against each other with no voids in the thermal blanket. In another more common method, furring strips are built out from the wall on 2-foot centers, and the foam is fitted between the strips. But this creates gaps in the blanket since 2 inches of wood are about R-2.

In the miscellaneous tips department, it's important to get the mesh as tightly stretched as you can to ensure that the stucco coats (there are usually at least two of them, sometimes three) will be flat and not wavy. Stuccoing wasn't something that this do-it-yourselfer wanted to do, so I hired it out to a contractor who could do it right and fast. You'll find that there is a wide selection of colors available for the finish coat, and you can have a smooth or rough finish or something in between. As mentioned, windowsills and doorstops have to be built out, but not

sists of a nylon insert that is set into a ¼-inch hole drilled through the foam into the masonry. Then an equally long nail is driven through the center of the insert to set it. The assembly has a shear strength of about 200 pounds per fastener, and that's more than enough to carry the load of

(continued on next page)

necessarily window or door frames. If you think you might have more windows than necessary on the north, east or west sides of the house, this is definitely the right time to "erase" them. If you've got a southern exposure, don't insulate it. That wall is your ticket to solar space heating. If it's blocked from the sun, there is no ticket, so you might as well insulate.

Depending on how much of the work you do yourself, a job like this will cost $3.00 to $4.50 per square foot, and you can expect that with an R-16 increase in the thermal resistance of the wall, that cost will be paid back in energy savings in eight to ten years, depending on the climate and on what kind of energy is used for space heating. In my house about 1800 square feet of masonry wall was covered (and so were five windows). At about 100 pounds of brick and mortar per square foot, this job "moved-in" about 90 tons of thermal mass. The walls are a lot warmer now because the insulation has reduced conductive heat loss by about 80 percent. With that the house is a lot less shivery in winter and a lot less sweaty in summer.

Joe Carter

References

Langa, Frederic S. 1980. Insulate on the outside. *New Shelter,* April 1980, pp. 41–47.

Michels, Timothy I. 1979. Results: the retrofit of an existing masonry home for passive space heating. In *Proceedings of the 4th National Passive Solar Conference,* ed. Gregory Franta, Newark, Del.: American Section of the International Solar Energy Society. pp. 564–567.

house, you can think about adding masonry over wooden floors or adding it over interior partitions.

Tile, brick or stone can be installed over wood floors by laying a bed of mortar or using a thin-set cement or mastic. The highest-mass, lowest-cost floor we've seen was built using concrete block wall toppers. These 1½-inch-thick concrete rectangles were laid over a wood floor, and with or without masonry paint they make a surprisingly good looking mass floor. They add about 15 pounds per square foot to the floor. A high mass floor can also be added by pouring a slab of concrete, and by laying polybutylene pipe in the floor, the slab can be heated using a solar water heater or woodstove water coil.

If you have some nice hardwood floors and don't want to cover them up, you can get the needed mass by adding stone, concrete, block, adobe or brick as a new facing over existing walls. If a new wall is being built, you can make it a mass wall if the floor joists can be reinforced to bear the added load. An easy-to-build mass wall is a surface-bond concrete block wall. Simply stack up the standard concrete blocks and then trowel on the surface bonding in accordance with the manufacturer's instructions. For added mass the cells of the blocks can be filled with sand. A natural stone wall is more difficult to build but can be very good looking, as can new or recycled red brick. Artificial stone is easier to work with because it is uniform in thickness and lighter weight. It is usually glued to the wall and then grouted. Most artificial stone is concrete-based with a surface finish that looks like stone, although you always know it's artificial.

Tanks, Drums, Cans and Culverts

In many cases, water is the best type of mass to add. The wide range of possible containers

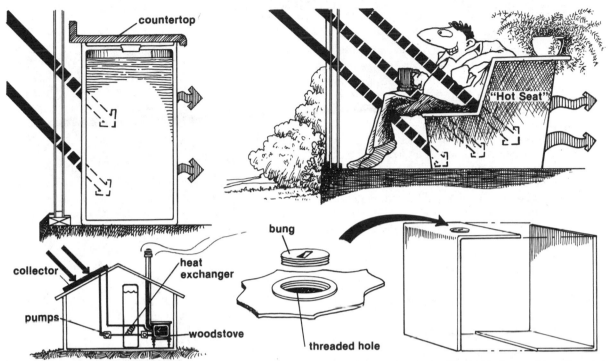

Figure III-57: Among many possible uses for custom-made steel tanks are the water-filled "kitchen cabinet" and the "hot seat" where people can heat their own thermal mass along with the water mass. At lower left is shown a more active possibility for storing heat in thermal mass that isn't exposed to direct sunlight. Water can be circulated between the tank and a flat plate collector on the roof, or it could be used to store some of the heat produced by a woodstove.

allows flexibility in fitting the mass in unobtrusively, and water stores the most Btu's per cubic foot. One of the best containers is a simple steel tank welded together out of plate steel. You can usually get a local welding shop to make up a tank to your specifications and deliver it to your door whence it can become a support for a counter, a half-wall room divider or whatever suits the situation. This and other metal containers should be painted inside and out with red lead primer and an acrylic enamel. Since a 2 by 3 by 6-foot tank weighs about 400 pounds, sloshing the paint around the inside can be difficult, but it's absolutely necessary to get complete coverage inside to avoid disastrous rust-out. Moving these tanks into place can also be interesting. For a 400-pound tank, you'll need dollies and plenty of ready muscle power when you're moving it.

Fifty-five gallon drums can be substituted for a steel tank. They are easier to move, and at $15 or less per reconditioned drum they're one of the cheapest storage containers around. If you stack them, be sure that you have access to the fill hole and that the hole is at the top before you strap, block or clamp them in place. Rectangular 5-gallon cans of metal or plastic are even easier to handle than drums and can be used in small spaces or where the house structure can't handle the concentrated weight of drums or a big tank. If you use recycled cans, thoroughly clean and paint them and test them for leaks before bringing them inside the house. A water pipe or culvert can also be used as a container for water mass. (See the box, "Converting Culverts to Containers.") Plain large-diameter pipe (irrigation pipe) can often be found used at prices that are con-

(continued on page 449)

Converting Culverts to Containers

With a little imagination and some ordinary steel culverts, the do-it-yourselfer can create attractive water walls for storing solar energy that can do double duty as room dividers, partitions, parts of furniture, book shelves, and whatever your imagination comes up with. When properly placed in a house, these solar pillars, unusual as they are, can look as much at home as columns on the Parthenon. Culverts are helical steel tubes, and when painted dark colors, installed vertically and filled with water, they absorb and store solar energy when the sun shines on them. In summer, the cylinders should be shaded so they only absorb internal house heat and thereby help to keep the indoors cool.

One of the first homes to use culverts in this way was the Dickinson House in Chico, California, which was designed by Living Systems of Winters, California. It features a large skylight and south-facing windows that allow the sun to penetrate culverts that are placed against the north wall of the living room. The culverts are painted a nice blue color that blends with the room and makes them very much a part of it. The Thigpen/Hunt residence in Davis, California, also designed by Living Systems, has seven 16-foot-tall water-filled culverts stationed directly behind a two-story window wall. Painted dark brown, this array is a strong but not overpowering element in the living room. These are examples of new homes, but it's certainly possible to install culverts into existing buildings. One example is the culverts that were installed at the Rodale Press North Street Design Department offices. Twelve culverts were used not only as thermal mass, but also as the structural support for office work space dividers.

Nearly everyone who has used culverts as containers would probably agree that the best culvert to use is made by Armco, Inc, which has offices throughout the United States. The Armco culvert has a welded seam that makes it watertight. Other manufacturers use rolled seams that are fine for life under a road but not so good

Photo III-66: In this new solar house, these culverts serve as a room divider. Needless to say, they have a striking, though not overwhelming, appearance.

when absolute watertightness is a must. If you don't have a listing for Armco in your phone book, write to Armco, Inc., Construction Products Division, P. O. Box 170, Middletown, Ohio 45042 or telephone: (513) 425-6541 for the address of the nearest distributor. These gal-

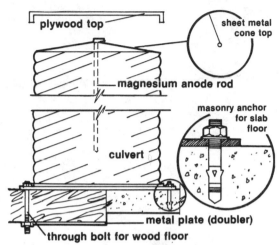

Figure III-58: Making a container out of a culvert starts with welding on a base plate to seal an end. The cone top supports the anode rod, which is available at plumbing supply outlets. The plywood top makes a flat surface for plants or other decoration.

vanized culverts come in 12-, 15-, 18- and 24-inch diameters and in stock lengths of 14, 20 and 24 feet. If you can't locate the Armco brand, all is not lost. It is possible to line a culvert with a big 6-mil polyethylene plastic bag that has been heat-sealed to make it waterproof. The researchers at Rodale's Design Department recommend seeking out a local swimming pool liner company and having the liner custom-made, or a packaging company might be able to do the same thing. Another excellent liner material is a synthetic rubber called Hypalon, which could be custom-fabricated into a bag that would last forever.

It is not difficult to securely install a culvert if you have a concrete slab floor. Bob Flower at the Design Department suggests welding an 18-inch-square steel plate (⅜ inch thick) onto the bottom of the culvert. Holes are drilled in each corner of the plate, and then the unit is secured to the floor with ⅜-inch expansion bolts that are 6 to 8 inches long. In homes with suspended (wooden) floors, the metal plate can be bolted into the floor joists, but keep in mind that mounting culverts on a suspended floor will probably require reinforcing the joists to carry the additional load. To ensure a rustproof seal at the base of the culvert, pour in a quart of catalyzed epoxy, and let it harden before adding water. David Bainbridge, founder of the Passive Solar Institute, suggests placing a steel plate below the joists and through-bolting to it. Or, he says, it might be easier to build up a small slab beneath the floor to mount the culvert on, if there's not too much space between the ground and the floor. He also recommends wrapping the base of the culvert in insulation if the floor under the culvert isn't insulated. For people living in earthquake-prone regions, he cautions that installations should include very secure cross-braces (at the top and bottom of the cylinder) that attach to a strong partition. The top of the culvert can be insulated by floating a disk of Styrofoam on the water. Evaporation can be minimized by topping the culvert with a metal cap.

In his *Passive Solar Energy Book* (Emmaus, Pa.: Rodale Press, 1979), architect Edward Mazria recommends that water walls be located so they receive direct sunlight between 10:00 A.M. and 2:00 P.M., and the sun-exposed surface of the culverts should be painted a dark color. A good rule of thumb is to try to use about 7½ gallons (1 cubic foot) of water for each square foot of south-facing glass. A resourceful person can install water-filled culverts into a home so that they are both aesthetically pleasing and functional, and in the dead of winter when they are radiating free heat, these solar pillars can look downright majestic, especially when you know, and feel, what they're doing for you.

Margaret J. Balitas

Photo III-67: Welded tanks that are custom-sized will give you the freedom to do exactly what you want with your water thermal mass. Here they replace kitchen cabinets as a counter support. Taken a step further, a sitting bench-tank could be placed in front of the counter to make a place for people as well as thermal mass.

Photo III-68: These honey can shelf supports are the same as those used in the previous water-wall project. Small containers like these can be stacked in any number of ways to create an interesting as well as a thermally effective moved-in mass.

siderably lower than the cost of new culvert.

You can combine thermal storage with gardening, using a planter as an effective and attractive way to add mass to a sunroom. It can provide the added benefits of early starts for garden vegetables or be a source for leafy greens in winter. The choice of container is up to you; your taste and budget will determine what will work best for you. The simplest and least expensive planter is a garbage can or two set on large casters to make moving easier and to facilitate cleaning. A large wood planter can also be built of exterior plywood or boards. These can be octagonal, rectangular or whatever. The wood will reduce the effectiveness of the soil as thermal storage, however, as the heat transfer through wood is relatively slow. Or you might simply decide to build sturdy wood shelves and use cut-off wine jugs or water bottles for the planters. A four- or five-shelf planter would allow you to grow a variety of plants in a small space and provide considerable thermal mass.

Whatever type of mass you choose will improve the comfort of your living space and probably lower your auxiliary heat bills, too. Thermal mass is what makes passive solar possible, providing the thermal flywheel without finicky machinery and complex controls. If you plan to enlarge your glazing to let in more sun or just want to take better advantage of solar gain that's now being wasted, move in the mass.

David Bainbridge

References

Bainbridge, David; Corbett, Judy; and Hofacre, John. 1979. *Village homes' solar house designs: a collection of 43 energy-conscious house designs.* Emmaus, Pa.: Rodale Press.

Browne, J.S.C. 1966. *Basic theory of structures.* London: Pergamon Press.

Parker, Harry, and Hauf, Harold D. 1975. *Simplified engineering for architects and builders.* 5th ed. New York: John Wiley & Sons.

AIR HEATING
E. COLLECTORS

Photo III-69: This array of air heating collectors was site-built and custom-sized to fit the available wall space. When direct-gain solar windows aren't the desired solar solution for a frame house, the next logical option is to go with collectors like these.

Hot Air

Sometimes history has a nice way of repeating. A few decades ago forced-air central heating systems were developed for the residential market, and it didn't take long for them to become one of the most popular. Heating air and moving it through ducts and registers was simpler and cheaper than heating water and moving it through pipes and radiators. The early systems used natural convection to deliver hot air from huge furnaces equipped with fat duct systems. They looked like robots. Then blowers were added; everything became streamlined, and forced-air systems took the lead in home space heating. Nowadays solar air heating collectors are one of the most popular solar space heating systems. They're simple and relatively low in cost. They can be hung out from windows or hung onto walls; they can occupy entire walls or roof pitches. They can be run by the sun-driven force of natural convection or by electrically driven fans and blowers. Being as simple as they are, hot air systems are just right for do-it-yourself retrofitting. Beginners can begin with a window box or something called a TAP; these are weekend projects that'll bring instant rewards. Hot air systems are forgiving. If a big collector array has a leak, there's no threat of puddles or water damage, and leaks can be fixed with caulk and tape and no panic. Gas furnaces, electric furnaces, solar furnaces —they all do the same thing, but these days, or from now on, solar hot air takes the lead in economical home heating.

Photo III-70: To get a better tilt angle for his collector, the author built an extension duct to raise the window box off the shallow roof pitch. Window boxes are typically light and easy to move around, and if the need or desire arises, they can be seasonally stored. (Photograph courtesy of David Chesebrough.)

PROJECT

A Window Box Room Heater

In the world of air heating collectors, this is the simplest of the simple. Hang a collector out one of your south windows, and you've got solar space heating. Natural convection circulates room air through the unit, which consists of single or double glazing, a blackened aluminum absorber and some insulation. Cool inlet air flows through the back of the collector, which is separated from the absorber with insulation. Then the air passes between the glazing and the absorber, where it's heated, and finally outlets back into the room. It's a daytime heater, a weekend project, a good place to start with solar retrofitting.

As an architecture student in Pittsburgh, Jim Halpern had been involved with solar design, but it wasn't until he saw a magazine article describing window box heaters that he realized his own front windows were a natural for a simple solar retrofit. On a whim, he constructed a window box unit to try out the idea.

Halpern's main design problem was adjusting the collector plans to accommodate the setback of his window from the south wall. An insulated horizontal duct was the answer, allowing the collector to sit at a 47-degree angle on a porch roof that was pitched at 32 degrees, not a good angle for winter collection. The arrangement works well and is so easy to build that Halpern is aiming to organize community workshops because he's certain that anyone can learn to build them.

The window box itself is a single-glazed, flat plate collector. The sides were made with 1 × 8 lumber, and all other wooden parts were cut from a sheet of ⅜-inch CDX plywood. (Halpern later realized that it is possible to cut the pieces for the entire collector from a single sheet of plywood.) The pieces were joined in simple butt joints secured by wood screws that Halpern sealed with Sikaflex, a dense urethane sealant that is comparable to silicone, but cheaper.

Halpern lined the sides and back of the collector with pieces of ¾-inch Styrofoam rigid insulation, but because Styrofoam degrades at 180°F (82°C), a better choice for insulation is fiberglass or rigid polyurethane or isocyanurate foam. Another piece of rigid insulation divides the warm and cold air channels. It's important to minimize heat gain into the cold air channel, and the foam divider serves this purpose.

The aluminum absorber can be scrap sheet stock, corrugated roofing or siding. Both sides are painted with flat black paint. (See the following project, "The TAP: A Modular Air Heater," for procedures on painting aluminum.) The edges of the absorber are sandwiched by the side wall insulation, and it's supported in the middle with a piece of wood ripped from a 1 × 4. The foam divider is also supported longitudinally in the same way. The glass glazing rests on the edges of the collector box with silicone caulking used to seal between the wood and the glass. Then the perimeter is capped with standard metal or wood corner molding that's pressed over a strip of butyl tape (see figure III-59).

Because Halpern's collector was going to be hung out over a shallow-pitched roof, instead of being run at a steep pitch to the ground, he needed a duct to increase the pitch of the collector and to connect the collector with the double-hung window. The duct was made with exterior plywood and lined with rigid foam. To keep the cold and warm air channels separate, the duct also has a foam divider. The duct was secured to the collector with plywood gusset pieces and wood screws. After sealing all of the exterior seams with urethane caulk, he painted the collector black, and in bright yellow he added the word "EXPERIMENTAL" to catch his neighbors' attention. It has!

The free end of the duct was placed inside an open triple-track storm window, and the bottom of the collector was secured to a 1×2 strip fastened to the porch roof. The duct was sealed to the window opening with Styrofoam strips. When the storm window was closed, all the seams were caulked to make a total seal. It's very important to have an absolute seal between the window box and the window or infiltration losses will negate solar gains. When the collector is not in operation, the double-hung window closes in front of the duct opening to reduce building heat loss. All in all, construction time totaled about eight hours, and Halpern didn't use anything more than hand tools and a circular saw.

Operation and Performance

This collector is a *backpass* design wherein the heated air flows behind the absorber. Other designs are *frontpass* (flow between the glazing and the absorber) and *doublepass* (flow on both sides of the absorber). Sunlight striking the absorber plate heats the air in the channel behind it, and this air rises out of the collector and into the room. As the heated air rises, it pulls cooler room air into the bottom channel through the inlet opening. This natural convection cycle continues as long as the sun shines on the collector. Halpern says there is enough available energy to drive the convection cycle even on bright hazy days, although the output temperature is lower. Intermittent cloudy days with patches of blue sky produce an on-again, off-again heat output. On heavily overcast days, the collector sits idle.

A nice feature about the window unit is that it is self-dampered, which means that the unit can remain open to the room throughout winter, al-

though to cut conductive heat loss and infiltration around the collector, Halpern keeps the double-hung window closed until the sun rises. At night, or during cloudy periods, the cool air settles to the bottom of the collector and all circulation stops.

Measured output temperatures have been in the range of 100 to 140°F (38–60°C). The calculated heat output averages 500 to 550 Btu per square foot of collector per day for the Pittsburgh climate, so during a heating season, this collector produces about 140,000 Btu per square foot. A seasonal output of about 1,250,000 Btu represents the following yearly savings using January 1980 energy cost figures from the Pennsylvania State Energy Office: natural gas, $6.30; fuel oil, $10.20; and electricity, $17.60. If you have an interest in payback times, those savings, considering Halpern's cost for materials and assuming a 12 percent yearly rise in fuel costs, come to roughly 7.5 years, 5.3 years and 3.4 years, respectively, for users of natural gas, fuel oil and electricity. Halpern's small 9-square-foot unit cost $70 ($7.77 per square foot) for all new material with $25 for glass being the costliest single item. Salvaged glass could lower the cost considerably. The 3-foot duct on Halpern's unit also inflated the price. The nice thing about window boxes is they can be custom sized to fit any window, and the length can be varied to suit your situation.

Overall, Halpern is pleased with his unit, and he plans to install three more in the remaining windows. The problems that did crop up were more matters of proper materials, operation and workmanship than weaknesses in the design. The channel dividers slipped because they were not secured properly; there were cracks in some of the sealed joints, and the wood warped. The Styrofoam insulation melted and shrunk, and in some places separated from the collector, decreasing efficiency. This is why other insulations are recommended. It's not only important to use insulation that will stand the heat inside the collector, but to use adhesive that's compatible with the kind you do use. Some combinations simply don't work. Duct tape or small nails will help secure pieces of insulation that might otherwise work loose.

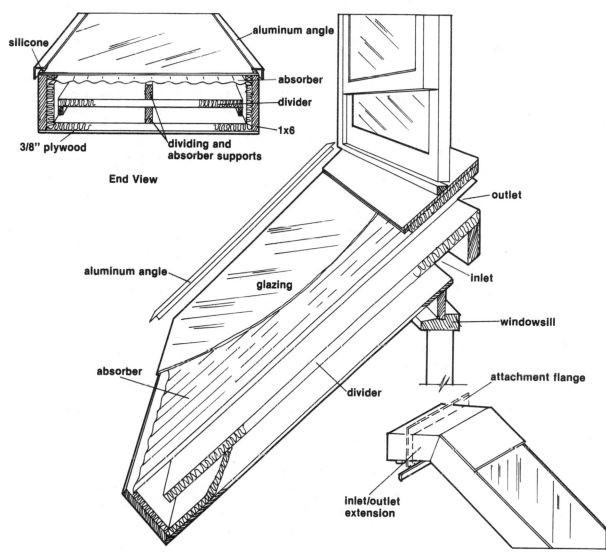

silicone

aluminum angle

absorber

divider

1x6

3/8" plywood

dividing and
absorber supports

End View

aluminum angle

glazing

absorber

divider

outlet

inlet

windowsill

attachment flange

inlet/outlet
extension

Figure III-59: Because the window box heater is a passive system, it operates at a generally higher temperature than would a fan-powered collector. Thus insulation is most important for reducing heat losses when the system is operating, since the collector itself doesn't store heat. Insulation is used to create a thermal envelope around the inside of the entire box. The divider itself is made with rigid foam to limit heat loss from the hot air channel to the one carrying cooler inlet air.

Materials Checklist

1×8 wood or plywood strips
³⁄₈" CDX plywood
corrugated aluminum roofing panel
¾"–1" rigid foam, foil-faced isocyanurate
glazing: FRP or glass

aluminum angle
foam weather stripping
butyl tape
silicone caulk

Your Window Box

Since collector area is highly variable in a design like this, the dimensions of passive air heaters must be carefully sized to allow adequate air flow from the inlet to the outlet. One researcher from New Mexico has developed a sizing system for this design. Scott Morris has been studying passive systems and natural convection for several years, and he has graciously contributed the following guidelines.

Morris recommends a collector tilt of latitude plus 10 to 15 degrees. A shallower pitch will work, but since a chimney effect provides the air flow, a steeper angle is better than a shallow one. The best orientation is of course due south with allowable departures as far as 30 degrees from true south.

The collector should be at least 5 to 6 feet long. The gap between the glazing and the absorber should be ¾ to 1½ inches. The depth of the upper and lower air channels (see dimensions CH in figure III-60) should be equal and not less than 2 inches. A depth of 3 or 4 inches will produce a slightly better air flow rate, especially for a long collector. The inlet and outlet openings (IO and OO) and the channel

crossover at the bottom of the divider (CC) should be about one-and-one-half times the depth of one channel. In general, the cross-sectional area of a channel should be at least 1/20 but no more than 1/10 of the exposed absorber surface area.

The collector can be made wider than the window and then narrowed down so that the vent openings still fit in the window frame, but this construction can cause reduced flow rates and reduced efficiency. If you try this design, use strips of wood in the channels to angle the air flow into the vent openings to prevent pockets of air stagnation. A small fan can also be used to boost flow rate. It should be sized to provide a flow rate of about 2 or 3 cfm per square foot of absorber surface.

The most common material for the absorber plate is corrugated steel or aluminum roofing painted flat black on both sides. Corrugated metal, which has more surface area than flat sheet stock, should always be mounted so that air flow is parallel with the channels. The gap between the glazing and the absorber (dimension GA) should be between ¾

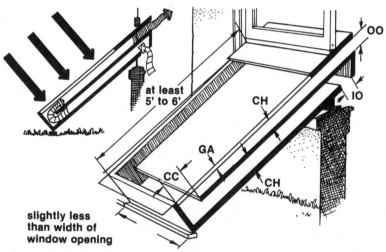

Figure III-60: The measurement codes in this illustration are described in the text. Whenever possible, size vents and air passages on the generous side of whatever your calculations show, the point being that with more space there is less flow-reducing friction between the air and solid surfaces.

and 1½ inches.

In moderate climates, greater efficiency is possible with a doublepass design in which air passes both above and below the plate. In cold climates, the backpass design we've described is best. Be aware that air flow above the absorber plate can lead to dust building up on the inner surface of the glazing and on the absorber, which can reduce heat gain. A solution would be to mount the glazing on a tightly fitting, removable frame that allows access for cleaning. Single glazing is usually sufficient in either design, but in climates colder than 7000 heating degree-days double glazing should be used. Because of the high temperatures inside the collector, the inner glazing should be tempered.

The top edge of the glazing should not be higher than the bottom of the intake opening. If it were higher, the cold night air in the warm air channel could gain enough extra head to develop a reverse air flow that would force cold air into the room through the inlet vent. It's called reverse convection or reverse thermosiphon. If it seems that cold air is coming in at night, first check the perimeter seal where the collector box enters the window. If that's tight, you may be getting some reverse convection, so cover the outlet with a piece of foam.

Exterior-grade plywood, ½ to ¾ inch thick, is sturdy enough for the sides, and ¼-inch plywood is used for the bottom. The collector will be exposed to the elements, which rules out materials like Masonite or particle board that might disintegrate. Paint all exposed wood surfaces with a good oil-base paint. Insulate the sides and bottom to R-5 or better with fiberglass or rigid foam. The channel divider should also be a rigid foam insulation. The collector is so simple in actual construction that we won't go into the construction steps. Refer to figures III-59 and III-60.

Installation

The collector should fit into a double-hung window with about ⅜-inch clearance on all sides. The perimeter is sealed with foam weather stripping and if necessary wood or rigid foam filler pieces. The window sash lowers onto the collector and weather stripping. A bead of caulk around the outside of the perimeter finishes the seal, but check this seal monthly to make sure it hasn't been disrupted. You can improve the looks and durability of your installation by making a flange to go around the collector where it meets the exterior of the sash. Closed-cell foam strips between the flange and the window sash will seal the joint as you tighten screws through the flange to the sash.

Carefully seal the gap between the upper rail of the raised sash and the glass of the top sash by stuffing the gap with fiberglass insulation. Do this even if you have a storm outside the window. Awning or casement windows of course present special problems with this kind of installation, and it may or may not be worth the bother to you to get into all sorts of sealing hassles with these window types. When the installation requires removal of a fixed storm window, use an interior storm window kit with clear plastic glazing to provide a second glazing layer above the collector outlet. One last point to remember about a ground-mounted installation is that the lower end of the collector must be held off the ground; otherwise, moisture can rot the wood. On a roof installation, place spacers under the collector and the nailing strip to allow water to flow past.

Next to direct solar gain and living in the desert, this has got to be the simplest route there is to solar space heating, and it's certainly one of the cheapest. If you build it yourself, collect all the materials and give yourself a Saturday to build and install. The already low cost of materials is discounted 40 percent by Uncle Sam's tax credit, and you begin reaping solar Btu's from day one. It's a good deal.

David Chesebrough

☼PROJECT

The TAP: A Modular Air Heater

The next step up on the air heating collector ladder is the TAP, or the Thermosiphoning Air Panel. It's a neat, clean module that is fastened to a vertical wall and is coupled with the house via inlet and outlet ports that are cut through the wall. It's a window box heater without the window, and it operates the same way: A glazed box with an aluminum absorber and a backpass air flow design delivers heated air by either natural or forced convection. TAPs are another weekend project, and by building one at a time or several at once you can line up as many as you like along a south wall. But if you want to cover an entire wall, then you may be a candidate for the next rung, the full wall system described in the next project.

Think of the TAP as a unit space heater that works independently of any other heating system in the house, quietly delivering heat to its appointed rooms whenever the sun is shining. These 3 by 7-foot modular collectors are easily home-built for vertical mounting onto the outside of an existing south-facing wall, where they supply warm solar-heated air to the adjacent rooms without controls or motors and without thermal storage. At night and during overcast days, the collectors do nothing, but since they're mounted on the outside of an existing insulated wall they don't increase heat loss at night, as do insulated direct-gain systems.

Like the window box heater, a TAP is the most basic of air heating collector systems, consisting of a black metal absorber plate built into a wooden frame that carries single- or double-glass glazing. The airspace between the glass and the absorber plate is sealed from any air movement. Air flow is in the 2-inch space between the absorber plate and the back of the TAP.

Since TAP is a totally passive collector, natural convection is the essential ingredient to its operation. When the sun hits the absorber plate, it heats the air behind the plate, which then rises in the collector and enters the room through the upper vent. Cooler room air from near the floor is drawn in through the lower vent; it heats up in the collector and rises, continuing the cycle as long as the sun shines.

There are all of two moving parts in the whole system: a manual damper and a paper flap attached to the lower vent grille that acts as an automatic one-way damper to prevent reverse thermosiphoning. The manually controlled damper in the upper vent grill is used to shut off the heater in summer. The simplicity and ease of construction of TAPs make them an attractive retrofit, especially for frame houses that don't have a large south-facing wall area.

Ruth Stigers and her two sons live in the Berkshire mountains in western Massachusetts (42 degrees north latitude). In January of 1979, Ruth noticed a headline in the local newspaper: "Free Solar Heaters to be Offered to Area Residents Under Federal Solar Demonstration Project." Her reaction: "If it works at all, I'm for it!" She applied to the program and by September of that year, the south wall of her house had been retrofitted with five thermosiphoning air panels.

The panels were designed and built by the Center for Ecological Technology (CET) a local non-profit research and technical assistance organization that wanted to get several of these systems installed to see how they would perform in the field. Ruth agreed to take daily temperature readings and make a few other measurements that would show how well the TAP worked. The south wall of the Stigers' single-story residence is oriented 4 degrees east of truesouth. The most-used spaces in the daytime—kitchen, dinette, and bathroom—are on the south side of the house, and that's where four TAPs were installed to heat the kitchen; one more was installed to heat the bathroom.

Locating the panels was a logistical feat. In the kitchen, all the collector vents had to be placed so as not to interfere with the kitchen cabinets. In the bathroom the tub was in the way, making it necessary to install the lower inlet vent to the adjacent bedroom. Cool air is now drawn from the bedroom, heated in the TAP collector, and discharged through the upper vent located in the bathroom. This installation was generally typical of the problems encountered in planning, designing and installing a TAP retrofit. By following the steps outlined in this project you should be able to anticipate similar problems and come up with a design that will work for your house.

Is a TAP right for your home? The evaluation of any solar system usually asks at least two questions: How much heat does it produce, and how much fuel does it save? A general rule of thumb for TAP performance in the Northeast is that it should deliver approximately 100,000 Btu (equivalent to about 1 gallon of oil) per square foot of collector per heating season. Ruth Stigers' twice-a-day data collection showed that temperature rise through the collector varied from 10 to 60°F (6–33°C) on sunny days

Photo III-71: The TAP module can be located so as not to interfere with existing windows or with stationary objects, like built-in cabinets, on the inside. Fortunately the interior layout in this installation allowed placements between all the south-facing windows. Painted to match the house color, these units are relatively unobtrusive additions.

with outlet temperatures as high as 130°F (54°C). On bright cloudy days, the temperature rise was usually 10°F or less. Ruth has never complained about overheating although her kitchen has gotten up into the mid-70s on some days, warmer than she normally kept it with oil heat.

An interesting note about overheating: Several TAP retrofits have produced room temperatures in the mid- to upper-80s on sunny, cold February days. However, in talking with the occupants of those houses it became clear that overheating is in the eye of the beholder. To an engineer, indoor temperatures above 72°F (22°C) constitute overheating, but to homeowners who have been struggling to heat their houses to 68°F (20°C) against the severe Northeast winters, 85°F (29°C) can be downright heavenly —especially when it's free heat from the sun, magically pouring in through what used to be a cold exterior wall.

Ruth used about 200 gallons less oil in the year following the retrofit, equal to about 2 gallons of oil saved per square foot of collector for her 110-square-foot system. Some of that savings was due to the fact that the 1979–80 winter had 11 percent fewer heating degree-days and more hours of sunshine than average. But despite the weather-related savings, it is safe to figure that the rule of thumb of at least 1 gallon of oil saved per square foot of collector per heating season is a realistic, if not conservative, expectation for a TAP used in the Northeast. As an added benefit, several TAP homeowners have found they were just as happy not heating the rest of the house as long as they have their warm solar rooms to work and relax in. Often they would shift their daytime activity to the solarized rooms, and it was possible to allow the rest of the house to "idle" at cooler temperatures. As one homeowner puts it: "Sunny days mean much more to me now. When I wake up and see that it's going to be sunny, I know I'll be able to turn my furnace off all day."

To find out whether a TAP will work for you, evaluate your south wall with the help of a compass, a tape measure and some graph paper. If

Figure III-61: The inlet and outlet vents in the TAP are offset to promote a more comprehensive air flow in the collector, thereby increasing heat collection. As with the passive window box perimeter insulation is a must for getting the highest possible air temperature at the outlet.

the wall orientation is within 30 degrees of true south, your site is okay. Next check for shading. Are there obstructions like wide roof overhangs, trees, hills or houses that shade the winter sun from the collector site? Look at the wall itself. What is it made of? In building your system, you will be cutting two vents through the wall and nailing a wooden frame onto the outside. If the wall is made of stone, brick or some other masonry material, you're probably headed for a mass under glass project described in the previous section (III-D). Wood-frame construction is the most appropriate for TAPs. You'll have to find out just how the wall is made. It may have clapboards, wood shingles, board and batten, asbestos or asphalt shingles, aluminum or vinyl siding, or stucco. Whatever the type, you will have to remove it to mount the collector directly onto the sheathing underneath. If there isn't any sheathing, it will have to be added.

Now measure the length and height of the south wall and draw the wall to scale on the graph paper. Measure the size and location of all windows and doors and draw them into your sketch. Now you can pick out potential collector locations. Each collector will require a clear wall area 38 inches wide and 86 inches high. Locate all the possible spots for the installation, and draw them in on a sketch. Examine the wall section you're considering for oil tank filler pipes, water spigots or anything that would make it difficult to remove the siding or cut through the wall or attach the collector.

Now evaluate possible locations from inside the house, deciding first of all what spaces the TAPs will be heating. Since the TAP is a daytime heater, its most sensible application is to heat daytime activity spaces like the kitchen, playroom or den. Bedrooms are probably a second choice, while storage rooms, guest rooms and other seldom used spaces should be the lowest priority. In Ruth Stigers' house, one TAP is used to heat the bathroom.

The next step is to check for obstacles on the interior wall in the same manner as you did for the exterior wall. You are going to be cutting two 6 by 14-inch vent holes through the wall. One vent will be as close to the floor as possible and the other will be near the ceiling. Look for obstacles. Usually there is no problem with the upper vent, but there may be radiators, plumbing or electrical outlets near the floor that will be in the way. Is there a woodstove nearby

that might interfere with proper operation of the collector? Kitchen cabinets are a special situation. You can install TAPs behind kitchen cabinets as long as there is free air flow through or under the cabinets in the kick space, but don't count on using long ductwork because the thermosiphoning force is too weak to overcome the increased air flow resistance created by long ducts.

But if long ducts are the only way that you can have your TAP, you do have the option of making it into a FAP, or a forced convection air panel. Simply by adding a small fan with a thermostatic control, you gain the ability to direct heated air farther away from the collector, or draw it in through a longer air duct. The fan is mounted on the collector outlet, and the thermostatic control totally automates operation. Recent studies have also shown that the FAP can have an efficiency advantage over the TAP without too much added cost. In the following construction steps we'll work in FAP details wherever they come up.

Photo III-72: In this variation on the TAP design an exhaust vent is included across the top of the collector, and an intake vent is concealed under the bottom. This collector uses FRP glazing instead of glass, and since it can degrade when exposed to high temperatures, the collector must be vented during warm weather.

Materials Checklist

two 10-foot 2×6's
two 10-foot 2×2's
two 10-foot 1×6's
one 8-foot 1×6
16d galvanized nails
6d galvanized nails
6d aluminum nails
34″ × 76″ single- or double-pane tempered glass
aluminum absorber plate
½″–¾″ rigid foam, foil-faced isocyanurate
foam gasket
2½″ × #10 wood screws
staples
6″ × 14″ vent grilles (one operable)
duct tape
frisket paper (nonadhesive)
neoprene setting blocks
reflective foil builder's paper
four tubes clear silicone caulk

paint or wood sealer
flat black paint
carpenter's glue

From a sheet metal shop:
flashing: 39½″ × 6¼″
two 6″ × 14″ metal duct boots

FAP hardware:
small 50–100 cfm fan; 4–5-inch diameter
    equipment or computer cooling fan
snap-disk thermostatic switch (which closes
    on temperature rise at 90°F, 32°C), or other
    suitable thermostatic switch
manual override switch (standard light switch
    or any type of single pole, single-throw switch
    rated to carry at least three amps)
miscellaneous wiring

Construction Steps

The materials for constructing an 18-square-foot TAP cost around $150, or $8.33 per square foot of glazing. The FAP option adds about $25 to $30 for the cost of electrical hardware. The design described in this project uses 34 by 76-inch double-pane sliding glass door replacement panels, which are relatively inexpensive because they're so common. You can easily modify the design to use 34 by 96-inch or other glass sizes or to use single-pane glazing.

The absorber plate should be corrugated to help relieve expansion and contraction stresses as well as to increase the heat exchange area. You can buy corrugated aluminum roofing panels and paint them yourself, but make sure the paint will withstand frequent temperature fluctuations from below freezing up to 200°F (93°C) or higher. Both sides of the plate should be painted black for maximum absorption and reradiation into the air channel.

Painting the absorber isn't just slapping on the paint. Proper surface preparation and priming are essential for good paint adhesion and long life. Surface prep begins with a solvent such as paint thinner or acetone to remove residual oils deposited during manufacturing. This is followed by a wash-down with detergent and then surface etching with a strong base (caustic solution). The first coat of paint is a primer coat, so use a standard paint made for just that purpose. For the flat finish coat an oil or acrylic-base high temperature (300 to 400°F, 149–204°C) paint is preferably sprayed on to obtain an even, thin coating. If you use a roofing material with a baked enamel finish, any glossiness must be lightly sanded away or dulled with liquid sandpaper prior to applying the finish coat. Heed these practices! They've been learned the hard way by having to strip off glazing and sandblast and repaint blistered, flaked-out absorbers. Neoprene setting blocks for setting the glass are available from any glass dealer. The rest of the materials list consists only of lumber, caulk and some sheet metal, all widely available and relatively inexpensive. One last item is the frisket paper (non-adhesive), used to make the one-way damper in the lower vent. This is a very light silk-screen paper available in art supply stores. It's a durable material that can withstand sun exposure and high temperatures. If you can't find it, there are several alternatives in the form of other thin film plastics that are used in solar applications. All these materials should be locally available.

Building a TAP doesn't require any special skills other than basic carpentry ability and lots of care. You'll need standard carpenter's hand tools and a circular saw. A table saw or radial arm saw is nice but not necessary. There is some sheet metal bending involved in fabricating the vent sleeves and

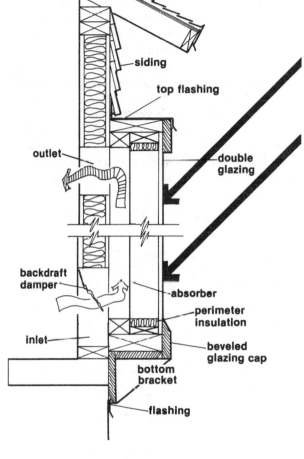

Figure III-62: As with any collector the TAP should be made as airtight as possible to keep out the weather and keep in the hot air. You can test for tightness by sending smoke through the unit and seeing if or where any escapes to the outside. Caulking is the primary means of air sealing, so use a good quality variety such as silicone, which won't dry out after years of exposure.

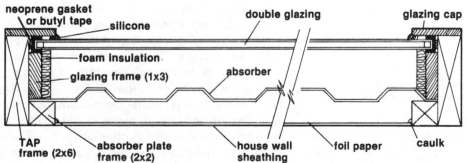

Figure III-63: This is a cross section of the TAP across the top. Since air flows behind the absorber plate, it's important to make a good seal between the plate and the 2×2's that make up the absorber plate frame. It's also important to have a reflective surface over the building sheathing behind the absorber to limit heat transfer into the insulated wall. It has also been found that if unprotected wood is exposed to high collector temperatures, it can be seriously degraded.

flashing that may be best done by a pro at a local sheet metal shop. For $5 to $10 you'll save yourself lots of grief and end up with a much neater job. The entire retrofit can be done by one person working alone, but two people can easily build a TAP in a day, and building two at a time shouldn't take much longer. By the way, of the four people who built Ruth Stigers' TAP, none had ever built one before and only one had any previous carpentry experience.

The TAP module with 34 by 76-inch glazing is built with the pieces described in the following cutting list. If you're thinking mass production, now's the time.

	Number of Pieces	Lumber Dimensions	Length
collector box	2	2×6	79½″
	2	2×6	34½″
absorber plate supports	2	2×2	76½″
	2	2×2	31½″
glazing supports	2	1×2	76½″
	2	1×2	33″
glazing caps (rip down a 1×6)	2	1×3	79½″
	2	1×3	32¾″ (bevel one to 45)
support braces	2	1×6	37½″
	3	2×6 triangles	---

1. To locate the collector, measure the distance between your floor and ceiling and transfer both levels on your exterior wall. Choose the installation area for your unit within those marks. The unit should not be above the level of the ceiling as the outlet vent must of course be below the ceiling. The lower vent should be as near to the floor as possible to permit the coolest room air to flow into the panel. The collector itself may extend a little below the floor if that's necessary to make it fit.

2. Proceed with the removal of the siding by marking two vertical lines on the house wall corresponding to the length and width of the collector box plus ¼ inch. Make sure to add 6½ inches to the length of the collector to accommodate the support brace. Wear goggles when cutting and use a blade that you're ready to sacrifice to any nails you might run into. Cut completely through the siding but not into the sheathing. After all four cuts are made, remove all the siding inside the cut lines.

You may for some reason not want to remove the exterior siding, in which case you do have the option of building in a back for the collector. The inside edges of the 2×6 collector frame will have to be rabbetted before they're assembled to carry a ¼-inch sheet of plywood. With this variation it is necessary to assemble the four sides and the back before mounting it on the wall. You'll also have to cut the inlet and outlet vent openings. To ensure airtightness, run a bead of silicone along the rabbetted edges before nailing down the plywood back.

3. The bottom bracket is made with the two 37½-inch 1×6 boards and the three 2×6 triangles. Nail the two 1×6's to each other at right angles using exterior glue along the 37½-inch seam, and nail support braces at each end and in the middle (see figure III-65). Using a level to find a horizontal line, place the bracket assembly on the wall and screw it into wall studs and/or the sheathing. Take the prefabricated piece of flashing (figure III-64) and slip it in under the siding at the top of the cut-out area.

4. Begin the construction of the TAP frame by sealing any knotholes in the 2×6's. Glue and nail together the 2×6 perimeter pieces to make the basic box. The top and bottom pieces must be fastened in between the sides. Measure the diagonals at opposite corners to make sure the unit is square. When the diagonals are equal, nail on a temporary cross-brace to stabilize the frame during mounting. Place the unit on the support bracket, then toenail through the side into the studs and the sheathing. Check the diagonals once more, and if they're equal, nail the other side.

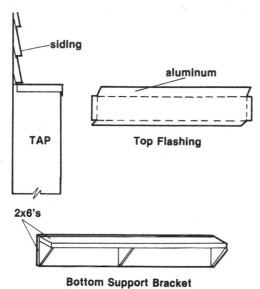

2x6's

Bottom Support Bracket

Figure III-64: The top flashing is made from a single piece of sheet metal. You can go to a local metal shop to have the bending done professionally, or you can fabricate a simple bending brake with pieces of wood. The design of the support bracket can certainly be varied to suit the look and location of your own TAP. You could, for example, use metal brackets that would eliminate the need for the triangular pieces.

A small shim between the frame and the support bracket may be helpful for fine squaring adjustments. Also, check the width of the unit for any bowing in the middle. Maintain the 37½-inch outside dimension and make adjustments by toenailing on either side of the 2×6 uprights.

The house wall itself may be bowed in or out a bit, and this can cause minor problems. Gaps of up to 1½ inches in 7 feet may be dealt with by splitting the difference, leaving equal gaps at the top and bottom. This is then covered by the 2 by 2-inch collector plate supports, which can be bent to conform to the wall surface. It's important to make sure that the collector box itself is flat, or there'll be problems when the glazing is installed.

5. Install the absorber plate frame by nailing the 76½-inch 2×2's inside the collector frame. Nail them first to the wall and then to the frame. Place the 31½-inch 2×2's at the top and bottom and nail them in place.

6. The positioning of the intake and exhaust vents is next and requires careful consideration. The upper vent must be as high in the TAP as possible while remaining at least 1 inch below the ceiling level. The lower intake vent must be as close to the floor as possible but must be installed 1 inch above the floor or the baseboard trim to accommodate the vent cover. Ideally, the vents should be located at diagonally opposite corners of the unit, for optimum air flow, but this is dependent upon the spacing of your wall studs. Don't cut a load-bearing wall stud unless you're prepared to support it properly with a little mini-header running between two adjacent studs.

Locate wall studs by finding nails in the sheathing, and mark the location of the vents, using your floor and ceiling lines as a reference point. The possibility of electrical cable should be considered before cutting into a wall, so set your circular saw only to the thickness of the sheathing and cut along the lines. Remove the cutouts and any insulation, and check inside the wall for wires or plumbing. The 6 by 14-inch vents should be placed horizontally, if possible, although a vertical placement will work almost as well.

Once vent holes are cut through the sheathing and any clearance problems are resolved, cut through the interior wall. From the exterior, drive a nail or drill your interior wall surface at the four corners of your vent opening. Go back inside, find the holes, and draw lines between them. Cut out the rectangle from the inside using a small keyhole saw or utility knife, if the wall is faced with Sheetrock, or use a circular or saber saw for paneled walls.

Once the holes are cut and clear of obstructions insert the prefabricated metal 6 by 14-inch duct boot so that the ½-inch flanges butt against the exterior sheathing. If you're going the FAP route, you might have an opportunity to fish some wiring up through the wall, through the inlet vent opening—a nice way to conceal the wire. Be sure to use wire

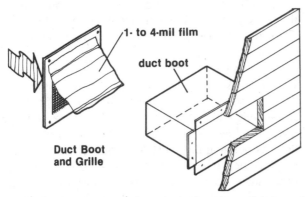

1- to 4-mil film

duct boot

**Duct Boot
and Grille**

Figure III-65: The duct boot can also be custom fabricated at a local metal shop or bought from a ductwork supplier. The grille that covers the boot should be made with hardware mesh or other open material that will present a minimal barrier to the thermosiphon air flow.

with insulation that can resist high temperatures in the 200 to 300°F (93–149°C) range. Figure III-66 shows the simple schematic you should plan to use for the FAP.

7. Use silicone or other high quality caulk to seal the seam where the absorber plate supports (2×2's) meet the sheathing. Cut the reflective builder's foil to fit the area inside the frame and staple it to the sheathing. Cut away the foil covering the ducts and tape the edges firmly to the boot with 2-inch duct tape.

8. Next comes the installation of the absorber plate. Nail a 12-inch 2×2 vertically onto the sheathing in the center of the frame. Fasten the end closure gaskets to the top and bottom 2×2 absorber supports. Most metal roof systems include some kind of die-cut gasket that mirrors the cross section of the roof panel. This is needed in the TAP to seal the top and bottom of the absorber to ensure a dead airspace between the absorber and the glazing. If a gasket isn't available, thick foam strips can do the job.

Lift the plate into the frame and make sure there is a ¼-inch gap around the entire perimeter between the plate and the collector frame. Nail the plate to the 2×2's, using aluminum nails with an aluminum collector plate. Nail about every 4 to 6

inches around the perimeter and twice into the 2×2 block in the center. Paint the nail heads with matching flat black paint and caulk around the perimeter with silicone.

9. Prepare to install the glazing supports by scribing a line around the inside of the collector frame at a depth that equals the thickness of the glazing you're using, be it double or single pane. This is easily done by setting your combination square at the appropriate measurement and using it to guide your pencil around the frame. Take the 1×2 glazing support pieces and nail them to the collector frame, carefully following the pencil line you've scribed. Attach the rigid foam perimeter insulation to the exposed faces of the glazing support pieces. The foil facing can be painted black if desired.

10. Place the neoprene setting blocks on the bottom of the TAP frame. These blocks cushion the weight of the glass. Thoroughly clean the inside of the glass, and with a helper lift the glazing into the frame. Make sure there is a ⅛- to ¼-inch space on all sides between the frame and the glass, and caulk the seam with silicone. Nail on the 1×3 glazing caps, making sure that the piece with the 45-degree bevel is on the bottom where it will shed water properly. Use silicone to caulk the seam between the cap pieces and the glass and between the cap pieces and the frame.

11. Nail the top flashing over the top cap piece and bend down and nail the 1-inch flanges to the sides and front of the collector. Silicone any areas around the flashing that may present leakage problems, especially along each side where the clapboards meet the flashing. Caulk the sides of the unit where it meets the siding and apply a final coat of paint or sealer to the TAP.

12. The rest of the installation involves placing the vent covers on the interior walls. The upper operable vent can simply be screwed into place with careful positioning so that it opens and closes freely.

The lower cover incorporates the one-way damper and requires a little more work. Place a 1-inch border of duct tape on the back of the vent cover over the louvered area. Cut a piece of frisket paper or light 1- to 2-mil plastic to cover the vent,

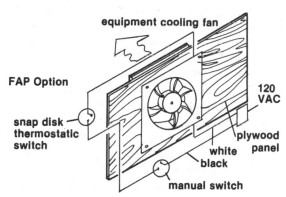

equipment cooling fan

FAP Option

snap disk thermostatic switch

120 VAC

plywood panel

white
black

manual switch

Figure III-66: In making a FAP out of your TAP, it's important that the fan be mounted on a piece of wood or plywood that completely covers the outlet opening. Otherwise it will actually draw some of its air from the room instead of the collector and less heat will be delivered.

and tape it along the top edge of the louver so that the tape functions as a hinge. The flap must be totally free to open into the duct when the vent cover is in place. Hold the cover in place and blow through louvers onto the paper. It should flutter open and fall back smoothly against the grille if it's been properly positioned. If it's a sunny day, the TAP will be operating and the flap won't return to a closed position. Close the upper vent to stagnate the unit and repeat the flap test, making any necessary adjustments. Screw the vent cover in place and once more check it for proper operation. If the height of the vent opening is more than the thickness of the wall, it may be possible for the paper or plastic flap to hit the back of the absorber plate and not become fully open. It could also impede air flow to some extent. One solution is to make a double flap that is split horizontally, thereby halving its reach.

To make your TAP into a FAP, the small fan

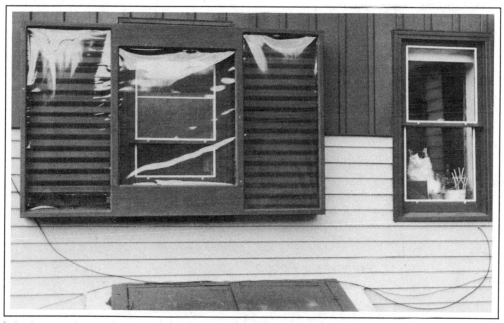

Photo III-73: This is another variation on air heating collectors that has two absorber plates on either side of a window feeding hot air through side vents and then through the window, which is open at the top. The window itself has an additional glazing layer which creates a sort of mini-manifold for the two "sidebar" collectors.

is mounted to a plywood panel sized to fit into the outlet vent. It's important to cushion all fan mounts with rubber gaskets to minimize vibration that increases noise. The panel itself should also be "shock absorbed" for the same reason. The snap-disk thermostatic switch is mounted at the top of the collector in the air passage, and the manual override is mounted at some convenient point of access in the room, perhaps near the fan. It will still be necessary to provide some means of closing the top vent in summer to prevent unwanted heat gain.

Operation

As you might suspect, operating the TAP is largely a matter of benign neglect. During winter the top operable vent is always left open, and the lower one-way damper prevents nighttime heat loss caused by reverse convection. At sunrise a convective flow is initiated, and the damper flap lifts up to supply inlet air. You'll notice that by around noon the flap reaches its highest lift, which correlates with the time of maximum solar input. Toward afternoon the air flow begins to slow down as solar gain diminishes, until finally the collector shuts down for the day. If you want to add protection against nighttime conductive heat loss through the vents, make a couple of insulation panels with rigid foam. These, of course, will require once-a-day removal and insertion. The FAP is also self-operating by way of the thermostatic switch. When the snap disk reaches 90°F (32°C), it closes and activates the fan. Depending on the weather and strength of the fan, the fan may cycle (go on and off) repeatedly or run continuously.

During summer, early fall and late spring (or whenever you don't need space heating) close the top vent to stop hot air from entering the room. And that's it. With proper weather sealing and maintenance, the unit will last for years, making hay while the sun shines, with little scrutiny required to keep it going.

Ned Nisson and Anastas Pollock

References

Center for Ecological Technology. 1980. *The story of a passive solar retrofit: TAP.* Pittsfield, Mass. Available from the Center for Ecological Technology, 74 North St., Pittsfield, MA 01201. Telephone: (413) 445-4556.
Fifty color slides and text of step-by-step instructions for building a thermosiphoning air panel.

Hughes, Ron. n.d. *Simple solar air heaters.* Available from New Life Farm, Inc., Drury, MO 65638.

Trutek, Inc. 1980. *Solar air heater.* Available from Trutek, Inc., P. O. Box 9068, Wichita, KS 67277.

Wolf, Ray. 1981. *Solar air heater.* Emmaus, Pa.: Rodale Press.
This book gives step-by-step instructions for building and installing a solar collector that mounts either vertically or horizontally onto solid masonry or standard frame walls. The collector, made of commonly available materials, requires simple carpentry techniques to construct. A complete tools and materials list and operating instructions are included.

Hardware Focus

Duct Grilles

Hart & Cooley
500 E. Eighth St.
Holland, MI 49423
(616) 392-7855

Hart & Cooley Catalog SA
Steel and aluminum registers, grilles and diffusers, all in a wide variety of sizes.

PROJECT
Walls of Warm Air
with Large Collector Systems

Here is a way to convert an entire wall into an air heating collector system, or half of a wall, or a third. Whatever area you'd like to devote to solar space heating, a large collector can be built to fit any size and shape. Like the little FAPs discussed in the previous project, this is a forced-air system, and because it is you can send solar-heated air any which way you want, up, down and sideways. And like any furnace it can be connected with an existing duct system, or new ductwork can be installed to deliver heat to distant rooms. These systems are big, but they're not complex. Planning is important because the ultimate design depends primarily on the wall in question. In this project we'll see how two do-it-yourselfers came up with different designs for different applications.

By the looks of early spring, winter in northern Nebraska must be as tough as the best of them. The muddy brown landscape has been laid low by snows and bitter winds, and farmers can only await warmer weather before bringing out their tractors and seeds. So winter is a time that many in these parts just want to "get through." Snow- and mud-bound farmers pass the time tending livestock, going over equipment and, of course, keeping warm.

It never used to take much to stay warm, but lately the prices of fuel oil and propane have been climbing above the cornstalks. The current 20 to 50 percent annual increases in energy costs make a farmer's net gains, if he's lucky enough to have any, noticeably slimmer, enough to make saving energy worth thinking about.

Enter the Small Farm Energy Project (SFEP) and their solar-wall air heating collector. SFEP (P. O. Box 736, Hartington, Nebraska 68739) is a nonprofit assistance group formed to help small farmers, already threatened by taxes, agribusiness and inflation, to reduce farm energy use. The wall collector is one of several solar and conservation options that SFEP has made available to farmers. It has proven itself to

be effective without being expensive.

We'll look at two different collectors installed by farmers on their houses to show the flexibility of the system. When you come to planning your own system, these examples should show that you're not "locked in" to a single configuration. The collector at Paul Phelps' house is the smaller of the two at about 120 square feet, and it provides direct heating from collector to house. The other system was built by Ken and Jan Stark and has nearly twice the surface area, 220 square feet, of the Phelps' system. The Stark collector is large enough to require a more roundabout air flow pattern than the Phelps' straight-through design, and also to warrant the addition of a heat storage component, in the form of small stones, to make more efficient use of the larger amounts of Btu's produced.

Before Paul Phelps built his collector, he knew what he wanted in a solar system: "Something that doesn't look funny and isn't sophisticated. You know, the worst place to have sophisticated machines is on a farm." Eight years ago he traded being an industrial engineer in Chicago for 480 acres and a run-down farmstead. He's been "busier than heck

Photo III-74: The Starks' collector shows what can be done when the goal is to cover most or all of the available south-facing wall area. The basic design of this collector can be applied to just about any wall size or shape, and similarly the way the hot air is handled can be customized to the house and the owner's needs.

ever since,'' and he loves it, seven days a week.

 With the farm's original old, gray farmhouse sagging serenely into oblivion, Paul's first big project was to build a new home. His wife, Wilma, was chief designer, and she allowed her pencil and ruler to be guided by her intuition. The result was a house built with a broad south face and a lot of window area for passive space heating. ''When we come around to the winter solstice,'' said Paul, ''the sunlight makes it all the way back to the fireplace wall. And that's a few ton of stone getting heated.''

 The solar-wall collector was added in the fall of 1978, and it easily passed Paul's ruling against complexity and weird looks. Simple it is. A look at its cross section shows it to be basically the same back-pass air heating collector shown in the previous project. In this case, however, the forced air system moves air horizontally, conforming to the horizontal shape of the collector.

 As the photograph of the Phelps' collector shows, there wasn't much space available for locating the collector, since much of the south wall had already been spoken for in windows and doors. Paul's aim for simplicity was held intact by not having to build the collector around any windows, although the building's heating load could certainly use the output of the extra collector area.

The horizontal layout of the collector called for a simple, end-to-end air flow pattern from the inlet across to the outlet. As with the TAP module, this collector is coupled with the house by way of holes cut through the frame wall. In the Phelps' collector the cool air inlet is located in a little-used, and therefore unheated, bedroom. The outlet vent penetrates the wall and then is run through the first floor into the family room. This is where Paul wanted to put the heat. The heated air is distributed throughout the family room, but it doesn't stop there. Because of the open interior plan of the house, some of that

warm air rises by natural convection up into the rest of the house via a wide staircase.

The blower is located at the collector outlet so that room air is "sucked" rather than pushed through the collector. There is good reason for this configuration: The collector operates at a negative pressure, and were there any air leaks to the outside, the cold air would be drawn into the unit where it would be heated. If the blower were at the inlet, the collector would be pressurized, and any leakage would send valuable hot air to the outside. It's a small but important detail, and hopefully you won't have to depend on it to get the most heat from your collector. The goal of construction is really to end up with zero leakage, no matter what direction, and that goal is realized by thinking "seal, seal, seal" every step of the way in construction.

Materials Checklist

2×4's and 2×2's for perimeter framing, baffles, blocks and internal glazing supports	glazing: FRP, glass, etc.
reflective foil builder's paper	miscellaneous nails
corrugated aluminum sheet roofing material	gasketed roofing nails or screws
thick, closed-cell foam strips	wood or aluminum flat battens and corner pieces
flat black paint	squirrel cage blower
silicone caulk	remote bulb thermostat
aluminum pop rivets	operable registers or grilles

How It's Built

The basic construction of the collector starts with the perimeter framing that defines the collector area (see figure III-68). We'll call this the absorber support. It is ripped to a certain thickness that is determined by some simple design calculations that are presented later in the project. Basically, in any forced-air system the size of the air passages can't be too big or too small. Nor should the blower be under- or oversized. By calculation and trial-and-error SFEP has come up with sizing rules of thumb that help to ensure optimum collector output. The vents are cut between wall studs, and in the Phelps'

system the outlet was opened to the lower floor by cutting through the stud-wall sole plate and the rim joist below it. The absorber used was corrugated aluminum roofing material, and before it was fastened to the absorber support a single horizontal baffle was nailed to the wall. The baffle helps to "organize" the flow of heated air and literally keep it on track. As we'll see in the Starks' system, baffles are vital to directing the air in several directions so that there are no heat-wasting dead-air pockets.

The sheets of corrugated roofing material are cut so that the corrugations run perpendicular to the direction of air flow. The thought here is that more turbulence is created if air is blown across the

Photo III-75: The Phelps' new home originally had a generous area of south-facing glass for direct-gain space heating. To collect more solar energy the horizontal air heater was built to fit the only available space along their deck. The job was somewhat easier and cheaper than adding more windows.

ripples of the absorber, and this turbulence increases the extraction of heat. This detail differs from the TAP design, where the goal is to minimize resistance to the gentle, natural convection air flow. Because more than one sheet of roofing is used in these larger collectors, all the seams between sheets must be sealed to preserve the integrity of the backpass air flow. After both sides of the absorber sheets are painted black (per the instructions given in the previous project, "The TAP"), they are fastened one by one to the

collector and to each other. A high-temperature caulk, such as silicone, and aluminum pop rivets are used to seal corrugated edge to corrugated edge. Where a corrugated edge lies over the absorber support, there are a couple of possible ways to make a good seal.

In some areas it's possible to buy strips of wood that are "wavy cut" so as to mirror the shape of the corrugations. These strips can be used with silicone to seal the edge. Some metal roofing systems

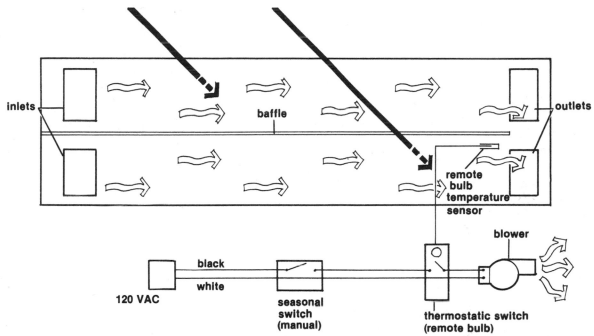

Figure III-67: This schematic shows how the Phelps' collector is wired for automatic operation with a manual override switch for seasonal operation. Because there is one horizontal baffle, there must be two inlets and outlets in the collector itself, though they are actually served by one inlet and one outlet vent on the room side.

incorporate a die-cut foam "closure strip" which, like the wood strip, conforms to the cross section of the roofing material. If neither of these two items is available, you can use thick, closed-cell foam strips to get the job done. One source might be recreational vehicle stores, which sell these strips for sealing camper tops to pickup trucks.

  With the absorber sheets in place, a second run of perimeter framing is nailed on outside the first. We'll call this the glazing support. It is ripped to a width that creates about a 1½-inch gap between the glazing and the absorber depending on the calculated thickness of the absorber support. The seam between the absorber and the glazing support is also caulked, and so is the seam between the glazing support and the wall of the house (seal, seal, seal!).

  Both the Phelpses and the Starks used a single layer of FRP to glaze their collector, although

just about any kind of glazing could be used, such as double-wall extruded acrylic or glass. As a semirigid glazing, FRP needs internal as well as perimeter support, and this is achieved by fastening internal glazing supports *over* the locations of the baffles that are on the underside of the absorber. For example, the Phelps' collector has one horizontal baffle running the length of the collector. After the absorber was put in place, the glazing support was screwed (not nailed; that deforms the corrugations) through the absorber to the baffle. The thickness of internal glazing supports is such that they are flush with the perimeter glazing support.

  The Phelps' collector was made a little less than 4 feet wide so it could be glazed with a single 4-foot-wide roll of FRP. (With its high impact strength FRP was a good choice for the Phelps' collector given its proximity to their much-used deck.) First, the leading edge of the roll is tacked to the collector with small ring-shank nails. A bead of silicone is run between the glazing support and the FRP. Then the glazing is unrolled and tacked in one direction only to ensure that it lies flat with no built-in ripples. The tacking is a little tricky because it's best to predrill the FRP instead of just nailing right through the material. Predrilling helps to avoid creating little fractures that might become bigger fractures. The same fastening is done to the internal glazing support(s), although there's no need for silicone between the support and the FRP. If the fastening points on the internal glazing supports are not to be covered with a batten, as are all the perimeter edges, then a gasketed roofing nail or screw should be used in place of the ring-shank nails. The collector perimeter is capped with a wood or aluminum corner. A bead of silicone is run between the glazing and the corner, and the fastening is done through the other face of the corner into the face of the perimeter glazing support. If the top edge is exposed to weather, flashing is essential. It should probably be incorporated during the framing (see figure III-68).

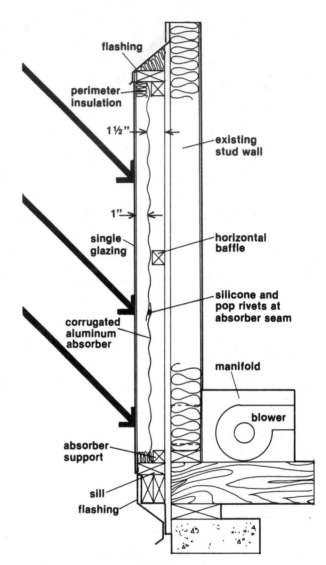

Figure III-68: The basic construction of a large wall collector is relatively simple in terms of its top and side cross sections. As with the TAP unit sealing up any possible air leak points is important, and in a backpass design like this it's important to make a tight seal between the absorber and the house wall. It may be more desirable to locate the blower out of the living space, in which case the collector outlets could lead down under the floor to a concealed manifold, with standard outlet registers cut into the floor of the space to be heated.

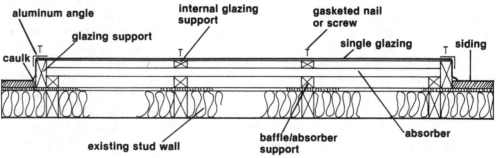

Adding the Juice

That's the collector. Now it has to be powered, and as the wiring diagram (figure III-67) shows, there's not much to it. Power is brought through a manual switch (used for seasonal control) to the thermostatic control and the blower. Using the right blower is a matter of a little calculation, based on the guidelines that are presented shortly. Basically, the more collector area you build, the bigger the blower has to be. There are a few options for the thermostatic control. With a $35 remote bulb thermostat (Honeywell T675A or equivalent) the sensing bulb can be bonded or mechanically affixed to the absorber plate near the collector outlet. The thermostat switch box can then be located near the blower so it's accessible. With an adjustable thermostat the "blower on" temperature setting can also be changed for fine-tuning the system. If, for example, the setting is too low, the velocity of the blower-driven air will make the air feel cooler than it really is. If the setting is too high, heat is wasted because the blower doesn't come on soon enough. It's also possible to vary the speed of the blower by installing a special speed controller.

Five dollars, on the other hand, gets you a simple "snap-disk" thermostatic control, which is simply a switch controlled by temperature. It is also mounted at the outlet, but it isn't adjustable. Instead, it is available in preset switching temperatures; a 100 to 120°F (38–49°C) switch-on (the switch closes on temperature rise) is desirable for these collectors.

Another important control to have is operable registers at the collector inlet and/or outlet, whichever is located low on the absorber. At night during cold weather cold air will want to "drop" out of the collector into the house, possibly drawing warm room air in the collector to replace it. Without vent control, such "reverse thermosiphoning" can steal back much of the day's collected heat. These registers can be manually operated or spring-controlled louvers that open by the force of the blower.

Because maximum control over the collected solar heat is the name of the game, it's necessary to have insulation in place behind the collector.

If not, some of the heat radiating from the back of the absorber plate will migrate through the wall into the room. That may not be so bad during cold weather; all that's being sacrificed is an element of control and therefore some efficiency. During the warm months, however, the collector can heat up if it's not shaded, and without some thermal resistance (insulation) between it and the room there will be some unwanted interior heat gain. It's been assumed here that the building's stud wall has already been insulated. (You're not adding solar heating to an uninsulated house, are you? Tsk, tsk.) But if for some strange and unusual reason the collector wall isn't insulated, the insulation should be added either to the stud-wall cavity or placed behind the collector when it's built. Rigid foam can be used to save space. It should be covered with thin plywood and reflective foil builder's paper. No insulation should be left exposed directly to the absorber plate; rigid foam can degrade and fiberglass will get picked up by the moving air, not healthy.

These larger collectors will also need to be ventilated during warm weather when heat isn't being extracted for space heating. It's likely that the collector wouldn't be totally shaded by overhangs on tree leaves, and there will be enough diffuse and reflected (and a little bit of direct) solar gain to heat up the collector considerably. FRP glazing in particular is susceptible to degradation at temperatures over 200°F (93°C), which can be reached in a *stagnating* (unventilated) collector. All that's needed are a couple of small, screened openings, an inlet and an outlet that can be sealed shut during the heating season and left open when it's warm.

The Starks' Collector

We've talked about how the design of this collector can be varied to fit any situation presented by a wall; now we can show its flexibility by looking at the design of Ken and Jan Starks' system. As the photograph shows, it's clearly a more ambitious undertaking: It's almost twice the size of the Phelps' collector; it's built around, not over, windows, and as we'll see the air flow pattern is rather complex. But we'll begin at the beginning and see how this design evolved.

In a sense, the beginning was a propane-fired, forced-air space heating system. The fuel was starting to cost too much for a winter's worth of comfort, and the Starks wanted to, had to, do something about it. Their first steps were a complete insulation job for the house, walls and ceiling, and this improvement alone cut their heating bill in half. When they heard that the SFEP could help them build a low-cost solar heating system, they were ready, and so was their house.

The size and layout of this system was again primarily dictated by the available wall space the owners wanted to devote to the collector. The Starks were looking at building a 16 by 17-foot collector (220 square feet of net collector area when the window area was subtracted) to almost totally cover their largest south-facing wall. Because of its size, the fact that it had to be built around the windows to preserve interior ventilation in summer, and the fact that both the inlets and the outlets were to be placed at the bottom of the collector, the air flow pattern could not be a simple straight run. The pattern had to be "shaped" by baffle placement to force air through the entire collector to prevent wasteful dead-air hot spots. As figure III-69 shows, the baffles direct air up and down, with a little sideways flow as well. The east and west halves of the collector essentially mirror each other, so we can discuss one half to show how the whole thing works. The three inlet vents draw cool air from the basement, and the three outlets deliver heated air to the living room on the first floor. The blower is installed in a manifold that connects the six outlet vents on both halves of the collector.

The horizontal lap siding on their wall presented a small problem when it came to installing the perimeter absorber and glazing supports. In older houses that don't have sheathing, the supports can be custom-cut to match the profile of the lap siding. In houses that do have sheathing a groove can be cut in the siding wide enough to accommodate the com-

bined width of the glazing and the absorber supports (3 inches when 2×2's and 2×4's are used). In both cases the edges must be thoroughly caulked.

The Starks' lap siding also presented an overly irregular surface for the backpass air flow, and it had to be smoothed out. This was done with discarded sheet aluminum press plates tacked over the siding. In other cases the siding could be removed if there is sheathing underneath, or it could be covered with a thin, cheap wall paneling that would be covered with reflective foil builder's paper. The reflective surface helps to resist the conduction of heat into the stud wall, where it isn't as useful relative to having it delivered directly to the living space.

Whenever possible the baffles were fastened to the existing wall studs. As with the Phelps' collector the internal glazing supports are fastened through the absorber and into the baffles. The stock width of the glazing material being used should also be considered when planning the layout of the baffles and the internal glazing supports. This is why some single blocks are shown in figure III-69 in line with some of the baffles. They are there to provide a fastening point for the internal glazing supports. Note too that in places where the baffles form right angle corners, a hole is drilled into the horizontal baffle to allow some of the air to "leak" through. This is another way of preventing dead-air pockets.

Relative to the size of their house—about 900 square feet of heated floor area—the Starks' collector is a large one. The combination of a tight, insulated building and large collector area could lead to daytime overheating, in which case excess heat would either have to be endured or "dumped" outside—wasted, in other words. Another possible solution: thermal storage. If the collector area exceeds 20 to 25 percent of the building's heated floor area, a heat storage component can be effectively used to store heat for use at night or during cloudy days. The common storage medium for hot air systems is 2- to 4-inch diameter stones stored in an insulated bin and connected with ductwork to the collector and, if there is one, to the existing furnace duct system. This component somewhat increases the complexity of the system, and it would require the aid of someone experienced in air handling systems.

As figure III-70 shows, the addition of thermal storage brings with it the need for additional thermostatic and air flow controls, all of them automatic. These controls allow for more than the single on-off operating mode of the direct-heating storageless system. If the house thermostat calls for heat during a sunny day, hot air will bypass the rock bin and enter the house directly from the collector. When the house is sufficiently heated, collector hot air is diverted through the rock bin where it flows

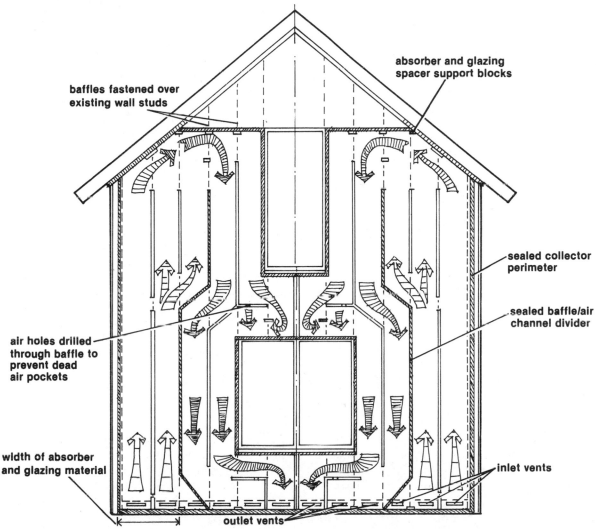

Figure III-69: The layout of the Starks' collector is a good example of what has to be done in a retrofit situation to design and build the collector relative to the location of the existing wall studs, windows, roof pitches and so on. Some attention should also be given to how the layout will look when it's done. The job, after all, should enhance the look of the house.

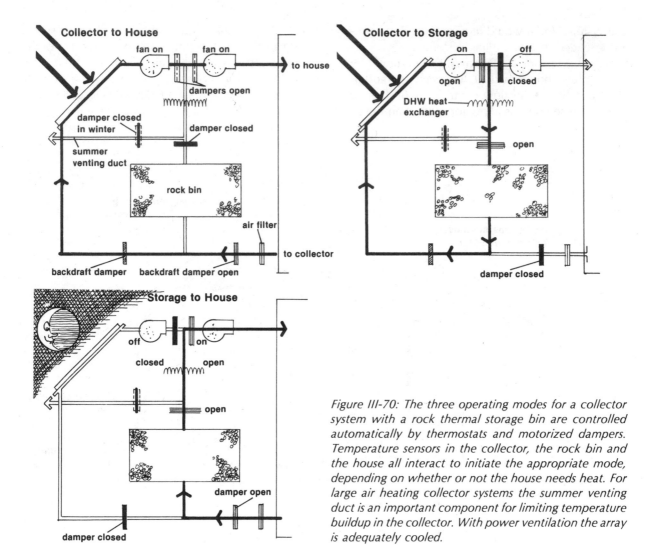

Figure III-70: The three operating modes for a collector system with a rock thermal storage bin are controlled automatically by thermostats and motorized dampers. Temperature sensors in the collector, the rock bin and the house all interact to initiate the appropriate mode, depending on whether or not the house needs heat. For large air heating collector systems the summer venting duct is an important component for limiting temperature buildup in the collector. With power ventilation the array is adequately cooled.

through the spaces between the stones and pebbles, heating them up. Thus the rock bin becomes "charged" with heat and available for service in the third operating mode: The house calls for heat; no heat is available from the collector (nighttime or a heavily overcast day); automatic dampers and the storage blower are actuated to draw heat from the rock storage and distribute it through the heating ducts. It's a bit on the complicated side, but it's not exotic. Every bit of hardware comes right off the shelf, but as mentioned it will take the experience of a specialist to put it together.

Cost and Performance

Anyone who's taking a serious look at solar heating is obviously looking for heat at a lower cost

Photo III-76: Here is yet another variation on the air heating collector theme: Build a pair of hinged panels to go in front of a roll-up garage door. When the sun shines, up goes the roll-up so the collectors can deliver their forced-air heat—not a bad idea for those garages that also double as home workshops.

than that provided by their existing system. The solar wall collector comes out looking pretty good on that score. The Starks' collector had a materials' cost of about $575, or $2.60 per square foot, and the rock storage with the additional controls and ductwork was estimated to cost again as much as the collector, bringing the total cost to about $5.25 per square foot of collector area, which in the realm of solar heating rates the title "low cost." But low cost for what

benefit? Ken Stark notes this: "Before we put up the collector, we added insulation around the house, and cut the propane bill in half. The solar heat has cut it by another 35 to 40 percent." The SFEP calculates about the same by comparing the home's heat loss of 400,000 Btu (through an average January day at 15°F, −9°C, average outside temperature) with the collector's average clear day heat gain of 175,000 Btu. The solar contribution is calculated to be about 43 percent, with the balance made up by the propane furnace. At that rate of heat production the system is likely to pay for itself in eight to nine years if the cost of energy continues to be so upwardly mobile.

Of course, the more sunshine there is on the collector, the sooner the system's cost is recovered. Snow can be an energy booster. When it's covering the ground in front of the collector, it reflects more sunlight onto the absorber. SFFP estimates that snow creates a 10 to 20 percent increase in the energy received by the collector.

With no heat storage mechanism, the Phelps' system cost about $2.25 per square foot for 120 square feet of collector, with the help of some recycled framing materials. Dub that system "very low cost." To evaluate this system the people at SFEP looked at its annual heating potential and came up with the wide range of 5.2 to 10.7 million·Btu per heating season (given the vagaries of available sunshine). The Phelpses are buying electricity at 3¢ per KWH, below the national average of 5¢ per KWH, but even with minimum solar gain they still are saving about $50 a year and paying the system off in about six years. At the current price of #2 fuel oil, annual savings in the $55 to $115 range are possible. The Phelpses are pleased with their retrofit, so much so that they've lately added a 66-square-foot collector to another part of their south wall. With that addition the Phelpses can look forward to a full 25 percent solar contribution to their cold weather comfort. These are the kinds of numbers that make solar heating sense, and for these Nebraskans they add up to a lot of valuable hot air.

Some Design Rules of Thumb for Large Air Heating Collectors

Getting ready to build? The following guidelines have been developed by the SFEP and other groups working to optimize air heating collectors. These calculations are designed to give prospective solarizers a much-better-than-ballpark set of numbers for sizing the collector area, the air passage, the fan, the vent openings and, if it's contemplated, the size of the rock thermal storage component.

1. To build a collector that operates without the need for storage, don't let the collector area exceed 25 percent of the house's heated floor area, if the house is tight and reasonably well insulated. Otherwise you may have overheating problems, which is really heat wasting. In very cold climates (over 6500 heating degree-days) this limit can be exceeded, although there will of course be limits on available wall area.

2. For large, "complicated" collectors, baffle layout should be such that no single "air run," the distance between an inlet and an outlet, exceeds about 24 feet. Larger collectors such as the Starks' are divided into two separate halves or zones, though both are powered by a single fan. A multizone collector can also have outlets opening onto different parts of the house. For example, a collector can be divided horizontally to heat rooms on different floors.

3. The blower-powered air flow rate should equal an "actual" 2½ cfm per square foot of collector at sea level, and 3 cfm per square foot at an altitude of 7000 feet, because of decreased air density. "Actual" flow is the blower's rated capacity (in cfm) less the effects of resistance to air flow (called static pressure drop) caused by the friction of moving air against solid surfaces. SFEP has measured the average pressure drop through these larger collectors to be from 0.3 to 0.5 inch on an instrument that measures pressure change in "inches in water" in a column. (It gets complicated; just take us at our word.)

Most blower cfm ratings are given for a range of pressure drops, so choosing the right unit won't be difficult. SFEP recommends direct drive blowers for small collectors (less than 200 square feet) and belt drive units (squirrel cage) for larger systems to allow for experimentation and fine-tuning with different pulley sizes. The goal for both of these is to produce the minimum temperature rise through the collectors —usually 15 to 20°F (8–11°C)—while still maintaining an outlet air flow that feels warm. The cooler-running collector is more efficient because it transfers that much more heat to the house (and less back out through the glazing).

For example, say that the Starks' 220-square-foot collector created a pressure drop of 0.5 inch of water because the air flow pattern is fairly convoluted. At 2½ cfm per square foot, about 550 cfm are needed for efficient operation. A shaded pole blower with blower-wheel diameter of 9 inches is rated at 865 cfm in "free air" operation, i.e., with no flow-reducing ducts connected to it. But at a 0.5-inch pressure drop (known in the trades as *specific pressure,* SP) this 1/6-hp blower is rated at 760 cfm. That's a little on the fast side, but by equipping the blower with a standard speed controller the blower speed can be tuned to get just the right outlet air temperature. Note, however, that not all blowers will accept a speed controller. Ask your supplier for one that will.

4. The depth of the air passage (the gap between the absorber and the existing wall) is determined by the thickness of the absorber support framing. The passage is sized so the blower can maintain a minimum air velocity, which in this case is 800 feet per minute (fpm). First the calculated actual cfm (2.5 cfm times collector square feet) is divided by 800 fpm to get the area (in square feet) of the air passage cross section. The passage depth is then found by dividing the cross-sectional area by the total width of the collector air passage in one direction of air flow.

For example, in the Phelps' straight-through collector the total passage width is 45 inches and the design air flow is 300 cfm. The area of the air passage is therefore: 300 cfm ÷ 800 fpm = 0.375 square feet, or 54 square inches (0.375 × 144). The air passage width is 54 square inches ÷ 45-inch passage width = 1.2 inches. The Phelpses settled on an air gap of 1¼ inches.

The Starks' collector has a variable width of 11 to 16 feet due to the presence of windows. The average width was figured to be 13.5 feet, but because the air flows through the collector in two directions, up and down, the actual collector width in *one* direction is half the average width, or 6.75 feet (81 inches). At a design air flow of 550 cfm the calculation proceeds: 550 cfm ÷ 800 fpm = 0.6875 square feet or 99 square inches. The air gap width is: 99 square inches ÷ 75-inch width = 1.32 inches. A 1½-inch actual gap assures that the air flow won't be restricted. It's always better to be generous, not stingy, with these calculations.

5. The collector inlets and outlets must be a size that is at least equal in area to the airway (between baffles) they serve. For example, the Phelps' two airways measured about 43 inches, with a 1¼-inch gap. With one inlet and one outlet vent, each opening had to be about 43 × 1.25 = 54 square inches or 6 by 9 inches. The Starks' airways were generally 16 inches wide with a 1.5-inch gap. That means an area of 16 × 1.5 = 24 square inches. Thus, each of the 12 vent openings was cut to measure 4 by 6 inches.

6. When ductwork is needed to get a more extensive distribution of solar heat, the design velocity for ducting is 800 fpm. Thus, a 200-square-foot collector designed for a 500-cfm flow rate would need ductwork with a cross-sectional area of 500 cfm per 800 fpm or 0.63 square feet. A 10 by 10-inch duct could be used, or one that was 6 by 15-inches. It is possible to build a duct onto an exposed basement ceiling using the floor joists and subfloor for three sides and thin plywood for the fourth side. Sealing air leaks with caulking and duct tape is crucial, along with providing insulation.

7. A rule of thumb on storage sizing calls for using at least 50 to 60 pounds of rock per square foot of collector. Working with Btu's, the specific heat of rock is such that 1 cubic foot stores about 20 Btu for every 1°F it rises in temperature. In the case of a 40°F (22°C) rise, a cubic foot would store 20 by 40 or 800 Btu. Let's say for example that the collector output is 150,000 Btu per day. In order to store half that much heat, 75,000 Btu/day ÷ 800 Btu/cubic foot of rock = about 94 cubic feet of rock needed, or about 9400 pounds of the stuff. That's roughly 3½ cubic yards; with 1 cubic yard weighing about 2700 pounds. The storage bin also should be proportioned for minimum surface area to minimize storage heat loss. The most practical shape for this is the simple cube.

It is stressed again that incorporating storage into the collector system is no simple task, and we've by no means included all the information needed to do the work. But there are some excellent references listed that give complete information on thermal storage for air heating collector systems. Foremost among them is a construction manual titled *Model-TEA Solar Heating System,* published by Total Environmental Action, Inc. of Harrisville, New Hampshire.

Joe Carter

References

Nicholson, Nick. 1977. *The Nicholson solar energy catalogue and building manual.* Ayer's Cliff, Quebec: The Ayer's Cliff Centre for Solar Research.

———. 1978. *Harvest the sun.* Ayer's Cliff, Quebec: The Ayer's Cliff Centre for Solar Research.

———. 1978. *Prototype Canada.* Ayer's Cliff, Quebec: The Ayer's Cliff Centre for Solar Research.

Small Farm Energy Project. 1980. *Small farm energy project: final report.* Available for $5 from Energy Project, P. O. Box 736, Hartington, Nebr. 68739. Telephone: (402) 254-6893.

☼PROJECT

A Roof-Integrated Collector System

We've had them hung out from windows, attached to walls like oversized dominoes, and covering entire south walls. So what else is new? Maybe your roof is a place for an air heating collector system. With this design the collector actually replaces a portion of the roof, and it uses the existing rafters as the primary framing for the absorber bays and the air passages. This is a forced-air system, and like the large vertical wall system, new ductwork can be installed for hot air distribution and the return (cool) air supply, or the collector can be connected with existing ducts. Steep, south-facing roofs are likely to have a lot of "convenient" solar surface area, convenient because they're not cluttered with doors and windows, nor shaded by bushes and buildings. You can start clean and develop a large system that will blow a lot of heat your way.

Rockland is near the midpoint of Maine's jagged coastline, at the northern limit of the region marked "temperate" on most climatic maps. Annual heating degree-days in this region average around 7400, with winter daytime temperatures commonly in the 10s and dipping to −10°F and lower overnight; add the often fierce winter winds, and you have a climate that's a challenge to any solar addition.

For Sharon and Francis Merrow, however, adding a solar space heating system wasn't the primary goal of their home improvement plans. In fact it was somewhat of an afterthought to their desire to have a bigger kitchen, a garage, more storage space and more living space. But with the addition they had planned, there was a large, steeply pitched south-facing roof that was just begging to be solarized. So instead of roofing the roof the Merrows added the air heating collector system in combination with some skylights for daylighting and direct-gain heating. The system collects heat with 230 square feet of collector area that is built nearly flush with the 60-degree roof of the addition. There is no rock bin for thermal storage, so the system delivers all its heat to the house during the day, substantially reducing the day-time heating load.

From a bird's-eye view the building forms a T, and the solar addition extends east along the top of the T over ground that was once occupied by a large barn. The addition added 270 square feet of floor space to the original kitchen, another 430 square feet of garage and shop space, and 414 square feet of storage and living space in the peaked room behind the collector wall. The garage and the storage room above it, which comprise about half the addition, are uninsulated and not heated. The blower and ductwork are located in the heated half of the upstairs living space, which is insulated and weather-sealed.

To briefly review the collector design, it is essentially a series of "cells" that are divided from each other by the 2×6 roof rafters that are placed on 24-inch centers. Each cell measures about 24 inches wide by 9 feet long. The cells are backed with plywood and insulation, and the absorber consists of black-painted insect screen that is bunched into waves or loops to increase surface area. There are 15 cells in the array, 2 of which have no backing or absorber. These are the skylights that light and heat the upstairs room. For this cold climate a double glazing of FRP was used over the collectors and the skylights. A ½-hp blower pushes air through the collector up to a manifold-plenum that connects all the outlets of the cells. Hot air is delivered down to the kitchen through an insulated duct, and the cooler return air enters another manifold-plenum that connects all the inlets of the cells. A differential thermostat controls the blower. Whenever the collector is 15°F (8°C) hotter than the downstairs temperature, the blower is switched on.

The collector system provides only daytime heating to the addition and to part of the original house, which is backed up by a combination wood/oil furnace. Since additional living space at a reasonable cost was the main goal of the project, it was an easy decision not to add storage mass and the attendant controls. They would have greatly increased the cost of the solar modification. The 270 square feet of glazed area (collector plus skylight) provide considerably more heat than is required by the addition even

during the dead of winter. Had this space been isolated from the rest of the house, it would easily overheat by noontime, but since surplus heat is distributed into the house, overheating is prevented; overall daytime heating requirements are reduced, and the need for thermal storage was eliminated. However, the area of a south roof on a medium-size (1500 square foot) house can easily equal 25 to 40 percent of the heated floor area. If the entire roof were solarized, thermal storage would certainly be necessary.

A Simple Collector

Two factors that must be balanced in the design of a simple collector are the amount of surface area available to absorb and transfer solar energy to the air moving through the collector, and the resistance to that air flow posed by a large surface area and a restricted air passage. Collectors have been

Photo III-77: It's hard to heat a big old house economically, especially when it's out on the Maine coast, but the Merrows attacked the problem with a thorough weather sealing and insulation program, followed by the air heater built into the roof of their addition.

built using corrugated and honeycombed metal sheets, metal lath or screen and metal strips, each with different effects on the flow of air. Collectors that use insect screen or expanded metal in some configuration of loops, ripples or multiple layers offer a great deal of surface area, but they also require larger blowers to overcome the resistance to air flow. If the air flow is too slow, the collector will run hotter, which means that more heat is lost back through the glazing by reradiation. The owners of this installation considered using a corrugated metal absorber, but they rejected the idea because of higher material costs for less absorber surface area and settled instead on using insect screen in a doublepass air flow configuration.

The plywood and screen design was attractive to the Merrows because they saw they could build the collector system themselves. Although they enjoyed the encouragement of a brother-in-law who worked in the solar field and helped with the overall design, neither of the Merrows had any prior familiarity with solar heating, or even extensive carpentry experience. But the only contracted assistance they required during the construction came when the old barn had to be torn down and when the foundation for the addition had to be formed and poured. Other than that, the Merrows happily point out that the additional heat contributed by the collectors comes as the result of a system built entirely on-site and without professional assistance.

As was mentioned, the collectors were built between the 2 × 6 roof rafters on the south side of the addition. In this case the rafters were located on 23¼-inch, instead of the standard 24-inch center, to accommodate the width of the FRP glazing over the collectors. With this rafter spacing each 4-foot-wide by 9-foot-long strip of FRP would cover two cells and still overlap each neighboring sheet by about ¾ inch.

Once they had framed the addition, the Merrows prepared the rafters for carrying the plywood backing and the absorber. They set 1 × 2 furring strips, along both faces of each rafter, parallel to and 2¼ inches below the top edge, which has been built up with a 2 × 2 to raise the glazing above the plane of the roof. These strips support the plywood collector backing. The length of the furring strips of course depends on the length of the collector. As figure III-74 shows, the plywood backing stops short of the floor and the roof ridge to leave room for the two manifold-plenums. The plywood backing panels were covered with reflective foil builder's paper after they were placed between the rafters. The foil has a dual purpose here: to reflect heat back into the air passage and to protect the plywood from degrading in the high-temperature environment of the collector. It has been found that wood fibers can actually break down over time if they're routinely exposed to the high (above 200°F, 93°C) temperatures that are possible in a collector. This isn't usually a severe or dramatic problem if the collector is ventilated whenever the sun is shining, but the foil facing is an ounce of prevention.

Thirteen 21½-inch-wide and 8-foot-long plywood panels were cut from sheets of ½-inch CDX plywood. The strips of aluminum insect screen were cut 10 feet long and 21 inches wide. They were made longer so there could be some bunching for increased absorber surface area. The screen was painted on both sides with flat black paint, and then each strip was stapled to a plywood section, with the excess length divided evenly to form shallow, broad waves of blackened absorber material.

After the plywood backing was positioned, the inlet and outlet manifold-plenums were built. Figures III-72 and III-76 show general views of how the manifold-plenums generally relate to the absorber and plywood backing, and figure III-73 shows some detail of how the manifolds are attached and sealed to the rafters. Of course every attic isn't going to be like the one we've drawn, but these representations should show enough of what's going on so you'll be able to work this plan into your application.

The manifolds at both ends of the collector essentially unite the inlets and outlets of the collector; they're also the transition from the collector to the supply (outlet) and return (inlet) ducts. The two manifolds are made with thin plywood, paneling or Masonite strips that are fastened to the rafters at the back

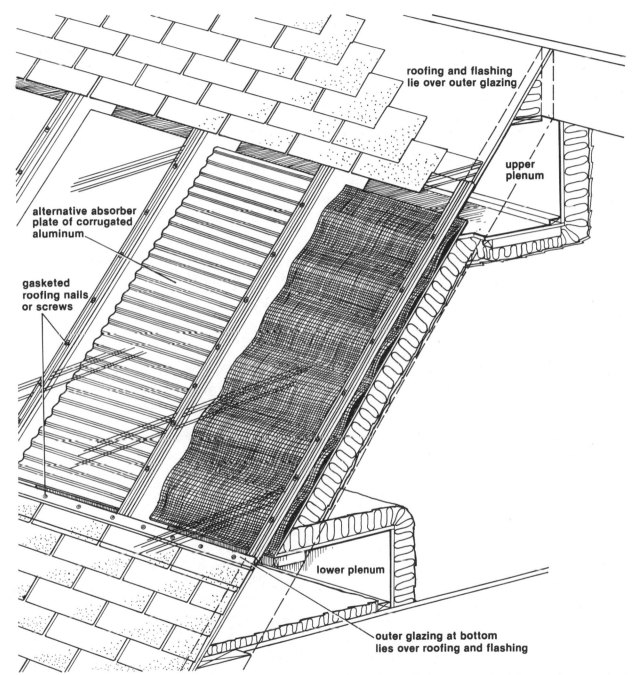

roofing and flashing
lie over outer glazing

upper
plenum

alternative absorber
plate of corrugated
aluminum

gasketed
roofing nails
or screws

lower plenum

outer glazing at bottom
lies over roofing and flashing

Figure III-71: There are definite advantages to integrating an air heating collector system into the roof structure: It's cheaper and more energy-efficient than putting collector boxes on the roof. The Merrows' installation made good use of blackened aluminum insect screen for the absorber but an alternative is to use corrugated aluminum roof panels, as in the previous project.

of the collector. The strips are 16 to 24 inches wide, varying with the space you have available above and below the collector. Where necessary, the strips are notched so they can fit around the rafters and butt up tight to the plywood backing panels (see figures III-72 and III-73).

Sealing out air leaks is of utmost importance in maintaining collector efficiency. After the manifolds are assembled and put in place, all seams must be sealed with caulk and standard duct tape. At some point in the construction of the collector, the plywood backing must be sealed to the rafters (to the 1×2 supports) using duct tape and caulk. Otherwise this will be a major source of hot air *ex*filtration from the collector. It's also been found the air can even leak through imperfect or knotty plywood laminations! The foil facing will help to limit this kind of leakage, but it's also a good idea to be selective in the plywood you use.

The two inlet and outlet openings between each collector cell and the manifold are as wide as the spaces between the rafters. The other dimension of these openings is the space between the upper or lower ends of the plywood backing panel and the existing roof sheathing or other solid obstruction. This gap should be about 6 to 10 inches, depending on the length of each cell, with wider gaps being made for longer cells. In the Merrows' 8-foot-long collectors the gap is 6 inches.

Glazing

To prepare the rafters for the glazing, glazing support members are installed between the rafters at both ends of the collector. For good support when the glazing is being fastened, 2×4's should be used turned face up so as not to block either the inlet or the outlet. The top edges of the rafters should also be checked and planed or sanded free of significant bumps and waves. Then 2×2's are nailed along all the rafters, thereby raising the new ''rafter'' edge above the plane of the original roof. The 2×4 glazing support is installed flush with this new edge. Because of the low daytime temperatures in their climate zone, the Merrows glazed their collector with a double layer of FRP glazing. In warmer climates single glazing is an acceptable design.

With the FRP, which comes in 4-foot-wide

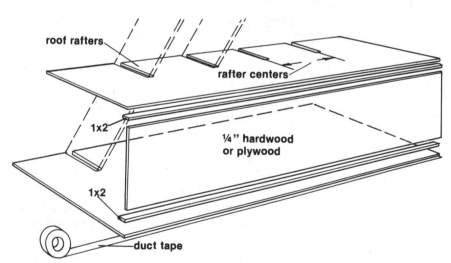

Figure III-72: The manifold-plenums are made with thin plywood or paneling that is fastened at the corners with 1×2 or 2×2 strips of wood. The corners are then sealed with duct tape. Prior to assembly the panels are notched to conform with the rafter spacing.

rolls, it's important to use the right attachment method or the mistakes will be all too apparent in the finished product. First, the top edge of the glazing is tacked down at the top of the collector cells (4 feet covers two cells); then it's unrolled for the full length of the cells to check for proper alignment. Slight errors can either be trimmed with thin snips or taken up with the overlap of the next roll. If there are major errors, the top edge should be picked up and repositioned. After the alignment is checked, the FRP is

tacked to the rafters using small ring-shank nails, with the nailing points being predrilled. Nail in one direction with the nails spaced about every 6 inches. At overlap points and around the perimeter of the entire array, run a bead of silicone to make the air seal.

The insulating air gap between the first and second glazing layers is created by nailing 1×2's along all the rafters and the glazing support members using widely spaced 6d nails. Then the second glazing layer is attached using the same method described for the first layer. The finish work for the glazing involves attaching battens and flashing the perimeter of the array. The battens can be wood or aluminum, and with either material a gasketed roofing nail or screw is used. It should be long enough to reach at least an inch into the rafter. Before applying the fastener drill a hole that is slightly larger than the shaft of the fastener down to the top edge of the rafter. This will help to relieve any possible expansion or contraction stresses that might occur in the FRP with wide daily or seasonal temperature swings.

The top and sides of the collector must be flashed with aluminum coil stock that extends at least 12 to 16 inches from the collector perimeter. At the bottom of the collector the glazing can actually extend over the roofing, with flashing being used only under the first glazing layer. Also, it isn't necessary to use a batten across the bottom edge, as a wood batten especially could impede rainwater drainage. Along the bottom edge just use the gasketed fasteners straight through the glazing. See figure III-74 for

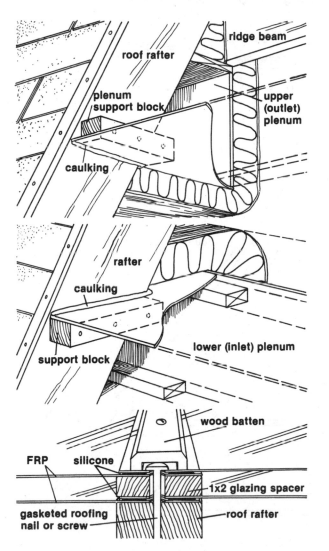

Figure III-73: The upper and lower plenums are attached to the rafters by way of small support blocks that are nailed into the sides of the rafters. After the plenum panels are notched and assembled, as in figure III-72, they are slid over or under the support blocks and then screwed to the blocks. All seams should be carefully caulked to stop air leakage. The lower detail shows a simple fastening system for two layers of FRP glazing. For good appearance the gasketed nails are covered by a wood batten that can be milled to shape on a table saw.

some of these finishing details. Of course, the original roofing should be replaced back to the perimeter of the collector.

Air Handling

The rest of the air handling system leads away from the manifold-plenums. These are the sup-

ply and return ducts that lead to and from the living space, and the design of this part of the system will be customized to where you want to distribute (supply) hot air, and from where you want to supply (return) cool air to the collector. For this you may do well by asking for advice from a heating, ventilating and air conditioning (HVAC) contractor, who will be able to recommend duct paths, sizes and outlet and

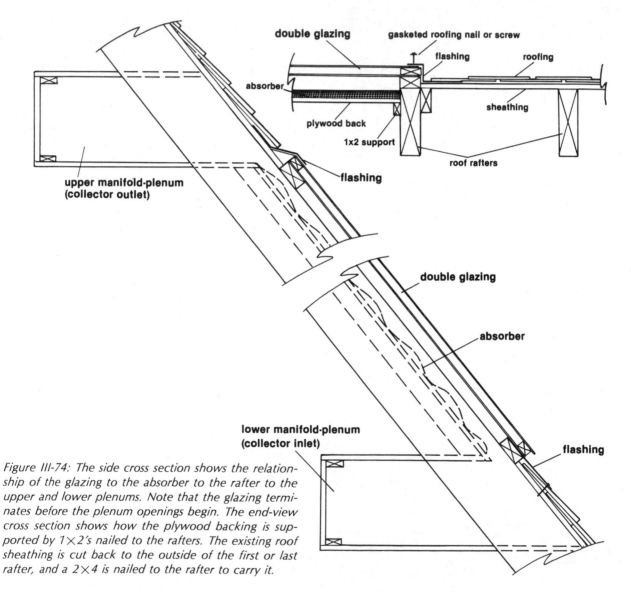

Figure III-74: The side cross section shows the relationship of the glazing to the absorber to the rafter to the upper and lower plenums. Note that the glazing terminates before the plenum openings begin. The end-view cross section shows how the plywood backing is supported by 1×2's nailed to the rafters. The existing roof sheathing is cut back to the outside of the first or last rafter, and a 2×4 is nailed to the rafter to carry it.

inlet locations. If you're planning to integrate your solar furnace with your existing furnace, you'll definitely need professional assistance, particularly because more control components will be needed.

A couple of design guidelines: It's important to maintain a fairly uniform rate of air flow through each of the collector cells. Otherwise some cells may be overcooled, and some may be overheated. The general practice for ensuring this uniformity is to put the supply and return ducts at opposite ends of their respective manifolds, as shown in figure III-76. The location of the blower in either duct is optional. As

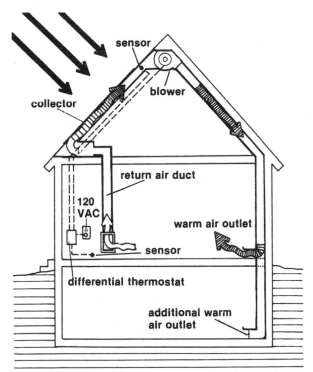

Figure III-75: In a typical collector system without rock thermal storage the air handling system is relatively simple. The return air duct leads to the inlet plenum. It should be located away from the warm air outlet to ensure that there won't be any mixing of the two. When integrating a collector system with an existing forced air furnace system, your best move will be to work with an HVAC contractor, especially when it comes to figuring out the thermostatic controls.

was mentioned in the previous project, putting the blower on the outlet (supply) duct maintains a negative pressure in which air leakage results in cold air infiltration rather than hot air exhaust from the collector. Since the blower will run hotter when it's in the outlet duct, it should be checked and lubricated on a regular basis. If there are space limitations to having an "outlet" blower, then it can be installed on the inlet (return) duct.

The way this system is designed, it's a ready candidate for night heat loss by reverse thermosiphoning. Warm house air could rise up through the outlet duct and enter the collector, where it would cool and become denser, and then "fall" back into the house through the inlet duct. Thus it's necessary to have some kind of manual or automatic control in these ducts or at the room inlet and outlet grilles, so the collector can be "shut down" at night and during cold, sunless days.

Ductwork can be home-built using the same kind of thin plywood or Masonite construction that was used to make the manifold-plenums. Of course, standard metal or preinsulated ductwork can also be used. Retrofit ductwork doesn't have to look bad. Round metal ducts running through the living space can be quite good-looking if they're painted with either bright or subtle colors. If total concealment is the goal, it's also possible to use the cavities formed by interior stud wall partitions. The wall section to be used would have to be opened up and cleared of any obstructions. Then the cavity can be fitted with a rectangular duct made of rigid insulation (a standard item in the HVAC trade). It's important that all ductwork be thoroughly insulated to at least R-6 to R-8 (2 inches of fiberglass) in order to have complete control of the heat produced by the collector. Otherwise conductive losses through the duct will reduce the outlet temperature. For the same reason it's even more critical that the duct system be completely sealed, or there will be warm air leakage into places you're not trying to solar heat. We again refer you to the expertise of an HVAC contractor. Some solar contractors strongly recommend putting the finished collector system to a smoke test using

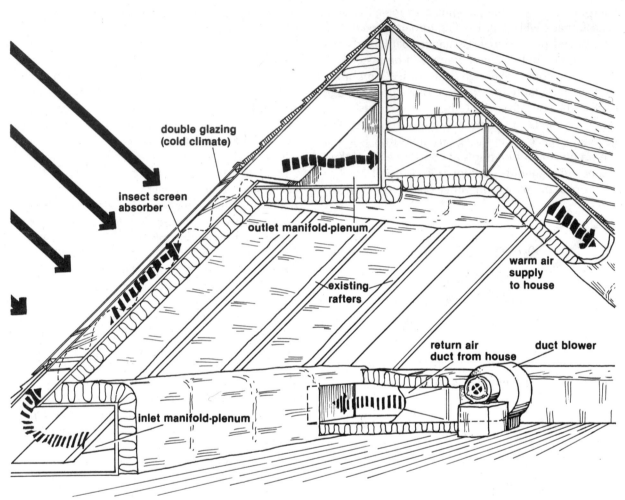

Figure III-76: To ensure a comprehensive distribution of the forced air through the collectors, locate the inlet and outlet at opposite ends of the array. If the attic is unheated, it's very important to insulate all ducts with at least 2 inches of fiberglass.

standard smoke bombs (available from Superior Signal Co., Inc., Spotswood, New Jersey 08884). Any invisible leaks that show up can be attacked with caulk and tape.

Another design element that should be considered is summertime venting of the collector. Since there's no space heating load, there should be some way of exhausting heat from the collector. The temperatures that can be reached in a stagnating collector (above 200°F, 93°C) can shorten the life of plastic

glazings and also negatively affect any wood that is exposed inside the collector. Also, if there is a living space behind the collector, there will be considerable heat gain into the room even though the back of the collector is insulated. An idea: Put an exhaust leg into the outlet duct to provide a natural convection escape route for collector hot air. If the blower has a variable speed control, it could be run at a low speed to boost the ventilation rate. For something a little more exotic, the collector could be used as a big

Photo III-78: The space behind the collector contains the air handling system: the two manifolds, the blower and the ductwork that leads to and from the heated living space. In the future the Merrows plan to convert this space into a usable room, so they planned ahead and installed a skylight, instead of a collector, in one of the rafter bays.

solar chimney that would draw warm air out and cool air into the house. (See the project "A Solar Chimney" in Section V.)

Cost and Performance

The low cost of this installation is one reason why the Merrows are now true believers in do-it-yourself solar retrofitting. The total they spent on materials for the collector, ductwork and controls came to about $135 more than the $632 bid they got from a roofing contractor to shingle the same roof. (Of course, they didn't count their own labor into the cost of the solar system.) And the contractor's estimate did not include the cost of skylights that open

onto the room behind the collector, which for the Merrows are an added bonus.

Along with their solar improvement, the Merrows have wisely attended to the basic conservation needs of their not-too-conserving 19th century house. Exterior walls have been insulated and weather stripping has been installed in a concerted effort to plug up the many heat leaks that were seemingly built into the old place. Since the Merrows use wood as their primary heating fuel, it's difficult to estimate accurately how much wood is saved by conservation and solar heat, but their use has certainly dropped at least a couple of cords per season, from an original use of five to six cords. The collector has run smoothly, delivering air heated as high as 105°F (41°C), plenty warm for a forced-air system.

Materials Checklist
(for the Merrow's collector system)

½" plywood: seven 4×8 sheets @ 8.95/sheet	$ 62.65
⅛" untempered Masonite: six 4×8 sheets @ 4.95/sheet	$ 29.70
495 feet of furring: ¾" × 1½" (1×2)	$ 19.74
silicone caulking, 12 tubes @ 3.50/tube	$ 42.00
aluminum insect screen: 42 yards of 24-inch-wide screen	$ 33.00
fan: ½-hp Dayton blower	$118.00
differential thermostat	$ 94.00
FRP glazing: three 4' × 50' rolls	$368.00
total cost	$767.09

Your Collector

If anything should be stressed about this kind of system, it is that the design is eminently flexible. Some of the basic rules must be followed, but size of course can be varied greatly. Increasing the length of the cells to gain more collector area is certainly possible if you've got the room. A larger cell, however, requires a larger fan to push air through the panels fast enough to keep the collectors operating at the minimum temperature. A larger blower of course uses more electrical power, which diminishes the overall benefit of the solar heat because of increased energy costs. Thus a more desirable way to increase collector area is by having more cells of shorter length, if that's possible.

Another option allows placement of both supply and return ducts along the bottom of the collector. In this configuration the existing rafters are sheathed on both the top and bottom edges, which forms the outlet duct. To form the collector, 2×4's are nailed over the plywood along the top of the existing rafters. Air will thereby flow from the inlet manifold up through the collector and enter the duct formed by the existing rafters. From there it will flow down to the outlet manifold, which can be located above or below the inlet manifold. Needless to say, air can be moved hither and thither pretty easily, and when you come to designing your own system you certainly don't have to be locked into a specific design. In retrofitting especially, the design's got to fit the available spaces.

The Merrows started with brand-new roof framing and built the system into their new addition. But there's not really much difference between exposed new rafters and an old roof that can be ripped off to make way for solar heating. The system works for the Merrows because it's essentially simple and low in cost for all the heat it delivers.

Erika Morgan

References

National Aeronautics and Space Administration. 1978. *Installation package for a solar heating system.* Report no. DOE/NASA CR-150876. Washington, D.C. For sale by National Technical Information Service, Springfield, VA 22151.

Temple, Peter L., and Adams, Jennifer A. 1980. *Model-TEA solar heating system.* Harrisville, N.H.: Total Environmental Action.

Hardware Focus

Complete Air Systems for Space Heating

Contemporary Systems, Inc.
Rte. 12
Walpole, NH 03608
(603) 756-4796

Deltair Solar Systems
Rte. 2, Box 53D
Chaska, MN 55318
(612) 474-4182

International Solar Technologies, Inc.
R.R. 2, Box 321
Plainfield, IN 46168
(317) 272-2996

Research Products Corp.
P. O. Box 1467
1015 E. Washington Ave.
Madison, WI 53701
(608) 257-8801

Solar Development, Inc.
11799 E. 30th Ave.
Aurora, CO 80010
(303) 343-8154

Solar Farms Industries, Inc
P. O. Box 242
Stockton, KS 67669
(913) 425-6726

Solar, Inc.
008 Sunburst Lane
Mead, NE 68041
(402) 624-6555

Solaron Corp.
1885 W. Dartmouth Ave.
Englewood, CO 80110
(303) 762-1500

Sunflower Energy Works, Inc.
P. O. Box 85
Goessel, KS 67053

Sun Wise, Inc.
40th St. and River Drive N.
P. O. Box 7005
Great Falls, MT 59406
(406) 727-5977

Tritec Solar Industries
P. O. Box 3145
711 Florida Rd.
Durango, CO 81301
(303) 247-8497

Differential Thermostats for Air Systems

Dan-Mar Co., Inc.
R.R. 2
Box 338B
Wikel Rd.
Huron, OH 44839
(419) 433-4479

Deko-Labs
P. O. Box 12841
University Station
Gainesville, FL 32604
(904) 372-6009

Solar Heating System Controller, TC20
On differential fixed at 15°F (8.3°C); off differential fixed at 5°F (2.8°C).

Model TC-3 Temperature Controller
On differential fixed at 10°F ± 1°F at 135°F (5.6°C ± 0.6°C at 57°C); off differential fixed at 5°F ± 1°F at 135°F (2.8°C ± 0.6°C at 57°C).

Dencor, Inc.
2750 S. Shoshone
Englewood, CO 80110
(303) 761-2553

Solar Controls Systems, Model 915 or 916
On differential 20°F ± 2°F (11°C ± 1°C); off differential 10°F ± 2°F
(5.6°C ± 1°C).

Hawthorne Industries, Inc.
3114 Tuxedo Ave.
West Palm Beach, FL 33405
(305) 684-8400

Fixflo Control H-1503-C
On differential 18°F (10°C); off differential 6°F (3.3°C).

Honeywell, Inc.
Residential Division Customer Service
1885 Douglas Dr. N.
Minneapolis, MN 55422
(612) 542-7500

Solar Control Panel
Adjustable on/off differentials from −10 to 40°F (−5.6°C to 22.2°C); factory
set for 18°F (10°C) on differential and 3°F (1.7°C) off differential.

JBJ Controls
P. O. Box 383
Idaho Falls, ID 83401
(208) 522-2200

Model 80A Self-Contained Air System Control
On differential 15°F (8.3°C); off differential 5°F (2.8°C).

Natural Power, Inc.
New Boston, NH 03070
(603) 487-5512

Series S25 & S26 Differential Thermostat
On differential set anywhere between 3°F (1.7°C) and 25°F (13.9°C); off
differential adjustable between 0 and 22°F (12.2°C).

Rho Sigma, Inc.
11922 Valerio St.
North Hollywood, CA 91605
(213) 982-6800

RS 106 Differential Fan/Pump Control for Space Heating System
On differential 20°F ± 3°F (11.1°C ± 1.7°C); off differential 3°F ± 1°F
(1.7°C ± 0.6°C).

Robertshaw Controls Co.
Temperature Controls Marketing Group
100 W. Victoria St.
Long Beach, CA 90805
(213) 636-8301

SD-30 Solar Commander
On differential 15°F ± 5°F (8.3°C ± 2.8°C); off differential 5°F ± 3°F
(2.8°C ± 1.7°C).

West Wind Electronics, Inc.
P. O. Box 1657
Durango, CO 81301
(303) 588-2275

Temperature Differential Switch
On differential adjustable from 0 to 9°F ± 1.8°F (0 to 5°C ± 1°C); off
differential factory set at 1.8°F ± 0.9°F (1°C ± 0.5°C).

Willtronix
1927 Clifton
Royal Oak, MI 48073
(313) 399-9557

WD-1 Differential Temperature Thermostat
Operating temperature is adjustable with a constant differential of 19°F ±
4°F (10.6°C ± 2.2°C).

Wolfway Product Consultants, Inc.
R.D. 1, Box 1135
Tamaqua, PA 18252
(717) 668-4359

Solar Controller Differential Thermostat
On differential preset at 15°F (8.3°C). Off differential preset at 3°F (1.7°C).
Can be adjusted within a wide range. Available in assembled units or kit
form.

Air Handlers

Airmax, Inc.
P. O. Box 129
White Oak, TX 75693
(214) 836-2785

Contemporary Systems, Inc.
Rte. 12
Walpole, NH 03608
(603) 756-4796

Delta H Systems
Rte. 3
Sterling, CO 80751
(303) 522-4300

Helio Thermics, Inc.
1070 Orion St.
Donaldson Ind. Park
Greenville, SC 29605
(803) 299-1300

Helitrope General
3733 Kenora Dr.
Spring Valley, CA 92077
(714) 460-3930

Hot Stuff
406 Walnut
La Hara, CO 81140
(303) 274-4069

Park Energy Co.
Star Rte. Box 9
Jackson, WY 83001
(307) 733-4950

Solar Control Corp.
5721 Arapahoe Rd.
Boulder, CO 80303
(303) 449-9180

Solar Development, Inc.
11799 E. 30th Ave.
Aurora, CO 80010
(303) 343-8154

Tritec Solar Industries, Inc.
P. O. Box 3145
711 Florida Rd.
Durango, CO 81301
(303) 247-8497

F. SOLAR GREENHOUSES

Photo III-79: For most houses an attached greenhouse is very unique in design and function, yet it is probably the most popular of the major space heating retrofit options.

Solarizing
for Food and Heat

The Northeast Solar Energy Center has estimated that even a modest-sized solar greenhouse can cut a family's fuel bill by 5 to 20 percent and produce up to $300 worth of food a year. Considering its versatility as a heat and food producer, as well as being a nice place just to sit, an attached solar greenhouse is an ideal retrofit project that starts with a sunny building spot and needs only common building materials to take it to completion. You can build a greenhouse that will conform in design, size and function exactly to your needs and to the demands of your climate. "Solar greenhouses are not limited to any one style, size or use," says the Energy Center, "but they are all designed with the sun and other natural elements in mind." This section will examine in detail one successful retrofit project, and you'll also see some examples of design variations for this multi-use solar addition. Other articles will pass along more of the experience that has been gained with greenhouses in recent years: how to plan them, how to operate them. They've come a long way from the tack-on heat-mongers that used to represent the state-of-the-art in home greenhouses. The solar greenhouse has been bred to be an energy-efficient, attractive, versatile and flexible structure. Fed by the sun and guided by its owner, it heats, stores heat and grows things and gives people a unique living space where they can be with their meditations, their flowers, their family and friends. It's a good spot.

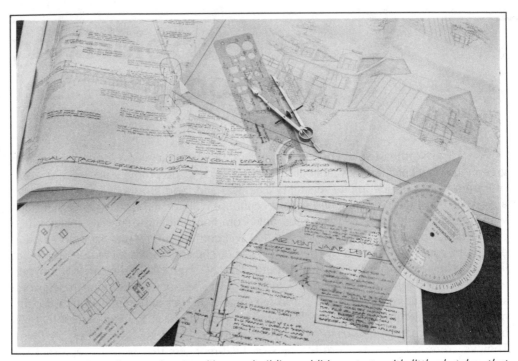

Photo III-80: Greenhouse planning, like any building addition, starts with little sketches that come straight from the imagination, and proceeds to the final working blueprints. Every planning step is an opportunity for change; even blueprints are flexible.

Planning and Designing Solar Greenhouses

You can tailor it to fit your needs

Many people consider solar greenhouses as the solar option par excellence because it not only can provide heat to a house, but food and added living space as well. It can provide these benefits in different proportions according to the owner's needs and priorities, and this is the key to a successful design: deciding what you want from your greenhouse. There is no sense in building a food production center if it remains fallow because a busy life-style keeps you away from it. Likewise, much of a greenhouse's daily heat gains won't be retained if the movable insulation isn't used. A greenhouse can be many things: a passive heat collector, a relaxing sunspace, a "foodspace," but it won't achieve its maximum potential at the hands of an overly "passive" owner.

There are three basic phases in the life of an attached greenhouse: planning and design, construction, and use. They must all be well-considered before anything really happens, so in a sense they're all part of the planning process. Since the construction naturally reflects the design, and the design anticipates the intended use, it is imperative that some important decisions be made at the start of the planning and design phase. You can ask yourself some questions about just what you want from this addition. There are no right or wrong answers, only your own answers and, along with other guidelines, you can follow your answers toward your ultimate design-construction-use plan. It's also true that the use you plan for now won't eliminate change later on. Even when you have your greenhouse cast in concrete, wood and glass, it still has flexibility. You can "re-model" your use of the space to fit changing needs, and you can certainly remodel the structure itself if the need arises.

The primary decision asks the question: How do I most want to use the greenhouse? There are any number of options.
- I want to raise as much food as possible—year-round.
- I want to use the greenhouse to augment my garden.
- I only want to raise orchids (or tomatoes, cactus).
- I just want a warm space where some houseplants can grow.
- I don't care if I grow anything; I just want to read the morning paper in a sunny, warm room—preferably one that heats itself.
- I want a warm place where the kids can play.
- I'll probably only spend the evenings there.
- I don't care as much about using the space as I do about getting some of that good, free solar heat.
- I'm just looking for a tax-break (credit) and trying to fulfill my social obligations by

helping to reduce the need for oil imports.

Some of these options may be far away from your plans, some very close, but we could probably find a large group of people who would pick any given answer.

When you find your answer to that question, where do you go from there? To keep things as simple as possible, we'll say that your answer puts you somewhere on what we'll call the heat-horticulture continuum, which contains all the options for using a solar greenhouse. If you put yourself on the heat end, your aim would be to use the space to get maximum heat for your house (all other factors being secondary). If you put yourself at the horticulture extreme, your primary interest is in growing plants and food (again with all other factors being secondary to that goal). Somewhere in the middle of this line there is also "living space," which draws from both ends of the heat-horticulture continuum. Most people will find their interests in the "wide middle" that involves a little bit of everything in varying degrees. This makes a lot of sense because an attached greenhouse really does have a lot to offer. Heat management is fairly easy, and even the most committed horticulturalist can't really ignore it; living space is easy too (you'll need a chair); horticultural management, as you might suspect, takes the most work, but in dollar value returned a greenhouse's food production can in fact outstrip its heat output. Perhaps our continuum is an uphill one.

But what are some possible consequences of learning toward one side or the other? If, for example, food production is your primary goal, you will probably glaze one or both of the end walls; plants like the morning sun, and they need as much light as they can get in winter. You'll probably have to sacrifice some heat gain to the house in order to keep the plants happy. An emphasis on horticulture may influence where you locate the addition, and how big you make it. Your "payback" calculation will include the value of the food produced along with the heat gained, which may influence your decision about the size of the space.

Outside of your anticipated use of the greenhouse, there are a number of other factors that affect the final outcome. Site conditions are foremost among them. Factors like orientation, available wall and ground space, what's in the wall or the ground, the presence of trees—these all influence design and function. Budget is certainly an important factor. It will influence your sizing of the greenhouse because the cost of building additions usually varies in almost direct proportion to the square feet of floor space that are built. It also affects your choice of materials. Glazing materials vary a lot in cost and quality, and so do framing materials, glazing insulations, and air handling equipment.

With many layers of factors, confusion might seem to be the overriding factor, but taken one step at a time a greenhouse isn't hard to understand in all of its parts. That's what we're going to do in the next few pages. We'll cover some of the basics and some of the details that make solar greenhouses what they are. These are guidelines with flexibility, and you should use them to stay within the right ball park in terms of design and construction so as not to make gross errors. On the other hand you're not going to create a finely tuned sports car either. That's the beauty of passive solar systems: They're effective, and they're forgiving. Spend some time with planning. It's the best time to give your imagination free reign, because changes only cost you in pencils and paper. The more carefully you examine the options, the more likely you are to uncover problems that later on could mean wasted time and materials. You might also get a better handle on your answer to that seminal question: What do I want? To that end you are urged to read through the entire greenhouse section, particularly the last essay, which talks about greenhouse gardening. If you've been thinking heat, maybe horticulture isn't such a great unknown.

The Site

Whether you want plants galore or thermal gain, the greenhouse should face as close to due south as practical and receive direct sunlight for at least part of the day (at least four hours). You'll want it to attach to the house cleanly so it won't look awkward, though an angled attachment is possible if the house faces too far from true south. If the corner of the house happens to split true south down the middle. you can consider developing a wraparound design that puts part of the greenhouse on the southeast wall and the other part of the greenhouse on the southwest wall. It's important to be aware of the daily and seasonal shading patterns of nearby trees and buildings; check the drainage at the site or sites you're considering and look for buried pipe or cable where you plan to dig the foundation.

Although true south is the optimum orientation, greenhouses that face as much as 35 degrees east or west of south can still be viable heat and food producers. The more it's oriented to the east or west, though, the more it makes sense to glaze the opposite side wall to gain extra usable heat and light. For example, glaze the east wall of a southwest-facing greenhouse. Plants need all the light they can get, especially in winter, and a glazed side wall can help considerably when the orientation is less than optimal. If you have more than one choice for orientation, calculate the total heat gain for all of them based on one or more apertures and glazing tilt combinations. Then you can see if the solar heat gain in the coldest months will match your requirements. (Use the calculations in "Finding Your Heat Gain" in Section I.) If your south wall is on the far side of the garage, for example, then you'll have to develop a long air duct system to deliver heat to the house, which is perfectly possible. You may also be moved to remodel the garage into a living space, with the greenhouse as the heat source. If your south wall is close to a sidewalk, lot line, or your favorite shade maple, other issues are raised, such as how to protect the glazing from passersby and how to be assured that your neighbor won't build something that will block the sun.

In placing the greenhouse along a house wall, it's a good idea to build over one or more windows or doors to facilitate heat exchange between house and greenhouse. Sometimes, especially in a horticultural greenhouse, it may be necessary to supply the greenhouse with house heat to protect plants on the coldest nights. Placement also affects access: Locating the greenhouse over a house door will of course allow direct access from the house, which is useful during cold weather. People won't have to go outside to get in and this reduces infiltration heat loss caused by opening both a house and a greenhouse door to the frigid outdoors. In snow country take particular care that snow and ice that are shed from a pitched roof won't damage the greenhouse.

Photo III-81: For this retrofit the site presented some pretty tough obstacles. True south was right off the corner of the house, and at the same corner there stood a big shade tree. The owner's solution was to build an L-shaped greenhouse with enough southeast and southwest glazing to compensate for the reduced solar gain.

In terms of the house floor plan, a greenhouse placed against living, dining or kitchen areas will make a nice expansion of the living space. Placement next to a kitchen is desirable since food grown in the greenhouse is readily available to the cook.

Finally, keep an eye out for the mundane realities such as the location of utility service entrances, meter boxes, phone lines, water pipes and septic tanks or leach fields. Examine the house's structure at possible points of attachment to ensure that the studs, beams or joists are strong enough to support the added load.

Thermal Performance, Sizes and Shapes

The factors that affect heat production are solar collection, heat storage and heat transfer, and these factors just happen to match the guidelines that the Internal Revenue Service uses to determine if a solar system qualifies for the federal solar tax credit. In other words, if your greenhouse addition is designed to provide usable heat by collecting, storing and transferring it to the living space (including the greenhouse living space), then by definition it is a bona fide system.

Photo III-82: Along with its other benefits (a place for thermal mass and growing beds) a greenhouse kneewall is a perfect candidate for exterior insulating shutter reflectors of the type shown here.

There are several options for the shape of a solar greenhouse. The amount of glazing used and the extent to which it is tilted, or not tilted, make for a wide variety of possibilities for the way a greenhouse looks. Glazing configuration also affects performance, as evidenced in the research of solar designer Bill Zoellick, which is summarized in the box, "A Comparison of Different Greenhouse Shapes." As was mentioned, there isn't likely to be much of a difference in the per-square-foot cost of a small or large greenhouse, so the main deciding factors for size are budget and available area. A larger space will of course cost more initially, but it will also collect more heat and produce more food. If heat collection is your prime emphasis, it would be difficult, practically speaking, to build too large a greenhouse, i.e., one that would always overheat the main house. In northern latitudes, for example, every square foot of greenhouse glazing will provide heat for about 2 square feet of heated floor area. If a greenhouse were half as big as the house, it would certainly be a construction project of considerable magnitude.

As a general rule of thumb the length of the greenhouse should be about one-and-one-half to two times the width to ensure that sunlight will penetrate all the way

(continued on page 506)

A Comparison of Different Greenhouse Shapes

In this study Oklahoma solar designer Bill Zoellick modeled the performance of four glazing configurations for greenhouses with the same floor area (160 square feet) and the same floor-to-ceiling height (10 feet). All of the designs shown in figure III-77 have the same opaque roof configuration with a shed roof extending 5 feet from the house wall at an angle of 15 degrees. Design 1 has 7.3 feet of vertical glazing (116.8 square feet, with no subtraction for opaque framing members for the purposes of the model) and 5.2 feet of roof glazing at 15 degrees (83.2 square feet) for a total of 200 square feet of glazing. The glazing angles and dimensions of the other designs are shown in the illustration.

It's clear that the vertical glazing design incorporates the most square feet of glazing, though it's not necessarily the best design in terms of thermal performance.

The performance calculations shown in the table are for a latitude of 36 degrees, which would typically have relatively mild winters and very warm summers. At this latitude the need for limiting gain in summer is just as important as getting maximum gain in winter. Although these calculations are pegged to this particular latitude, they are still a useful reference for greenhouse performance at other latitudes, because the numbers are based on solar gain for a single "clear day" with no cloud cover. At

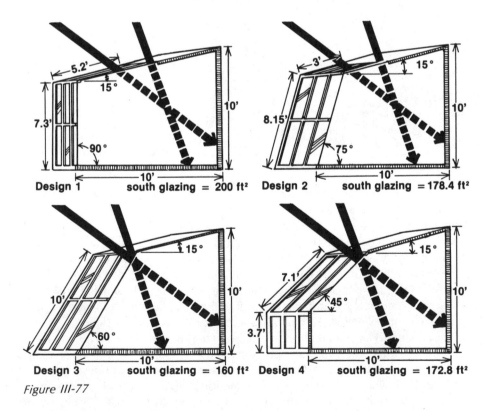

Figure III-77

Total Clear-Day Btu Gained through South-Facing Glazing					Btu Gained per Square Foot of South Glazing per Day			
	Jan. 15	Apr. 15	July 15	Oct. 15		Jan. 15	Apr. 15	July 15	Oct. 15
Design 1	274,900	254,700	228,100	285,000	**Design 1**	1375	1274	1141	1425
Design 2	274,500	253,800	213,000	284,500	**Design 2**	1539	1423	1194	1595
Design 3	272,800	248,800	203,200	280,800	**Design 3**	1705	1555	1270	1755
Design 4	274,800	249,300	211,200	282,200	**Design 4**	1590	1443	1222	1633

higher latitudes the solar altitude is lower, which tends to favor vertical and steeply pitched glazing. At lower latitudes the reverse is true. It should be noted that the numbers in the table represent performance for a "naked" greenhouse that has no movable insulation or summer sun screen. Both of these components would somewhat affect real performance. Also, neither glazing heat loss nor the effects of ventilation are included in the results. Again what we're looking at here is pure solar gain data for one sunny day.

Interestingly enough, all the designs perform about the same on January 15. However, design 3 has 20 percent less glazing area than design 1, but it collected only about 1 percent less solar energy than design 1. Since glazing can be the costliest material in a greenhouse, design 3 looks to be one of the most economical. It also does best at limiting gain in summer. It collects 11 percent less energy than the "hottest" design, number 1, and about 4 percent less than design 4, which has the second lowest gain on July 15. Both designs 3 and 4 have shed roofs that are completely opaque, which accounts for their reduced summer gain.

Design 4 has about 12 more square feet of glazing area than design 3, and a more involved framing plan with the glazed kneewall, but it comes in second in winter heat gain. For the horticultural greenhouse in particular, a glazed kneewall improves headroom, and allows placement of a planting bed at the vertex of the kneewall and tilted glazing, and thermal mass right behind the kneewall. The numbers are significant indicators of greenhouse performance, but they should always be looked at in relation to other factors such as cost of construction and the usefulness of the enclosed space.

There are of course a number of possible variations to these basic designs. Design 4 could have a kneewall, and the glazing could be tilted at 60 degrees. Design 2 (a close third in winter performance) could also have a 60-degree glazing tilt. Both of these variations would register better performance at higher latitudes. When you come to working out your own design geometry, performance will certainly be an important consideration, but so will appearance. If you prefer a design that may have performance disadvantages in winter or summer, worry not, because you can still employ glazing insulation, shading, and ventilation controls to balance minor performance deficiencies with a design you like.

to the house wall. Also, a width much greater than 10 feet can require somewhat more substantial roof and support structure, which increases costs without adding appreciably to heat output.

To maximize winter heat gain and the amount of light that reaches the back of the greenhouse, a rule of thumb for the glazing tilt angle is your latitude plus 15 degrees. However, a consequence of this glazing angle is that it increases the amount of direct gain in the "marginal" seasons, spring and fall, when overheating can be a problem if adequate shading or increased ventilation isn't provided. In the summer months vertical glazing is more self-shading than tilted glazing, and it is more easily shaded with a curtain or overhang. You don't want to overdo the summer shading if you are gardening seriously, because plants need a lot of light in summer to maintain peak growth throughout the year.

There is also the matter of the *kneewall*. Even slightly tilted glazing cuts down on headroom and therefore on readily usable floor space. A vertical kneewall essentially "jacks up" tilted south glazing, and in so doing headroom is increased and the southern-most strip of floor space is more usable. With a 3- to 4-foot-high kneewall, for example, a planting bed or countertop could be installed right at the break between the tilted and vertical glazing. Thermal mass in the form of water-filled barrels can then be placed under the bed to receive direct gain through the kneewall. It's a nice combination.

Glazing Options

Side wall glazing is a variable you can decide on. To glaze or not to glaze depends on how you plan to use the space and what the orientation is. As has been discussed, a southwest- or southeast-facing greenhouse would benefit by having side wall glazing on the east or west walls, respectively. Your desire for more privacy may run counter to your interest in side wall glazing. Also, opaque walls are already insulated, while a glazed wall would have to be included in the nighttime glazing insulation routine. Also, east- and west-facing glazing will admit more low-angle summer sunlight if it's not shaded, though compensation for that is increased ventilation. Finally, side wall glazing isn't an all-or-nothing deal. You can glaze the front half and make the rear half solid.

The greenhouse roof presents more options for glazing. To glaze or not to glaze depends on a number of factors. First of all it isn't assumed that you must build a shed-type roof. The tilted south glazing can be run up to the house wall, though this invites overheating problems unless some interior or exterior shading is provided. With a shed roof design common practice is to glaze just the front part and permanently insulate the back part. To determine where the line between glazing and roofing should be, you can use the sun path charts (see Section I) to study solar angles at different times of the year.

The sun reaches its highest noon altitude around June 21, though August may be the hottest time of the year. The lowest noon altitude is on or about December 21, though January or February may be the coldest month; they might also be somewhat sunnier than December. If you want to maintain maximum heat collection from the higher angle February or March sun, the opaque section of the shed roof would be

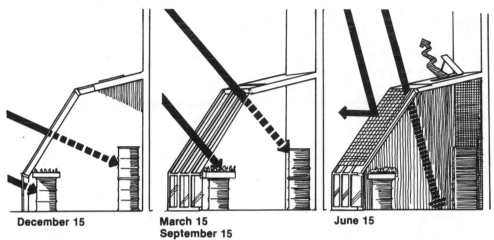

December 15   **March 15**   **June 15**
                         September 15

Figure III-78: As with any solar design, greenhouse design considers solar altitude angles for the whole year. In winter and mid- to late fall and spring the goal is to have maximum sunlight penetration into the greenhouse. But this goal must be tempered by the need to at least partially shade the greenhouse in summer. In the panel at right the opaque roof section extends out far enough so that summer sunlight cannot reach the back wall or the thermal mass if it is placed along the back. Further shading is achieved with a translucent curtain.

narrower than that of a roof that was designed to begin limiting solar gain after December 21. You can determine the time when the roof will begin to shade the greenhouse by varying the width of the opaque section. This will also have consequences in summer. A narrower roof section will of course admit more high-angle sunlight and increase the need for shading (movable curtain) and ventilation. A rule of thumb here is that at noon on the summer solstice, or a few weeks after the solstice when it's hotter, the opaque roof should not allow direct sunlight to reach the house wall or any thermal mass that might be up against the wall.

Other shading considerations have to do with controlling gain in summer. As with any solar glazing, deciduous trees can be a shading friend. Movable insulation can double as a summer shade although for the horticultural greenhouse it is considered better practice to use a translucent shade to limit gain without completely eliminating light.

Any of the common glazings discussed in "All about Glazing" (Section III-A.) can be used in greenhouses. You can select for cost, looks, longevity, ease of handling and privacy or view. For many reasons our preference has always been to use glass. It doesn't degrade from exposure to ultraviolet radiation; it's dimensionally very stable under wide temperature swings; it's available in relatively low-cost, double-pane, tempered units, and it's good looking, has high consumer acceptance, and needs minimal maintenance when it's installed properly. However, FRP glazing has the one advantage of diffusing sunlight and distributing the energy more comprehensively throughout the room.

In all but the mildest climates greenhouses should be double glazed, to reduce day and night heat loss. An effective lower-cost alternative to double glass is a glass-film combination (film on the inside). Double-wall, extruded acrylic glazing gives two glazing layers, and since it doesn't provide a clear view, it's a good choice for privacy when a greenhouse is built in tight quarters.

Insulation

Greenhouses can collect tens of thousands of Btu's on a given day, but the trick is to retain them as long as possible. When you have a tightly sealed greenhouse, heat retention is made possible with insulation and thermal mass. For foundations, rigid insulation (Styrofoam) is best, and it can be located inside or outside the masonry foundation. Inside the foundation it helps to isolate the dirt from heat loss. If remote rock storage is located in the greenhouse floor, insulation inside the foundation is preferable. Exterior insulation includes the mass of the foundation in the thermal dynamics of the greenhouse. A greenhouse with a slab floor should have exterior insulation or insulation between the edge of the slab and the foundation.

For the walls and roof, insulating with fiberglass and an inside vapor barrier of 6-mil polyethylene is common practice, although rigid insulation is as effective if fire protection can be assured (foam insulations give off poisonous fumes when burned). The vapor barrier must be virtually hermetic because in the moist greenhouse environment a small gap could leak enough water vapor to reduce the value of insulation and even cause dry rot in wooden members. One way to minimize the risk of moisture buildup is to use a closed-cell rigid foam insulation, e.g., Styrofoam, instead of fiberglass batts.

Glazing Insulation

A very important insulation component is movable insulation in the form of insulating shutters and curtains. It isn't needed if the greenhouse is used only as a daytime heat collector, but it is highly beneficial in the horticultural greenhouse and the "living space" greenhouse. It can reduce conductive heat loss through double glazing by 50 to 80 percent, depending on the R-value of the insulation. That represents a significant reduction in overall heat loss.

Night insulation is an emerging technology. There are a number of options—from homemade, pop-in rigid panels to the electrically powered Beadwall system, in which millions of tiny beads of polystyrene foam are alternately blown in (night) and evacuated (day) from between two layers of glass spaced 3 to 5 inches apart. An important concern with night insulation is to install something you will actually use, because if you don't in fact pop in the panels, they do no good. There are three basic options for simple installations. 1. Roller shades (manufactured or homemade) can be as simple as a double layer of reflective Mylar or a sandwich of fabric and an insulating

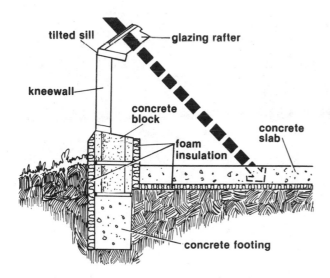

Figure III-79: There are a couple of options for placement of foundation and floor slab insulation. Insulation under slab should be laid over a gravel base that's been covered with sand. Exterior insulation will have to be protected with siding or cement stucco. To date the best known class of rigid foam for this purpose is extruded polystyrene (Styrofoam, Foamula 2).

material. Two factors greatly affect their performance: the insulating properties of the materials and the care taken to seal edges against convective air flow between the glazing and the insulation. At least three of the four edges must be pretty airtight. This can be achieved by using hinged slat of the type used with the insulating roller shade in Section III-B. Velcro or magnetic strips are also effective edge sealers. 2. Thermal curtains: Some commercially available varieties have sufficient insulating value, but again you've got to provide a tight edge seal. 3. Pop-in panels made of rigid foam are handled twice a day. They require storage when not in use, which may be a problem in a small greenhouse. Extruded styrene (Styrofoam) and polyisocyanurate foam (Thermax) are both good choices.

Some people have experimented with night insulation located outside the glazing, particularly on the kneewall. Hinged panels are opened by day so that the inside surface can function as a reflector to bounce more sunlight into the collector. (See the project, "Exterior Shutter-Reflectors for a Greenhouse," found later in this section.)

Thermal Mass

Do you need thermal mass for heat storage? The heating load of the house during the winter daylight hours is about one-third of the 24-hour total. If you are primarily concerned with providing heat to the house, you can assume that the house can comfortably absorb most of the surplus heat from the greenhouse without overheating. This operating routine essentially obviates the need for thermal mass heat storage. In many cases though, there is a need to store heat to prevent overheating in the greenhouse, to minimize day-night temperature swings and to help maintain comfortable temperatures at night for the "lived-in" greenhouse.

Photo III-83: Instead of putting in big water barrels for thermal mass, consider using smaller containers, which can allow for more flexibility before, during and after the greenhouse is put into operation.

The materials commonly used to store solar energy in greenhouses are rock, man-made masonry or water. Given their ability to hold heat, rocks generally require about two-and-one-half times the volume of water to be able to store the same amount of heat. This naturally has design implications. The greenhouse may easily have room for a barrel of water; but does it have room for two-and-a-half barrels of rocks? On the other hand, rocks don't leak, although the properly prepared water container can be relied on for leakproof service. (See "Moved-In Mass" in Section III-D.) A masonry wall can be effective, but expensive to build. It's better to use an existing masonry wall. A concrete slab can be insulated along the edges and underneath to be an effective thermal mass. Rock bins require additional excavation and the use of fans to move heated air across the stones.

In sizing the mass, you can use between 4 and 7 gallons of water per square foot of south glazing, depending on the severity of the climate, the available sunshine and the amount of heat you wish to store (a function of the greenhouse volume and the length of time it will be heated). The nice thing about water mass is that the amount is easily increased or decreased. For rock bins, assume 1 to 3 cubic feet of 1- to 1½-inch stones for every square foot of glazing.

The greater the amount of mass, the smaller will be the temperature fluctuations in the greenhouse, which is a good quality in the horticultural greenhouse. In some

Photo III-84: Fifty-five gallon drums are by now the classic thermal mass container for greenhouses. In this greenhouse they are stacked three high, which is considered to be a safe limit relative to the crush strength of the drums in the first course.

climates, a greenhouse without thermal mass can experience day-night temperature swings as wide as 70 degrees, which can be very hard on plant health. Overheating is also a waste of heat since heat loss is greatly increased. By sizing mass to keep the temperature swings within a moderate range of around 30 to 40 degrees, you will provide a more comfortable environment for both plants and people.

The rate at which heat can be stored is a function of the surface area of the storage mass that is exposed to direct sunlight and the temperature difference (delta T) between the surface and the interior of the storage mass. The greater the delta T and the greater the surface area exposed, the more rapid is the transfer of heat into storage. To speed heat transfer to storage, you should plan to charge a rock bed with the warmest air available in the greenhouse, which will be at the peak of the structure. A common approach is to automate this function by installing a standard differential thermostat to turn on a fan when the delta T between rock storage and greenhouse peak reaches a certain predetermined point. The fan pulls hot air through a duct and discharges it into the rock bed. (See the project, "A New Hampshire Greenhouse," following this essay.)

Mass performs most efficiently when it is placed in direct sunlight, which isn't hard to do in a greenhouse. Large amounts of mass, such as a wall of water-filled cans or barrels, should be located at the rear of the greenhouse and be painted flat black (or a flat, dark color) to increase absorption. If clear glass or semiclear plastic bottles are used for water storage, the water can be made dark with dye. Planting should be planned so that tall plants and leafy crops don't significantly shade the storage.

Heat Transfer

In a greenhouse there are four basic means of heat transfer that can be used singly or in combination. 1. Direct light transmission (direct gain) occurs through a transparent opening in the residence/greenhouse common wall. 2. Direct air exchange occurs between the greenhouse and the residence via vents and other wall openings such as windows and doors. 3. Conduction can occur through the residence greenhouse wall if it's an uninsulated masonry or frame wall. 4. Heat can be transferred to remote storage (a rock bin) and be used later to heat the greenhouse or the house.

Neither the direct gain nor the conduction modes are ever likely to be adequate as the sole heat transfer mechanisms, which means that an air exchange mode must be employed. This can involved either natural or forced convection or both, though experience has shown that forced convection is almost always necessary to limit overheating, even when thermal mass is present. Vents are needed to move heat into the house and also to exhaust heat (in summer) from the greenhouse.

In the house heating mode existing doors and windows can be used to control natural convection. To boost the flow of heat a window fan rated at about 200 to 400 cfm should be sufficient. The fan operation can be automated with an inexpensive cooling thermostat (line voltage type). A fan should be placed as high as possible to pick up the warmest greenhouse air.

For heat exhaust and plant ventilation, a greenhouse exterior door (intake) coupled with an exhaust vent can be very effective. For such a system to operate successfully by natural convection, the difference in height between the two vents is the key item. We have had success by having 20 percent of the glazing area filled with vents (intake plus exhaust). Good vent locations are the south kneewall (intake) and the top of either or both side walls (exhaust). A screened exterior greenhouse door (intake) can also be a useful intake and exhaust vent. Screening the vents to keep out bugs and small animals requires that you increase the area of the vents by about 30 to 40 percent since the insect screen itself impedes the natural convection flow. Plant ventilation requires not only keeping optimal greenhouse air temperatures (avoiding overheating) but also a regular flow of air around the leaves.

Natural convection is adequate for air movement in most properly designed attached greenhouses, but an exhaust fan will allow for reduced vent area. A 1000-cfm fan will be adequate for most greenhouses. Fans used to stir air around plant leaves should be placed at or near plant level. Care must be taken to keep fan installation from leaking water if the fan is installed through exterior wall or roof, and it will be important

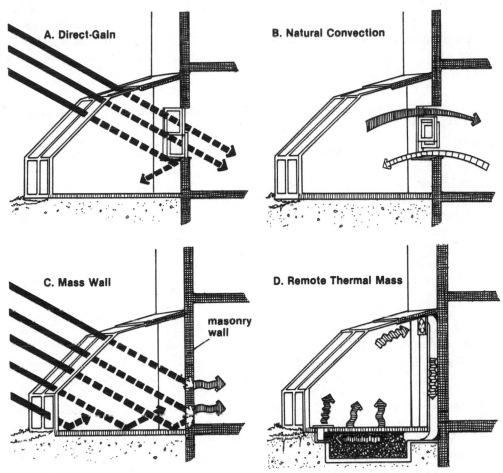

Figure III-80: Four basic options for greenhouse heat management include purely direct-gain heat to the house (top left), natural or forced convection from greenhouse to house (top right), thermal storage and conduction through a masonry wall (lower left) and remote thermal storage with redistribution to the greenhouse or the house. The latter system, being the most complex, is described in detail in the project that follows.

to seal the fan opening in winter to stop infiltration heat loss.

The storage and transfer of heat that is collected in a greenhouse creates an energy system. How you manage the heat of course depends on your goals, but there are three main management options. 1. The greenhouse can be designed to heat itself completely day and night, requiring thermal storage in the greenhouse. In this mode the house will receive only surplus heat, i.e., heat that is available beyond the capacity of the thermal mass to absorb and contain it. 2. The greenhouse can operate in cycle with the house, providing most of its daytime heat to the house, then drawing on auxiliary (house) heat at night, when necessary. 3. The greenhouse can be operated so that surplus heat is stored in a thermally separate (isolated) rock bin that can feed either house or greenhouse, as required. This offers maximum flexibility. You can have it one way in December and another in March, but it also increases the expense and complexity. After a while you'll get a feel for the way your greenhouse operates and you'll fall into a simple heat management routine.

Construction

When you've thought and schemed until you can't squeeze another inch of floor space or south glazing into your plans, you are probably ready to build. Even now, though, your plan isn't necessarily etched in stone, because as soon as you pick up a shovel or wield a hammer, you may start to see things differently. Some changes will be minor, but others could be more involved, and as with any major project, exciting new possibilities sometimes reveal themselves during the construction phase.

The first construction step is site preparation: salvaging shrubs or sod, rerouting any pipes or cable that might be in the way or would be exposed by the excavated foundation. The soil excavated for the foundation should be retained for backfilling after the structure is built.

In freezing climates, it's important to place the footings below the frost line, whether you use a continuous concrete wall foundation or concrete piers. In frost areas, insulate the foundation all the way to the frost line, either inside or outside the foundation. For that portion of the foundation that is above ground, 2 or 3 inches of rigid insulation are fastened on the outside and coated to prevent damage. White beadboard has been found to be used by carpenter ants for nesting purposes, but blue Styrofoam hasn't been known to have this problem, and it's a better insulator anyway. Exterior coatings that are applicable over rigid insulations include: Thoroseal, Block Bond, Pressure Wall, Easi Mix, Hydrasite 60. Another coating option is to use a hard, corrugated asbestos sheathing called transite, which is available from masonry suppliers.

The junction of foundation and the wooden sill is, over time, a critical area because of the potential for moisture buildup that could lead to wood rot. Treat the sills with copper napthanatex as a preservative (never creosote or penta, which harm plants), and countersink and plug the holes of the anchor bolt sill fasteners to keep them from rusting.

Framing is usually done with 2×6 stock for both the side wall and roof rafters, and for the rafters that carry the glazing. Figure III-81 shows some basic framing plans for a greenhouse. It is of critical importance that the construction at all points be tight against air leaks, downright hermetic in fact. Buy yourself a case of high quality caulk so there'll always be some on hand to shoot into nooks and crannies.

Although designs can vary, a normal attachment procedure is to attach a 2×6 ledger board across the length of the residence wall that will be covered by the greenhouse. This should be secured with lag bolts into the existing house studs at the level desired for the bottom of the greenhouse roof rafters.

Side-wall framing of the greenhouse is standard. The side-wall stud closest to the residence should be attached to the residence wall stud. If the greenhouse is designed in a standard modular length, both side walls should run into residence wall studs. The spaces between the house wall and the end stud should be caulked to prevent infiltration. This should be done at the framing stage and again after the greenhouse siding has been applied.

Where the greenhouse roof meets the residence roof, standard wall flashing is required. Normal flashing procedures apply: Remove a length of house siding, slide an

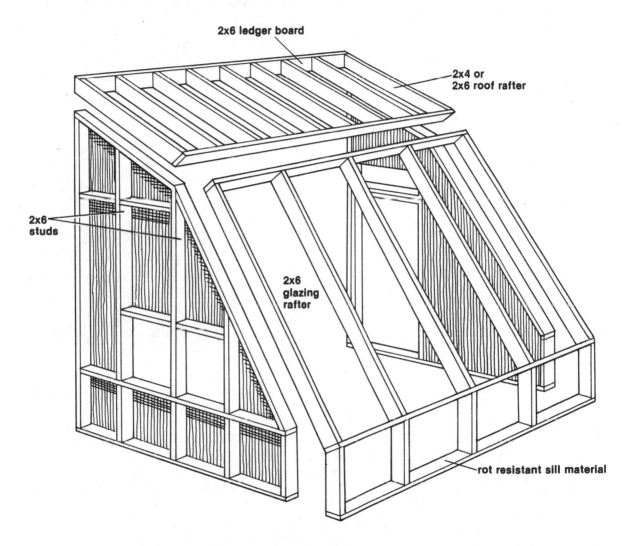

Figure III-81: The basic framing of a greenhouse involves standard carpentry practices. The side walls are framed with 2×6's so they can carry more insulation. The opaque roof section can be framed with 2×8's or 2×10's to carry even more insulation. Note that where the glazing rafters meet the kneewall jackstuds, the sill is tilted so that water running down the inside of the glazing (from condensation) doesn't puddle and ultimately rot the sill.

Photo III-85: No one ever said it wasn't okay to build a greenhouse upstairs if there isn't enough room or sunlight downstairs. When it comes to adding thermal mass, the support structure, in this case the porch roof, must be thoroughly capable of bearing the extra load, or it must be reinforced.

8- to 14-inch-wide piece of flashing to the wall, replace the siding, lay the other edge of the flashing down on top of the greenhouse solid roof and cover with roofing material and sealant.

We can't instruct you in all the basics of carpentry and greenhouse framing, but it's a very straightforward sort of undertaking for someone with basic carpentry experience. There is of course plumbing and electrical work that has to be planned and built in, though in small- to medium-sized greenhouses these components won't overly complicate matters. A greenhouse is one of several possibilities for a solar addition, more of which are presented in Section IV. The whole idea of a solar addition is one that requires careful planning. It's not just more living space; it's an energy system that is integrated with the existing building, and in the case of a greenhouse it's a specialized horticultural space that is probably unlike any other space in the rest of the house. This essay is a start in the planning and design of an attached solar greenhouse, and the articles that follow

Photo III-86: The planning and design stages of your greenhouse ought to be the fun that leads to the work of construction. The work can be fun too. You can organize the bigger parts of the project to be a solar barn raising, a gathering of some handy friends who can help you put the walls up and the roof down. (Photograph courtesy of The Habitat Center, Walnut Creek, Calif.)

in this section will provide a lot more useful information, but don't limit your research. The following reference section is intended for you to expand your studies so you can find out more about the state-of-the-art of the attached solar greenhouse. Greenhouses have gone through a lot of design evolution in the past few years, and the more you know about it all, the more complete your own design palette will be.

Dennis Jaehne and Sheldon Klapper

References

Clegg, Peter, and Watkins, Derry. 1978. *The complete greenhouse book: building and using greenhouses from cold frames to solar structures.* Charlotte, Vt.: Garden Way Publishing Co.

DeKorne, James. 1978. *The survival greenhouse: an eco-system approach to home food production.* 2d ed. Culver City, Calif.: Peace Press.

Ecotope Group. 1979. *A solar greenhouse guide for the Pacific Northwest.* 2d ed. Seattle, Wash.

Fontanetta, J., and Heller, A. 1979. *Building and using a solar-heated geodesic greenhouse.* Charlotte, Vt.: Garden Way Publishing Co.

Geiger, Rudolf. 1965. *The climate near the ground.* 4th ed. Cambridge, Mass.: Harvard University Press.

Hayes, John, and Gillett, Drew, eds. 1977. *Proceedings of the Conference on Energy-Conserving, Solar-Heated Greenhouses.* Marlboro, Vt.: Marlboro College.

Hayes, John, and Jaehne, Dennis, eds. 1979. *Proceedings of the Second Conference on Energy-Conserving, Solar-Heated Greenhouses: solar greenhouses: living and growing.* Brattleboro, Vt.: New England Solar Energy Association.

Hix, John. 1974. *The glass house.* Cambridge, Mass.: MIT Press.

Jensen, Merle, ed. 1976. *Proceedings of the Solar Energy Food and Fuel Workshop.* Tucson, Ariz.: Environmental Research Laboratory, University of Arizona.

Kasprzak, R. 1977. *The passive solar greenhouse and organic hydroponics: a primer.* Flagstaff, Ariz.: R.L.D. Publications.

McCullagh, James C., ed. 1978. *The solar greenhouse book.* Emmaus, Pa.: Rodale Press.

McDonald, Elvin. 1976. *How to build your own greenhouse.* New York: Popular Library.

Nearing, Helen, and Nearing, Scott. 1979. *Building and using our sun-heated greenhouse.* Harborside, Maine: Social Science Institute.

Seeman, J. 1974. *Climate under glass.* Geneva: World Meteorological Organization.

Smithsonian Science Information Exchange. 1979. *Solar greenhouse.* Washington, D.C.

Wolf, Ray. 1980. *Rodale's solar growing frame.* Emmaus, Pa.: Rodale Press.

Yanda, Bill, and Fisher, Rick. 1979. *The food and heat producing solar greenhouse.* Santa Fe: John Muir Publications.

PROJECT

A New Hampshire Greenhouse

By now there may be 10,000 attached solar greenhouses in this country, give or take a few thousand, and there's probably not a carbon copy among them. They all have their quirks and details, but they all share a lot of the same features. This project describes a greenhouse that is pretty representative of current designs. It's a 10-foot-wide by 20-foot-long addition with a kneewall. The central element of its energy system is the rock thermal storage bed located under the greenhouse floor. It is charged by hot air drawn from the greenhouse ceiling and blown through the rocks. The advantage of the "remote" heat storage is that it saves space in the greenhouse. There's plenty of room for growing, plenty of room for living.

The village of Alexandria is nestled into the foothills of New Hampshire's White Mountains, a place where winters are simply quite cold. This is where Peter and Madelyn Brown live, and in order to live more comfortably and with more self-reliance against cold winters and rising food and fuel bills, they added a solar greenhouse to their old white clapboard "cape." As far as the Browns are concerned, the wedding of traditional architecture and solar design has come off quite well. Their home's colonial roots remain undisturbed yet updated by the solar addition, which provides heat, food and just a nice place to sit.

Like many people, the Browns had their "energy awakening" during the first oil embargo of 1973, and they made their first energy investment in insulation for the walls of their venerable New England foursquare. But, anticipating the solar addition, they didn't insulate a 20-foot-long section of their south wall where the greenhouse now sits. This way solar heat can more easily be conducted through the frame wall. Besides providing heat and shelter for the Browns and their plants, the greenhouse also houses a small overhead water tank that is solar-heated to provide hot showers in a very pleasant greenhouse environment.

In 1978 the Browns spent about $1500 for materials to build their greenhouse, which was re-

cently appraised as being worth at least $5000. That's what an addition like this could cost if it were hired out to a contractor, but the Browns went at it in a different way, a way that saved money and gave solar building experience to a group of homeowners, all of them potential greenhouse builders. The Browns saved by using framing materials produced from their own backyard sawmill, certainly a viable option if you've got access to timber and a way to move it around. They probably saved the most, however, by having the project built as part of a solar greenhouse workshop sponsored by the New Hampshire Audubon Society. The Browns provided all the construction materials and prepared the greenhouse foundation, and the workshop's volunteer labor force did the rest.

"There were 23 people who showed up here one Saturday morning, in the rain by the way," as Peter Brown recalled the event. "They came from as far away as Montreal and eastern Massachusetts," noted Madelyn, "and some of the Massachusetts people were part of an inner-city group planning to erect greenhouses on vacant lots in low-income neighborhoods." They all came to gain first-hand experience in building solar greenhouses, experience they could take back to their homes and communities. Under the sponsorship of organizations like the Audubon Society, state and federally funded public

Photo III-87: The Browns' greenhouse shows how a solar structure can be blended with the existing house by matching siding, paint colors and roof materials. Proportions and glazing tilt and roof pitch angles are other design elements that should be carefully considered.

interest groups, and other organizations, such workshops and their offshoots have built literally thousands of greenhouses all around the country. It's the old-time barn raising redefined for the solar age, and the Browns' experience typifies the good that comes of cooperation.

With the foundation, the rock thermal storage and the concrete floor already in place, the work crew completed the rest of the greenhouse in just two days. That is testimony not only to "speed in numbers" but as well to the simplicity of the greenhouse structure. Naturally, for a workshop to succeed as well as this one did, all tasks must be well organized: There should be a job for every worker or pair of workers; all the materials must be ready at the site, right down to the screws, nails and nitbits; and there should be a leader to organize and direct every step of the construction. (If the workshop strikes you as a way to go, you'll find a couple of references in the bibliography to help you get it together.)

As a regular construction job it would be good to have a minimum work crew of three or four able-bodied souls. A fair estimate would be that such a crew could build the whole thing from foundation to finishing touches in two to three weeks, for a greenhouse of 200 to 300 square feet of floor area. For the weekend owner-builder anywhere from two to three months or more should be allotted, depending on the weather, the help and so on. But by any count the task is not so huge or complicated as to be beyond the capabilities of someone with basic carpentry, plumbing and electrical skills. There should be at least one person in the crew with that sort of experience. There is of course more than one way to build these things, but the Browns' retrofit is a good example of what is being done coast to coast.

The Foundation and the Rock Bed

The Browns' greenhouse is actually a hybrid system, combining passive direct-gain heat collection with a system of ducts and a small fan for distributing the heat collected in the greenhouse to the rock storage bed or to the living space. The rock bed, located under the slab floor is known as a re-mote heat storage component; that is, it's removed from direct contact with solar radiation, which is why the ducts and fan are used to move heat around. Since this storage is integral with the foundation, it must be planned and developed right at the beginning of the project. The alternative to remote storage is to place thermal mass, usually water in containers, in the greenhouse space. Water barrels can be used both to support growing beds and to keep them warm.

To start, the Browns hired a local earth-moving contractor to excavate a 10 by 20-foot pit on the south-facing grassy slope adjoining their house. The hole was dug 4½ feet deep, the depth required to contain the rock bed and the overlying floor slab. An important design detail is to have the greenhouse floor finish out a couple of inches below the floor level of the existing living space. After the pit was excavated, the Browns poured a concrete footing (8 inches wide by 8 inches deep) around the perimeter below the frost line (4 feet below grade in this re-gion). Then they laid up a foundation wall of concrete blocks bringing the top of the foundation about 1 foot higher than the planned finish floor level of the greenhouse. This 1-foot extension puts the wooden sill plates that much more above grade and keeps them and the subsequent framing drier.

They insulated the entire interior of the block wall with 2 inches of Styrofoam to keep the heat from conducting sideways from the rock bed through the masonry foundation and into the earth. Then they leveled off the bottom of the earth pit and insulated it the same way. A 6-mil polyethylene vapor barrier was placed around the inside of the foundation and over the floors, completely covering the insulation board. It was also folded over the top of the wall to protect the insulation during construc-tion. The insulation is a critical component for below-grade, thermal-mass systems. Earth gives very little insulating protection to the valuable stored heat. "We wanted all the heat right up here," Peter ex-plained, indicating the concrete floor of his solar

greenhouse.

After insulating, the base of the rock bed was prepared for an air handling system. Peter then laid standard 8 by 8 by 16-inch hollow concrete blocks on the polyethylene sheet at the bottom of the pit, with their hollow cores aligned and running par-allel to the ground, forming channels to carry the warm air. He left spaces between the blocks to aid air circulation and separated each row by 6 to 8 inches, covering the entire floor of the pit with rows of blocks. As will be described, the channels formed by the blocks carry the return forced-air flow after the hot air has passed through the rocks, which lie over the blocks.

The blocks were covered with a 10 by 20-foot piece of livestock fencing with a 6-inch-square mesh, and the fencing layer was covered with strips of ½-inch mesh "hardware cloth." The fencing and hardware cloth support the stones, which were the next addition to the storage bed. Peter carefully poured 6 cubic yards of 1-inch stones over the screening, bucketful by bucketful, so as not to upset anything. (It's important that the stones be washed free of dirt and dust, which could foul the forced-air system.) He covered the surface of the wire with an 8-inch layer of stones, and stopped to add the ple-num and return ducts of the air-spaced 4-inch holes cut into opposite sides (not the bottom or top). One end of the plenum was outfitted with a wooden "right-angle elbow" intended to extend up through the rock storage bed and protrude above the con-crete pad into the greenhouse itself. From there it would be connected with the draw-down duct that extends to the peak of the greenhouse. The last foot of the other end of the plenum was tapered to create a bottleneck effect to ensure that the heated air would in fact reach the far end of the plenum.

The 4-inch diameter holes on each side of the plenum were outfitted with PVC pipe that spanned the entire rock bed. You can use factory-perforated drain-field pipe for this job, but Peter pre-ferred to start with solid pipe and decide for himself where to drill the holes, thus gaining a little more control over how the air would be distributed. He drilled a row of 1-inch holes—4 inches apart—on

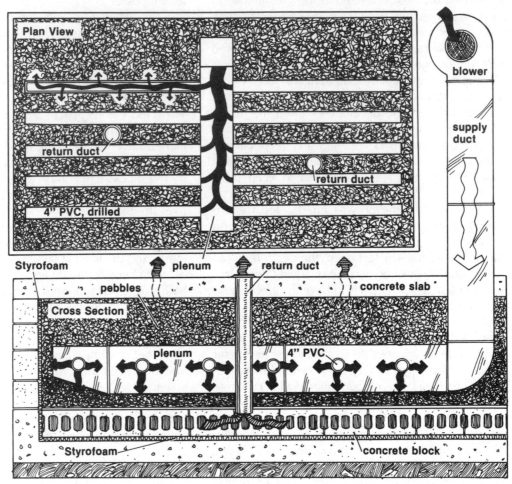

Figure III-82: Two views of the rock thermal storage system show how the blower feeds the manifold and how the manifold feeds the perforated PVC pipes that extend into the rock bin. The concrete blocks under the rocks provide a necessary plenum for the return air to collect and flow back into the greenhouse through the return ducts.

each side of the PVC pipe sections. He then drilled ½-inch diameter holes, again 4 inches on center, along what would be the bottom of each pipe. One end of each pipe was capped with a standard PVC fitting. Thus prepared, all 12 pipes were fitted into the plenum, with the open end stuck in the box and the undrilled surface facing up. The entire assembly was carefully bedded with more crushed stone, and the

rest of the crushed rock was added to bring the top of the rock bed to a point just covering the pipes and plenum. The rock bed was then covered with another sheet of 6-mil polyethylene, and over this was poured a 6-inch concrete slab. When the concrete had dried, the Browns painted the slab with a standard black masonry paint to enhance its ability to absorb the sun's rays.

Framing and Glazing

After the foundation work comes the framing, glazing, insulating and general finishing. The south wall of the greenhouse starts with a 2-foot-high masonry kneewall (the top of the foundation) topped with a framing sill that carries 2×6 glazing rafters set at 60 degrees from horizontal, a pitch that represents the optimal angle for wintertime heat collection at this latitude.

The kneewall and the east and west above-grade foundation are insulated with 2 inches of Styrofoam to the outside. The foam is then covered with fastened stucco. As was discussed in the preceding essay, the kneewall could be a framed and glazed wall section, rather than masonry.

The side walls are framed with 2×6's, and the opaque roof section is framed with 2×8 rafters. Both the studs and the rafters are placed on 2-foot centers. This is rather beefy construction for the likes

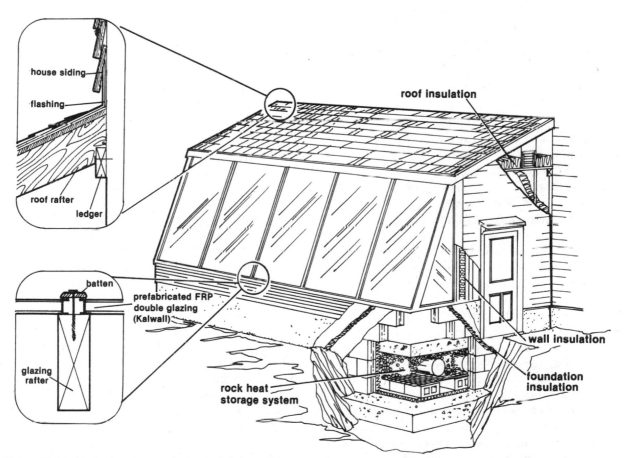

Figure III-83: The primary attachment of the greenhouse to the existing wall is by way of a 2×4 and 2×6 ledger lag-bolted to the studs. Removing one or two courses of siding above the greenhouse roof rafters simplifies the job of flashing. The Kalwall double-layer FRP glazing panels are easily fastened to the glazing rafters with a simple batten and gasketing.

of a greenhouse, but it allows for thicker insulation to be placed in the cavities between studs (6-inch fiberglass, R-19) and rafters (8-inch batts, R-27). Even the greenhouse door is insulated, in this case with 3 inches of Styrofoam (R-16). The back wall of the greenhouse (common with the house) remains uninsulated. Otherwise the framing involves standard carpentry procedures with careful attention to sealing all possible points of cold air infiltration. The east wall includes a triangular glazing section, and a glass door for morning light. All nonglazed framing is sheathed and finished with siding that matches the house siding. An intake air vent is built into the lower part of the west wall. The screened door opening serves as the exhaust vent.

To glaze the south wall the Browns used prefabricated, 4×8-foot, double-glazed panels (glazed with FRP) built into extruded aluminum frames (made by Kalwall). These frames are fastened to the glazing rafters with screws, and a weather sealing trim strip covers the seams between panels. It's a simple glazing system.

The air handling system was completed by extending the 12 by 12-inch plenum protruding through the concrete floor to a point about 6 feet above the floor, where it forms a tee. A rotary duct fan powered by a ½-hp motor is mounted on one side of the tee, and on the other a duct leads into the house, with a simple plywood hatch separating the house from the duct. When the blower is operating and the hatch is open, heated air flows from the top of the greenhouse through the top part of the tee and directly into the house. When the hatch is closed, the blower forces hot air from the greenhouse down the plenum and through the rock bed, where it gives off its heat to the cooler rocks. The air travels to the bottom of the storage containment (where the coolest air is) and back up through the return ducts (the 4-inch solid PVC pipes). An adjustable thermostatic switch located inside the greenhouse automatically turns on the blower when the air temperature inside the greenhouse is warm enough, typically over 85°F (29°C) for direct house heating.

Operation

Things start to happen in the Browns' greenhouse shortly after sunrise. Growing more intense by the minute, the sun's rays pass through the glazing and start to warm the slab floor and the wall of the house. On any frigid-but-sunny winter day, this absorbed solar energy soon creates enough temperature difference between the air and the rock bed to trigger the blower.

When they are home, the Browns open the hatch so the blower forces heated air into the living space, letting direct space heating take precedence over heat storage. When they do want to maximize heat storage, they close the hatch connecting the duct with the house, forcing warm air back into the rock storage bed. "We can warm the rock storage bed up to 60 to 70°F in five to six hours on a sunny winter day," explains Peter. The effect of this warm thermal mass is to keep air temperature inside the greenhouse above freezing—even during the coldest winter nights.

There is an informal strategy involved here, one planned around the family's day-to-day comings and goings. The Browns' idea is to maximize solar space heating whenever they're at home and to heat the rock bed when they aren't or when the coldest weather endangers greenhouse plants. Their primary heat source is wood burned in two woodstoves, with an old oil burner as a reliable but unpopular backup to the wood and solar systems. After gaining some experience in managing heat from the greenhouse, the Browns have developed a plan that better distributes heat through the house. By opening a door from the bedroom to a hallway leading to the second floor and closing another door to the den, heat is directed by natural convection upstairs to the children's rooms. The air flow continues back downstairs via another corridor, into the den and out to the greenhouse again, a complete loop. The establishment of a whole-house air circulation pattern is a good way to make the most of solar heat distribution, and the aspiring greenhouse builder will do well to try to

Photo III-88: The air handling system delivers heated air either to the rock storage under the greenhouse floor slab or directly to the house, with the switch being made by a simple hatch cover located at the outlet to the house. The short column at right is one of the return air ducts (from rock storage back to the greenhouse space).

establish this sort of total circulation, either by using the house as it stands, as the Browns do, or by making appropriate modifications.

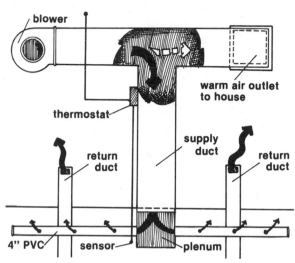

Figure III-84: This schematic of the Browns' air handling system shows how the blower can deliver warm air directly to the house or to the rock bin simply by opening and closing a hatch over the house duct. When the temperature difference between the greenhouse air and the rock bin is great enough, the blower is actuated automatically by the thermostat.

Performance

The Browns built their greenhouse without major problems, and they are committed to getting the most they can from it. But how well does it work in northern New Hampshire? Is it really equal to the challenge? "It works," says Peter. "We had a lot of cold weather here last February, and one night the temperature dropped past 20 degrees below zero. It was 50 here in the greenhouse," he said. "And on a sunny winter day I can sit at my desk, or Madelyn can use the bedroom, and we don't need any other heat, even on the coldest days. And we've never had to use anything other than solar heat to keep the plants alive either." Peter Brown states that the daytime heat output from the greenhouse keeps the den and children's rooms at a comfortable 65 to 75°F (18–24°C).

Some solar greenhouse designs include

earth berms on the sides to provide additional insulation, and many greenhouses are equipped with some type of shutter system used over the glazing for additional nighttime insulation. But, in theory at least, a well-designed solar greenhouse with sufficient heat storage capacity should maintain above-freezing temperatures throughout the year without night insulation. The Browns' experience supports the theory.

Over the past winters, Peter, Madelyn and their children have also enjoyed fresh salads all winter long, raising endive, spinach, tomatoes and parsley as well as several types of lettuce in greenhouse pots and beds. Peter planted a late winter crop of melons and winter squash, while Madelyn raised begonias, ferns, ivies and other ornamentals in hanging pots. Of course, a solar greenhouse's food production can only be as good as the gardeners who manage it, so it's important to cultivate your indoor gardening skills. A solar greenhouse is a unique, miniature ecosystem with different requirements from an outdoor garden.

Many solar designers agree that the payback on a solar greenhouse installation will come faster with food production than in heating value alone. Certainly, the economics of the basic solar greenhouse are sound. A payback period of four to eight years for the value of food and heat produced can be attained if serious effort is put into food production. Outside of food and heat, there is of course the added value of sun-filled living space that the greenhouse can add.

"We've always had some sort of garden, but we were never terribly good at it," Madelyn explained. "The solar greenhouse really gave us a green thumb. It's like having an extra room, too. We'd always talked about adding a screened porch, but actually this space is much more useful. The kids come out to play and do their homework in here during the winter."

The Browns also learned how to make use of their solar addition during the summer. "By turning on the blower system when it gets hot in here"

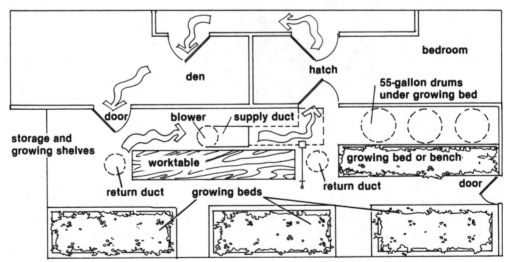

Figure III-85: Often the floor plan of a retrofit greenhouse is greatly affected by the arrangement of doors and windows on the existing house wall. The Browns had to allow for two doorways opening into the greenhouse, so they put a worktable between them and arranged growing beds along the south side.

Photo III-89: For the interior design of their greenhouse, the Browns combined plant-growing facilities with their solar-heated shower and, as shown in the preceding photograph, a "family table" for meals, homework and get-togethers.

six hours on a sunny winter day, I know that we could charge additional mass as well. Sometimes during the winter we'll have four or five cloudy days in a row. More thermal mass would provide more heat to carry us through those times." To get that extra thermal storage capacity, Peter has added a row of those familiar water-filled 55-gallon drums along the north wall of the greenhouse. By expanding the unit's thermal storage capacity, these drums further minimize temperature fluctuations while providing a handy support for a growing bed.

On Sizing and Design

The 10 by 20-foot dimension for the greenhouse was arrived at because it made for a good blend with the size and shape of the house. A larger greenhouse might well have overpowered the humble cape, while a smaller structure would have been less useful in all regards—for heating, growing and pure enjoyment. A 10 by 16-foot solar greenhouse will usually fit well alongside most average-sized houses and can produce enough food to keep a family of four in salad and other greens all winter long, as well as providing a good chunk of the house's space heating.

With many options to consider, a solar greenhouse addition should be thoroughly planned before the picks and shovels come out. You want to create a workable system, not just a structure, and you want to make a good integration with the existing house. As Peter Brown concludes: "You are limited only by your imagination in designing a greenhouse that suits you. We like to spread the word that duplicating a greenhouse like this can be done easily. The most difficult part was excavating the pit and pouring the foundation. The most intricate part was installing the air manifold. Still, it's pretty cut and dried. Anyone can do it."

Peter explained, "we can cool 100°F air down to 60° after it's passed through the rock storage bed."

Now that the greenhouse has operated for a couple of years, the Browns can see that a couple of changes could make good operation better. "If I were doing it over again, I would add about 10 percent more crushed rock," says Peter. "Because the present thermal mass can be charged in just five or

Michael Harris

Hardware Focus

Greenhouse Plans

Applied Technologies, Inc.
Solar Energy Division
Lyndon Way
Kittery, ME 03904

The Solar Greenhouse
Two sheets of construction drawings. Greenhouse size: 7'5" (H) × 6' (W) × 6'6" (L).

Cornerstones Energy Group
54 Cumberland St.
Brunswick, ME 04011
(207) 729-0540

Cornerstones Solar Greenhouse
Plans and materials list for greenhouse 16' or 21' × 10'. Rock storage under floor. Includes plans for thermal shutters.

Ecotope Group
2332 E. Madison St.
Seattle, WA 98112
(206) 322-3753

A Solar Greenhouse Guide for the Pacific Northwest
Step-by-step instructions for designing and constructing a solar attached greenhouse and a window box greenhouse. Text, drawings and materials lists.

Garden Way Publishing Co.
517 Ferry Rd.
Charlotte, VT 05445
(802) 425-2171

Solar Room
Plans for freestanding or attached greenhouse of variable dimensions.
Build Your Own Window Greenhouse
Plans for double-glazed greenhouse. Area over 75 cubic ft. Adaptable to any window.

The Lightning Tree
P. O. Box 1837
Sante Fe, NM 87501
(505) 983-7434

An Attached Solar Greenhouse
By W.F. and Susan Yanda; 10' × 16' greenhouse as example; management tips and planting schedule included.

Maine Audubon Society
Gilsland Farm
118 U.S. Rte. 1
Falmouth, ME 04105
(207) 781-2330

Dual Mode Lean-to Greenhouse with Gravel Storage and Maine Vocational Region Ten Hybrid Solar Greenhouse
Designed by Charles Wing. Hybrid greenhouse size: 10'8 (H) × 11'9 (W); lean-to size: 13' (H) × 12'2" (W). Blueprints include notes on materials.

John Muir Publications, Inc.
P. O. Box 613
Sante Fe, NM 87501
(505) 982-4078

The Food and Heat Producing Solar Greenhouse
By Bill Yanda and Rick Fisher. This book contains complete construction details for building a solar attached greenhouse. Includes a list of materials and tools for building a 10' × 16' greenhouse.

Sierra Club Books
530 Bush St.
San Francisco, CA 94108
(415) 981-8634

The Integral Urban House
Contains working plans for window greenhouse suitable for double-hung or casement windows that open in; also, working plans for solar greenhouse (8' × 16'); may be lengthened by 4' increments; manually operated vents; steel drums for thermal storage; gravel floor. Drawings, materials lists and construction procedures included.

Solar Applications and Research, Ltd.
3683 W. 4th Ave.
Vancouver, B.C.
Canada V6R 1P2
(604) 733-5631

Solar Sauna
Box 466
Hollis, NH 03049

Solstice Publications
P. O. Box 2043
Evergreen, CO 80439

Weather Energy Systems, Inc.
39 Barlows Landing Rd.
Pocasset, MA 02559
(617) 563-9337

Four Seasons Solar Products Corp.
672 Sunrise Highway
West Babylon, NY 11704
(516) 422-1300

Green Mountain Homes
Royalton, VT 05068
(802) 763-8384

Pacific Coast Greenhouse Manufacturing Co.
8360 Industrial Ave.
Cotati, CA 94928
(707) 795-2164

Solar Resources, Inc.
Box 1848
Taos, NM 87571
(505) 758-9344

Solar Technology Corp.
2160 Clay St.
Denver, CO 80211
(303) 455-3309

Sun-Ray Solar Equipment Co., Inc.
4 Pines Bridge Rd.
Beacon Falls, CT 06403
(203) 888-0534

Solar Reliant Greenhouse
Blueprints and siting and operations guide; typical width of greenhouse 8′ to 10′ minimum; length selected is multiple of glazing unit; space allotted for water storage; double or triple glazing; air lock entry; insulated reflector/ shutter for glazed roof section closes at night or during summer heating conditions.

Solar Amenities Module
Construction plans, materials list and cost estimates for 12′ × 24′ sunspace combining a solar greenhouse, solarium, hot tub, sauna and deck space.

Three sets of plans available: Free-Standing Solar Reliant Greenhouse; 10′ × 20′ Single-Story Attached Passive Solar Greenhouse; Two-Story Attached Greenhouse. Very extensive detail.

Sun Haus Greenhouse/Solarium
Detailed blueprints and construction and operation manual; size: 10′ (H) × 8′ (W) × from 12′ lengths. Double glazing; storage in floor.

Greenhouses

Add-A-Room Greenhouse
6′ × 8′ or 6′ × 12′; double polycarbonate glazing or triple glazing with polycarbonate and glass; aluminum frame; air circulation system; rock storage available.

Solar-Shed Sunspace
8′6″ × 13′8½″, 16′9½″ or 19′10½″. Kit includes directions, double-glazed glass in wood frames and all building materials except thermal mass, roof shingles, insulation and optional fan.

Prefabricated redwood and glass or fiberglass greenhouses designed to customer's specifications.

Solar Room
8′ (H) × 9′ (W) × 9′ to 39′ (L) in 3′ increments. Galvanized steel ribs; air inflated double glazing of heavy, plastic film; redwood frame; air circulation active or passive; can be dismantled in summer.

The Soltec GreenRoom
10′ × 12′ × 12′ expandable by 4′ sections; roof glazing: Exolite double-wall acrylic sheet; wall glazing: tempered thermopane glass; redwood frame; automatic ventilation system; options include thermal storage and movable insulation.

Sol-Arc Solarium
13′ 7½″ (L) × 8′ (W). Extra length in increments of 26″. Wood, glass and aluminum construction; double-glazed. Sol-Arc collectors on roof of greenhouse provide heat for domestic hot water while providing shade in summer and insulation against heat loss in winter.

A Two-Story Greenhouse with Passive Water Heating

Harry and Bertha Riley were thrilled when the Lancaster County Solar Project offered to build them a two-story, lean-to greenhouse for their row home in Marietta, Pennsylvania. The Rileys' house interested the Solar Project (a local community action group) because it had a good southern exposure on the rear wall (facing the backyard). At the first-floor level a basement dug into a slope had a foot-thick foundation wall that could provide a healthy supply of thermal mass. The lean-to design with its tapering cross section would provide ample floor space for growing beds while the height of the two-story design would encourage natural convection of warm air into the house. The greenhouse would also be roomy enough to hold a breadbox passive water preheater.

Our plan interested the Rileys because

Photo III-90: The Rileys' greenhouse has brought a big change to their lives. No longer must they shut themselves in during the long winter and hope they can afford to buy enough food and heat. The greenhouse gives them both, and they're very happy to be able to continue their gardening through the winter.

their fixed income was being severely strained by winter heating costs, even though they used relatively low-cost natural gas. They are also avid gardeners and more than eager to use the 156-square-foot growing space. We all felt we had an irresistible chance to make a graphic statement about appropriate technology at a site that was a scant eight miles downstream from the disabled Three Mile Island nuclear power plant.

The floor above the basement cantilevered about 5 feet beyond the south foundation wall. Originally this space contained a storage room and a bathroom, both adjoining the kitchen. We removed the rear wall of the storage room and angled the greenhouse glazing from ground level to the roof of the storage room. Thus the greenhouse encloses a small platform (the old storage room floor) that carries the solar preheater tank. We added a window to the kitchen wall for better air circulation and weather-stripped the door between the kitchen and the platform, since the greenhouse isn't insulated at night. A 4-inch concrete slab was poured at ground level to provide a floor and additional thermal mass. Now it's filled mostly with raised growing beds.

As the photograph shows, our design does not incorporate any opaque roof section; the glazing runs right up to the vertical wall. We framed a 2-foot kneewall with 2×6's, 24 inches on center, on the east, south and west sides. The unglazed west wall was framed above the kneewall with 2×4's, 24 inches on center. We used 2×6's on the south and east walls to support the glazing. All the solid wall sections were sheathed with ½-inch plywood and insulated with rolls of R-11 and R-19 fiberglass. The inside is finished inside with ½-inch plywood. The 2×6's for the tilted glazing were placed 48 inches on center with 2×4 crosspieces between them for extra support. Vents at the top and the bottom of the south glazing use the crosspieces for support, along with the vertical members and the kneewall or header. The Solar Project now uses verti-

cal framing 24 inches on center without the crosspieces instead of 48 inches because this holds the FRP glazing flatter. The only other support used was a 4×4 post under the platform to help support the 62-gallon water tank.

After staining the frame, we applied the glazing following the same procedure as described in the box "Glazing for a Two-Story Porch," found earlier in Section III-C. Rolls of Glasteel for the inside layer and FRP on the outside were used. The 1½ by 4-foot vents are double glazed with FRP on a flanged frame that

Photo III-91: Inside the greenhouse the batch-water heater sits on the balcony, leaving the entire floor available for growing beds.

(continued on next page)

resists water penetration. Raised flashing above the hinges deflects rain without interfering with the operation of the vents. The top vents are operated with a simple rope and pulley system from the platform. The vents also have screens for summer ventilation without insects.

For the breadbox water heater we bought a new, glass-lined pressure tank, painted

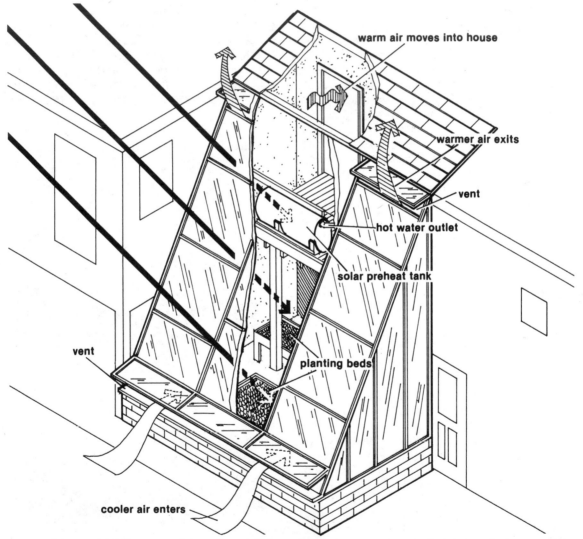

warm air moves into house

warmer air exits

vent

hot water outlet

solar preheat tank

planting beds

vent

cooler air enters

Figure III-86: In the Rileys' greenhouse the solar domestic hot water preheat tank sits on a platform over the ground-level planting beds. Vents at the top and bottom of the tilted glazing allow air to flow through the space preventing summertime overheating. In winter warm air collects at the top of the greenhouse and flows into the kitchen through a door and a window that open onto the water heater platform.

it black, and mounted it on its side, cradled on three 2×6 supports. Foil-faced urethane insulation placed under the tank between the supports reflects more sunlight onto the tank. The plumbing hookup is pretty basic. It involves a bypass loop from the gas water heater's cold-water feed line to the bottom of the preheat tank and from the top of the tank back to the water heater. All pipe lines were insulated. We installed a thermometer in another threaded opening in the top side of the tank, and all other openings were plugged. Then we built a light frame of 1×1's to cover the tank. The sides nearest the house and the west wall were insulated with foil-faced board and covered with thin plywood on the outside. The front, top and east sides of the frame were glazed with a clear vinyl film. Even without movable insulation the tank delivers water at a year-round average of 90 to 100°F (32–38°C) to the 40-gallon water heater, whose thermostat is set as low as possible. The solar hot water tank has cut the Rileys' gas consumption by 20 percent.

The greenhouse temperature doesn't fall below freezing unless the temperature outside is several degrees below zero. The greenhouse temperature has not dropped below 28°F (2°C) under any conditions, so there is no chance that the water tank will freeze in Lancaster County's climate. In winter, the greenhouse heats the entire house for five to ten hours a day during reasonably bright weather. Heating management is aided by Bertha's cat (neé "Solar"), who insists that the door to the greenhouse be open in the morning if the sun is shining.

The Rileys get all their spring plant sets, including tomatoes, peppers, eggplant and cucumbers from seed in the greenhouse and supply the neighbors as well. Bertha grows an indoor vegetable crop from fall through January, then begins spring production by early March. As soon as the ground thaws and the danger of frost is past, the Rileys dig up every square foot of backyard not needed for walkways and fill it with seedlings from the greenhouse. They grow

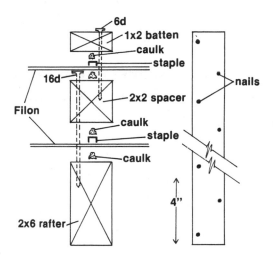

Figure III-87: Using the FRP glazing material, the glazing attachment is relatively simple. The first layer is tacked over a bead of caulk, and a 2×2 spacer is nailed through the glazing to the rafter. The second layer is tacked on, again with liberal use of caulk, and the exterior batten is nailed on through the 2×2 spacer. The cross section of the greenhouse framing shows vents at the top and the bottom, protected from water leaks by weather-stripped flanges and the raised flashing at the top of each vent. Note that the sill is tilted so that water won't puddle on it. In a greenhouse the inside glazing layer usually picks up a lot of condensation, which ultimately runs down to the sill.

a staggering amount of produce, and the greenhouse alone yields many pounds of food annually. It has made the house much more comfortable by supplying about one-half of the heating load. In summer the convection from breezes rising through the upper vents helps cool their rooms. Bertha and Harry are now building up credit rather than going into debt when they pay their winter gas bills on the budget plan. At the Solar Project we consider this greenhouse one of the most successful we've done.

M.R. Carey

"A greenhouse is only as good as its operator."

Quoth the author.
After it's built, the work begins

Once you've built the greenhouse of your dreams, the die is cast. You've created the shell of a system that can provide you with great quantities of heat and food, but only if you attend carefully to the realities of its operation. This new-breed passive solar greenhouse needs to be managed by simple daily routines, so that its valuable potential can be realized. Solar greenhouses are neither thermally nor horticulturally quite like their all-glass ancestors, and they aren't at all like your backyard garden. Thus solar greenhouse operation begins with an understanding of the nature of the beast and of your responsibilities to it. Once you work out a few routine tasks, your own experience will enable you to refine the procedure for superior results. This essay will take a look into this phase of your greenhouse project with the aim of getting you started in the right direction.

Getting heat from greenhouses is certainly less involved than creating and maintaining a food growing environment, but as you learn from experience, the horticultural workings of your greenhouse will become clear. The successful greenhouse operator will maintain a balance between factors that affect space heating and comfort, for plants and people. "Balance" is not meant to imply treading a thin line between these factors. Horticultural operation that is "imbalanced" will certainly yield some crops, but a well-tuned, truly productive greenhouse is by no means a distant goal, nor does it require a great deal of gardening expertise.

Some of the early greenhouses of the 19th century were passive solar structures that incorporated many of the conservation and solar heating techniques recommended in this book. Insulated walls, movable insulation and wood-burning stoves as back-up heating systems were common features of these greenhouses, but as inexpensive fossil fuels became available, traditional designs came to depend on automatic controls to stabilize interior temperature and humidity. These active systems were wasteful, because they exhausted surplus heat during the day and added fossil fueled heat to the greenhouse at night, with no effort to conserve the sun's energy.

Food production is no longer economic in these oil-heated dinosaurs, but passive solar greenhouses can take full advantage of available sunlight to provide an optimum environment for plant growth. Recent greenhouse designs, which are energy-efficient and small enough to attach to houses of just about any size or shape, can use the surplus solar energy they collect to heat the house, while providing a pleasant living

Photo III-92: There aren't any seedlings up yet, but this greenhouse garden has already been started. The beds are made; the soil has been composed, and the tilling is underway. So begins a journey further into self-reliance.

space. The most noticeable impact of these new greenhouses is on the life-style of the people who own and operate them since they really require an active participation. Thus, the greenhouse owner should be involved from the very beginning of the design process, right through to planting and harvesting. This kind of preplanning generally hasn't been done, either by people interested in building solar greenhouses or by organizations promoting them. This has resulted in some frustrated and disappointed owners who didn't realize how much time and energy was required to create a flourishing greenhouse environment until after it was built.

Operational Responsibilities

The temperature in a passive solar greenhouse space is controlled by the routine manipulation of vents, windows, shades, insulation and thermal mass rather than by purely automatic controls. Some of the tasks, however, can be automated. If you want to use the area primarily as a heat collector and living space, where human comfort and

TABLE III-8 OPERATING STEPS FOR A PRODUCTIVE PASSIVE SOLAR GREENHOUSE

Task	Frequency	Time Required	Comments
Opening and closing vents	daily	5 min	to cool greenhouse; depends on season and weather
Opening and closing windows and doors to house	daily	5 min	to convectively heat house
Installing and removing night insulation	daily	15–30 min	necessary during heating season
Installing sunshade	seasonal (summer)	2 hours	makes greenhouse cooler; do not reduce light levels below plants' requirements
Plant watering	weekly	1–2 hours	depends on planter box size, soil quality, greenhouse temperature and type of plant
Plant maintenance	daily	10–15 min	examine plants and soil for pests and disorders; prune and thin plants
Applying pesticides	monthly	2 hours	apply appropriate pesticides as a last resort
Weeding	weekly	1 hour	know difference between seedling and weedling
Transplanting	seasonal	1 day	requires care and should be done late in the day
Mixing planter bed soil	seasonal	1 day	should be done each time a seed or transplant is planted
Adding vegetable food scraps to greenhouse compost bin	daily	10 min	mix scraps in thoroughly
Hand pollination	daily	30 min	only required in closed ecosystem greenhouse

convenience is of primary importance, active controls such as thermostatically controlled fans, back-up space heaters and motorized insulation systems may be worthwhile. This automation allows temperature control to closer tolerances, but it can be an expensive convenience.

If the primary function of your solar greenhouse is to produce vegetables, herbs, fruits and houseplants, your responsibilities are more demanding. In addition to the thermal controls there are added horticultural considerations. Plants can withstand wider temperature swings than would be comfortable in a living space, so a horticultural greenhouse does not have to be held to as narrow a "plant comfort" temperature range.

But there may be several plant varieties involved, each with its own requirements for sunlight, soil composition, humidity, plant companions and pollination. Here the operator must be able to integrate the horticultural and thermal operations into an overall scheme. If your solar greenhouse will serve as both a social and horticultural space, plant care must still take precedence.

The additional tasks necessary for horticulture typically require anywhere from one-and-a-half hours per day (averaged throughout the year) for intensive gardening, to as little as one-half hour per day for a more modest effort. Naturally, during certain times of the year, when tasks such as planting or seasonal maintenance are required, two or three full days may be needed to finish jobs that can't be spread over longer periods. Once you are happy with the way your greenhouse is operating, you'll most probably have developed a simple management routine that can really be more pleasure than chore.

Environmental Balance

The key to a comfortable and productive solar greenhouse is balancing the interdependent factors of transmitted solar energy, temperature and moisture, with the dominant factor being temperature. Heat management of course begins with a sunny day, and on a typical winter day with an outside temperature of 20°F (−7°C), a well-designed greenhouse is able to collect about three times the amount of solar energy necessary for daytime space heating of the greenhouse and plant photosynthesis. This means, for example, that a well-designed and well-built 10 by 20-foot, single-story greenhouse in Denver, Colorado, or in Washington, D.C., could capture as much as 220,000 Btu of excess energy on a sunny day, about the same amount of energy available in 47 pounds of dry white oak firewood, 220 cubic feet of natural gas or 1½ gallons of fuel oil. To fully realize the benefits of your greenhouse operation, the excess collected solar energy must be used for daytime heating of the adjacent house or stored in thermal mass within the greenhouse or the house. It is possible to transfer about 60 percent of this available energy into thermal storage, leaving a balance, in this example, of 88,000 Btu in the form of hot air, which should be moved into the house or vented outside (depending on the season) to maintain appropriate greenhouse temperatures.

In the daytime, energy enters the greenhouse in the form of solar radiation that is converted to heat and contained by the glazing. The heat is transported primarily by convection (hot air movement), but conduction is significant in transferring the solar energy deeper into water or masonry thermal storage mass. At night the dominant form of heating is usually infrared radiation from the storage mass surfaces (at 70 to 90°F, 21–32°C) to the plants or people. For an object to receive direct radiative heating from the warm storage mass surface, it must be visible from that surface, due to the fact that radiation travels in straight lines. However, if there is sufficient energy in storage to keep the surface temperature above 90 to 100°F (32–38°C), then convection loops can also provide useful heat to the adjacent house. These barely perceptible warm air currents rise gently to make contact with plants and people or with the cold glazing or night insulation surface, where they give up their heat, then sink downward. These simple phenomena are what make the passive solar greenhouse work to keep itself and its occupants warm, be they flora or folks.

The precise thermal characteristics of passive solar greenhouses vary with design, construction, climate and orientation, but temperature differences of 60 to 70°F (33–39°C) can generally be maintained between inside and outside through a winter's night, if night insulation is used. However, if cloudy, cold weather persists for several days, a back-up heater or house heat may be required to keep the greenhouse above 55 to 60°F (13–16°C). A greenhouse that does not incorporate movable insulation will cool off rapidly, but through a single night it should still maintain temperatures that are 25 to 35°F (14–19°C) above the outside temperature. Even if the greenhouse is thermally isolated from the adjacent house, it will seek a temperature somewhere between the house temperature and the ambient temperature. Nevertheless, night insulation for attached greenhouses is an important design element.

Contrary to what you might expect, it isn't frost or freezing you should be most concerned about, but rather excessively high temperatures, because overheating is actually the number one problem in a passive solar greenhouse. Still, in a properly designed greenhouse, temperature can be easily maintained in the desired range with proper ventilation, shading, and placement of thermal mass.

Sun Screen

Since daytime overheating results from the conversion of solar radiation into heat, a first step in preventing this problem would simply be to reduce the amount of solar radiation reaching the interior of the greenhouse. But there are limits to how much the solar radiation can be reduced without negatively affecting horticultural and thermal performance. If the level of radiation is reduced too much (for instance by more than 70 percent), the remaining light will be insufficient to support photosynthesis of such plants as tomatoes, and very little energy will be available for thermal storage.

If the spectrum of solar radiation is altered significantly by a colored glazing, plants may suffer growth anomalies. To avoid this, the solar spectrum must be neutrally filtered by reducing all wavelengths proportionally, called "neutral density shading." This can be done most simply with a white cloth or plastic screen with holes of such size that the right amount of sunlight will be transmitted. The holes transmit all wavelengths equally and the white cloth reflects and scatters the blocked light back out of the greenhouse, limiting the potential for overheating. This screening material is sold at greenhouse supply stores, although cheesecloth will also work. The neutral density sun screen should be in place during the hot months and possibly during the late spring and early fall.

Ventilation

Even with shading, the passive solar greenhouse will still need ventilation during warm periods. In cool weather, convective ventilation to the house prevents the greenhouse from overheating while it warms the house, but during warm weather, when

Photo III-93: Sun shades will protect greenhouse plants in summer by limiting solar gain and temperature rise. Slatted shades, like the ones shown, and lightweight, white fabrics should be chosen for their durability in strong sunlight.

circulation to the house isn't needed, exhaust ventilation should be maximized. More ventilation will be required with exposed masonry heat storage than with water storage because masonry's heat capacity is lower and because it doesn't pull heat away from its surface as quickly as water does. Thus the masonry surface temperature is higher, which means it reradiates more heat back into the greenhouse than would a water storage medium.

A shaded passive solar greenhouse with water storage will require at least 6 to 10 air changes each hour to maintain proper temperatures. (One air change occurs when a volume of air equivalent to the volume of the greenhouse moves through the greenhouse structure, thus 10 air changes would be a volume of air 10 times that of the greenhouse.) The same greenhouse without neutral density shading could require 15 to 20 air changes each hour. A good rule of thumb where masonry wall storage is used is to increase the ventilation rate 40 percent more than for water storage. Ventilation can be accomplished passively by using wind-powered turbine vents, roof-mounted, boat-hatch or recreational-vehicle vents or large side vents, but in many cases electric fans with gravity dampers are best.

The outlet vents should be placed downwind near the top of the greenhouse to exhaust hot air, while any inlet vents should be upwind near ground level to help destratify the air. Locating vents on east and west walls promotes cross ventilation. To automate this system, simply install standard thermostatically controlled fans in the exhaust vents. Forced circulation of excess hot air into remote storage within or below the house improves the overall thermal efficiency of the greenhouse and does away with

Photo III-94: Even for a relatively small greenhouse a large exhaust vent is a good idea. The one above is a plug-in type that comes out completely when warm weather returns.

exposed thermal mass that can heat up in warm weather.

Manual vents are generally opened in the morning and closed just before sundown. Roof vents should be avoided because of the possibility of water and air leaks, but if you must use them to get enough ventilating area and keep your costs down, use boat-hatch vents or the sort commonly used in vans. If it doesn't leak on a ship or van, it probably won't on your greenhouse. There are many venting options, so be sure to choose the combination that suits your needs, your microclimate (in particular, prevailing wind direction) and your design.

Movable Insulation

Night insulation *must* be employed during winter to make full use of the solar energy stored during the day. Movable insulation decreases heat loss by 60 to 80 percent because the storage mass surfaces and plants do not "see" the cold surface of the glazing, which would otherwise transmit much of their radiated heat to the outside. This heat loss phenomenon, called "night sky radiation," occurs most rapidly on clear nights. The heat that does escape the insulating panels does so by conduction.

While movable insulation is primarily used to retain heat in the greenhouse at night, it may also be used as a sun shade during extremely hot days. However, when shading is used in this manner, don't cover more than 20 percent of the greenhouse glazing, and use insulating material that is protected from degradation by solar radiation. Movable insulation is usually put in place and removed manually, but automated systems are available.

Make your decisions concerning installation and operation of movable insulation prior to construction of the greenhouse because positioning insulation is the single most important and potentially the most time-consuming daily operation made during the winter. When selecting movable insulation for your greenhouse, look for high R-value (at least R-6), ease and reliability of operation and durability.

Passive Solar Greenhouse Horticulture

In the passive solar greenhouse, horticulture is an art. Plant propagation and growth take place in a balanced, man-made environment. The solar greenhouse environment is lush, and conditions are ideal for both plant and insect growth. The insect population must be balanced between pests—insects deleterious to plants—and the natural enemies of these pests. If the greenhouse is allowed to get too hot, greater than 90°F (32°C), such beneficial predators as ladybugs tend to vacate the space, and plants will transpire excessively. The result is weakened plants, inhibited pollination and reduced yields. Overheating also increases the pest population, further threatening the already stressed plants. When disequilibrium or stress is detected within the greenhouse environment, quick and decisive action must be taken. If plants grow too rapidly or become diseased, prune them. If pests multiply rapidly, check their spread by reestablishing a healthy environment and, if necessary, by introducing natural predators or even natural pesticides. If the soil begins to dry out or show nutritional depletion, nourish and water it.

The greenhouse is an ecosystem that can operate in either the open or closed mode. The open ecosystem occurs when the greenhouse vents, doors or windows are open, allowing air and insects from outside (if the openings aren't screened) to pass freely into and out of the structure. This mode is generally used during warm days or nights when the greenhouse is being cooled by ventilation. The closed ecosystem is used during cooler periods, when there is no air exchange between the greenhouse and the outside.

The open ecosystem allows natural pollination to occur from insects and sometimes the wind. Natural pollination saves time, but the open system also allows easy entry for pests. A closed system may be desired where hybrids or specialized plants such as orchids or violets are being propagated, and it can be created simply by covering vents with insect screen. In cool weather, it's necessary to operate the greenhouse in the closed mode for extended periods of time anyway, but artificial pollination is generally not a burden because the leafy vegetable crops usually grown at this time do not require pollination to produce. Generally pollination is only needed for "fruiting" vegetables such as tomatoes, peppers and eggplants.

Photo III-95: Many people use their greenhouse as a "season extender" to start seeds before they could survive outside. Timing is important because you don't want to get stuck with trays of crowded seedlings in need of transplanting before it's warm enough to move them out.

Sunlight for Growing

The effect of sunlight on plant growth is perhaps the least understood phenomenon influencing passive solar greenhouse yield. The color (wavelength) and intensity of the solar energy transmitted into the greenhouse has an important impact on both thermal performance and plant growth. This transmitted solar energy is the driving force of the greenhouse system, and some appreciation of the relationship between it and the greenhouse environment is essential. Plants generally prefer varying amounts of diffuse sunlight. This scattering is accomplished by using a diffusing glazing material such as fiberglass-reinforced plastic (FRP) or a sun screen, and with flat white paint applied to all surfaces except thermal storage walls or tanks.

Diffusion slightly reduces the intensity of sunlight at any one location, but it helps to scatter light into what would normally be shaded zones. For example, the planter boxes in zones #1 and #4 of figure III-88 receive enough light to sustain growth of certain plants even though they are actually shaded from direct sunlight during much of the year. Scattered sunlight also provides a softer, more comfortable environment for people enjoying the greenhouse as a social space. The use of diffuse glazing does of course block a direct view to the outside.

A plant uses primarily visible light for growth. The spectrum of sunlight is constant in outer space, but by the time it passes through the atmosphere, clouds, pollution, dust and the greenhouse glazing, the spectrum has been altered. Even white interior walls (from which most sunlight is reflected) have some effect on the spectrum. Most of the energy absorbed in all of these cases is in the ultraviolet and infrared regions,

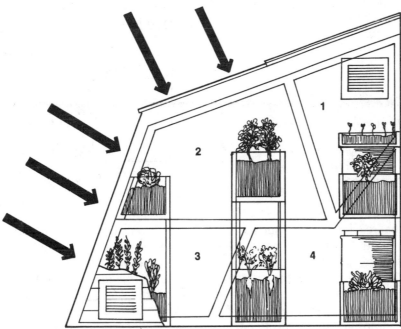

Figure III-88: Even in the relatively small confines of a greenhouse, there will be zones that have different lighting and heating characteristics. The four zones described by the author can be an aid in planning which plants should go where.

which is fortunate because the energy in these nonvisible regions of the spectrum isn't generally needed for plant growth.

However, if glazing materials or paints are used that do absorb much sunlight in the visible region, plant growth and yield can be severely hampered. This should not be a problem if greenhouse-quality glass or glazing materials such as FRP, acrylics and polycarbonates are used. Problems may occur if tinted glass is used for glazing or if colored paints are used for interior surfaces. The important thing to remember is that plants not only need sufficient levels of light, but also energy from specific wavelengths to sustain normal growth. If these requirements aren't met, growth may be slowed and symptoms such as spindliness and dwarfism may occur, symptoms that are easy to confuse with other horticultural problems.

Photosynthesis

Photosynthesis is the most important plant growth process for a greenhouse operator to understand. The essential ingredients are adequate amounts of sunlight, carbon dioxide, water and green plant matter (chlorophyll). About one-third of the solar

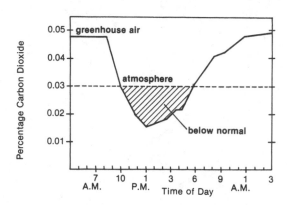

Figure III-89: *This graph shows the carbon dioxide content of greenhouse air at different times of the day. As you might expect, when photosynthetic activity is at its highest during the middle of the day, the least amount of CO_2 is present. As the author suggests, it's important to have some way of adding CO_2 to the air at this time in order to maintain optimum growing conditions.*

energy that reaches the plant surface is of suitable wavelength to support photosynthesis, but only about 1 percent of the total incident solar energy is actually used by the plant in this process. Photosynthesis can proceed normally for most plant types with only one-third to one-fourth of the maximum possible solar illumination. (Of the remaining incident sunlight, 10 to 20 percent is reflected by the plant surface and the remaining 70 to 80 percent is absorbed by the plant and converted to thermal energy, which promotes transpiration.) If the greenhouse is designed, constructed and operated properly, neither the amount nor the quality of solar energy should cause a problem in plant growth.

Figure III-89 shows the typical concentration of carbon dioxide in outside air and in a closed ecosystem greenhouse. By 10:00 A.M., carbon dioxide levels in the greenhouse are below outside levels and far too low to support satisfactory plant growth. This condition is not nearly as severe when ventilative cooling is being used, but can still occur. A remedy is to add compost to the greenhouse. The aerobic decomposition of organic matter in a 48-cubic-foot (4 by 3 by 4-foot) compost bin gives off as much as 2 pounds of carbon dioxide per day, much more than a fully planted greenhouse ever needs. It's very important to screen the compost from the greenhouse space so that pests in the compost won't become pests for your plants. Another solution to the carbon dioxide problem for closed, unventilated greenhouses is to use a carbon dioxide (CO_2) fire extinguisher. A few blasts around midday will provide ample carbon dioxide to support photosynthesis, and your fire extinguisher will go a long way at that rate of use. The midday timing is important because that is when plants need most of the carbon dioxide that they consume. Commercial greenhouses use CO_2 tanks with a very slow but continuous release.

Selecting Plant Type and Variety

Choose plant varieties that are well-suited for the passive solar greenhouse environment. Several varieties have been bred for commercial greenhouse production, and while these varieties aren't always suitable in passive solar greenhouses, they are worth a try. A certain amount of experimentation will be necessary to find plants that will do well in your sunspace.

Hardy varieties with broad temperature tolerances are best for planting during cold seasons or in climates experiencing predominantly clear, sunny days with wide daily temperature swings (Southeast and West). Varieties requiring lower light levels, such as greens and oriental vegetables, grow well during the winter in northern latitudes. Table III-9 provides information on some vegetables suited for greenhouse growing. Varietal information is also available from seed companies, universities, extension offices and commercial greenhouses. Owners of solar greenhouses are another source of valuable information. Keep records of each planting, noticing how each type grows and produces. Then use this information to help select varieties for your next planting.

TABLE III-9   GROWTH CHARACTERISTICS FOR GREENHOUSE PLANTS

Plant Type	Water Requirements	Germination Time (days)	Germination Soil Temperature (°F)	Growing Temperature Day	Growing Temperature Night	Plant Spacing (inches)	Plant Box Depth (inches)	Maturing Harvest Begins (days)	Light Requirements	Frost Tolerance
Lettuce (loose leaf)	moist	3–5	40–80	80	35	4–12	12	45–50	filtered	moderate
Spinach	moist	7	45–85	80	35	5–6	12	42–50	filtered	moderate
Swiss chard	moist	9	50–85	85	35	6–8	12	50–60	direct	tolerant
Chinese cabbage	moist	5–7	40–80	75	32	6–12	10	45–75	direct	tolerant
Tomatoes	quite moist	7	75–85	85	60	12	18	60–80	direct	very susceptible
Cucumbers	moist	5	80–95	95	45	12	18	60–70	direct	very susceptible
Peppers	moist	15–18	75–95	85	55	6–12	14	60–75	direct	susceptible
Broccoli	moist	9	70–80	75	50	12	18	55–60	direct/ filtered	tolerant
Cabbage	moist	7	70–95	75	50	12	18	60–90	direct	tolerant
Parsley	moist	9	50–80	90	32	1–2	6	60	filtered	susceptible
Mint	moist	20	50–85	90	45	6	6	60	filtered	moderate
Chives	moist	5–10	40–80	90	40	6	10	45	direct	moderate
Basil	moist	5–10	75–95	90	40	8	10	45	direct	susceptible

Soil Preparation

The best growing media for your passive solar greenhouse is probably a mixture of compost and local loam soil. A rich mixture of this type is usually not feasible for outdoor gardens because of the sheer quantity of compost necessary for large gardens, but for an attached greenhouse a small compost bin built into the greenhouse and fed with household organic matter and sawdust, leaves or grass clippings will provide ample volume.

Healthy soil consists of about one-half solid matter and one-half airspace. Good soil, therefore, is relatively lightweight and porous, allowing living organisms such as worms to breathe freely and permitting long-term watering of plant roots by capillary action in the soil. The solid matter in soil is the loam, made up of varying amounts of organic matter, silt, sand and clay. Loam soil suitable for high-quality vegetable production should have about 5 percent organic matter. The amount of clay, sand or silt determines the porosity or wetting ability of the soil and the capacity of the soil to store water. The larger the clay content, the more water will tend to stand; the more sand,

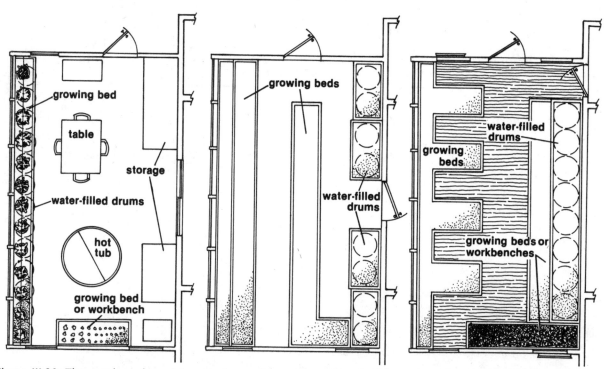

Figure III-90: The number of possible greenhouse floor plans will probably be as many and varied as the number of greenhouses, but the above ideas show some often repeated variations. When you make your own plan, remember that there are no rules against seasonal or long-term flexibility if you think your greenhouse activities are going to change from time to time.

the greater the tendency for water to run through the soil, leaching soil nutrients to the bottom of the planter box. Once good soil composition is achieved, it must be maintained, for as plants grow, soil nutrients become depleted, and the soil begins to pack down, losing its porosity. Regular addition of compost materials will correct this problem, but if compaction becomes severe, add 20 to 40 percent perlite, vermiculite or a bean-sized pebble to keep the soil loose.

Sixteen naturally occurring chemical elements in two groups are needed by plants for balanced growth: (1) trace elements, required only in the very small quantities contained in clay and compost, and (2) the three major plant foods, nitrogen (N), phosphorus (P) and potassium (K). Measurement of trace elements is difficult and should be done by your local agricultural office or county agent if you feel there may be a problem with your soil. You can, however, measure the soil's NPK values and its pH level. The pH measurement characterizes the acidity or alkalinity of the greenhouse soil, which is largely determined by the nitrogen, phosphorous and potassium ratio. There are many kits of varying complexity that can be used to make these measurements.

The use of pH as an indicator for monitoring your planter boxes or beds is generally sufficient for passive solar greenhouse operations. When you establish the appropriate NPK ratio with the soil test kit, see what the pH is at the same time. From then on you can simply monitor changes in pH to tell if there have been significant changes in NPK. Acidic soil can be corrected by adding small amounts of hydrated lime, limestone or hardwood ashes from your fireplace or woodstove. Alkaline soil can be adjusted by adding minerals that contain trace elements such as borax, manganese or gypsum, or by using oak bark or peat moss. Alteration of soil composition should be done in small stages over a period of several months, accompanied by soil tests.

Dirt doesn't complain when it's in poor health, and it takes a while for the neglect to show up in your plants. But good soil is the foundation of your horticultural program; give it the care it deserves.

Seed Planting and Germination

Germination occurs when everything is right for growth: soil condition, temperature, moisture and oxygen levels. These levels must be maintained during the germination period to sustain rapid and healthy growth.

Optimum soil temperature and approximate germination periods for several popular vegetables are given in table III-9. The temperatures in the table are soil temperatures, not greenhouse air temperatures, which are a little higher. To ensure correct germination temperatures, use both soil thermometers and maximum/minimum air thermometers for monitoring. The soil must be loose enough and just moist enough to permit the flow of oxygen into the soil and carbon dioxide out of it, which makes seed planting depth and soil composition important during germination.

Most vegetable plants should be started from seed each year. Perennials such as asparagus and most houseplants are generally propagated by transplanting new growth or cuttings from a mature plant.

Photo III-96: Because of the great temperature difference between the inside and the outside air, condensation will most always occur on the inside of the glazing. If it chronically collects on the glazing sill, it may rot wood members, but with a water catch like this the condensation is caught and returned to the growing beds.

Transplanting

You can use the greenhouse to extend the outdoor growing season by starting seedlings in it during late winter and early spring. Most seeds are planted very close together in flats or trays to conserve space in the greenhouse. When the seedlings reach 1 to 2 inches in height, they can be transplanted into permanent beds or individual pots, in the greenhouse or out to a garden plot. Seedlings germinated in a passive solar greenhouse generally will experience less shock after transplanting than seedlings raised in your house or a traditional greenhouse, but there will be some. Once the greenhouse seedlings overcome this initial shock, they should grow more rapidly and yield produce in a shorter period than an identical plant raised out in the garden.

To minimize shock, transplanting should be done late in the evening or on a damp, cloudy day. Place each seedling in appropriately spaced holes about twice the size of the seedling root base and line these little holes with compost and water. Then if peat pots are used, simply place the entire container into the hole, being careful to cover any part of the peat with good loam soil to reduce moisture wicking away from the peat moss.

Deep rooted crops such as beets and turnips and fleshy vegetables such as melons, squash and cucumbers are among those plants that are not well-suited to transplanting because they may suffer undue shock. Fleshy vegetables, which require a large growing space, should not be grown in passive solar greenhouses unless they can be supported vertically. The new varieties of bush cucumbers, however, grow well in a greenhouse.

Watering

It's essential to strike the proper balance in watering your plants. Too little water stresses the plants and stunts growth; too much water eliminates airspace in the soil, damages roots and can promote stem rot, fungus and other plant disorders. Plant respiration requires large quantities of water, and since the respiration rate is increased by higher temperatures, a lot of water is required by plants in a greenhouse. For example, a mature tomato plant can require 20 to 25 gallons of water during its productive greenhouse life of approximately six months, about 3 or 4 quarts of water at its base every three days. See table III-9 for the recommended soil moisture for other popular vegetables.

The cycling air of the open-ecosystem greenhouse and the limited depth of soil in the planter boxes tend to accelerate water evaporation from the soil. As the greenhouse air temperature goes up, the capacity of the warming air to hold moisture also increases, raising humidity at the expense of soil moisture. The ideal relative humidity for most vegetable plants ranges from 50 to 70 percent, but the wide variations that can occur during the day in a passive solar greenhouse do not pose a serious problem to most plants. It's not hard to reduce this diurnal humidity cycle by keeping air temperature below 80°F (27°C), maintaining proper moisture in the soil and mulching planter boxes well.

Since water needs will vary with plant types and seasonal weather conditions, the most effective method for watering is to establish a schedule for checking your plant beds, adding water only when needed. A simple test for soil moisture content is to take a pinch of soil from below the surface of your planting bed. If it's just wet enough to hold together, it's usually wet enough to support growth of mature plants.

The ideal water temperature for greenhouse irrigation is around 70°F (21°C). Most tap or well water is considerably colder than this and should be warmed by running the water through a black hose or other inexpensive solar collector placed in direct sunlight before and during watering. This will raise the temperature of the cold water on its way to the plants.

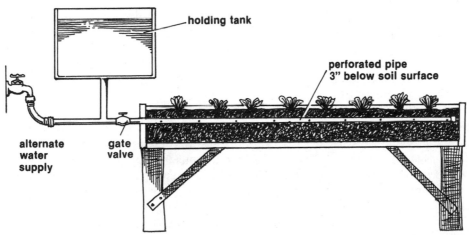

Figure III-91: With a watering system like this you can trickle water to the roots of your plants, a very efficient way to water. When filling the holding tank, close the gate valve; when it's time to trickle, open the valve just a little to create a small flow through the perforated pipe.

To improve effectiveness of plant watering, a subsurface irrigation system can be made by placing a perforated hose or tube in the soil no more than 3 inches below the surface of the soil in the planter box. Connect the hose to a container mounted just above the tube level. This water container can be raised to obtain the appropriate water pressure and flow rate in the growing box. The hose perforations should become more frequent the farther they are from the container. For the first 5 feet of tube, space the holes at 8- to 10-inch intervals, and at the 5- to 10-foot length increase the frequency to 4- to 5-inch intervals. With this method, precise amounts of water can be "leaked" into the planter box without compacting the soil or promoting fungus or stem rot. Domestic water from your house can be introduced directly into the subsurface system using a standard hose bib spigot. To water, just open the hose bib so that it barely flows, creating a kind of underground drip irrigation that's good for plant growth.

Pest Control

There are many types of insects that will thrive in the lush habitat of the passive solar greenhouse. Insects such as aphids and spider mites are deleterious to plant life, yet there are other insects such as ladybugs and lacewings that are beneficial as predators of these pests. Bees and other flying insects help pollinate fertile blossoms. A fourth group of insects, mosquitoes and common houseflies, is attracted to the environment, but usually just passes through. If there are bugs in the greenhouse, there are bound to be bugs in the house, unless you stop the migration before it starts by screening openings between the house and greenhouse.

You can help nature establish an acceptable balance within the greenhouse ecosystem, one where the loss of vegetables or plants due to pests is acceptable. A first step is to prevent overheating, which drives off or kills beneficial insects. If the pest population becomes unacceptable, external controls may be required. Remember that these methods can also affect the beneficial organisms in the greenhouse, and if the balance is grossly disturbed, long-term imbalance can result. Thus controls should be applied systematically, in small doses and with the greenhouse balance constantly in mind. Three approaches to pest control are: companion planting, increasing the predator population (insects and beneficial parasites) and using certain short-lived organic-base sprays and powders.

Companion planting must be practiced when planting or transplanting begins. The purpose of companion planting is to restrict the mobility and propagation of the pest population. Plant small patches or rows of plants that repel pests in between rows or patches of susceptible plants. Increase predator population by introducing wasps, ladybugs or praying mantises and by maintaining ideal conditions for their reproduction. The pesticides recommended for pest and disease control are not considered strong, but many are still poisonous and should be applied with the same precautions as for stronger commercial pesticides. Combinations of these remedies will usually help establish an acceptable insect balance. For more detailed information, refer to other sources listed at the end of this essay.

Plant Disorders

Plant disorders result from disease or environmental stress, reducing yield or even killing the stricken plant. Problems may result from viruses, bacteria and fungi, or such environmental factors as soil deficiencies, overheating and overwatering. Unfortunately, they are often difficult to detect and identify in time to save a diseased plant. Plant diseases that have developed far enough to show noticeable symptoms will usually claim the life of the plant or plant part and may spread to other similar varieties. Thus a diseased plant or plant part should be removed from the greenhouse as soon as symptoms are detected, and attempts made quickly to determine the cause of the problem.

The growth of bacteria and fungi is encouraged by excessively high humidity, very wet soil and high soil temperatures (greater than 80°F, 27°C). Bacteria causes stem and leaf rot and blocks the plant's vascular system, causing wilt and loss of natural color. Fungus generally shows up on leaves or stems as a visible powdering or rust-colored mildew or as yellow spots on leaves. This mildew spreads throughout the veins of the plant and blocks the plant's nutrient flow. Other fungi live in the wet surface of the soil causing "damping off" of young seedlings. Viruses can cause quick death and are usually the most difficult disease to control. Mosaic viruses destroy chlorophyll and interfere with photosynthesis and the plant's food supply. This results in dwarf plants, excess foliage (instead of more root or fruit growth) and yellow leaf curl.

The best treatment is prevention. Simply stated, if the passive solar greenhouse is operated properly, disease should be minimal except where disease is introduced by seedlings or plants brought in from other greenhouses. Proper watering and temperature control coupled with healthy plants are the basis for preventing plant disease. Maintain the soil within the proper pH range for the plant types you're growing. After each harvest, remove all crop residues and rotate the crops (plant different varieties in a given plot). If you experience excessive problems with plant disease, select varieties that are more disease-resistant and buy only disease-free certified stock from reputable dealers.

Floor Plan

The floor plan for a horticultural greenhouse should accommodate as much growing space as possible while still allowing sunlight to reach each planting area. You will also want to avoid plant arrangements that shade other plants. If planting is to be done directly in an earth-floor section of the greenhouse, as shown in plan 3 in figure III-90, the floor slab must of course be poured accordingly when construction begins. When planning locations for water faucets, sinks or a hot tub, always think in terms of shading, convenience and possible modifications to the foundation, floors or walls.

Planter Boxes

To get the most planting area in a limited greenhouse space, you'll probably need planter boxes, which can be made of masonry or wood. Make your planter 3 inches deeper than required for the plants you will grow in it. Once the planter boxes are in

TABLE III-10   GUIDELINES FOR FOOD PRODUCTION USING A PASSIVE SOLAR GREENHOUSE

Crops in the Solar Greenhouse

Crops	Planting Dates	Harvesting Period
Staggered Planting		
Carrots	February through April; August through November	year-round
Cucumbers	July; mid-March	October, November; June
Swiss chard, lettuce, spinach	January; early August through November	March, April; October through April
Oriental vegetables	September through December	November through April
Peppers	June, July; early March	October through December; June
Radishes, turnips	September through December	end of November through April
One-time Planting		
Broccoli	August, September; mid-February	November through January; mid- to late-May
Herbs	February, March; October, December	year-round
Peas	August, September	November through January
Tomatoes	June, early July; mid-February to mid-March	September through early December; early June

Greenhouse Plants to Transplant Outdoors

Crops	Greenhouse Planting Dates	Transplanting Outdoors	Harvesting Period
Spring			
Broccoli, cabbage, kohlrabi	end of February, March	April, May	July, August
Eggplant, peppers, tomatoes, lettuce	end of March, April	mid-May, beginning of June	July through September
Leeks, onions	February	April	July, August
Fall			
Brussels sprouts	mid-June	September	October, November
Broccoli, cabbage, cauliflower	end of June, beginning of July	September	October

SOURCE: Information for this table was provided by Malcolm Lillywhite and also by Stephen Ganser, Charles Kauffman and Eileen Weinsteiger, research staff members of the Organic Gardening and Farming Research Center, Maxatawny, Pa.
NOTE: Designed for 40–42° north latitude. Adjust starting dates to your area.

place, mix the soil and fill the boxes to the very top because in time the soil will pack down exposing the rim of the box, which keeps water from overflowing. If the boxes are wood, line them either with a waterproof liner or a coat of marine-grade waterproof paint. All boxes should be sturdy and painted white on the outside.

Planting Schedule

Besides coordinating greenhouse production with other available food supplies, growing schedules must take account of seasonal influences. Tomatoes are generally available from June until early fall from your garden, and you might expect to be able to grow tomatoes in your greenhouse during the winter, when they aren't available elsewhere, but because tomatoes require long sunny days for productive growth, they don't produce well in winter even in the most temperate greenhouse. In fact, tomatoes use so much space that they are of questionable value in small attached greenhouses, unless you are growing in high altitudes or altitudes where the climate may be too cold or have too short an outdoor growing season for tomatoes or similar plants. A more productive use of your greenhouse during the winter is to grow hardy vegetables that have low light requirements and a high yield per square foot. These vegetables include most greens, such as spinach, Swiss chard, lettuce and oriental vegetables, and sometimes cabbage, leeks, peas and carrots as well.

TABLE III-11   SAMPLE OPERATION AND MAINTENANCE WORKSHEET

Date	Weather Conditions	Planting	Germination	Indoor Temperature Min	Max	Outdoor Temperature Min	Max	Soil (planter) Temperature Min	Max	Relative Humidity	Comments	Corrective Action	Begin Harvesting
2/1/80	cloudy	black simpson lettuce	---	65	55	35	15	63	59	60	good	---	---
2/20/80	partly cloudy	bush cucumbers, tomatoes	Swiss chard	67	55	38	14	64	58	58	good	---	Swiss chard
3/1/80	sunny	carrots, basil	tomatoes	75	56	36	12	68	60	55	soil drying out	mulched	Bibb lettuce
3/15/80	sunny	transplanted tomatoes	---	78	57	37	12	70	62	55	black simpson infested	threw out infested plants	---

NOTE: Dates taken at random. Actual worksheet should be a daily log.

Photo III-97: After all is said, your greenhouse will become just what you want it to be: all for heat and space, all for food or heat or a bit of all three. In this greenhouse food plants, houseplants, stereo speakers and sitting places combine to make it a most pleasant spot.

Personal preference and climate are important in developing a planting schedule. After you have a couple of growing seasons under your belt, you'll probably be well on your way to establishing your own unique horticultural and thermal practices. Table III-10 gives planting and harvesting schedules for a number of vegetables that do well in greenhouses.

Monitoring

An aid to establishing an operational plan is to monitor and record as much about your greenhouse as you can when you are first beginning to use it. Record indoor and outdoor air temperatures using a maximum/minimum thermometer, ground temperature using a soil thermometer, daily humidity, planting dates, plant varieties and locations, notes on plant health, vegetable yield, pests, and anything as you work toward optimizing its performance. Records are hard data that you can always refer back to for making evaluations, comparisons or changes in your operating plan. A sample data sheet that can be used for record keeping is given in table III-11. Begin keeping good records immediately, including information such as date planted, variety, plant location, germination time, plant health, yield and anything else you think is important. Now you are ready for both you and your greenhouse to grow together.

Malcolm Lillywhite

References

Abraham, George (Doc), and Abraham, Katy. 1975. *Organic gardening under glass: fruits, vegetables and ornamentals in the greenhouse.* 1975. Emmaus, Pa.: Rodale Press.

Arthurs, Kathryn L., ed. 1976. *Greenhouse gardening.* Menlo Park; Calif.: Lane Publishing Co.

DeKorne, James. 1978. *The survival greenhouse: an eco-system approach to home food production.* 2d ed. Culver City, Calif.: Peace Press.

Halpin, Anne. M., ed. 1980. *Rodale's encyclopedia of indoor gardening.* Emmaus, Pa.: Rodale Press.

Jordan, William H., Jr. 1977. *Windowsill ecology: controlling indoor plant pests with beneficial insects.* Emmaus, Pa.: Rodale Press.

Klein, Miriam. 1979. *Biological management of passive solar greenhouses: an annotated bibliography and resource list.* Butte, Mont.: The National Center for Appropriate Technology.

————. 1980. *Horticultural management of solar greenhouses in the northeast.* Newport, Vt.: The Memphremagog Group.

McCullagh, James C., ed. 1978. *The solar greenhouse book.* Emmaus, Pa.: Rodale Press.

Taylor, Kathryn S., and Gregg, Edith W. 1969. *Winter flowers in greenhouse and sun-heated pit.* New York: Charles Scribner's Sons.

Yepsen, Roger B., Jr. 1976. *Organic plant protection.* Emmaus, Pa.: Rodale Press.

PROJECT

Exterior Shutter-Reflectors for a Greenhouse

> *This shutter does double duty on a greenhouse kneewall. By night it insulates, and by day it lies open to reflect more sunlight into the greenhouse, always a welcome commodity. If water thermal mass is placed behind the kneewall glazing, the shutter is a particularly effective, if not necessary component, because it helps to reduce heat loss from the mass and also boosts the solar gain to it.*

Faced with the task to design and build a total night insulation system for a 32 by 14-foot greenhouse, we had to come up with different solutions to suit different parts of the structure. That meant shutter-reflectors for the south-facing kneewall and the side walls, and roll-up quilts for the tilted south-facing wall.

The tilted south wall consisted of eight 4 by 12-foot glazing sections running at a 45-degree slope. Since this area was too large for rigid shutters, we decided to use a commercially available product, Window Quilt, despite its relatively high cost. Using these shades for such a long, sloped span required some modifications. The units are designed for vertical runs over standard windows and are raised by means of a cord-and-roller system. They are lowered by releasing tension on the cord, allowing a weighted rail at the bottom of the shade to pull the curtain down by the force of gravity. In our case, the 45-degree glazing tilt prevented this from happening, but the cure was simple: We removed the weights from the bottom rail and installed a secondary pull-down cord running from the bottom of the shade to a pulley at the bottom of the tilted glazing. With this alteration, the shades work smoothly. (The following box describes an insulating curtain system that can be home-built.)

Next came the job of building the shutters for the various vertical glazing sections. In this greenhouse the kneewall consisted of eight glazed sections, each measuring about 45 inches on a side. The glazed sections of the side walls were trapezoidal, square on the bottom and angled at the top, following the slope of the glazing tilt. The southernmost section was about 8 feet high, and the section next to it went up to about 12 feet at the high point of the trapezoid. This shutter design is suitable for even these large glazed areas.

The shutter is basically a three-part lamination: ¼-inch tempered hardboard (Masonite) on the outside, a 1-inch-thick Styrofoam core (which could be 2, even 3 inches), and a reflective surface on the inner face. Problems arose when we made some test laminations, because we couldn't find an adhesive that would adequately bond together these different materials. We finally discovered a flooring adhesive called "The Perfect Putdown" (made by H.B. Fuller Company of Palatine, Illinois) that would not dissolve the Styrofoam and would bond it fairly well to the aluminum and the hardboard, although not quite well enough that we could rely on it completely. (Dow-Corning, which makes Styrofoam, now has a glue for it.) Along with the adhesive, we decided to encase the whole package in a rabbetted wood frame that would firmly clamp all the edges together.

For this frame we used carefully selected fir

Photo III-98: In their daytime role the shutters become reflectors in their open position. In this installation the kneewall shutters rest on a permanent run of pipe to keep them off the ground, but supports could also be built onto the shutters themselves to save yard space.

2×2's. On one side we cut a rabbet ½ inch deep by 1⅛ inch wide, knowing that we could compress the sandwich a little to make a tight fit. We then cut the frame pieces to length and performed the necessary joinery work for the corners. For good appearance we made a mitered slip joint at each corner, but this was probably more work than it was worth. A much simpler splined miter joint would have looked just as good, and the simplest reliable joint would be a corner lap, glued and screwed or doweled. It's important to make the joints strong, because the shutters have to take a lot of stress. When the framework was finished, we cut the hardboard to size and slightly chamfered the outside edges so they would slide more easily into the frame. The same hardboard pieces also became the templates for cutting the Styrofoam and aluminum sheets. For this job we recycled aluminum plates that were discards from a local offset printer.

Using a notched trowel, we applied an even layer of adhesive to all surfaces, including a liberal coating inside the rabbet of the 2×2 frame. The corner joints of the frame were evenly coated with resorcinol, a waterproof glue. The various parts were then assembled, using bar clamps to pull them together and square them up.

The end panels proved a little more complicated to construct because of the angled joints that had to be made and the general unwieldiness of their large size. We decided to include additional 2×2 braces at 4-foot intervals to make a strong, unified framework.

After the glue dried, we cleaned up the wood frame with a plane and sander (a good time to fine-tune the shutter's fit into the glazing section), and with paint thinner we washed the aluminum clean of residual ink and caulk. The woodwork was treated with Cuprinol-20 wood preservative and given two coats of an oil stain to match the rest of the green-house woodwork. The hardboard was painted with a bright blue alkyd (exterior) enamel, the choice of the architects that we first balked at but later agreed was a very pleasing color detail.

To understand how we mounted the shut-ters on this greenhouse, it is necessary to understand our glazing system. The glazing material is a double-wall extruded acrylic panel that is normally mounted with aluminum clamping bars provided by the manu-facturer. To complement the good looks of the greenhouse's frame, we devised our own clamping system using red oak battens and ⅛-inch-thick neo-prene gasket strips. Together they form a flexible seal

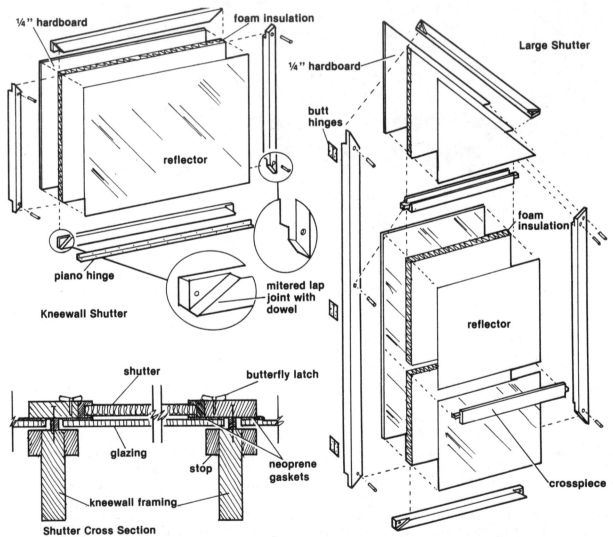

Figure III-92: *Because these shutters are exposed to the elements, selection of durable materials is important. The wood frame members should either be thoroughly painted or naturally rot resistant. The exterior panel should be equally weather resistant.*

with the glazing material, which is necessary because the acrylic expands and contracts somewhat with temperature change.

Since it is essential that the shutters form a tight seal when they are closed, we decided to use this same neoprene gasket for the secondary function of sealing the shutters. Rather than trimming the neoprene flush with the edge of the 3-inch-wide battens, we let it hang out ½ inch on each side. By sizing the shutters so they would fit between the battens, they make a nice, tight seal against the extended lip of the neoprene. There are, of course, other routes to the weather-seal solution using standard weather stripping materials. See figure III-92 for a couple of ideas.

The eight kneewall shutters were mounted with piano hinges along their lower edge, and they're held in a closed position with simple butterfly twist latches, the kind that are used to hold storm windows in place. When opened, the shutters lie on a horizontal rail made with 2-inch galvanized pipe with vertical pipe sections set in concrete piers poured directly into the ground. The height of the rail was planned so that the shutters would lie about 5 degrees below the horizontal plane so that rainwater would run off. This is also a suitable angle for effective reflection of winter sunlight into the greenhouse.

On the east and west sides, the larger shutters to the rear of the greenhouse are hinged with standard butt hinges so that when open they too increase direct gain through solar reflectance into the greenhouse. To do the same with the shorter shutters in front would only block sunlight from the rear glaz-

ing, so we mounted these shutters with standard storm window hardware (hangers and butterfly latches) so they could be removed daily and stored inside. This was a rather unwieldy operation, but we could see no alternative.

The overall system is quite successful. With our gasket system all the shutters seal tightly and operate smoothly. They are weathering well and look good either opened or closed. A good way to get a feel for the effectiveness of shutters is to make "the touch test." On a cold winter's night leave one shutter open and from inside the greenhouse put one hand on the surface of the open glazing and the other on its shuttered neighbor. You'll notice quite a temperature difference.

You can also look at some numbers. Depending on the kind of insulation you use, you'll end up with anywhere from an R-6 to R-10 level of insulation with the shutters in place (including about R-1.8 for the double glazing). This improvement represents a 65 to 80 percent reduction in conductive heat loss through the shuttered glazing sections. The R-3.4 Window Quilt raised heat resistance of the tilted south wall to R-5.2, reducing heat loss through that section by about 60 percent. It's clear that the difference between a naked greenhouse and one with night insulation is one that's going to mean having much more heat and light available without having to build a bigger one.

Tom Wilson

A Sliding Thermal Curtain

Bill Ginn's attached solar greenhouse was helping to keep his 1837 Cape Cod farmhouse warm during the winter, but he was worried that the cold Maine nights would send the temperature in his greenhouse below freezing, damaging his plants. Since he already had a thermal storage system, the only thing missing was night insulation for the greenhouse glazing.

To get it, Bill made his own insulating curtains and strung them along heavy wire close to the sloped glazing. During the day, the curtains are tied out of the way at the peak of the greenhouse, and at night they slide down the guide wires to cover the windows.

The sides and bottom of the curtains have Velcro strips spaced every foot, so the curtain can be pulled taut and fastened to other strips along the glazing perimeter. After using this system for a season, Bill realized that some cold air was infiltrating between the Velcro strips. To perfect the seal, he added wood battens attached to spring-tensioned cabinet hinges along the sides and bottom. Now the Velcro holds the curtain in place until he snaps down the battens, completely sealing the curtains.

The curtains themselves are handmade from quilting material, a vapor barrier and two layers of decorative fabric for the exterior surfaces. The Fiberfill II quilting material is about ¾ inch thick and is usually available in fabric shops. Other possibilities for insulation include Polarguard or more expensive Thinsulate, which has a higher R-value. The various layers are simply tacked with thread (like a real quilt) to keep them together.

The vapor barrier is important in preventing water vapor from condensing inside the quilting and destroying its insulating value. For that reason, the vapor barrier must be located inside the quilting. If you use polyethylene, it should be at least 6 mils thick to ensure long life and better resistance to water vapor. Bill used Astrolon, a reflective-surfaced material devel-

Photo III-99: In smaller greenhouses manual operation of the insulating curtain is done without difficulty, but with larger curtains some sort of drawstring system would probably be needed. The same system of wire tracks and washers could also be used with a summer shading curtain.

oped by NASA. The Astrolon also reflects some radiant heat back into the greenhouse. If you don't mind the looks of the Astrolon, omit the fabric layer, leaving the Astrolon exposed to the greenhouse interior for better thermal performance. Never expose it to direct sunlight, however, because it is degraded by ultraviolet light. Bill made his first curtains that way, and within a year the Astrolon had simply vanished.

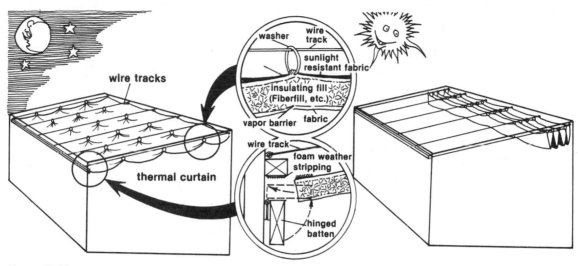

Figure III-93

The support wires, spaced every 2 feet along the 10-foot width of the glazing, are fastened to the highest point in the greenhouse (next to the house wall) and strung along the sloped glazing down to the top of the kneewall, about 9 feet. Securing the wires to the curtain required stitching on metal washers with heavy thread. The washers are sewn only at one point, so that when the wire is threaded through them, the curtain hangs below. The wire is pulled as tightly as possible so the curtain won't sag.

To remove the curtain in the morning, it's pushed back up the wire tracks, folding like an accordion. At the top, it is tied in place with cord or fabric ties. The size of the folds is determined by the spacing of the washers along the wire; spacing them every 24 or 30 inches works well.

Once he installed the curtain, Bill noticed a big difference in the nighttime temperature of the greenhouse, as much as 20°F (11°C) warmer. Based on comparison with similar insulating curtains, he estimates their insulation value at about R-4, although this could be improved with the use of Thinsulate and exposed Astrolon. But for his purpose—simple freeze protection—the curtain is fully adequate as it is, all for a modest investment of time and money.

Hardware Focus

Night Insulation for Greenhouses

Automatic Devices Co.
2121 S. 12th St.
Allentown, PA 18103
(215) 797-6000

Greenshield Systems
Rolls from spring-type roller to motorized roller following shape of house. Blanket has 3 parts: clear polyester, shading cloth and thermal blanket.

SOLAR
IV ADDITIONS

Photo IV-1: Sunspace additions give a house a whole new dimension in a living space that projects to the outdoors yet still maintains thermal comfort. On a sunny winter day people will bask, even bake, in the direct solar gain, happily oblivious to the chill they can see but can't feel.

Solarizing the
Expanding House

Additions usually come later. After most or all of a house has gotten the remodeling once-over, industrious homeowners look for new frontiers and more living space. There are of course normal growth patterns: Tots become teens, and *everybody* needs more space. There are semiemergencies too, like triplets. The old place no longer fits the clan. And last but not least there is the bottom line: The American habit of moving up to that better place has become very difficult to finance, and more and more people are finding out that the best way to improve their lot is to improve their lot.

Nowadays the cost of that improvement is enough of a liability, and there's no need to incur big energy debts as well. With the cost of energy going the way it's going, solar design and energy-efficient construction are perfect accomplices to home expansion schemes. No matter where an addition is located, energy efficiency should be as primary a concern as looks and size. Insulate like mad; make it tight as a bucket. If there is even the slightest ray of solar exposure available to the planned addition, it would be a missed opportunity not to include solar gain in the final outcome. A building addition is somewhat like new construction, and it's been proven that solar heating for new buildings, particularly passive solar heating, is in the not-too-long run more cost-effective than just about any other form of space heating. (The one exception is probably firewood that you cut yourself, if your labor is free, but is that how you want to spend your summer?)

The solar addition is a transformation. It brings light and heat; it can heat more than itself; it can be a space like no other, a magnet for family and friends. "Sunspace" can be any space: a solar kitchen, a solar bedroom, a solar breezeway, den, bathroom, greenhouse, spa. You can have it your way, and if you can share it with the sun, it'll be all the better.

563

Photo IV-2: The Welsh Addition

A Sunny Family Room

With the increasing interest in home improvement versus new home buying, the residential addition is no longer a rare luxury, but a typical (albeit major) expenditure for a family in need of more space. The reasons for wanting to build an addition on your house vary widely, but often they arise from the feeling of being crowded by growing children who need living space of their own or from the parents' need for a private room to which they can retreat. An addition also creates increased energy demands for the heating and cooling of the added rooms. But simply by using adequate amounts of insulation along with energy-efficient construction techniques, that increase can be held way down. And with a little help from solar design the addition can be an energy asset that helps to reduce the total building energy demands.

In 1978 the Welsh family wanted to add a family room, and they asked me to work up a design for the project. Their house, a lovely and well-preserved Pennsylvania brick was built about 1830 in Coopersburg, Pennsylvania, and the most recent addition to it was a kitchen and bedroom built around 1880. As well as having a second family room, the Welshes also wanted to put in a first-floor bathroom for their guests and for their own convenience, and they wanted a new entry foyer to shift the two entrances away from the center of the living room and the kitchen and make a single entrance.

Fortunately the logical place for the addition was on the south side, and the obvious design approach was for direct-gain solar space heating incorporating the existing brick wall as some thermal mass. Distribution of the heat that would be collected and stored in the addition required a design that allowed for a natural convective circulation for heated air to move into and share its energy with the house.

To gain the benefits of natural convection, the height of the new room had to be increased from the original design. This not only increased the area of the existing brick wall to be enclosed, for more thermal mass, but it more importantly connected the second floor with the addition, allowing the convective loop to work. Heated air rises and enters the second-floor bedroom, and as it gives off its heat, it circulates down the stairwell, returning through the kitchen back to the addition to be reheated. The added height of the room did cause some increase in expense, about 20 percent, but this was further justified by adding an 8 by 15-foot loft-sitting area adjacent to the master bedroom. The double benefit of gaining the loft space and the improved thermal performance made the increase in the project cost acceptable to the Welshes. Certainly the change was critical to turning the addition into a greater energy asset for their heating budget.

For auxiliary heat in the addition, an airtight woodstove was placed in a back corner where it heats the addition, and via the same convection loop, the loft, the master bedroom and the kitchen.

As a further energy conservation effort, the combination boiler/water heater was supplemented with a small electric water heater, connected in series between the

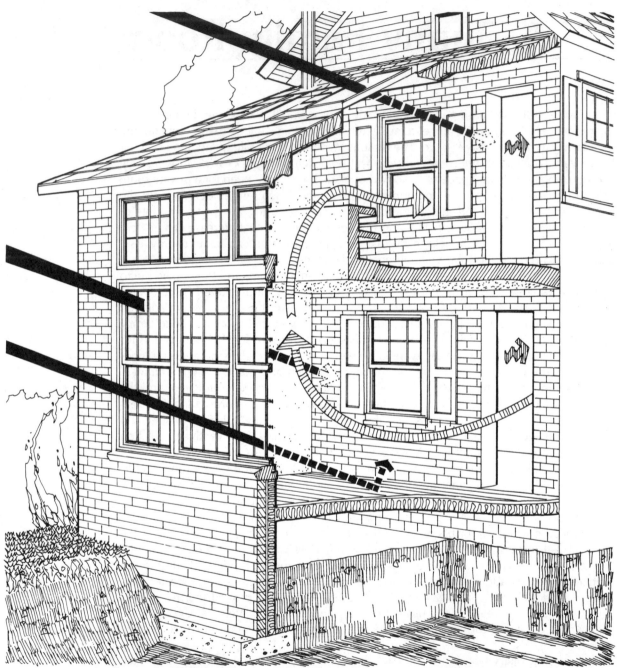

Figure IV-1: In designing this addition the architect made full use of basic solar heating principles: Direct gain strikes the original brick wall (now thermal mass) at the back of the addition, where heat is stored and ultimately radiated into adjacent rooms. As the air in the addition is heated by the sun, it rises by convection and flows into the main house. In summer the operable skylights allow excessive heat to be exhausted outdoors.

boiler and the house. This reduces oil consumption in the summer because the boiler is shut down, and only the electric water heater is used. The efficiency of burning oil just for domestic water is down in the 15 to 25 percent range, and even though electricity is more expensive than oil, this simple modification creates a new reduction in energy use, with about a three-year payback on the cost of the water heater.

As shown in figure IV-1, the addition is built on conventional concrete footings and foundation walls with an excavated crawl space for future access. Rigid Styrofoam insulation is installed below grade covered by a plastic vapor barrier for moisture control. Wall construction is a brick veneer (for aesthetic reasons) over a frame wall of 2×6 studs, which are filled with 6-inch fiberglass batt insulation, and insulating sheathing is placed between the brick and framing. All told, the walls have a heat resistance of about R-25. The floor is also insulated with 6-inch batts. The roof was framed with 2×12's to carry 9-inch batts (R-30). Where the roof framing was headed around the roof windows, it was carefully and thoroughly drilled to allow air flow across the top of the insulation to prevent condensation from trapped moisture. All the windows used are double glazed, and the roof windows—added for increased light and sunshine in the loft—are also

Photo IV-3: How nice it is to be able to gaze into the winter sunlight, all the while enjoying its warmth.

operable for summer ventilation. Summer ventilation is also improved by convection, using the operable roof windows for outlets and the lower floor windows as inlets for cooler air.

The solar aperture consists of three double-hung windows with three fixed windows above, for a new glazing area of 81 square feet. The thermal mass enclosed by the addition is 158 cubic feet (19,436 pounds) of existing brick and 12 cubic feet (1500 pounds) of new brick at the corner where the woodstove is located, making a total mass of about 11 tons. The frame construction is finished inside with white plaster to reflect more light onto the masonry and to maximize natural daylighting.

The combination of the due south orientation and the availability of the existing masonry wall for thermal mass were the essential ingredients that favored designing for solar heat gain. Also, changing the design before construction to promote whole-house heat circulation made the addition not only a solar collector for itself, but part of the original house as well. And not incidentally, the visual height of the family room is welcomed by the Welshes, as the rest of their house has lower, flat ceilings. This contrast provides a comfortable spaciousness and represents a positive meshing of energy-and people-related design.

Charles Klein

A New Sunspace
for an Antique House

You might not expect a solar addition to fit in with the look of a two-story post-and-beam colonial, but with the careful design work of architect Harrison Fraker and project designers Peter S. Brock, Larry Lindsey, and William Glennie of the Princeton Energy Group, it does. While the angular 500-square-foot, lean-to-style sunspace contrasts with traditional lines of the house, it has also been made to blend as much as possible. The tilted (60 degree) glazing consists of totally clear double glass to minimize the visual separation between the house and the addition by allowing a clear view of the original south wall. The glazing is trimmed with unobtrusive dark bronze-colored aluminum that is more eye-pleasing than eye-catching. The east end of the addition is both sided and painted to match the antique blue-gray tone of the original house. An old Japanese maple spreads its delicately leafed branches over the sunspace, as if the original house and landscape had never been changed. But the changes are quite apparent to the owners, for this sunspace now supplies 70 to 80 percent of the house's heating needs, as well as being a rather elegant living space.

To accomplish this energy savings, the owners asked the Princeton Energy Group to design an addition that would promote a comprehensive convective flow of air that would bring most of the house into thermal contact with the sunspace. Harrison Fraker produced a design that required some remodeling of the second floor to get the kind of air circulation they wanted. Upstairs the bathroom was moved to create a hallway that connects to the back stairs on the north side of the house. This created a complete loop for convective air exchange between the house and the sunspace.

To begin the sunspace the builder dug a 36 by 15-foot perimeter for the foundation. He used 8-inch concrete block for the foundation wall, with a 4-inch block at the top to accommodate 1½ inches of rigid urethane insulation and to provide bearing for the slab floor. The original porch floor was small—24 feet by 7 feet—and only 4 inches thick. To add thermal mass, the builder poured an additional 4 inches of concrete over the original slab and 8 inches of concrete around the perimeter of the original porch to extend the floor area to the foundation walls. An extra 2 inches of concrete were poured in the zone along the south edge of the slab where the water drums were to be located. Slate was set in the concrete because of its good looks and dark, heat-absorbing color.

The addition was framed with 2×6's and 2×8's as roof joists, 2×6's in the side walls, and 3×8 rafters as glazing supports. The beam at the peak of the glazing is formed by a 4×8, which is supported along its length by 5-inch steel pipe columns.

The roof was sheathed with plywood roof decking that was mopped with hot tar and covered with an asphalt membrane in successive layers to form a built-up

Photo IV-4: A modern sunspace adds to the attractiveness and value of this post-and-beam colonial.

bituminous roof. Inside, two layers of 3½-inch fiberglass batts (R-26) were tacked in the 2×8 joist spaces. Sheetrock was attached to the joists, spackled, sanded and painted for a finished ceiling surface.

The east end side wall was sheathed in plywood, insulated with fiberglass batts, and fitted with an entry door. The exterior was finished with siding to match the rest of the house, and the interior was finished with Sheetrock.

The primary glazing assembly uses an exterior pane of ¼-inch tempered glass in 4×8 and 4×4-foot sheets, lap-joined under a compression seal, made with extruded aluminum battens and neoprene gasketing. Sheets of ¼-inch Plexiglas are attached to the interior edges of the rafters. They are simply butt-joined under wood battens and caulked with butyl. The total glazing area provides 500 square feet for direct-gain heating. An additional 110 square feet of vertical double-wall acrylic glazing covers the 3-foot-high south kneewall. Sunlight passing through the kneewall heats up 17 55-gallon water-filled drums placed underneath a counter covered with a black heat-absorbing plastic laminate. Additional thermal mass was added in the form of nine 18-inch diameter 10-foot Kalwall water tubes and three 12-inch diameter 7-foot tubes that stand along the back of the sunspace.

Because of its old-fashioned construction, the house itself has some of its own

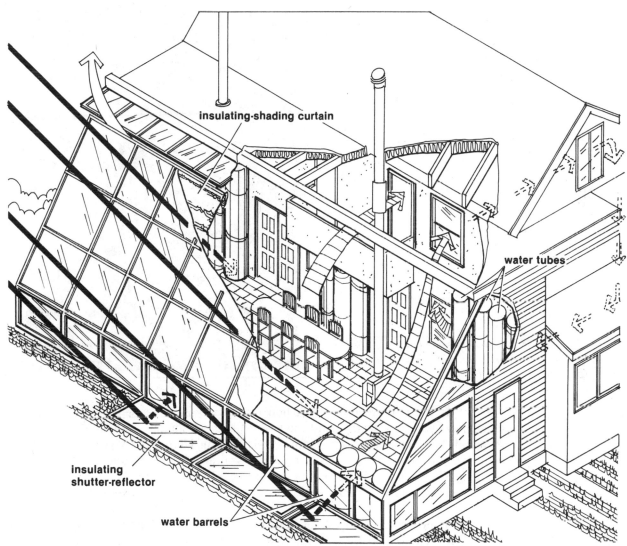

Figure IV-2: Creating a convection loop was essential to the success of this passive solar design, so the owners moved a second-floor bathroom to create an air passageway. Now the heat radiated by the water drums, Kalwall tubes and tile floor rises up to the ceiling and is shunted into the original house through a second-story window over the balcony. It travels through a hallway, down a stairwell and through another hallway to reenter the sunspace through a door or window on the first floor.

thermal mass built in. When the siding was removed from the south wall of the house, the builder discovered the wall was of old-fashioned "noggin" construction with half-timber framing. In this method of construction the spaces between the timbers were originally filled with brick and masonry rubble, thus making the heat capacity of the wall much greater than anticipated.

Photo IV-5: These homeowners use their solar addition not only as a greenhouse, but also as a dining room, where they do lots of entertaining.

To keep the greenhouse addition a little cosier in winter, exterior insulating and reflecting shutters are used against the kneewall. These shutters are made with ¾-inch plywood, 1½-inch isocyanurate foam board and 5-mil aluminized Mylar reflective film. In the closed position they make an airtight seal against compressible foam weather stripping. Their effective insulating value is estimated to be about R-12. When open, they lie nearly horizontal to reflect about 15 to 20 percent more solar energy onto the water drums.

Some additional insulating value is gained from a curtain that spans the entire tilted glass section. It operates smoothly with a cable and track system made especially for the sloping roofs of commercial greenhouses. The curtain is made of aluminized Tyvek fabric, which has reflective properties, though its main value is in creating an airspace between itself and the inner glazing. The curtain is also an effective summer shading device, usually a must for tilted glazing.

Ventilation at the top of the sunspace also reduces summer heat buildup. At the top 26 inches of the tilted glazing is a continuous 36-foot-long vent, which is controlled by a standard operating mechanism manufactured by the Lord and Burnham Company for their prefabricated greenhouses. The 72 square feet of vent opening combined with the operable windows and doors at ground level ensure adequate warm-weather ventilation.

To predict the addition's thermal performance in the design stage, the Princeton Energy Group used a hand-held calculator program (PEGFIX) they have developed, as well as other simulation models. Their primary design parameter included maintaining an inside temperature range of 55 to 80°F (13–27°C) and according to the owners, the addition meets this criterion well. As a back-up heat source the owners added a wood-stove, which to date has seen very little use.

The owners are delighted to be able to cut down on the use of the oil-fired hot water that originally heated the house. Besides using the sunspace to grow plants, they enjoy using it as a dining room and family room. When snow has drifted high up against the kneewall, it is delightful to sit in an easy chair with a book and a cup of coffee, basking in the sun.

Amy Zaffarano Rowland

Reference

Brock, Peter D. 1979. *Recent greenhouse projects: a large residential addition and a novel passive solar greenhouse assisted heat pump system.* Princeton N.J.: Princeton Energy Group.

A Sunspace/Coolspace

As a boy I often visited my grandparents' house in central North Carolina. In summer, with its porches and shade trees, it always seemed cool and comfortable. In winter, though, it was either too hot or too cold, subject to the vagaries of woodstoves and fireplaces, and in later years gas or kerosene heaters.

When my wife and I moved into the same old house in 1978, it needed plenty of improvements, and energy efficiency was very high on the list. As much as we could, we wanted to make the house independent of the utility companies and be able to go away for a month in winter knowing that the stored warmth of the sun would prevent frozen pipes.

Those south-facing porches that I had enjoyed so much as a child seemed a natural site for a passive solar system that would not only provide heat in the winter, but continue to shade the house in the summer. We decided to enclose the upper and lower 24 by 8-foot porches to make a two-story sunspace addition that would decrease both our winter heating and summer cooling loads, and also moderate the spring and fall temperature extremes. With our own labor and the help of a carpenter, we've made the house more comfortable year-round than it ever was when my grandparents lived in it. The house itself, however, posed many problems for us along the way.

First, while the existing south wall offered enough glazing area for substantial solar heating, it also received the hot summer sun in the morning and, because it was set back from the west wing, afternoon winter sun was blocked. To gain shading for the porches in summer and allow the winter sun to enter throughout the day, we decided to site the window wall of the sunspace flush with the extended west wing of the house, which also seemed like a good design decision. This increased the floor area of the sunspace to about 480 square feet by enclosing an additional 12 by 24-foot space in front of the porches. With this design the window wall would be about 16 feet high and 24 feet wide.

Even with vertical glazing we needed a way to shade the window wall to prevent overheating in late spring, summer and early fall. A roof overhang of 30 inches was the obvious method of shading the top course of 34 by 76-inch double glass panels. To shade the lower window panels, we installed 24-inch-wide removable plywood shades mounted on permanent wood supports and shelf brackets that were attached to the framing. Hinged, insulated reflector panels direct the sun's light through the 34 by 38-inch thermopane panels at ground level into water-filled heat storage barrels behind them. When they are closed, they serve as insulating shutters that shade the glazing in summer and minimize heat loss from the thermal mass during winter nights.

The sunspace heats only the rooms opening directly onto it: an upstairs bedroom and bath, the downstairs living room, office and dining room. A woodstove heats the kitchen, dining room and back hallways, while the bedroom above us is heated by

Photo IV-6: Nelson Blue's solar addition echoes the lines of the porch columns and balcony rails on this stately old North Carolina house.

warm air rising through a floor vent. We heat the four west-facing rooms in the front of the house with portable electric or kerosene radiant heaters, when necessary.

Of course, without insulation and other weather sealing improvements the solar gains would have been of little or no benefit. We began work on a house that was drafty and entirely without insulation. The front wing of the house was built on support piers with an open foundation so that cold air circulated freely under the flooring, a problem we corrected with R-11 fiberglass laid in the joist spaces. We insulated the ceilings with R-19 fiberglass (6-inch). We removed all the exterior siding boards and installed R-11 fiberglass (3½-inch) and 15-pound felt. Using R-11 fiberglass, we also insulated the foundation walls at the rear of the house up to the floor line. We made the house more airtight with caulking and covered the inside of the window screen frames with 5-mil clear film, taping cracks with transparent tape. Now we've arrived at the point where infiltration losses have been reduced from over two air changes per hour to about one (in the solar heated part of the house).

Once the sunspace was built with some 380 square feet of south glazing, we had to contend with the problem of providing enough thermal mass to prevent overheating. The sun does not shine directly on the walls of the inside rooms, and the walls themselves are of frame construction, which contains little mass. To fill this need, we

Photo IV-7: A roof overhang shades the upper sections of glazing, plywood shutters are hung from wooden supports to shade the middle sections, and the insulated reflective shutters are raised against the kneewall to protect the addition from excessive insolation in summer.

installed six kerosene tanks (30 by 30 by 36 inches high) on the lower level of the sunspace with their sides exposed. They contain about 7000 pounds of water. We also placed 23 55-gallon steel barrels, filled with about 10,000 pounds of water, under a vented counter that was built behind the bottom run of glazing panels. The heat stored in the barrels and tanks is released by radiation and convection on cold nights and cloudy days, and our records show that this has prevented the temperature in the sunspace from dropping below 50°F (10°C) even during extreme winter weather.

The first-floor rooms that adjoin the sunspace maintain about the same temperature as the sunspace itself during the winter, although a portable heater is sometimes called for in the living room. Upstairs, the sunspace and adjoining bedroom are always slightly warmer than downstairs and never need additional heat. This passive heating system maintains a comfortable sleeping temperature of around 55°F (13°C) at night in the upstairs bedroom, which by day stays in the low 70s (21°C). These temperature swings have been no discomfort for us.

As for cooling in summer, the air movement provided by breezes or portable electric fans keeps the sunspace in the high 70s on the hottest days. With summer

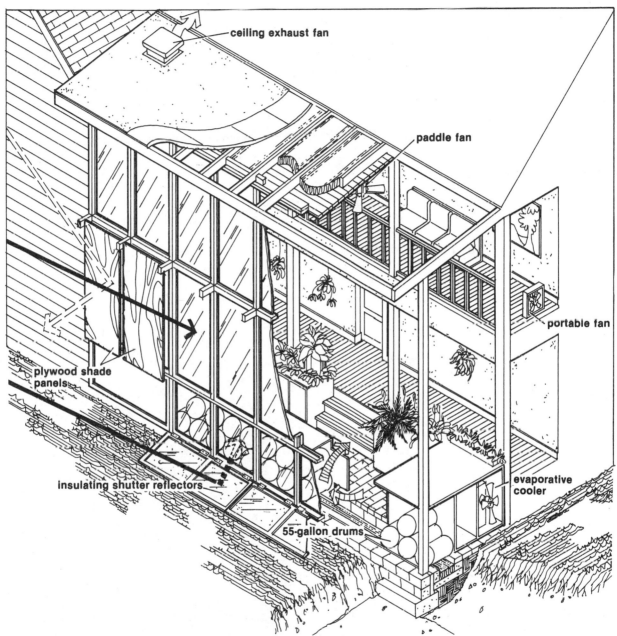

Figure IV-3: After insulating and weather sealing well, the Blues built a sunspace with plenty of glazing area. To store some of the direct-gain heat, they installed 55-gallon water-filled drums under a counter built behind the lower glazing section. The heat released by the barrels and tanks is sufficient to keep the sunspace temperature from dropping below 50°F (10°C) even in extreme winter weather. Convection brings heat to the rooms adjoining the sunspace and a paddle fan further promotes circulation of heated air. In summer plywood shade panels are mounted on the middle glazing section. An evaporative cooler is used occasionally when fan-powered ventilation is insufficient. With these controls the Blues can make their sunspace a coolspace as well, despite the hot southern sun.

shading and night ventilation the indoor temperature seldom exceeds 82°F (28°C). We pull cool night air into the sunspace with the help of the fan and a two-speed, 2000/3000-cfm evaporative cooler. Hot daytime air is exhausted by a 1500-cfm utility fan in the sunspace ceiling. We also make use of a 52-inch Casablanca fan suspended from the ceiling, which helps to flush hot air out of the sunspace at night. The moving air also increases the cooling effect of sweat evaporating from skin. In winter we can use the fan to pull warm air off the ceiling and blow it down to the floor.

In experiments made during the summer of 1980, circulating air through the cool basement and crawl space under the house markedly improved our ability to cool off the sunspace. The coolth obtained from under the house actually gave the same cooling effect as did running the evaporative cooler at full tilt, which meant that we needed less of that appliance. For several days in July 1980, the ambient temperature reached 100°F (38°C). On these days the first-floor temperatures reached 85°F (29°C), and the second floor got up to 89°F (32°C) by the late afternoon, without any cooling. By using our new-found cooling technique, we were able to keep the temperature down to around the low 80s. We've also found that using small portable fans to blow air directly over us when we're sitting or sleeping is an effective way to increase cooling comfort. With this method our bodies sense an air temperature that is around five degrees cooler

Photo IV-8: The south-facing porches have been trans-formed into a two-story sunspace complete with a paddle fan to keep the warm air circulating.

than the actual air temperature. It seems as though the key for cooling in humid climates is in drawing upon cool air sources and in keeping lots of interior air gently circulating.

We were able to build our addition using standard building practices and materials. The glazing for the whole window wall is double-pane tempered glass. For the middle and upper courses the units are standard 34 by 76-inch patio door replacement windows. The lower course uses 34 by 38-inch windows (another standard size) in front of the water barrels.

The framing for the glazing is 2×6's that are 36 inches on center. The roof framing is 2×6's set 24 inches on center with a ½-inch plywood deck and composition roofing. The east end wall is also framed with 2×6's 24 inches on center and finished with wood siding that matches the rest of the house. All the unglazed wall and roof sections are insulated with 6 inches of fiberglass.

Inside, the ceiling is finished with acoustical tile. The floor between the counter and the porch is brick over sand for the area behind the counter, which adds a little more thermal mass. The footing is 8 by 8 by 16-inch concrete block, which is all that is necessary in this mild climate. The barrels under the counter are supported on each end by 8-inch concrete blocks running lengthwise, forming a 20 by 8-inch duct space. An inch-and-a-half of Styrofoam insulation (R-7.5) is used on the inside of the stem walls and under the blocks supporting the barrels.

The cost of the addition, not including my labor, was $4000. This works out at about $14 per square foot for the two-story, 288-square-foot addition (the floor space between the porch and the window wall). Of that cost, the glass came to just over $1600. Interestingly enough, to that 288-square-foot increase in living space we also add 384 square feet more usable living space gained by enclosing the two porches. This makes for a total addition of over 670 square feet with one large window wall, which brings the cost down to about $6 per square foot.

In our 3393 heating degree-day climate our auxiliary heating costs included one cord of wood ($80), 50 gallons of kerosene for two portable heaters ($45), for a total cost of $125. We estimate the cost of running fans and the evaporative cooler to be about $10 a month, or about $50 for the whole cooling season. Thus our total space conditioning cost is about $175 a year for our home, which has 1400 square feet of heated and cooled living space in addition to the 480-square-foot sunspace. We figure that in normal, sunny, cold-weather conditions, we're getting about 75 percent of our heating energy from the sun while in very cold weather the fraction drops to about one-half solar and one-half auxiliary heat.

We find our sunspace to be a delightful living area. It is pleasant in summer and winter, comfortable when the sun shines and an effective buffer zone, with its stored mass, when it is cold and cloudy. The lowest winter temperatures inside are always about 30°F (1°C) above the lowest outside temperature. In summer the high temperature inside is at least 10°F (6°C) cooler than the peak outside temperature. We think our purpose has been accomplished.

H. Nelson Blue

V NATURAL COOLING

Photo V-1: Air conditioning without air conditioners isn't all that new. This grape arbor has been around for years shading a west-facing porch from the devilish rays of the summer sun. Along with the shading, vegetation also provides an evaporative cooling effect.

If you can't stand the heat....

If you can't stand the heat, you can get rid of it. Better still, you don't even have to let it in, and you can unplug the air conditioner, which is the villain of the story. Introducing natural cooling: Just as a house can be upgraded to be self-heating, it can also be made better at self-cooling. The first steps involve resisting heat gain from the summer sun, the enemy. The techniques for shading are many, and insulation for walls and ceilings is not only a friend in winter. When heat does get in, ventilation comes to the rescue. Natural and forced convection are used to flush out the heat and bring in cooler air from allied suppliers, namely basements and crawl spaces; even the earth can be enlisted as a source of *coolth* in one of the more advanced cooling schemes. Other schemes make use of radiation, evaporation, dehumidification and combinations thereof to provide effective cooling and to eliminate or at least minimize the need for switching on the power-hungry air conditioner. In this section we take an all too brief look at some of the basic and some of the exotic approaches to natural cooling. It's a big subject, and we'll definitely devote a whole book to it in the not-too-distant future, but with this once-over it should become clear that air conditioning without air conditioners is very much in the realm of the possible.

Photo V-2: On this new solar house the grape arbor was built during house construction. As the deciduous grapevines mature, they'll cover the overhang and fill in over the south glass. When cool weather returns, the vines drop their leaves to clear the way for direct-gain solar heating.

Air Conditioning without Air Conditioners

As the green tide of spring spreads slowly north, people's thoughts turn from staying warm to keeping cool in the hot summer days ahead. For many the rites of summer begin with the ritual Uncovering and Cleanup of the Air Conditioner, followed by the Setting of the Thermostat and the Plugging-In. After duly initiating the cooling season, the attention shifts to chaise lounges or lawn mowers, or to any number of repairs to the ravages of winter. The cooling system will be left to defend the house against oppressive heat and humidity, and it will run and run and run as long as it takes to get the job done. It gets expensive. In many parts of the country a summer's worth of cooling with an air conditioner is more expensive than winter space heating, but there are all kinds of techniques for house cooling that don't involve a single kilowatt-hour. When more cooling is needed, low-power fans are used instead of power-hungry air conditioners. In this essay we'll follow two basic themes of natural cooling. The first is reducing the amount of summer heat that gets into a house, and the second is providing additional interior cooling when heat gain can't be adequately resisted. By implementing some of these methods, you may be able to give your air conditioner a summer vacation instead of a summer job.

In the previous two sections we've presented all sorts of ways to add solar space heating, but it's important to keep in mind the need for cooling because increased solar heating capability can easily contribute to an increased cooling load. The area of your solar glazing should be tuned to the actual heating demand and not just expanded to cover all the available solar surfaces. To prevent summer heat buildup from direct, reflected and diffuse gains, space heating systems should be designed to vent solar-heated air directly to the outside and/or to block solar gain with overhangs, awnings, landscape shading and other shading techniques.

There are also simple management techniques, such as controlling windows, that you employ without making any drastic changes. Building insulation is another friend of natural cooling. In summer the heat source is of course outside the house, and insulation can greatly reduce conductive heat gain through walls, roofs and windows. In some areas, natural convection or fan-powered ventilation with cool night air, in conjunction with heat-gain reduction, can provide all of the cooling necessary for an energy-efficient house. If these techniques don't give you enough cooling comfort, there are ways to minimize the energy used by air conditioners. There is also something called an evaporative cooler, which uses less power than most air conditioners.

As you read, keep in mind the information on human comfort presented in "The Sun in Your Place" in Section I. Keeping cool doesn't necessarily mean keeping the air temperature at 72°F (22°C) or even 78°F (26°C), but involves the relationship between air temperature, humidity, air movement and mean radiant temperature (the temperature of surrounding surfaces), as well as the nature of your own indoor activities. Many of the techniques described here increase comfort by increasing air movement or reducing the mean radiant temperature and not just by reducing air temperature. With a lower mean radiant temperature, drier air and a gentle breeze, even at 84°F (29°C) you can feel comfortable.

Hot Weather House Management

Before you try cooling techniques that require major modifications, see what you can do to reduce the cooling load in your house just as it is. You may find that with wise management you can somewhat reduce the cooling load without any need for major changes. The most effective of these procedures involve controlling solar gain, ventilation cooling and minimizing interior production (internal gains).

Ventilation can provide the most powerful cooling in many cases. When the night air temperature drops, open windows and doors to let breezes cool the inside. This coolness or "coolth" is stored by the interior mass of the house (Sheetrock, tile floors, etc.). If there isn't much breeze, ventilation can be increased by using fans, the fan on an air conditioner or the furnace blower in the duct system if there is a "fan only" setting.

Ventilation helps not only by cooling the inside of the house but also by increasing the sweat evaporation rate from your skin and thus your sense of comfort. For example, a small, quiet portable fan in the bedroom can be aimed to blow over you while you sleep.

Solar heat gain through windows and sliding glass doors can be significantly reduced if the drapes, shades or blinds are closed when the sun is shining on them. Most effective are light-colored, opaque and insulated drapes and blinds. For maximum protection they should be well sealed at the top (with a valance), bottom and sides. If they don't have a sealed valance at the top, studies have shown that window coverings can actually increase heat gain by setting up a convective loop in which heat that builds up between window and shade is transferred to the room.[1]

Curtains and shades are often inset in the window frame, and in that location good side and bottom seals can be developed. Drapes, blinds or shades on east windows should be closed in the early morning; windows should be covered by mid-morning, and west windows should be shaded by early afternoon. If it is very hot and bright or if everyone is away during the day, close all the drapes, blinds and shades first thing in the morning and later open only the south and east ones in the evening, until the sun sets. (You might recall this is the exact opposite of the winter routine when you open the drapes on the east in the morning, south at midday, and west in the afternoon.) The savings from this simple activity can be significant. Studies made in Davis, California, have shown that interior shades can reduce room temperature 16°F (9°C) or more below that of a room with bare windows,[2] while another study at the Illinois Institute of Technology found that plain white roller shades could reduce total heat gain through the

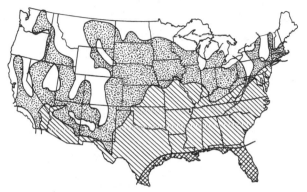

☐ **solar control and radiative (night sky) cooling**
▨ **solar control ventilation**
◩ **solar control, evaporative and radiative cooling**
⊠ **solar control, evaporative cooling, dehumidification**

Figure V-1: This map shows some of the cooling strategies that can be used in different climate zones.

window by nearly 50 percent.[3] For very hot rooms, putting aluminum foil on the glass will make an even bigger difference.

Internal heat gain—the heat generated by people, lights and appliances—can make a significant contribution to an increased cooling load. Again, tailoring your household activities to be compatible with the season, and careful use of electrical equipment (lights, appliances) will limit this gain. Rather than baking or cooking meals that require use of the oven or range burners, create meals that can be prepared with no cooking or at least a minimum of stove "on time." If you are reaping a steady supply of fresh fruits and vegetables from your garden, you have the foundation for superb summer meals that don't need heat. Low-energy meals also reduce the electric or gas bill. If you do use the stove top or oven, plan to use it at night, if you can, so the added heat can be carried off by the night breezes and ventilation. Save the fancy cooking for winter when every bit of the added heat can be put to use.

The refrigerator and freezer both give off heat. If you can, put the freezer in the garage or in a room that can be closed, although some freezers can malfunction if they're kept in unheated spaces, and the manufacturer's warranty might be voided, so look over your owner's manual. Generally speaking, refrigerators and freezers should not be put in unheated rooms that drop below freezing. You can also help these appliances operate more efficiently by moving them a little farther from the wall, keeping the coils clean and opening them as little as possible. When you replace a freezer or refrigerator, get an energy-efficient one and buy the smallest unit that will meet your needs. The most efficient refrigerators use as little as 50 KWH of electricity per month instead of the norm of 200 to 300 KWH.

If you have the space, consider drying your clothes on a clothesline in the summer instead of in the dryer—reducing the cooling load of the house and saving the energy that the vacationing dryer doesn't use. If you must use the dryer, make sure it's properly vented to the outside with dryer vent hose to avoid the discomfort of added summertime humidity. Spending time outside while the weather's fine will also reduce heat gain from people, the tv, stereo and lights that you might otherwise use inside. Just by properly managing your house, you can greatly reduce the need for air conditioner cooling. You can go a lot further with natural cooling, however, by employing some or all of the following options.

Solar Control

In the broadest sense, solar control is a year-round practice, and most of this book is devoted to controlling solar heat in cold weather. But the sun, which provides such excellent heating in winter, becomes an unwelcome visitor in the summer. By using the appropriate solar controls you can block the summer sun yet still allow solar access for winter heating. Solar control for cooling relies on factors that are equally common to solar heating: orientation, shading, insulation, color and material, and landscaping.

Orientation

Your home's shape and orientation will play a vital role in summer heat gain and the ease of solar control. As you learned in the introduction, the best orientation for solar heating is with a rectangular house with its longest dimension facing due south and containing most of the windows. The north side should have the fewest windows. This shape will also stay coolest in the summer and is most amenable for solar control techniques (see table V-1).[4] However, a poorly oriented house offers the greatest room for improvement because it is so vulnerable to heat gains, so even if you have a tall house facing west, don't despair; there is much you can do to keep cool no matter what the building shape.

The sun rises a little north of east in summer; by noon it is very high in the sky and in the evening it sets to the north of west. During a summer day the sun's energy is concentrated on the east and west walls and windows while the south and north receive considerably less. Tables V-2 and V-3 show how significantly exposure or aspect is related to wall and window heat gain.

East and west windows receive almost twice the solar energy as do the south windows and nearly three times as much as the north windows. For walls, it's clear that increasing insulation has the greatest effect on reducing heat gain. The actual impact on your house will depend on its design and construction, local climate, and on the solar exposure of both the house and the site. The eastern sun may overheat the house early in the morning, requiring air conditioning all day. Or the house may be hit by the western sun in the afternoon, the warmest time of the day. The answer to most of these orientation-caused problems is solar control.

Insulated walls do much to resist conductive heat gain. Windows, on the other hand, provide little protection against conductive gain, but this isn't the primary concern with windows. Direct and diffuse radiation are by far the greatest contributors. Thus the first steps toward window control should be with shading, whether by overhangs, wing walls, exterior shades, interior drapes, blinds, shutters or tinted or reflective films.

TABLE V-1   COMPARISONS OF POTENTIAL SUMMER OVERHEATING FOR HOUSES OF DIFFERENT SHAPES AND ORIENTATIONS

Shape and Orientation	Interior Temperature (°F) Rise above Ambient
Dome	47
Tall broad model, east/west	47
Gable roof facing east/west	42
Cube facing northeast/southwest	36
Cube facing north/south	34
Two-story house facing north/south	26
Two-story white-roof house facing north/south	14

SOURCE: L.W. Neubauer, "Shapes and Orientations of Houses for Natural Cooling," *Transactions of the American Society of Agricultural Engineers,* vol. 15, no. 1 (1972), pp. 126–128.

NOTE: The models used in this study were closed, black metal boxes in order to standardize conditions. So when reading the numbers, it's important to bear in mind that people do not, of course, live in closed, black metal boxes. If all the above building shapes and orientations were part of real, open, well-insulated and ventilated houses, the temperature rise would be nowhere as great. These temperature data are used simply to show that certain building shapes are more liable than others to heat up.

TABLE V-2   SOLAR HEAT GAIN THROUGH WINDOWS AT 40° NORTH LATITUDE ON JUNE 21 (SOLSTICE)

Window Orientation	Btu/ft²
N	484
NE, NW	894
E, W	1200
SE, SW	1007
S	622

SOURCE: American Society of Heating, Refrigerating and Airconditioning Engineers, *Handbook and Product Directory* (New York: ASHRAE, 1972).

TABLE V-3   SOLAR HEAT GAIN THROUGH SOLID WALLS AT 40° NORTH LATITUDE ON JULY 21 (AVERAGE AIR TEMPERATURE 83°F)

	Btu Gain per Square Foot of Wall Area			
Wall Orientation	8" Brick, Uninsulated	Wood, Uninsulated Dark Paint	Wood, Uninsulated Light Paint	Wood, Insulated Dark Paint
N	760	380	367	190
NE, NW	820	410	384	205
E, W	864	432	393	216
SE, SW	855	427	388	214
S (shaded by overhang)	717	358	335	174

SOURCE: Clifford Strock and Richard Koral, eds., *The Handbook of Air Conditioning, Heating and Ventilating* (New York: Industrial Press, 1979).

Shading

The high angle of the summer sun makes the use of overhangs very attractive for solar control, particularly on a south wall. Extending the roof may not be cheap, but it is effective and permanent, while providing better weather protection for the south wall and windows. The required overhang extension for south windows depends on the months when shading is required, the height of the window and the latitude. Table V-4 provides a simple method for determining the needed overhang.

On some houses an overhang can be added along the south roof pitch simply by extending the rafters (by bolting extensions onto the rafter tails) and extending the roof. When planning such an extension, be aware of how low the new roof edge will be. You may want to build the extension at a slightly shallower pitch than that of the existing roof. If the total overhang can't be practically extended, or if you need to shade south windows on the first floor of a two-story building, you can build a simple overhang or add canvas or aluminum awnings. Beware of fixed awnings, however, because usually they're hung so low as to block direct gain in winter as well as in summer.

Unfortunately there is not a practical overhang design that works well for windows facing more than 15 to 20 degrees east or west of south. To be effective, an overhang for a 4-foot-high southeast window 1 foot below the eave would have to be 10 feet long to provide adequate protection. Wing walls or fins are effective in blocking gain into windows that face east or west of south. These "vertical overhangs" are extension partitions or "mini-walls" that can be made of materials identical or similar to the existing siding. They also can be made like attached fence sections with a light, airy appearance that won't dominate the appearance of the wall. In general practice, these wings are extended as far as overhangs.

Photo V-3: A simple and nice looking patio overhang greatly reduces the solar gain to the west wall of this house, and it keeps the patio a little cooler too. The slats could also be crisscrossed to increase the shading effect.

TABLE V-4   Finding the Overhang Required for Summer Shading

North Latitude	F factor*
28°	5.6–11.1
32°	4.0– 6.3
36°	3.0– 4.5
40°	2.5– 3.4
44°	2.0– 2.7
48°	1.7– 2.2
52°	1.5– 1.8
56°	1.3– 1.5
Horizontal extension of overhang = window height F	

SOURCE: Edward Mazria, *The Passive Solar Energy Book* (Emmaus, Pa.: Rodale Press, 1979).
*Find the F factor in the table according to your latitude. The higher values will provide 100 percent shading at noon on June 21, the lower values until August 1.

To define an adequate degree of shading, it's helpful to use the term *shading coefficient* in describing the effectiveness of shading techniques on a numerical scale. A window fully exposed to the sun has a shading coefficient of 1 (maximum heat gain), while a window fully shaded by an overhang has a shading coefficient of 0.2. It is not 0 for an overhang because diffuse and reflected radiation can still get in. By contrast, an exterior vertical shade hung close to the window can have a shading coefficient as low as 0.1. The shading coefficient is a convenient way to compare the effectiveness of the different methods of solar control.

Windows that face east or west, or that face south with no overhang, may require exterior shading devices to provide an economical, attractive and effective means for shading them. Why not just use interior curtains or drapes you may ask? The answer is simple. Once the sun enters a window, heat has effectively entered the house, and although its impact can be reduced with an interior shade, some heat trapped between the shade and the window will still pass into the room. The simplest and cheapest method for exterior shading is often to hang a canvas or bamboo shade outside the window, supported by the roof overhang or an exterior valance. A space for venting should be

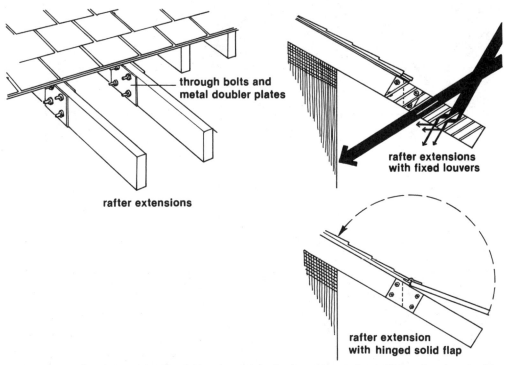

Figure V-2: Rafter extensions should be through-bolted to the existing rafter ends using doubler plates for maximum strength. Fixed louvers are one way to control solar gain in different seasons, while another possibility is to fasten a hinged panel to the extension. It can be put down onto the extension during warm weather.

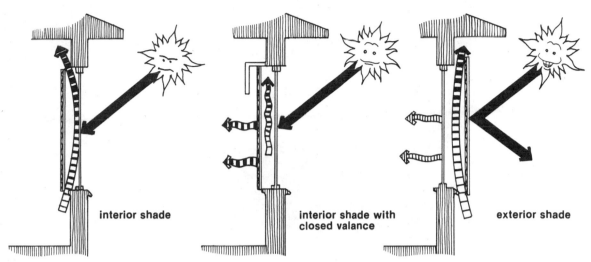

interior shade

interior shade with closed valance

exterior shade

Figure V-3: As the illustration shows, exterior shading is the most effective for stopping heat gain. Most of the heat built up between the glazing and the shade convects away. The second best option is to have an interior curtain with a closed valance at the top to limit convective heat flow into the room.

left at the top of the support to prevent buildup of a pocket of hot air. This kind of system provides good protection for a modest investment. The shade can be rolled up in the evening and, when fall comes, it can be taken down and stored, just like the time-honored winter storm window routine. Where wind is a problem, it is necessary to provide a ground anchor, preferably with an elastic shock cord, to prevent flapping and possible damage to the shade.

Louvered shade screens can be a more expensive means of exterior shading than a roll-down awning, but they are also more convenient—no need to roll them up or down, and although they do have some filtering effect on incoming light, you can see through them clearly. Shade screen, which consists of many tiny louvers that also serve as an insect screen, can be used in standard screen frames that snap onto the windows. Several companies manufacture louvered metal shade screens with various louver angles and shading coefficients. (See the "Hardware Focus" section at the end of this essay.) The shading coefficient is typically about 0.2, the same as a full overhang. The two drawbacks are sensitivity to damage (kids, animals, low-hanging branches) and cost. Bronze shade screens, for example, can cost over $5 a square foot. They may be best used on east- and west-facing second-story windows that are removed from possible sources of damage.

More recently several companies have started manufacturing shade screens that use flat fiberglass webbing instead of louvers. This material is much cheaper (around $1 per square foot) and provides good shading (the shading coefficient is about 0.3). They are much more resistant to damage than metal shade screens, and they may be the best alternative in many cases. Naturally, both types of shade screens should be taken down

Photo V-4: Exterior shading for west- and east-facing walls is the most effective for reducing heat gain because it blocks sunlight before it passes through the window. Interior shades block sunlight effectively, but they can also trap heat between the shade and the window.

in the winter on south-facing windows to allow as much solar radiation to enter the house as possible.

Large louvers may also be used for a very effective and attractive fixed or movable exterior shade. Unfortunately, they are expensive unless you build them yourself. Vertical louvers are best when installed over east and west walls; horizontal louvers are best for the south. Shading coefficients of 0.2 to 0.3 are typical values for well-designed exterior louvers.

Freestanding walls and fences may also be used for exterior shading and can be particularly effective on the critical west windows and walls. They can also help prevent side yards or courtyards from becoming heat traps. They should be louvered or built with slats to reduce adverse effects of wind loading.

Is your porch rather furnacelike in summer? A roofed porch, particularly an east- or -west-facing one, can be shaded with homemade vertical sun screens that look quite nice. The screens are made up with a "weaving" of diagonally crisscrossed wood strips, e.g., plaster lath, set into a frame. These units can be installed permanently or seasonally; they'll still admit enough light so the porch won't be dull and dim.

The ultimate design of a total exterior shading retrofit may well include a mix

Photo V-5: Shade screens block sunlight, but they don't block the view. The photograph on the left shows the view through the screen at eye-level, while the one on the right shows how light is blocked when it comes in from a high (summer sun) angle.

of the options discussed here, depending on the many variables of orientation, window placement, available space (for wings and fences), and certainly on how the retrofits will affect your home's appearance.

Shading with Interior Drapes and Shutters

As noted, an interior shading device will do little to reduce heat gain unless it's sealed tightly to the window frame. If drapes, shades or shutters are well designed, however, they can provide effective control over direct gain and reduce conducted heat gain. To begin, however, let's consider their value for controlling direct gain.

The most common window fixtures are, of course, venetian blinds, roller shades and drapes. These standard items provide shading coefficients only one-half to one-fourth as effective as overhangs or exterior shading. A white curtain or roller shade has a value of 0.4, a white venetian blind about 0.6. Darker colored drapes, shades and venetian blinds may have a shading coefficient of only 0.7 to 0.8, hardly worth utilizing for sun control.

A much better alternative is offered by special drapes and interior shutters made specifically for solar control. These all include most of the following features: highly reflective outside material, high resistance to conductive heat flow (R-value about 5), and tightly sealed edges (top, bottom and sides). Properly made and installed, these units have a shading coefficient lower than an overhang—down to 0.1 or less—which makes them at least twice as effective at cutting out heat gain. However, for that level of efficiency they will admit little or no light unless a small opening is left in them. This kind of window treatment is a perfect example of how hot weather controls interface with cold weather controls. In cold weather the same insulated curtains or shutters can be used at night to reduce building heat loss through windows, making them an effective year-round energy control.

Nowadays an increasingly popular material for solar control is tinted or reflective film applied directly to the window glass or attached to a shade-roller mechanism. Although these are widely advertised as *the* solar control treatment for windows, they aren't always worthwhile, particularly for south-facing windows. The most obvious problem with fixed films is that they are permanent and cannot be removed in the winter. They are marginally effective for hot weather shading, with typical shading coefficients of only 0.3 to 0.5 and little resistance to conducted heat gain. For east- and west-facing windows these films can be helpful because overhang requirements are so great for those orientations. Use the darkest or most reflective films that provide a shading coefficient of at least 0.3.

Insulation and Building Color

Just as window insulation is a useful control in both hot and cold weather, so too is building insulation about as important in summer as it is in winter. The major source of summer heat gain in most houses is the roof, and roof or attic floor insulation is very helpful in minimizing the conduction of heat into attics and second-floor rooms, particularly on houses with dark-colored roofs.

Although insulated ceilings significantly reduce the interior heat gain from a hot roof, it is often desirable to further reduce gains with roof treatments such as light-colored roofing material to reflect rather than absorb solar energy. This change can reduce roof temperature by 50°F (28°C) or more,[5] keeping the house cooler and extending the life of the roof. Certain roofing materials also naturally "run" cooler than others. Wood shakes and shingles are particularly effective in this way, followed by tiles and light-colored asphalt shingles. On the other hand, tarred roofs, metal roofs and asphalt roll roofing are among the hottest.[6]

A design sometimes used in the tropics that may be desirable for reducing heat gains in very hot climates is the double roof. The upper roof membrane should be light

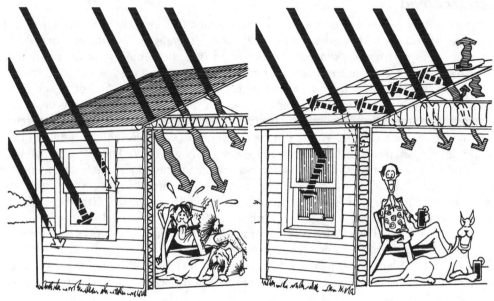

Figure V-4: By increasing building insulation, especially in the roof, you will minimize heat gain through the ceiling. The other treatments shown here are changing the roof material (when it's time to reroof) from a dark to a light color, increasing wall insulation and using window shades.

in color and thoroughly vented. The heat buildup between the two roofs sets up a convective air flow that exhausts hot air back to the atmosphere if the roofs are well-ventilated. This double roof idea meshes with another natural cooling option that uses the double roof at night to deliver cool air into the house. This is discussed shortly.

In a frame house the most critical walls vis à vis heat gain are the east and west walls, and they should be insulated and painted a light color (though you will certainly want to insulate all the walls to reduce cold-weather heat loss). If all the walls are massive (brick, cement, block, adobe or stone), exterior insulation and light color on the east, north and west walls would provide both summer and winter benefits by seasonally storing coolth in summer and heat in winter. Dark-colored massive south walls that are glazed over can capture the sun's energy and keep the house warm far into the night after outside air temperature has dipped below comfortable levels.

In fact, a stone, brick or concrete block building that has been covered with exterior insulation is likely to be a prime candidate for successful air conditioning without air conditioners, even in very hot climates. The large surface area of the mass walls helps keep the building cool by absorbing room heat. With vigorous nighttime ventilation, the cooling effect is enhanced even more, and those cool walls will be ready to absorb more heat during the following day. To overcome high interior temperatures, an evaporative cooler or air conditioner can be run during off-peak electrical use periods (night) to help the walls store additional coolth.

Landscaping

Proper placement of trees, vines and shrubs can provide very attractive and effective, but often overlooked, solar controls for roofs, walls and windows. Total shading from trees can have a shading coefficient of 0.2, the same as for an overhang! Properly placed deciduous trees will not block the winter sun if they are located mostly to the east and west of a building rather than directly to the south, where even bare branches might block valuable incoming winter sunlight. Deciduous trees are desirable for landscape shading because they automatically adjust for cool springs and hot falls. In early spring, the small leaves don't block much of the valuable sunlight, while in late fall the leaves will provide shade when there is still a need for cooling in many parts of the country. If a solid south overhang cannot practically be added, then landscape to provide needed shade on the large south windows. With an arbor, use a vine or plant that can be pruned back in the fall to allow nearly full sun through. Grapes, wisteria, morning glory and hops are all good candidates. Keep the structural members of the arbor as small as possible so they won't cast shadows in winter; wire works well, or you can make an arbor that is removable in winter.

The full landscape plan for a house might include: tall trees for roof shade, deciduous vines for shade on an arbor to the south of the windows, dark ground cover

Photo V-6: These arbors are retrofits, of a few decades ago. The structure is formed of bent and straight pipe sections that are covered with a mesh that carries the grapevines.

Figure V-5: This landscaping plan shows much use of deciduous trees for summer shading. Shadows cast on the ground as well as on the house itself help to cool the air around the house. Shady spots are also effective for evaporative cooling.

to the south to reduce ground reflection of light and heat, hedges to the east and west to shield the east and west walls and windows, and shrubs or small trees to the northeast and northwest to block late evening and early morning sun in the hottest period of the summer.

For even greater cooling, a hedge can be used as a living evaporative cooler. In Indio, California, a full hedge is laced along its top with a soaker hose that thoroughly wets the hedge. The evaporative cooling effect provides 10 to 15°F (6–8°C) cooler air for the patio and a cooler source of ventilation air for the house. Shaded, wet lawns can also contribute to a cooler building environment, especially compared to the way that concrete or gravel surfaces can heat up. Naturally, these methods work best in dry climates.

Not only does landscaping cool the house and yard, it also makes a quieter, cleaner and more satisfactory environment. In Sacramento, California, cutting down street trees in one neighborhood resulted in a 10°F jump in ambient summer air temperature and considerably increased cooling loads in adjacent homes.[7] Landscaping for cooling requires plenty of watering, trimming and leaf raking—but add up the energy, environmental and aesthetic benefits, and you've got a bargain in the making.

Ventilation

Ventilation with cool night air is a simple method of cooling, and it may be all that is required in many areas after the cooling load has been reduced as much as possible. By venting warm inside air and drawing in cooler night air, you can do much to keep a house cool during the day. As discussed earlier, natural ventilation with open windows and doors may provide sufficient cooling effect, but if it doesn't, you can modify the house to provide more effective ventilation by using the "thermal stack effect." Because of natural convection, warm air wants to rise, and when it does it is replaced by cooler air. The necessary modifications range from simply adding new operable windows or vents to more substantial changes that cause an increased rate of cooling air flow.

If natural convection ventilation proves insufficient, it may be desirable to use fan-powered ventilation with window fans or a forced-air duct system, or to add a whole-house fan. As you read through and evaluate the possibilities that can be applied to your house, keep in mind the two most common concerns with natural ventilation: security and allergies. Security can be maintained by installing stop locks and fixed or locked grilles, screens and vents, or by providing other blocks to open windows and large vents. Allergic reactions caused by the introduction of mold, fungi spores, pollen, dust or smog are best dealt with by using a filtered fan and by adding filters to intake vents.

In some areas, though, night ventilation will provide little benefit and may actually be undesirable. If the nighttime temperature doesn't drop below a dry 85°F (29°C), the best strategy is to put the most effort into resisting daytime heat gain. If the humidity is high with night temperatures remaining over 80°F (27°C), then night ventilation may not be as valuable as dehumidification. Much depends on the local climate and on the construction of your house and its inherent thermal performance.

To take an extreme case, if you live in an uninsulated house with a dark roof and the interior temperature at night is 95°F (35°C), then letting in even 90°F (32°C) outside air will be desirable (at least to help you be comfortable enough to think about adding insulation and light-colored roofing). On the other hand, if you've added exterior insulation to your concrete block house, painted the roof white and weatherized carefully, you're more in control of interior conditions. That's really the point of all energy upgrading—to gain control of outside conditions for both heating and cooling your home. Once you're in charge, common sense will most often tell you what to do.

Using Wind

Wind creates a zone of high pressure on the windward side of a house and a zone of low pressure on the leeward side. You can easily make use of these pressure differentials to cause air to flow through the house. There must of course be a cross flow since without an outlet the room is like a cave. Tests have shown that ventilation in a room with no outlet may be less than one-half that of a room with an inlet and an opposite-side outlet. For best wind flow through a building, provide an inlet and outlet

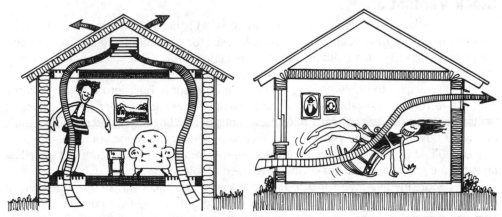

Figure V-6: Roof and wall vents will greatly improve ventilation cooling. On the left the floor registers are coupled to a cool basement or crawl space so that as the hot air leaves through the gable vent, cooler air is drawn in. At right an intake vent is placed in the path of the prevailing summer breezes, which can provide both ventilation and evaporative cooling effects.

in each room, and make the downwind outlet larger than the inlet. The wind flow will do the most good if it contacts people (evaporative cooling), so a low inlet vent coupled with a high outlet will maximize the cooling effect.

One of the simplest retrofits for improving natural ventilation is the addition of louvered vents or hinged panels either in partition walls or in or above interior doors that are kept closed at night. These vents should be sized approximately 2 by 2 feet, or as space permits. They may reduce your sound privacy a bit, so don't rush right into a retrofit without evaluating its impact. A solid, hinged panel or transom in a wall or over a bedroom door could be a solution to possible loss of privacy. An as added benefit, this kind of ventilation will also improve the distribution of solar heat collected in cold weather.

You can also replace fixed windows with ones that are operable or add new windows (on the south) or vents (on the east, west and north). Casement windows are often useful scooping in the wind. Make sure that the casement is hinged on the down-wind side of prevailing summer winds.

Opaque vents can usually be installed for somewhat less cost than a window, and they will gain less heat in summer and lose less heat in winter. Standard attic vents can be ganged together to make a larger, weather-resistant, bugproof and secure vent. They should be sealed and insulated in winter to minimize heat loss. Adding a bay window with a horizontal sill vent can provide excellent ventilation that won't be adversely affected by high winds or rain. The sill vent is also less noticeable than vertically mounted ones.

Stack Ventilation

For those who live in low-wind areas, wind is not the only method of natural ventilation. The thermal stack effect can also be used for ventilation cooling if outlet vents are provided up high in a building and inlet vents down low. This stack effect ("thermal chimney") is particularly effective in multi-story houses and can be boosted with a fan.

Typically, the only cooling of this type commonly built into a house is in the attic where vents in the gable allow hot air to escape, drawing in cooler air through soffit vents. If your attic doesn't already have these, consider adding them, and possibly a static or a wind-powered turbine vent as well. Good ventilation can reduce your attic temperature by 20°F (11°C) or more, which lowers the mean radiant temperature of the ceiling facing the living space and makes the area seem cooler.

More involved work like adding ridge vents, a cupola vent, an operable skylight or a roof vent can easily be added when you reroof (using light-colored roofing!), although adding them to an existing roof won't pose excessive difficulties as long as standard sealing and flashing practices are observed. Figure that you'll need to devote about 1 percent of the attic or top floor area to exhaust vent area, with an additional 1 percent of that area used for soffit vents.

Again, no matter what kind of exhaust vents you add that open into living spaces, always provide a tight-fitting insulated panel to close off the unit in cold weather. Winter ventilation of unheated attics should also be maintained to disperse water vapor that could otherwise condense and degrade insulation.

One of the ways to increase thermal stack ventilation is to provide a pathway for air from the cool basement or crawl space into the house, up to the attic and out. This could involve little more than screened, operable vents added in the floor and ceiling in several places, using the same rule of thumb for sizing as with attic ventilation (or 10 square feet each for inlet and outlet vents for a 1000-square-foot house).

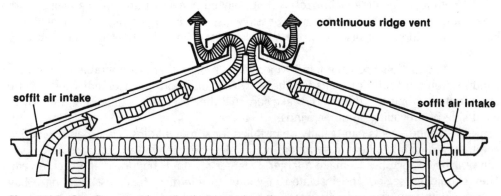

Figure V-7: Continuous ridge ventilation with soffit air intake is the best roof venting system available. As the roof heats up, a convective flow is initiated whereby heated air exhausts through the ridge vent, and cooler air enters through soffit vents.

You can do even more to increase the stack effect by building a solar chimney, such as the one described later in this section. The most basic solar chimney consists of a section of large diameter black pipe (8 inch minimum) extending a few feet up from the roof. For better results make a square chimney with east, west and south glazing and a black-painted north face. A wind-powered turbine vent is a suitable cap for the top. You'll also want an insulated inside shutter to close it off in winter. Again, remember to supply a cool air intake (basement, crawl space) along with a solar chimney. There won't be any benefit in using the stack effect only to pull in hot outside air. (See the project, "A Solar Chimney," later in this section.)

While the solar chimney is a relatively advanced passive cooling technique in as much as it should be preceded by the management techniques and building modifications discussed earlier, it also represents a fairly simple construction project. For even more cooling effect, you can install a "cool tube" to provide the primary air intake of the house. Several types of cool tubes have been built using such things as rectangular clay flue liners, metal culverts and sewer pipe. All of them involve a buried (4 or more feet deep) pipe or array of pipes with an outdoor air intake located in a shaded area. Air entering the house is first drawn through the tube by the flow induced by the solar chimney or a fan. The air is cooled by the earth (which is usually a constant 45 to 60°F, 7–16°C, at these depths), and in some cases it is also dehumidified by the time it enters the house. Although this system has been understood for a long time—similar systems have been used in Persia for thousands of years—experience with them has been gained only recently here in the United States.

One research team, the Princeton Energy Group in Princeton, New Jersey, has made instrumented studies of an actual cool tube made of eight, 40-foot-long, 6-inch diameter plastic pipes buried 4 feet deep. This array provided an average cooling effect of 8700 Btu per hour, equivalent to a small air conditioner. The researchers estimate that in an energy-efficient house of 2300 square feet, this cool tube array could meet 70 percent of the building's cooling load.[8]

This notion of coupling a cool air intake with a solar-driven exhaust can be applied to other solar space heating retrofits that serve as the exhaust driver. Consider the mass wall glazing retrofit used for passive space heating: It heats up in the summer by absorbing high-angle direct radiation and must be built with some venting capability to dump the unwanted heat to the outside. That of course implies a convective flow, and if the wall isn't heavily shaded in the summer, it may well be a solar chimney in disguise. So too might a vertical-wall air heating collector be the beginning of your natural cooling system. All of this speaks for advance planning that considers both a house's heating and cooling needs.

Before you invest your time and energy in solar chimneys and cool tubes, you should first consider using a little electricity to power one or more exhaust fans used in conjunction with added floor, wall and ceiling vents. Because air flow is forced, the fan option may work better than the more involved and subtle passive cooling effect. Fans do, however, use a fair amount of electrical power (though much less than air conditioners), an increasingly important consideration for the energy-conscious 80s.

Forced Ventilation

Forced-air cooling is at its simplest with a small fan pointed right at you. Increasing air flow past your body evaporates perspiration from your skin and keeps you feeling cooler. Portable fans can also be used to increase nighttime ventilation and cooling of a house. Place the fan so it will draw cool air into the rooms where you are sleeping.

One step beyond the portable fan is a "Casablanca" fan hanging from the ceiling. It can be used to help keep you cool (evaporatively) and to ventilate the house in summer, and it will also help mix the air for more uniform heating in the winter. Some models use only 40 watts on low speed. In the not-too-distant future, photovoltaic cells will be cheap enough to warrant their use in powering fans and other appliances. What a perfect combination: using solar electric power to remove excess solar heat.

For an even more powerful cooling system, consider a whole-house fan. The fan is installed in the ceiling to suck air out of the house and blow it out the attic vents, allowing cooler air to enter from a basement, crawl space or shaded side of the house. Such fans also play a role in increasing evaporation from your skin. The approximate capacity needed is indicated in table V-5. They can be thermostatically controlled to turn on when ambient temperature and humidity drop low enough and can be used with furnace filters placed in windows to reduce the amount of dust and pollen brought into the house. Whole-house fans are helpful and cost-effective even when used in conjunction with an air conditioner, in which case they should be set to kick on when the air conditioner goes off.

Moved-In Mass: the Coolth Connection

In many areas ventilation (up to a whole-house fan) with solar control will provide night comfort and daytime cooling for much of the summer, but for better cooling in an area with consistently high day and low night temperatures (and low humidity), added thermal storage may be helpful. Just as thermal mass stores solar heat for winter night heating, it can store the coolth of the night for use during a hot summer day. Night ventilation with cool air removes heat from the thermal mass by radiation and convection. The cool mass then acts as a sink for radiation from warm people and other internal heat sources during the day.

Thermal mass for nighttime cooling is most effective when it's widely distributed through wall, ceiling and floors, i.e., lots of surface area, because the heat exchange rate is fairly slow between air and mass. This contrasts with winter heating, when mass is most effective if it is directly in front of the south windows, where the concentrated solar radiation promotes more rapid heat exchange. You should definitely consider water storage over masonry for a retrofit with a limited area for thermal mass. "Moved-In Mass" in Section III-D describes in more detail some techniques and considerations for adding thermal mass.

Still more powerful natural cooling systems can be developed utilizing the effects of water evaporation and night sky radiation, or the seasonal storage of coolth in the form of ice or cold water. The most effective systems for retrofitting, and they are

TABLE V-5   RECOMMENDED FAN CAPACITIES

Floor Area of House (ft²)	Amount of Air Fan Should Deliver (ft³/min)	
	Cool Night Regions	Warm Night Regions
800	3000	6500
1000	4000	8000
1200	5000	9500
1400	5500	11,000
1600	6400	13,000
1800	7000	14,500

SOURCE: L.W. Neubauer and H. Walker, *Farm Building Design* (Englewood Cliffs, N.J.: Prentice-Hall, 1961).
NOTE: Table is based on air exchange rate of 1 air change per minute in warm nights, ½ air change per minute in cool night areas.

by no means simple retrofits, are those that combine evaporation and night sky radiation. An introduction to evaporative cooling and night sky radiation will help clarify how these systems work and how you might use them to help cool your house.

Evaporation

The evaporation of water can have a very powerful cooling effect, as you know from stepping out of a swimming pool on a windy day. The phase change from water to water vapor absorbs a great deal of heat (1000 Btu per pound of water evaporated). Evaporating a cubic foot of water will provide fully 60,000 Btu of cooling, a considerable amount indeed. With reasonably good air flow across the water surface, a small pond can evaporate 10 to 15 inches a month, a cooling rate of over 2000 Btu per square foot per day, or about 85 Btu per square foot per hour. A 500-square-foot section equipped with sprinklers for evaporative cooling can remove heat at the rate of over 40,000 Btu per hour. Roof sprinkling is a common application of evaporative cooling for reducing roof temperature. Wetting down roofs for cooling has long been used to cool and protect the flat, tarred roofs of commercial buildings, and the same technique can be used on houses to lower the temperature of the roofing material by as much as 30 to 40°F (17–22°C), which helps to reduce attic air temperatures by up to 25°F (14°C). Naturally, a cooler attic helps to maintain a cooler living space, especially in houses that have little attic insulation. A house with a well-insulated roof or attic floor will not gain maximum benefit from this technique, but nevertheless heat gain through the insulation will be somewhat reduced. See the box "Cool Roofs" in this essay, which describes a commercially available residential roof sprinkling system.

Night Sky Radiation

The effectiveness of night sky radiation depends on clarity of the night sky, low humidity and an air temperature drop of at least 20°F (11°C) from day to night. With these conditions, the cool night sky acts as an infinite heat sink for radiation from roofs and walls. The rapid chilling in the desert after a hot day results from night sky, radiative cooling. Even in more humid areas, cooling rates up to 50 percent of those in low-humidity areas are possible, which means that up to 200 Btu per square foot per day of cooling effect could occur on a clear night. Radiant cooling is definitely more effective in the drier areas of the United States, but combined with evaporation or with ventilation, it can be worthwhile elsewhere. Refer back to figure V-1 to see where in the United States night sky cooling will be most effective.

Work in Israel on night sky cooling has produced a system that may be applicable for some retrofits. This "Negev Desert" design features a second layer of roofing over the house's original roof. The white-painted sheet-metal roof is raised to form a plenum or airspace between it and the original roof so that as the metal cools at night, the plenum air is cooled and flows down through vents that open into the house.[9]

Combined Cooling Systems

Other more exotic cooling systems that use the combined effects of evaporation and night sky radiation are the roof pond, the cool pool and a heat exchanger coupled to a large body of water such as a swimming pool. All three options typically involve major modifications.

The roof pond involves building a metal roof support system that doubles as a ceiling for the room below and provides an excellent medium for radiant transfer. The ceiling supports an array of open water tanks or closed water bags that are in turn covered with an operable, insulated hatch-cover system. The roof insulation panels are drawn back at night to initiate cooling by night sky radiation, and if the water is not contained, by evaporation as well. The roof pond can also be used for heating by reversing the operation. The panels are open during the day to collect solar heat and closed at night; the heat radiates into the room through the ceiling. The cool pool concept is meant to be used just for cooling.

A more specialized cooling technique using evaporation and night sky radiation may prove useful for homes built on hot-water heated radiant slabs, where the coolth of an unheated swimming pool, pond or stream can be tapped. Pumping and filtering cool water through the slab keeps its temperature down and can provide sufficient cooling for the entire house by drawing room heat down to it and "flushing" the heat away via the circulating water.

Such a slab could also be cooled with water from a properly designed liquid flat plate collector that is run at night in the summer. It is best that the collector array be unglazed, e.g., a swimming pool heater, to allow adequate radiation to the night sky —since glass and some plastics are opaque to infrared radiation. Some plastic glazings, however, do not block infrared and may be suitable for a heating/cooling collector.

Figure V-8: The double roof helps with cooling both day and night. By day it works like a ridge vent system with soffit air intake, dumping heat buildup back to the atmosphere. At night the metal outer roof rapidly radiates heat to the night sky, and as the air between the two roofs cools, the cooled air is allowed to fall into the house.

Make an experiment out of it: Pump water through your collectors on a clear summer night and see what the temperature difference is from inlet to outlet. If the flow rate is 2 to 3 gallons per minute and the water temperature drops, say 5°F (3°C), the collectors are dumping 5000 to 7000 Btu per hour into the atmosphere, and it's taking those Btu's from the slab. That's a good bit of cooling when you consider that standard room air conditioners are sized to extract 5000 Btu per hour from 150-square-foot rooms and 10,000 Btu per hour from 300-square-foot rooms.[10]

Air heating collectors may also work for nighttime cooling if they are designed properly. Again, plastic glazing is more effective at radiating heat than glass, so consider summer cooling when you choose the glazing for an air heating collector. You will also want to include more mechanical and electronic controls, possibly reversing fans to allow the system to pull hot air away from high ceilings and return cooled air at floor level. An integrated system could provide both heating and cooling with no great increase in cost, but you'll do well to consult with a contractor experienced in solar design for such a dual-function system.

Evaporative Coolers and Air Conditioners

If the techniques we've discussed don't provide all the cooling you need, or if you're not ready to make major house modifications, you have the option of using a mechanical cooling system to make your house as cool as you'd like. There are products available that operate somewhat more efficiently than do their predecessors, and these

(continued on page 608)

Cool Roofs

A roof spray evaporative cooling system that helps keep a lid on roof temperature buildup is available for residential use from Spraycool, Inc., 890 Atlanta Street, Roswell, Georgia 30075. Telephone: (404) 922-4957. In this system, small amounts of water are sprayed onto the roof at timed and thermostatically controlled intervals. Heat is absorbed from the roof and dispersed into the atmosphere as the water changes from liquid to vapor. Spraycool claims this system can get rid of more than 90 percent of the solar heat absorbed by a roof, which can result in a 25 percent or greater reduction in a house's cooling load.

In summer roof and attic temperature can reach as high as 170°F (77°C), but if this heat buildup can be substantially limited, the amount of heat that can radiate down to living areas is also reduced. Although insulation slows the transfer of heat through a roof, it doesn't stop it. In fact, it is much harder to get rid of heat that has penetrated insulation, and this, according to the manufacturer, is why the system is useful even with insulated roofs. An additional benefit of keeping the roof cool is that lower temperatures can prolong the life of asphalt-based and other synthetic roofing materials.

In a typical residential installation, specially perforated PVC pipe is connected to a cold water supply and mounted along the roof ridge. In most cases, standard house water pressure is adequate for spraying. The controls for the system include a solenoid valve, a thermostat with a sensing element implanted on the roof and a timer that controls the spray intervals. The controls are usually set to keep the roof from getting above 90°F (32°C). When a temperature sensor signals the control panel that the roof tempera-

Photo V-7: Talk about unusual, a roof spray system in action is certainly a unique sight, but it may well become standard equipment for roofs in the not too distant future.

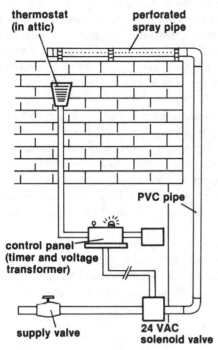

thermostat (in attic)

perforated spray pipe

PVC pipe

control panel (timer and voltage transformer)

supply valve

24 VAC solenoid valve

Figure V-9

ture is over 90°F, the solenoid valve opens, and the pipe system sprays a mist of water onto the roof. The system shuts off after 10 to 15 seconds and will come on again in about 10 minutes if the roof temperature rises over 90°F. The fine spray thoroughly wets the roof, but the system is sized so there isn't any excessive runoff or dripping.

In most residential applications, a single-zone system will cool the roof, but larger homes might require multiple zones. Spraycool is suitable for any type of roof and for any type of home including mobile homes, which may benefit from spray cooling more than most houses because they are usually such poorly insulated, metal-skinned hot boxes. Seasonal shutdown of the system simply involves turning it off and draining it in the fall to prevent freezing, and flushing it and turning it back on in the spring.

The amount of electricity needed to operate the system is negligible. Water consumption is also low. Commercial installations operating for a 10-hour day use between 1/5 and 1/10 of a gallon of water per square foot of roof per day, and residential systems generally use less. Water use will vary according to the size of the roof, the climate and the water pressure, but to determine roughly how much water a Spraycool system would use in your house, take a container and put it under the tap; turn it on full force for about 12 seconds, measure the volume and multiply by six to determine the amount of water that would probably be used in one hour (six spray cycles per hour). On the hottest summer days, the system will operate for up to about 10 hours (60 spray cycles), while on cooler days there will be fewer cycles.

Cooling energy savings will of course vary widely too. When a gallon of water evaporates from the roof, it takes about 8000 Btu with it, but that doesn't translate directly into an interior cooling effect. The Spraycool company claims that a 25 percent reduction in cooling load is possible. A big plus is that Spraycool, which is the only company offering a residential scale system, will put together custom-designed kits for do-it-yourselfers.

Margaret J. Balitas

are the ones to choose. There are two basic types of mechanical systems for adding cooling: the evaporative cooler and of course the air conditioner.

The evaporative cooler is the most efficient of the two, delivering in most cases comparable cooling capacity for one-quarter to one-half the energy cost of an air conditioner. There are two basic types. The direct evaporative cooler or "swamp cooler" draws outside air through wetted pads and blows the cooled air through the house. In recent years, the "swamp cooler" has been ignored in favor of the air conditioner, but it is effective, particularly in drier climates where the added humidity won't be a problem.

The second type is the indirect evaporative cooler. It is a direct evaporative cooler coupled to a heat exchanger. Typically, it draws air from the house through evaporative pads and through a heat exchanger, exhausting the warmed, humid air to the atmosphere. A second fan draws dry exterior air through a filter and through the heat exchanger, where it is cooled off and blown into the house. This is less efficient than a direct evaporative cooler, but it doesn't add humidity to the living space. It appears that this type of evaporative cooler has fallen out of production, but a system could be built by a skilled do-it-yourselfer.

In areas that are not very hot but where high humidity keeps indoor conditions out of the comfort range, a dehumidifier may be useful. A dehumidifier does use less energy than an air conditioner and can dry air out enough to cool by helping to increase evaporation from your skin, but in most cases an air conditioner is better, dehumidifying while it lowers air temperature.

Air Conditioners

Where temperatures remain high through the night or where, as in the Southeast, high humidity is consistently linked with high ambient temperatures, air conditioners may prove indispensable. Choosing an efficient air conditioner is vitally important, whether it's a first-time purchase or the replacement for an old clunker. Air conditioner efficiency is nowadays described by the Energy Efficiency Ratio (EER)—simply the unit's Btu rating (how quickly it removes heat) divided by its wattage. Thus the higher the EER, the less the unit will cost to operate. EERs for air conditioners range from 5 to over 11, and a unit rated at 10 is twice as efficient as one rated at 5, using half the electrical power to achieve the same cooling effect. Don't buy an air conditioner with less than an EER of 8 to 10. Even though the lower initial price of a less efficient unit may seem like a bargain, it isn't. The higher cost of an EER 10 air conditioner will be recouped with energy savings in the first two or three years of operation.

Air conditioners are of two basic types: air to air and air to fluid. Both are specialized, one-way heat pumps that move interior heat to the outside. Air-to-air heat pumps, the standard air conditioner, can achieve up to a 20 to 30°F (11–17°C) temperature difference between indoor and outdoor air, but the hotter it gets, the tougher is the job of cooling. So air conditioners are limited to achieving interior air temperatures in the 80s on days when the ambient air temperature goes over 100°F (38°C).

Water-to-air heat pumps are not standard items. They are designed to use cool source water circulated from a well, a swimming pool or from coils sunk in the earth. The temperature of the water from these sources is usually lower than the average summertime air temperature, perhaps 50°F (28°C) lower. So it's not hard to see that a water-to-air unit can be much more efficient than an air-to-air unit in dealing with summer heat. These special air conditioners are available from several reputable manufacturers. They will require access to a well, a lake or pond, a creek or a swimming pool.

Along with built-in efficiency, an air conditioner should be closely matched to the cooling load for best operating efficiency. If often happens that tradespeople size air conditioners using rather generous rules of thumb, figuring it is safer to oversize than undersize. This can result in an air conditioner that is 50 percent or more too large, particularly when installed in an energy-efficient house. You can assume that after applying as many solar control and ventilation cooling methods as possible, your house's cooling load will be cut by fully 25 to 75 percent.

Ideally, the unit should be on the north of the house or on a side with lots of shade. Failing that, it should be fully shaded with a canvas awning or wood shade, though you must be careful not to impede the unit's warm air exhaust with an add-on overhang. A window-mounted unit should be tightly sealed within the window frame to prevent any infiltration of outside air. With central air conditioning systems, check that ducts are sealed and insulated to make sure the cool air gets to you.

You can minimize energy use by keeping the air conditioner off when no one is home—it is cheaper to cool off the house starting one-half hour before you get home (by using a timer control) than cooling all day long. Also try to zone the house so you only cool the areas that really need to be cooled. You can also save by coupling the air conditioner to the whole-house fan so that the air conditioner goes off as soon as the outside temperature has dropped enough for effective ventilation cooling. Keep the thermostat at 78°F (26°C) or higher; every degree Fahrenheit that you lower the thermostat may add as much as 3 to 4 percent to your cooling bill. Keep the air conditioner

Photo V-8: The best place for an air conditioner is on the shaded side of a house. The unit is working to dump heat from the house via its exterior condenser, and if it's exposed to direct sunlight, it makes the task somewhat more difficult, reducing the efficiency of the operation.

in good condition. It's particularly important to keep the air filter and cooling coils clean.

If you think of them as cars, air conditioners can be gas guzzlers, or they can be Volkswagens, depending on the kind of unit you buy and how you use it. But better yet, transform your home into a summertime Volkswagen by implementing some of the techniques described in this essay and perhaps doing away with the need for any air conditioner at all. If we need to automatically flip on the air conditioner at the slightest threat of a hot day, then we're living by habits that work against the usefulness of overhangs, shade screens, improved ventilation and the like. Air conditioning without air conditioners is about modifying our habits as well as our homes, and the "coolest" success is in harmonizing the two.

David Bainbridge

Notes

1. B.G. Haynes et al., "Thermal Properties of Carpets and Drapes," Research Bulletin no. 68 (Athens: University of Georgia, 1969).

2. L.W. Neubauer and R.D. Cramer, "Solar Radiation Control for Small Exposed Houses," *Transactions of the American Society of Agricultural Engineers,* vol. 9, no. 2 (1966), pp. 194–195, 197.

3. A.J. Hand, "Insulating Window Shade," *Popular Science* (January 1979), p. 78.

4. L.W. Neubauer, "Shapes and Orientations of Houses for Natural Cooling," *Transactions of the American Society of Agricultural Engineers,* vol. 15, no. 1 (1972), pp. 126–128.

5. L.W. Neubauer, R.D. Cramer and M. Laraway, "Temperature Control of Solar Radiation on Roof Surfaces," *Transactions of the American Society of Agricultural Engineers,* vol. 7, no. 4 (1964), pp. 432–434, 438.

6. D.A. Bainbridge and M. Hunt, *The Effect of Roof Color and Material on Temperature* (Winters, Calif.: Living Systems, 1976).

7. J. Hammond et al., *A Strategy for Energy Conservation* (Winters, Calif.: Living Systems, 1974).

8. Kevin W. Green, "Passive Cooling," *Research and Design* (Fall 1979), pp. 4–9.

9. B. Givoni, M. Paciuk and S. Weiser, *Natural Energies for Heating and Cooling of Buildings,* report from the Building Research Station (Haifa, Israel: The Technicon, 1976).

10. The catalog of W. W. Grainger, Inc., 5959 W. Howard St., Chicago, IL 60648.

References

Olgyay, Victor, and Olgyay, Aladar. 1976. *Solar control and shading devices*. Princeton: Princeton University Press.

Reppert, Mary H., ed. 1979. *Summer attic and whole-house ventilation*. Washington, D.C.: U.S. Department of Commerce, National Bureau of Standards. no. 003-003-02089-3.

Hardware Focus

Attic Fans

Broan Manufacturing Co.
Hartford, WI 53027
(414) 673-4340

Broan Powered Attic Ventilators
Roof-mounted attic fan; thermostat is set at 100°F (38°C); models range from 930 to 1600 cfm.

Penn Ventilator Co., Inc.
Red Lion and Gantry Rds.
Philadelphia, PA 19115
(215) 464-8900

Breezecap Venter
Roof-mounted attic fan; thermostat is set at 95°F (35°C); 1310 cfm.

Evaporative Coolers

Evaporative coolers are relatively inexpensive to install and operate, but they are not suitable for every geographical area. They are generally not recommended for homes near the seashore or in other areas of high humidity.

Essick Manufacturing Company (see below) has a booklet entitled *Evaporative Air Cooling Handbook: A Manual for Contractors, Engineers and Architects* that is helpful in assessing the capabilities and limitations of evaporative coolers.

The companies listed below manufacture a variety of coolers. The range of sizes is indicated by listing the models with the lowest and highest cfm.

Champion Cooler Corp.
P. O. Box 886
1724 S. Scullie
Denison, TX 75020
(214) 465-1962

16 models. 2800 to 6500 cfm. 2-speed motor.

Dearborn Stove Co.
P. O. Box 28426
3000 W. Kingsley
Dallas, TX 75238
(214) 278-6161

6 models. 2400 to 4800 cfm. 1 or 2-speed motor.

Essick Manufacturing Co.
3215 Brown St.
Little Rock, AR 72204
(501) 372-7722

48 models. 2200 to 16,000. 1 or 2-speed motor.

Goettl Air Conditioning, Inc.
2005 E. Indian School Rd.
Phoenix, AZ 85016
(602) 957-9800

15 models. 3300 to 21,000. 1 or 2-speed motor.

Health Aire Corp.
27465 Pacific Ave.
Highland, CA 92346
(714) 862-0281

CemGlass Cooling System
Fiberglass construction. 4000 cfm. 2-speed motor.

McGraw-Edison Co.
International Metal Products Division
P. O. Box 20188
500 S. 15th St.
Phoenix, AZ 85036

2 models. 4000 or 4500 cfm. 2-speed motor.

Phoenix Manufacturing, Inc.
415 S. Seventh St.
Phoenix, AZ 85036
(602) 258-8483

14 residential models. 2000/1330 to 6500/4330 cfm. 1 or 2-speed motor.

Space-Aire Co.
169 N. Main St.
Lake Elsinore, CA 92330
(714) 674-1144

Space-Aire
3 models. 3000, 4400 or 5500 cfm. 2-speed motor.

United Electric Co.
P. O. Box 5148
500 Block Kell Blvd.
Wichita Falls, TX 76307
(817) 767-8333

9 models. 3000 to 5500 cfm. 1, 2 or 4-speed motor.

Williams Furnace Co.
14960 Firestone Blvd.
La Mirada, CA 90638
(714) 521-6500

12 models. 3200/2135 to 6500/4335 cfm. 1 or 2-speed motor.

Reflective Fabrics for Shades and Curtains

Duracote Corp.
350 N. Diamond St.
Ravenna, OH 44266
(216) 296-3486

Foylon 7001 & 7018
Drapery liner material. Foil on lightweight nylon scrim. 54" wide; 100 yds per roll.
Foylon Durashade 4413
Aluminum foil on one side, white vinyl fabric on other side. Designed for use in shades. 36", 45" or 54" wide; 60 yds per roll.
Foylon 7137
Aluminum foil on one side, black vinyl on other side. For use in greenhouse covers. 54" wide; 200 yds per roll.

King-Seeley Thermos Co.
Metallized Products Division
37 East St.
Winchester, MA 01890
(617) 729-8300

Astrolon

One layer of 0.00125 aluminized colored polyethylene and 1 layer of 0.00125 aluminized clear polyethylene, with both metal surfaces bonded to a glass scrim and embossed.

Roof Vents

HC Products
P. O. Box 68
Princeville, IL 61559
(309) 385-4323

Vent-a-System

Consists of Vent-a-Ridge, a continuous louvered opening along the peak of a roof that lets warm air escape from the attic, and Vent-a-Strip, continuous intake vents along the soffits parallel to the ridge. This is the most effective roof ventilation system currently available.

Sun Control Shades

The shading coefficients listed below are those claimed by the manufacturers on their product literature. Clear glass has a shading coefficient of 1.0. A shading coefficient of 0.25 means that 75 percent of the incident solar energy is rejected. Shading coefficients should be used as one guide to selecting shades. The shades listed below are for interior application unless indicated otherwise.

Joel Berman Associates, Inc.
102 Prince St.
New York, NY 10012
(212) 226-2050

MechoShade
Electro Shade

Roller shades; woven vinyl fabrics with varying degrees of openness; shades are single-layer, 2-layer or 2-layer with reflective film; shading coefficients range from 0.25 to 0.69 for various models and colors.

Gila River Products
6615 W. Boston St.
Chandler, AZ 85224
(602) 961-1244

Gila Insulating Shade

Roller shade; transparent polyester film with bronze-colored tint; shading coefficient: 0.33; 36", 42" or 48" shade mounts or custom sizes; side tracks are optional.

The Moore Co.
Marceline, MO 64658
(816) 376-3583

Moore Solarshades

Exterior operable, louvered aluminum shades; widths from 8' to 14'; primarily for commercial applications, but suitable for window walls, greenhouses, etc.; shading coefficient: 0.2.

MRS Interior Systems, Inc.
100 Marcus Dr.
Melville, NY 11747
(212) 895-4788

Prima Rolling Shade System
Motortex

Interior or exterior roller shades; manual or motorized controls; available in a variety of colors; shading coefficients range from 0.16 to 0.47 (for exterior applications).

The Plastic Sun Shade Co., Inc.
389–91 Union Ave.
Irvington, NJ 07111
(201) 373-8181

Solar Screen Co.
53–11 105th St.
Corona, NY 11368
(212) 592-8223

Sol-R-Veil, Inc.
60 W. 18th St.
New York, NY 10011
(212) 924-7200

J. P. Stevens
P. O. Box 1138
Walterboro, SC 29488
(803) 538-8045

Sun Control Products, Inc.
431 Fourth Ave. S. E.
Rochester, MN 55901
(507) 282-2778

Wind-N-Sun Shield, Inc.
P. O. Box 2504
131 Tomahawk Dr.
Indian Harbor Beach, FL 32937
(305) 777-3558

Sun Shades
Roller shades; acetate or Mylar; custom sizes; shading coefficients range from 0.16 to 0.51 depending on color; 8 colors; UV resistant.

Kool Vue Window Shades
One layer of metallized Mylar between 2 layers of mylar; roll-down shade; custom sizes; colors and shading coefficients: bronze (0.38), gold (0.25), gray (0.36), silver (0.27).

Sol-R-Veil
Roller shade; vinyl-coated fiberglass yarn; 60 sizes; smallest: 36"(H) × 18"(W); largest: 144"(H) × 100" (W); shading coefficients and colors: white (0.37 or 0.48), bronze (0.69), brown (0.74), for interior application; white (0.21), bronze (0.19), beige (0.21), gray (0.12), for exterior application.

Comfort Shade
Vinyl-coated fiberglass woven with rib in horizontal direction; colors and shading coefficients: white (0.62), gray (0.35), charcoal (0.37), bronze (0.35), bone (0.50); UV resistant.

NRG Window Shade
Roller shade on side tracks; aluminized polyester; colors and shading coefficients: charcoal/silver (0.15), bronze/silver (0.28); UV resistant; custom sizes.

Wind-N-Sun Shield
Drapery liner panels or roller shades. Aluminized polyester on one side. White vinyl on other. Shading coefficient: 0.01; UV resistant; sizes: 37¼" × 4', 55¼" × 6'; custom sizes.

Sun Control Films

Sun control films adhere to inner surfaces of windows to limit direct gain. The films are usually two or three layers of laminates made of metallized (reflective), transparent and/or tinted layers.

Dunmore Corp.
Newtown Industrial Commons
Penns Trail
Newtown, PA 18940
(215) 968-4774

Gila River Products
6615 W. Boston St.
Chandler, AZ 85224
(602) 961-1244

Dun Ray
Adhesive; metallized clear polyester film; silver; shading coefficient: 0.24; sold in rolls 36" or 48" × 25'; UV resistant.

Insul Film
Laminated polyester sheets; adhesive; reflective colors and shading coefficients: silver (0.30), gray (0.39), bronze (0.39), gold (0.38); UV resistant.

Madico
64 Industrial Parkway
Woburn, MA 01801
(617) 935-7850

Martin Processing, Inc.
P. O. Box 5068
Martinsville, VA 24112
(703) 629-1711

Metallized Products
2544 Terminal Dr. S.
St. Petersburg, FL 33712
(813) 822-9621

National Metallizing
P. O. Box 5202
Princeton, NJ 08540
(609) 443-5000

Plastic-View Transparent Shades, Inc.
P. O. Box 25
15468 Cabrito Rd.
Van Nuys, CA 91408
(213) 786-2801

Solar Screen Co.
53–11 105th St.
Corona, NY 11368
(212) 592-8223

Solar-X Corp.
25 Needham St.
Newton, MA 02161
(617) 244-8686

Sun Control Products, Inc.
431 Fourth Ave. S. E.
Rochester, MN 55901
(507) 282-2778

Reflecto-Shield
Adhesive; aluminum within polyester laminate; colors and shading coefficients: gray (0.25, 0.36), gold (0.24, 0.30), bronze (0.25, 0.36), silver (0.26, 0.31); UV resistant; sold in rolls 60″ × 100′ or in kits with precut sizes.

LLumar
Adhesive; metallized polyester laminate; colors and shading coefficients: silver (0.28), bronze (0.31), gray (0.30), gold (0.27); UV resistant.

Sun-Gard Window Films
Three polymer layers and 2 film layers; adhesive; colors and shading coefficients: silver (0.24), bronze (0.38), gray (0.36), gold (0.25), transparent silver (0.55); UV resistant; 9 precut sizes for do-it-yourselfers; also rolls of 100′ and 200′.

Nunsun
Two layers of polyester film with a layer of aluminum; adhesive; colors: bronze, gold, gray, silver; shading coefficients range from 0.20 to 0.49; UV resistant.

Plastic-View's Film-to-Glass
Two sheets of film with middle sheet aluminized; adhesive; colors and shading coefficients: gold/silver (0.25), gray/silver (0.36), silver/silver (0.27), bronze/silver (0.38), bronze/bronze (0.47), gold/gold (0.30), gray/gray (0.43); UV resistant; prepackaged sizes: 5′ × 2′, 3′, 4′ or 8′; roll form: 5′ × lineal ft.

E-Z Bond
One layer of aluminum between 2 layers of film; adhesive; colors and shading coefficients: bronze (0.38), gold (0.25), gray (0.36), silver (0.27); UV resistant; kit form: 20″, 40″, 60″ widths; 5′, 10′ lengths; roll form: 60″ × 25′; 24″, 40″ or 60″ × 100′.

Solar-X
Clear film with coating of Transolume metal; adhesive; colors: bronze/silver, silver, smoke/silver, gold/silver; shading coefficient: 0.24; UV resistant; sizes available: 18″ × 72″, 36″ × 78″, 48″ × 78″, 54″ × 78″.

Sun Control Film
Aluminized polyester; adhesive; colors and shading coefficients: silver (0.20), bronze (0.35), gold (0.26); UV resistant; custom-cut sizes; widest: 60″.

3M Co.
St. Paul, MN 55101
(612) 733-1110

Scotchtint P-19 Window Insulation Film
Adhesive; color: gray; shading coefficient: 0.23; UV resistant.

Vacumet Corp.
20 Edison Dr.
Wayne, NJ 07470
(201) 628-0400

The Eliminator
Adhesive; colors: silver and brown; shading coefficient: 0.25; do-it-yourself sizes: widths 14", 34", 48" and lengths 78", 156"; professional lengths up to 60" wide.

Sun Control Screens, Blinds and Shutters

The shading coefficients listed below are those claimed by the manufacturers on their product literature.

Hamel, Inc.
999 Airport Rd.
Lakewood, NJ 08701
(201) 367-7670

Haroscreen
Exterior roll-down screen; fabric or open-weave PVC-coated fiberglass; custom sizes; shading coefficient: 0.16; UV resistant.

Kaiser Aluminum
300 Lakeside Dr.
Oakland, CA 94643
(415) 271-3321

ShadeScreen
Exterior screens; louvered aluminum; colors and shading coefficients at 10° or 40° profile angle: green (0.411, 0.195); black (0.429, 0.139); widths 18" to 48"; 50' per roll.

KoolShade Corp.
P. O. Box 210
722 Genevieve St.
Solana Beach, CA 92075
(714) 755-5126

KoolShade Solar Screen
Exterior screens with small woven bronze louvers; aluminum frame; widths up to 72½"; standard color: black; shading coefficients: Standard Kool-Shade (0.232); Low Sun Angle KoolShade (0.15).

Joseph C. Maillard Enterprises
1233 E. Ramsey St.
Banning, CA 92220
(714) 849-3141

Faber Maximatic, Faber Solarmatic
Exterior aluminum louvers; maximum widths 13' × 13'; cord, rod or motorized controls depending on size; shading coefficient: 0.1. Expensive.

MRS Interior Systems, Inc.
100 Marcus Dr.
Melville, NY 11747
(212) 895-4788

Sundrape Vertical Blind System
Interior blinds with vertical louvers; available in various fabrics and widths; motorized system available with sun sensing devices; shading coefficients range from 0.39 to 0.74.

Nichols-Homeshield, Inc.
1000 Harvester
W. Chicago, IL 60185
(312) 231-5600 or (205) 345-2120

Nichols-Homeshield Blinds
Exterior aluminum louvers; custom sizes; shading coefficient: 0.17; numerous neutral colors.

Pease Co.
Ever-Strait Division
7100 Dixie Highway
Fairfield, OH 45023
(513) 867-3333

Phifer Wire Products, Inc.
P. O. Box 1700
Tuscaloosa, AL 35401
Toll-free number: (800) 633-5955

The Rolsekur Corp.
Fowler's Mill Rd.
Tamworth, NH 03886
(603) 323-8834

Sears, Roebuck and Co.

Serrande of Italy
P. O. Box 1034
West Sacramento, CA 95691
(916) 371-6960

J. P. Stevens
P. O. Box 1138
Walterboro, SC 29488
(803) 538-8045

Virginia Iron & Metal Co.
P. O. Box 8229
Richmond, VA 23226
(804) 266-9638

Penn Ventilator Co., Inc.
Red Lion & Gantry Rds.
Philadelphia, PA 19115
(215) 464-8900

Triangle Engineering Co.
P. O. Drawer 38271
Houston, TX 77088
(713) 445-4251

Pease Rolling Shutter V or VM
Exterior shutter of hollow PVC slats; motor-operated; rolls down on track; R-values when installed 1¾" outside single glazing: Model VM, R-1.76; Model V, R-2.47. Sizes to customer's specifications; UV resistant; shading coefficient: 0.1.

SunScreen
Exterior screens; woven fiberglass; brown; sold by rolls: 28" × 76", 32" × 76", 36" × 84", 48" × 84"; shading coefficient: 0.32. UV resistant.

Rolsekur Rolling Shutter
Exterior shutter of interlocking PVC extrusions or wood profiles; shutter manually or motor-operated from inside; UV resistant; custom sizes; shading coefficient: 0.1.

The Sears catalog lists a variety of sun control products.

Serrande Shutter
Exterior shutter of PVC or wood; all slats are ½" thick with reinforced steel; no limit to span; custom sizes; manually or motor-operated; shading coefficient: 0.1.

Comfort Screen
Interior or exterior applications; vinyl-coated fiberglass woven with rib in vertical direction; colors and shading coefficients: white (0.62), gray (0.35), charcoal (0.37), bronze (0.35), bone (0.50); UV resistant; widths up to 84".

Vimco Solar Shield
Exterior screens; woven fiberglass fabric; gray; shading coefficient: 0.33 (45° profile angle); widths 29", 33", 37", 49"; heights to 80"; UV resistant; can be rolled up and stored at head of window.

Turbine Vents

Turbine Ventilators
Diameters range from 6" to 30".

Wisper Cool
Turbine sizes: 12" or 14" diameters.

Photo V-9: Shade, shade and more shade: This was a primary aim in the retrofit work at the Houston House. It makes both the inside and the outside more comfortable through the hot Texas summer.

The Houston House

A Natural Cooling Retrofit

There has recently been more than a little conjecture about how 20 percent of all residences in this country could be solar-heated by the year 2000, primarily by the success of a vigorous effort in new construction. However, if one looks at some of the facts, that goal appears to be pretty optimistic, if not patently impossible. In 1977 there were over 82 million housing units in the United States. The percentage of net increase for the years 1970 through 1977 has averaged 2.6 percent per year. In recent years, though, the percentage increase of new construction has decreased. At the current 1980 construction rate, *all* houses built between now and 2000 would have to be solar-heated to come anywhere near the 20 percent projection. Clearly, this will not happen, and it means that the future solar home population of this country must involve much more than just new homes.

For most people the house they currently occupy will be the one they'll stay with during this period of rapidly increasing energy costs and high home mortgage costs. The most realistic approach to reducing residential energy consumption of course lies with retrofitting. A lot of the emphasis in retrofitting is on improving home heating systems, but the importance of employing natural cooling techniques should not be underestimated. It's a fact that many electric utilities in this country are "summer peaking," in that the heaviest demands for power come during hot weather, primarily because of the proliferation of air conditioners. Peak power is the most expensive, and as the peaks are pushed to higher heights, the need to build more generating capacity intensifies. For that we get coal and nuclear power plants.

To demonstrate the viability of natural cooling and passive solar alternatives, an 18-year-old residence in Houston, Texas, was retrofitted with features that put it more in tune with the local climate. Houston is located in the hot, humid Texas Gulf Coast, where the number of heating degree-days is about the same as the number of cooling degree-days. The climate in this region is characterized by high rainfall, mild, damp winters and hot, humid summers. The region has early spring and late fall growing seasons separated by two summer months (July and August) that are almost too hot for open air gardening. In summer the relative humidity often reaches 100 percent, even at night, and this is as important a factor as high temperature in making the weather seem uncomfortably hot. In studying the climate it was found that wind direction, temperature and humidity follow a strict seasonal pattern that can be used to reduce cooling loads primarily by using breezes and natural or forced convection. Winter winds come from the north, and summer breezes come from the south. This reversal of wind direction is a help in designing natural cooling and other energy conservation improvements.

The one-story brick house (the Houston House) was specifically chosen because of its high potential for accepting passive solar cooling modifications that wouldn't interfere with energy-efficient winter heating. The house has external dimensions of 50

Photo V-10: The combination of a tall fence and a roof overhang works to completely block the summer sun from reaching the courtyard. Other retrofit items are the turbine vents that pull heat out of the attic space.

by 60 feet and is built with brick-faced stud walls on a 4 to 6-inch-thick concrete slab that is covered by a hardwood floor. The floor plan is a U-shape with a den occupying the center of the U. The den has a high ceiling and a glass south wall that is shaded in summer by an added overhang and by a deciduous tree that stands to the south of the patio slab. With nature's perfect timing this tree loses its leaves in winter, allowing the sun to shine through and warm the den.·

None of the modifications that were made required major changes to the house, but taken together they produced significant energy savings in both summer and winter. We (myself and a team of coresearchers and friends) began work in August 1976, with the aim of having the house ready for the 1977 heating season. The first item was wall insulation, which was done from the inside by blowing in loose-fill insulation to fill all the 2½-inch-deep stud wall cavities. In some segments of the walls we found insulation, but the placement seemed to be more for sound deadening rather than for controlling heat loss or gain.

Next came perimeter insulation for the slab. We dug a 12-inch-deep trench around the foundation and faced the slab edge and the footing with 16 inch by 2-inch-thick styrene (Styrofoam). The insulation runs 4 inches above grade, and to make it less exposed a 2-inch layer of dirt was spread over the original grade. To completely cover the foam, dried grass clippings were piled up against the wall, with some clear space between the ground and wood parts of the house to prevent termite migration. This layer

Photo V-11: Instead of the usual pruning-for-neatness, the bushes around the Houston House were allowed to flourish for maximum shading.

provided some additional insulation value and would be kept adequately dry by the wide roof overhangs, some of which were extended (discussed shortly).

Next we insulated the (unfinished) attic floor with 6 inches of fiberglass. This was done as much to reduce cooling loads as to reduce heating loads. Summer comes early in this region and attics can easily heat up to 125 to 135°F (52–57°C). With an uninsulated floor a lot of heat is conducted through the ceiling. The fiberglass resists that conduction, but still the attic can heat up, so in order to reduce attic air temperature we added wind turbine ventilators to the roof. By cooling the attic air, the heat flow across the insulation into the house was further reduced. Instead of the 135°F high the attic now heats up to 5 to 10°F (3–6°C) above ambient temperature, which ranges between 90 and 100°F (32–38°C) in summer.

Infiltration control is another important improvement for both the heating and the cooling seasons. In summer it isn't just hot air that sneaks in, but the dreaded high humidity. Everywhere we could we sealed cracks in the building skin and caulked and weather-stripped around doors and windows. An unheated utility room was converted into an air-lock-type entrance. These improvements are really first priority, but we wanted to take care of some of the bigger jobs first so the house could be reoccupied sooner.

Solar control improvements included extending the roof overhang along the south side of the building, adding window shades and insulation and adding and improving bushes and trees around the house. Extending rafters and soffits is a simple enough task, and by roofing the extensions to match the existing roof material the change is only noticeable in reduced cooling bills.

The operation routine for the window controls is simple: closed by day in summer, open at night if the air temperature warrants ventilation cooling. Some windows use a combination of insulated drapery and a separate roller shade, the former being used in winter, the latter in summer. Landscaping improvements included the pruning of bushes and trees to promote new, thicker growth, especially on the east and west sides of the house. Temperature measurements showed that shaded wall sections were at least

Photo V-12: Simple, white window shades are known to reduce heat gain through a window by 30 percent. In rooms where there is no need for daytime light, windows can be shaded and draped to minimize solar gain.

20°F (11°C) cooler than sunlit sections, which means a significant reduction in conductive heat gain through the shaded sections. The final improvements at the Houston House included installing manual override switches on heating and cooling thermostats and reducing internal heat gain by venting the gas-fired water heater, clothes dryer and cooking heat to the outside.

The final proof of the value of any retrofit is of course in the reduction in the utility bills. A central air conditioner is still needed to maintain interior comfort levels, but the overall cooling was lowered by 66 to 75 percent. The heating load was cut in half. Thus the overall annual use (in terms of cost) is about two-thirds less than before the improvements. This is in part because the electricity used for cooling is much more expensive per Btu than the natural gas used for heating. We've found that the furnace and air conditioner just aren't needed as early in their respective seasons. The shading from trees and bushes are alone responsible for about a 10 percent reduction in the overall cooling load. By making fairly simple improvements designed to take advantage of the particular opportunities of the climate and the site, we created substantial energy savings with no compromise in comfort.

Arthur C. Meyers III

References

Meyers, Arthur C., III, and Way, George E. 1979. Passive cooling design approaches: a review of applicability for hot humid climates. In *Proceedings of the 4th National Passive Solar Conference,* ed. Gregory Franta, Newark, Del.: American Section of the International Solar Energy Society.

———. 1979. Passive solar heating and cooling in the hot humid gulf coast region. In *Proceedings of the 3rd National Passive Solar Conference,* eds. Harry Miller, Michael Riordan and David Richards, Newark, Del.: American Section of the International Solar Energy Society.

☀️ PROJECT
A Solar Chimney

This is a rather exotic natural cooling technique, but it's also a very simple one. A solar chimney (or thermal chimney) is a passive system that works by induced ventilation, which means that natural convection ventilation is boosted by the power of the sun. How so? The chimney itself is a collector with glazing on the east, south and west sides and a metal absorber inside. It's called a chimney because as well as being a collector it's also an air shaft that is coupled with the interior living space. Direct gain heats the absorber, and hot air convects out of the top of the chimney at a rate that varies with the intensity of the direct gain. Replacement air has to come from somewhere, and in this particular system cool air is drawn in from a crawl space under the first floor. That's it: Solar-induced ventilation speeds up natural convection ventilation and gives it enough power to pull in cooler, denser air. It's a bit of solar irony: a collector used for cooling.

Sacramento is like a baked-out sack of tomatoes in summer. You tend to sag a little as the temperature climbs beyond the 90s and into the 100s. The old-timers knew what to do with excess heat before the advent of air conditioners—dump it through the roof. On the roofs of the old fruit-packing houses, you can see rows of ventilating cupolas. In fact, the idea of cooling towers goes all the way back to ancient Persia, but nowadays they're only just beginning to reemerge as a viable natural cooling option.

A thermal chimney is best for cooling when it can pull in something besides warmed-over air, and the local climate obliges by providing cool nights that permit thorough ventilation cooling of the house, and especially of the basement and crawl space that are the daytime cool air sources.

We wanted to use the chimney to provide some relief during periods when summer heat would rise and accumulate at the top of the upstairs rooms, then creep down into the lower levels, making every room but the basement uncomfortably warm. The chimney is designed to work with the whole house so that every room could be included in the ventilation system, not a difficult task with our relatively simple floor plan. A cool air intake from the crawl space was placed at the center of the house just below the staircase to ensure that the coolest air, which is isolated from the warmer air at the perimeter of the crawl space, would be drawn in first. As the cool air is drawn into the house by the force of the induced ventilation, it spreads throughout each floor of the house as the warmer air rises and is exhausted through the chimney.

During the day, the cool air under the house, about 1600 cubic feet, is our primary source for replacement. There are also 1340 cubic feet of cool air in the basement room and 1730 cubic feet in the garage. These volumes aren't just drawn on once; rather, the coolest air is of course used first, and then the supply gradually warms up through the day. At night, everything cools down again.

A catalpa tree shades the south end of the upper roof during summer afternoons, so we placed the solar chimney high on the roof at the northwest corner of the house, where there is plenty of afternoon sun. Since the house is oriented 30 degrees south of west, both glazed sides of the chimney (south and west) are fully exposed to the sun during the afternoon, when the need for cooling is greatest.

We based the sizing of our chimney primarily on such things as ease of construction, cost, structural integrity and good appearance, figuring we could always add another tower if this one wasn't sufficient. (Since this idea is relatively new on the solar horizon, there hasn't yet been a great deal of research done toward optimizing chimney design. Therefore we can't give you a bunch of sizing or performance numbers.) The important variables for effective performance are the air flow rate generated by the tower, the total volume of the house and the size of the cool air source. To be sure the thermal chimney would pull air through the crawl space, not leaky windows, we sealed the window and door openings carefully with caulk and weather stripping. Before the chimney was built, we had already applied several other heat gain-reducing measures that greatly improve the ability of the chimney to do its job:

1. Ceiling insulation was increased to R-25.

2. Shade screens were installed on all windows exposed to direct sun during the summer.

3. A solar greenhouse shades and provides air circulation in front of the south wall in the summer.

4. An elevated solar water heater shades the south entry.

5. Outer walls were coated with white stucco.

We knew that a tall chimney would develop a stronger air flow, but we limited the height of ours to about 7 feet above the roof because this seemed to be about the structural and visual height limit. The perimeter dimensions were kept at 2 feet by 2 feet to make the best use of the 4-foot-wide FRP glazing we used. Only the top 4 feet of the south face of the chimney was glazed to prevent sunlight from shining directly into the house when the sun is high overhead in the summer. The west side of the chimney is completely glazed to catch as much sun as possible in the afternoon, when ventilation is needed most. The lower angle afternoon sun beams cannot enter the house through the west face. Finally, an 18-inch diameter Romlair wind turbine vent with a custom-made cap was installed on the top of the chimney because it seemed to provide the best combination of weather (rain) protection and wind-catching ability. It actually helps pull air from the chimney when a breeze starts it spinning.

Photo V-13: The importance of the turbine vent is that when the breezes come in the ventilation rate is increased. Often the hottest temperatures also coincide with windless days, which makes the solar chimney-turbine combination an effective duo.

Construction

To keep costs down, we built the chimney with standard materials and kept the overall design fairly simple. However, the installation did require careful planning and execution. First, we had to find an appropriate location in the ceiling to cut a hole for the chimney, so we went up to the attic to check the spans of the ceiling joists and roof rafters. We decided on a location that would allow for good tie-in of the chimney's 2×4 vertical members with minimal disruption to the roof structure. We cleared away the ceiling insulation and cut the hole through the ceiling and then through the roof. (No turning back now!)

We installed horizontal blocking between the joists and the rafters to frame the hole for the chimney, taking care to keep the blocking faces plumb. Next we cut the four plywood sides and nailed them to the blocking, in effect creating a small lightwell that connected the pitched roof with the horizontal ceiling. Finally, we nailed three vertical 8-foot 2×4's into corners of the hole and fastened more blocking between them at the top of the opening and at the top of the chimney. For the southwest corner we used a 2×2 upright to minimize shading of the absorber. This framing formed the basic structure of the chimney. We also secured a 2 by 4-foot sun baffle against an east-west ceiling joist that runs through the center of the chimney opening. To further ensure that no high-angle sunlight would be admitted into the room, all unglazed wall sections were sheathed with ½-inch plywood to make the box rigid.

We cut insulation board with a reflective foil face to cover the entire inside of the chimney. Then everything was painted flat black. The foil facing thus became the absorber. Flashing was carefully installed around the base of the chimney, and shingles that had been removed when the hole was cut were replaced. Lascolite (FRP) glazing was cut, fastened and caulked onto the south and west faces. The outside wood was painted white to match the house walls. The final step was setting the turbine vent on top of the chimney. Since we had a round turbine and a square chimney, we had a metal cap made at a local sheet-metal shop so the two could be joined.

Inside, a hinged, tight-fitting plywood hatch was built flush with the ceiling. It had a handle and latch for operations and weather stripping for a tight seal in winter. We caulked all cracks inside the chimney, and up in the attic we fastened insulation around

the chimney lightwell. To gain access to our cool air resources, we framed a 2 by 3-foot vent in the base-

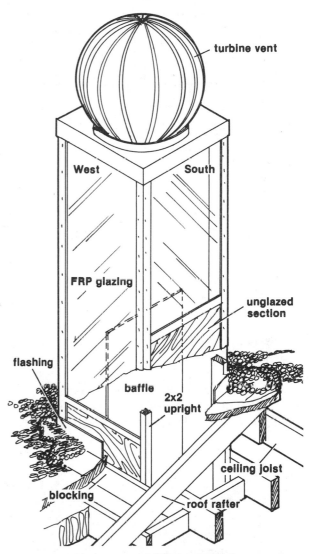

Figure V-10: *If you were thinking of installing a turbine vent anyway, maybe there's a chance for putting a solar chimney between it and the roof. Structurally it's a very simple affair, and it's also light in weight and won't require structural changes to the roof. Note that the light baffle inside the chimney faces south to block the high angle summer sun from penetrating the opening in the ceiling.*

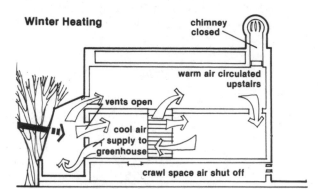

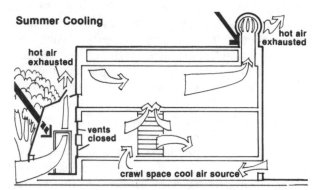

Figure V-11: The solar chimney installed at the author's house is part of a year-round heating and cooling management program that helps him to stay comfortable without expensive electrical air conditioning in summer and, thanks to the solar greenhouse, without using a lot of fuel in winter.

ment wall for a cool air intake from the crawl space. Another plywood hatch was built to cover the vent in winter.

Operation

When the weather turns hot, we open both the basement and the chimney hatches. We also open the greenhouse doors and vents to limit heat buildup on the south side of the house. A plywood vent cover is placed over the greenhouse-to-house inlet and outlet to stop natural convection of hot air from the greenhouse and to hold cooler air inside the house.

Daily operation of the system has a rhythm of its own. In the morning we close the chimney vent and all the windows and outside doors to hold cool air inside the house as the temperature rises outside. Around midday, when the upper floor starts to get warm, we open the chimney vent (still leaving all windows and exterior doors closed) to allow the chimney to draw in cool air from the crawl space. One to two hours after sunset, when the inside and outside temperatures are the same, we open all the windows and doors to allow better circulation of the cool night air. The solar chimney is left open through the night. In the fall, the basement vent to the crawl

space and the chimney vent are both closed tightly and the greenhouse vents are opened for natural convection space heating.

Performance and Evaluation

We estimate that the chimney is lowering the upstairs temperature by 5 to 10°F (3–6°C) on summer afternoons. Temperatures within the house now stay noticeably cooler than outside, and it is a rare day indeed when we have to resort to using a fan before sundown. But the fact that the basement usually stays several degrees cooler than the kitchen and upstairs rooms in the afternoon suggests that the chimney is not using all of the available cool air resource. Adding another chimney would increase the draw on the basement and crawl space air. We could also increase absorber area by installing blackened expanded metal. More absorber could increase the temperature inside the chimney, making for a faster air flow rate. A more radical design modification would be to tilt the chimney to a more favorable angle for receiving the high-angle summer sun, as shown in photo V-14.

The turbine vent increases the drawing power of the chimney when it's windy, but unfortunately the highest daytime temperatures usually coincide with the slowest breezes. However, the sea breezes usually pick up after sunset and help the vent do its job of exhausting hot air out of the upstairs rooms.

Your Place

The first step is to practice what was preached in "Air Conditioning without Air Conditioners" found earlier in this section. Then, if you need more cooling power, you can go further with an idea like this one. In planning the chimney, measure the amount of cool air you can draw. If you have a crawl space or basement and a multi-story house, then a solar chimney should work because the earth and basement walls represent an effective coolth storage mass, and the multiple levels enhance the convective flow.

Try to place the chimney over the hottest part of the house, as well as where it will get good sun and wind exposure. Locate the cool air intake vents to provide efficient use and distribution of the coolness, and be sure to provide good air circulation through all rooms and hallways.

Expenses

The simplicity of the chimney is reflected in its low cost. If we had to buy all the necessary materials new (which we didn't), they would have come to about $180. Considering the costs of buying and running an air conditioner, our natural cooling system represents a substantial savings. Although it doesn't have the cooling power of a bona fide air conditioner, the system works well and justifies our efforts. Obviously an air conditioner would provide cooler air, but the costs of operating one make this passive cooling option our preference.

John Burton and Jeff Reiss

Photo V-14: The two solar chimneys on this house are a possible improvement on the previous design. They were built with tilted glazing to reduce the angle of incidence from the high-angle summer sun, thereby increasing the gain, which could make for a faster convective air flow. The view in this photograph is from the northwest corner of this solar house.

VI THE WHOLE HOUSE

Photo VI-1: In this whole-house retrofit Californians Polly and Ken Haggard converted an old bungalow into something completely different. They built clerestories, extended roof overhangs, added south glass, added water and masonry thermal mass, installed a thermosiphon solar water heater and, among the other interior improvements, created a solar bathroom using glass blocks. A big old cast-iron bathtub was put back into service in a most unique way.

The Turned-On House

"It'll be a great place, if they ever get it built."

That's an old one about New York City, old but true: The place is being remodeled, repaired and rebuilt 24 hours a day. Maybe you can say the same about your place (leaving out the 24-hour thing). In your mind's eye you can see all the parts of your house and how perfect you'd like them to be. In the unwincing light of day, however, you can see the difference between your visions and your present state of affairs. Sigh. There's much to be done, but don't let it overwhelm you. The whole house is just the sum of its parts, and taken one part at a time the Big Job Ahead can be whittled down to manageable proportions. We, of course, hope you'll include a lot of solarizing in your whole-house plans.

In this section we present some fine examples of what it means to go all the way with an energy retrofit, all the way to the turned-on house. The owners of these houses have endured the toil and the dust, and the financing, to realize their own master plans. The dictates of architecture, site, climate and budget guided the final outcomes to unique solutions, and as these examples show, the turned-on house can be many things. A total retrofit is likely to include two, three, four or more different systems, each responding to a different need for a different part of the house. The versatility and flexibility of solar design makes it all possible.

These homes show us what the often-fantasized "home of the future" truly ought to be. They point the way toward a more sustainable household economy, one that replaces a costly dependence on "imported" energy and food with the independence of on-site production. In the decades ahead the designs and materials will undoubtedly change, but in this decade we know that it can be done.

Photo VI-2: The Feinour House

The Feinour House

There never really was a formal conspiracy among builders, architects and development planners to cover America with houses that were immune to the heating power of the winter sun. Semi-insulated, randomly oriented housing flourished through the expedient of cheap energy and a plain lack of awareness rather than any malign intent. Frustrating as this lack of foresight may now be for homeowners, that randomness did occasionally produce at least the opportunity for solar heating retrofits to be developed cheaply and effectively.

A case in point is Jean Feinour's small, comfortable house in the town of Blue Ball, Pennsylvania. There are doubtless thousands of houses that look just like it, with its low, open, cement porch under a roof supported by decorative iron corner posts leading to the front entrance, and three upstairs windows that nicely balance the porch and large living room window. What is fortunate about this place is that it has an excellent solar orientation, facing within 15 degrees of true south with no major obstructions in sight.

Unfortunately, all but one of Jean's south windows had been fitted with huge slat-metal awnings that covered fully half of the glass area and prevented any sun from entering for all but a very short time around the winter solstice. Despite the good exposure, Jean's windows were losing far more heat to the cold than they were gaining from the sun, and as heating oil prices rose, her limited income was being stretched to the limit.

In mid-1979, the Lancaster County Community Action Program's Solar Project devised a simple, attractive and cost-effective solar package for Jean's house, which included modifying the awnings to allow for summer shading only, building a glazed porch enclosure for additional direct-gain heating and building a small blower-driven air heating collector to provide supplemental heat to the living room. The estimated materials' cost for these projects was just $300.

First the building crew removed six awnings (four on the south wall, two on the west). No material costs were involved, but a lot of hard work was. After determining from sun angles that the finished shades should be 15 inches long, the excess overhang was cut off. It would have been easier and simpler to remove the awnings and get rid of them. But while this would have improved winter heat gain through the windows, it also would have eliminated the benefit of summer shading, trading lower heating bills for higher air conditioning bills with no overall reduction in energy use.

The enclosure we built for the porch entry was a simple, light frame wall of conventional construction, glazed with greenhouse-grade, fiberglass-reinforced plastic. This design was inexpensive with a cost of about $200 in materials for a 7 by 10-foot enclosure. To begin, the enclosure sill plates for the framework were anchored to the concrete porch with lag bolts and lead anchors over a thick bead of caulking to prevent drafts from coming in under the frame. Header plates were nailed to existing members in the porch roof.

All the enclosure framing was built with 2×2 lumber except for the west face, where we used 2×4's to carry a storm/screen door. The wrought-iron-style corner posts supporting the porch roof were left in place inside the enclosure, so the 2×2 frame wall only had to hold up itself and the glazing. All the studs are on 24-inch centers to accommodate the 4-foot-wide glazing material. A 2-foot-long 2×4 horizontal cross-brace was installed between the door frame and the existing wall to make a frame for a hinged vent at floor level. Another cross-brace was run between two studs high on the east wall to make a frame for an exhaust vent. These vents help to prevent overheating in the enclosure during summer by providing cross ventilation.

With the framing complete, the glazing went up quickly. First we spread latex caulk on the frame as a sealant, then put the glazing (FRP) into position and tacked it down with a staple gun. After caulking the joints between adjacent sheets, 1×3 battens were laid over and nailed into the framing, permanently securing the glazing. In addition to the outer glazing, we used Glasteel (a thin film) for the inner glazing and attached it in the same manner. Since the fiberglass is a diffusing glazing, we decided to create a totally transparent window in the middle of the south wall using a polyester film (Llumar).

We moved an aluminum storm/screen door from the house's front entrance and installed it in the porch doorway with perimeter weather stripping. The vent panels were framed with 1×2's, glazed with the FRP material, hinged to swing inward and carefully weather-stripped. All in all the porch enclosure was a very simple piece of work.

The third solar improvement was a 2 by 11-foot air heating collector mounted at a 60-degree angle just below the living room picture window. Two holes for duct fittings are cut through the back of the collector at the same end, about 1 foot apart from each other. A single baffle makes the fan-forced air travel from one end to the other and back again to the exhaust duct where it blows into the living room. Small ¾-inch holes for summer ventilation are also cut through the side of the panel on each end, just above the insulation board. We mounted small, bimetallic vents that are heat-controlled to automatically open and close a gasketed flap inside these vent holes. You can also rig a simple manual flap or cover plates for this as long as you remember to open them each spring to prevent panel stagnation and close them each fall to prevent heat loss. The absorber plate is a black corrugated aluminum trailer siding. A thermostatic switch is set to operate the blower when the collector heats up to 90 to 100°F (32–38°C). A single layer of FRP glazing was used in this collector, the edges of which were sealed with caulk and aluminum angle stock.

After we mounted the collector, the inlet and outlet ducts were run through the wall into the living room. Inside the living room, we connected the ducts to a small, rectangular box of stained wood which sits unobtrusively in a corner. A diamond-shaped vent in the upper half of the box delivers heated air to the room. The box is divided inside, with a small blower mounted in the bottom section to pull air from the room and push it through the collector. A switch on the box allows Jean to turn the system on, and then it works automatically when heat is available, controlled by the thermostat.

Figure VI-1: It took all of $800 in materials to bring about a dramatic change in the thermal performance of Jean Feinour's house. The simplest improvement was in turning back the metal awnings so that they no longer shaded the winter sun. The solar vestibule sends heated air in through the front door, and the air heating collector combines with the picture window to keep the living room cozy through cold, sunny days.

Finally, as an extra touch to the retrofit, the crew enclosed the east cellar windows with a second glazing to cut cold air infiltration into the basement. Heat loss and infiltration in basements can be much more of an energy liability than most people realize, and simple basement weatherizing can bring noticeable results.

The collector and porch enclosure were painted white to match the house siding, and they blend in surprisingly well with the rest of the exterior. In fact, the retrofit's appearance is so unassuming that when two of the Solar Project's staff members went to see it for the first time, they drove right by and had to backtrack to find it.

The awning modifications work perfectly and passively and now allow much more direct gain in winter, increasing both the family's comfort and the health of their houseplants. In the afternoon, the awnings on the west second-story windows admit sun as well, extending the time that the house can receive solar gain. The windows are still well shaded in summer, preserving the sole energy-saving benefit of the original awnings. One wonders just how many tens of thousands of fixed awnings could be modified in this way on houses all over the country.

Because we made no attempt to enclose any thermal mass, the porch enclosure does not provide the nighttime climate modification of a solar greenhouse, but it does do a powerful job of daytime air heating. Precisely because there is so little enclosed

Photo VI-3: Sunlight quickly warms the air in the enclosed porch in the morning, and Jean Feinour lets it circulate into the house when the weather is cool by opening the front door. Since there is little thermal mass in the porch, she closes off the space at night and relies on a 250-watt porch light to prevent plant freezing.

mass, sun shining through the 170 square feet of glazing has little to heat but air, and this heat constantly circulates into the house when the front door is open. Standing in the dining room on a sunny winter day, one can feel the warmth radiating from the porch.

Surprisingly, plants will thrive on the porch in a mild winter, although it was certainly not designed with winter horticulture in mind. As an experiment, Jean replaced the porch light bulb with a 250-watt heat lamp and found that it alone was sufficient to keep her plants from freezing overnight. Though it is doubtful that the same results will be obtained during a harsh winter, Jean is optimistically planning to raise some late-autumn vegetables.

Summer ventilation has been equally successful. The low, west-facing vent catches prevailing breezes, and the larger, east-facing vent allows heated air to rise out quickly. With screens in place in the outside door, the enclosure remains reasonably comfortable during the hottest weather, and it certainly doesn't add any heat to the house.

It is in winter, however, that Jean most appreciates the retrofit. Beams of sunshine warm the upstairs bedrooms, heated air from the porch gently circulates through the dining room, and the collector dependably blows heat into the living room. The entire front part of the house warms up quickly on bright days, and Jean has discovered that by cooking a hot breakfast to take the chill off the kitchen, she can often avoid turning the house thermostat above its low nighttime setting for the rest of the day.

Placing a hard dollar value on the savings that Jean realizes from her retrofit is difficult, because so much of the benefit is in the increased comfort of the house, rather than in the replacement of fuel oil needed to maintain a minimum bearable temperature. One financial fact is certain: At the cost of the retrofit, assuming a savings of 10 percent of Jean's fuel consumption, the installation will have repaid every dime in less than three years. And Jean swears that she is more than 10 percent satisfied with the system.

M.R. Carey

Photo VI-4: The Michels House

The Michels House

When we bought our 40-plus-year-old masonry duplex in 1976, my wife, Wanda, and I knew it would be expensive to heat, but we knew we'd have a lot of company since a large part of the housing in urban areas is comprised of this older type of masonry building. Because of their energy liability, these well-built but thermally poor houses are in danger of becoming prematurely obsolete, which puts the future of beautiful old places like ours in question. If they're not affordable because of high energy costs, will people just bite the bullet and suffer the expense, or will more and more of these houses just be abandoned to whatever fate ensues, including the wrecker's ball? Certainly our neighborhood wasn't threatened with demolition, but others probably are, with one of the causes being the cost of heating.

In our case, we had a real stake in correcting our house's thermal faults since we neither wanted to foot the bills nor leave the place. So it became a personal imperative to find solutions that it was hoped would also be useful for others faced with the same problem. With lots of research, it took about nine months to come up with a passive solar heating plan that would substantially reduce our heating bills. Energy conservation measures were clearly the first steps to take, and it seemed that a realistic goal would be to reduce the building's heating load by 60 percent.

Research done by the National Bureau of Standards indicated that when exterior masonry walls are insulated on the outside, they can provide a remarkable temperature stabilizing effect for a building's interior. It became apparent that there would be little or no cost savings between insulating outside or inside the exterior walls, and the added benefit of putting all that masonry inside a blanket of insulation made the outside option all the more attractive. More points for working outside: Insulating inside would have entailed stripping the walls of all their trim, pulling out electrical outlets, reframing all the windows and probably refinishing all the floors—all of which would be quite costly in addition to the basic insulation job. Also, we'd have lost interior square footage and would have had to suffer the mess and inconvenience of the job until it was totally done.

Insulating on the outside avoids all of these problems and simultaneously makes the building's mass available for the thermal mass storage that is generally desirable in a passive solar system. In our house this process put approximately 200,000 pounds of once thermally useless masonry inside the building's envelope. With a 15°F (8°C) interior air-temperature swing, this mass has a heat storage capacity of 660,000 Btu—enough for two average winter days of heat loss or one severe winter day. It was a pleasant surprise to learn that our humble abode had such an impressive heat-storage capacity. Our second important conservation project was to install insulating window shades. We chose shades made by Insul Shade of Branford, Connecticut, because one magazine's product review and analysis indicated that they offer the highest R-value for the money. Used over double-glazed windows they reach R-15, and they do seem to improve

Photo VI-5: An insulating drape hung over the big south window keeps night heat loss to a minimum.

thermal comfort. These shades have very good edge seals, and since the glass surface temperature drops really low behind the shade, condensation has occurred on both the outer storm window and the interior pane of glass. So far, however, we haven't found this condensation to be a problem.

We have also carried out all the standard but no less important conservation efforts of installing storm windows, R-19 attic insulation and replacing the old boiler with a unit that is 60 percent smaller than the original, making for much more efficient operation of the heating system. In the latter improvement, our old steam boiler had a rated output of 200,000 Btu per hour, but the house's reduced energy requirement didn't warrant such a huge rate of gas consumption. The replacement unit is rated at 60,000 Btu per hour, which is much closer to the house's rate of heat loss on the coldest nights.

The cost of each conservation effort was carefully weighed against its calculated benefit. The exterior wall insulation has a 9- to 11-year payback period, the longest payback of all our projects. This was calculated on a natural gas price of 40¢ per therm, which means that with a more expensive fuel such as oil or electricity the payback would come sooner. The R-19 attic insulation, on the other hand, will pay for itself in two years. Our window improvements, which included adding storms and the insulating shades, should make an energy-savings return on investment in six to seven years. The boiler replacement was a must-do improvement for which we haven't calculated a return on our investment. All these improvements were calculated to reduce the energy demand of the house by about 60 percent, and we feel that a 7- to 10-year payback period is a reasonable return for our conservation investment of about $8500, compared to leaving that money in a savings account.

After conservation we were ready to make use of our home's south wall, which we didn't insulate, as the appropriate next step toward further reducing our dependence on fossil fuels. Our dramatic modifications of that wall will, on the average, produce energy to satisfy approximately 50 percent of the heating load that remains after the conservation improvements. This will have the effect of reducing our annual heating consumption by fully 80 percent from the original heating load.

To develop as much useful passive solar collector area as possible, we took out a portion of wall on the second floor and replaced it with about 100 square feet of glass that reached all the way up to the cathedral ceiling. Below this direct-gain component, we built a small attached greenhouse that was thermally as well as visually connected with the second-floor glazing system (see figure VI-2). When all the remodeling dust settled, we had installed a passive heating system with about 275 square feet of collector area for about $4100, or $15 per square foot. For us the price was right, and we felt the addition looked good, too. Although the north, east and west walls were stuccoed over as part of the insulation job, the original brickwork on the south wall shows clearly through the glazing, preserving some of the original charm of the old place.

Our desire to make the best use of materials led to a fair amount of experimentation with the solar wall. In the first winter we experimented with single glazing for the second-floor glazing in order to gain an extra 12 percent of the available solar energy that would be lost in transmission through a second layer of glass. We felt that the use of window insulation at night would adequately compensate for the higher heat loss through single glazing. But when streams of daytime and nighttime condensation on the icy-cold glass proved to be too great a problem, we double glazed for the second winter.

We spent two winters experimenting with different glazing materials for the greenhouse, trying out Lascolite (FRP), 1-mil Mylar and a 7-mil material from the 3M Company, which generically is a laminated polyester-polycarbonate film often used in greenhouses that 3M calls #7410. The Lascolite and the #7410 each cost about 40¢ per square foot, the Mylar about 4¢ per square foot. We installed the Lascolite wrong. We tried screwing it to wood frames and then stretching it a little to make it as flat as possible. This caused too much stress, especially at very low temperatures (thermal contraction), and the Lascolite failed. Lascolite was also less desirable in this case because its light diffusing quality obscured any view from the first-floor windows, as well as the view of the house from the street.

The need for a tension-absorbing glazing led to the temporary use of 1-mil Mylar, which worked very well. But while the Mylar takes tension stresses well, it is not puncture resistant. It survived the winter exposure to the elements, but it did not survive summer storage when the greenhouse glazing frames were removed to prevent excessive heat gain. We kept poking things through it by mistake, which caused long tears.

Figure VI-2: The Michels house mixes conservation and solar improvements to optimize its energy efficiency. Added exterior, attic and window insulation performed a dramatic turnaround on the building's rate of heat loss, literally cutting it by more than half. The exterior wall insulation also transforms east, north and west walls into thermal mass that stores heat instead of dumping it outside. The mini-greenhouse feeds heat to both the upstairs and the downstairs, and the enlarged picture window upstairs provides effective direct-gain heating as well as bright daylighting.

The 3M #7410 film is both strong in tension and quite puncture resistant. It is also crystal clear, and for some reason dirt doesn't cling to it the way it did to the Mylar. For our purposes it is an excellent material, and the one we would recommend for a similar installation. We should mention too that lightness is an important factor with seasonal glazing systems, which makes thin films an ideal choice. You don't want to be lugging around ponderous double-glass frames, even just twice a year. Our greenhouse frames weigh in at about 15 pounds, and they're easy to handle.

The final solar improvement for the house is almost too small to notice. We glazed in our small porch with Plexiglas panels and hung a storm door at the top of the porch stairs to create a little solar airlock/vestibule. We don't depend on this retrofit for any heat contribution to the house, nor do we add any heat to it from the house, but the enclosure does stay somewhat warmer than the outside air. This reduces some heat loss from the house and, more importantly, it reduces cold air infiltration every time the front door is opened.

All things considered, technical, economic and aesthetic, this retrofit is a success. From an economic standpoint the solar retrofit costs were compared to the resulting energy savings, and the payback period came out to be ten years or less. We calculate an annual reduction in our gas use of about 450 therms for the solar improvements alone. In this particular instance the move to producing energy (with solar heat) was made at

Photo VI-6. The expanded window area floods the room with sunlight, creating an attractive environment for plants and people.

the point where the cost of options for conserving energy just exceeded the cost benefit of the solar improvements we made. Thus we rejected conservation improvements that had a payback period of more than ten years and put them into the list of possibilities for things to do *after* the solar retrofit.

Aesthetics is a criterion not often mentioned in the evaluation of a solar retrofit, but it was very important to us. Our quality of life has been enhanced by living in this passive solar home that has been created from the shell of an outmoded construction technique. The quality of the interior space, daylighting and our view of the world have all been greatly improved. What is the value of a more beautiful and comfortable home? To us, it's virtually priceless. There's a big old tree in our front yard, and for the first time we can see it with our expanded view from the living room. We love it!

Quite honestly, even if these remodeling improvements had been done without much regard for their thermal benefits, they could very well be justified on their aesthetic values alone. But you really can't do that anymore. Well, you can, but it's foolish not to combine the elegance of home improvement with the elegance of a more efficient home heating system. Needless to say, our improved environment is wonderful, and so are our reduced energy bills.

Tim Michels

Reference

Michels, Timothy I. 1979. Results: the retrofit of an existing masonry home for passive space heating. In *Proceedings of the 4th National Passive Solar Conference,* ed. Gregory Franta, Newark, Del.: American Section of the International Solar Energy Society.

The Doxsey House

Retrofitting is always a compromise between the ideal of achieving maximum solar efficiency and the reality of the structure's limitations. The house we (myself and a partner were coowner/contractors) bought in 1979 in Asheville, North Carolina, confronted us with limitations that were both financial and aesthetic. On the positive side the house was structurally sound, with a good, broad southern exposure and a rock-bottom asking price of $4000. On the negative side, it was derelict, with holes in the roof that measured 3 feet across and a long list of other roofing, heating, electrical and plumbing defects to correct.

Aesthetically, we were prevented from making any obtrusive solar additions by the fact that the entire neighborhood was listed in the National Registry of Historic Places. With signs of historical restoration springing up on every street, we felt that the addition of an attached greenhouse or very large expanses of glass would be inappropriate. Instead, we decided that the most cost-effective way to harness solar heat was to enclose porches on the first and second floors on the southwest side of the house, which actually faces about 20 degrees away from true south. The most appropriate and visually pleasing glazing system would be to use small panes of glass set in stained and varnished wood frames.

The structural work that was necessary to enclose the porches gave us the opportunity to build in the increased thermal mass the porches-turned-sunrooms would require in order to store heat and prevent large temperature fluctuations. Thus when we replaced the rotten floor joists of the downstairs porch, we added a beam and some posts and doubled up a joist that would carry a wall to the upstairs porch, all to handle the extra weight.

With both the upstairs and downstairs porches the additional thermal mass is supplied by floors made with recycled brick and by 8-foot-high Kalwall sun tubes filled with dyed water. The tubes were leveled in beds of sand, since the old floors themselves weren't quite level. They are 18 inches in diameter and spaced so that sun can also shine directly into each sunroom to keep the plants growing.

Since the house is a duplex, the solar modifications treat the two units separately. Downstairs, a stairwell running along the south wall of the house limited the amount of direct-gain glazing we could add. To compensate for this obstacle, we built a 76-square-foot air heating collector to serve the downstairs. The system consists of a 190-cfm fan that draws heated air through a series of baffles placed behind a black-painted steel absorber plate. A thermostat turns the fan on at 100°F (38°C) and turns it off at 90°F (32°C). We built the collector on-site using all off-the-shelf materials, including tempered-glass patio door seconds for the glazing. By choosing materials carefully, we were able to keep the collector cost down to about $300. We calculate that on an average January day the collector adds about 35,000 Btu to the house. The heated air

Photo VI-7: The Doxsey House

is ducted to the north bedrooms and some of the heat is absorbed by an interior brick chimney which acts as a thermal mass.

The essential modifications to the second floor included increased south glazing along with the water and masonry thermal mass placed on the porch (which is now a living room). Also, by removing a couple of interior partitions we have improved the circulation of heated air to other rooms. The total glazing addition to both floors amounted to 190 square feet added to about 60 square feet already in place. The porch conversions also added about 60 square feet of west-southwest glazing, and we have found that these windows collect a useful amount of solar energy in the late afternoon.

But while the solar modifications bring up to about 280,000 Btu of heat into the house on an average January day, it would be much less effective heat if a strong conservation program were not a part of the overall plan. Our initial approach to

Photo VI-8: Since a stairwell runs along part of the south wall, it wasn't possible to glaze the area in front of it, but it was possible to put the southern exposure to use with a wall-mounted air heating collector, made of tempered-glass patio doors, that provides hot air to the downstairs bedroom on the north side of the house.

conservation was basically to seal the house as tightly as possible with caulking and weather stripping. We sealed every crack and seam we could find, knowing that in an old house we could get to most but not all of them. We removed all the window trim and insulated around them in areas that blown-in wall insulation doesn't reach, such as where the window weights are located.

We gutted some exterior walls and installed foil-backed fiberglass insulation, while in other walls we had cellulose insulation blown in from the interior. The attic was insulated with 6 inches (R-22) of cellulose insulation, which we now think is too little even for our climate of about 4000 heating degree-days. We plan to add at least 4 more inches there.

Once the walls, floor and ceilings of a house are insulated, the largest heat losses occur from infiltration through doors and conductive heat loss through windows. We handled one infiltration problem by adding a two-door air-lock entry on the south side. This air-lock vestibule not only serves as a buffer zone but also collects enough heat on its own to furnish some to the apartments when the inner doors are open.

To cut down on heat loss through the windows, we installed insulating shades over the double glazing. The south and west sides of the house were fitted with Window

Figure VI-3: This whole-house retrofit was developed with consideration for both energy efficiency and historical preservation. Not wanting to make visually obtrusive solar improvements, the owners simply increased window area for direct-gain heating and added water thermal mass and an air heating collector. Big old frame houses like this one can present any number of retrofit opportunities, and with many options you can design to preserve traditional styles, to give the old place a face-lift or to do a little of both.

Quilt, a commercial insulating shade that has good edge seals and a pleasing quilted pattern. In conjunction with the double-glazed windows its effective R-value is over 5. Since the south and west sides of the house are the most exposed to passersby, using the same style of shade on all these windows unifies the overall appearance of the house, which was a concern given our location in a historical district.

Photo VI-9: Recycled brick and Kalwall sun tubes filled with dyed water provide thermal mass in the southwest corner rooms, both upstairs and down. Since the old floors aren't quite level, the water tubes were leveled in beds of sand and spaced far enough apart to admit plenty of sunlight into the rooms behind.

For other windows, home-built insulated Roman shades were made. The shade seals the window in the following manner: The top of the shade is attached to the top face trim, the sides are pressed against the side face trim by a wooden strip mounted with two spring-loaded hinges. The bottom of the shade rests on the windowsill. The shade itself consists of a poly-cotton fabric (outer layer), quilt batt (insulation) and 2-mil plastic (vapor barrier). String ties hold the batting in place and also support plastic rings on the back side through which the drawstrings pass.

Each kitchen has an under-the-counter water heater, a refrigerator and stove, all of which provide sufficient waste heat to keep them warm. Locating the water heater in the kitchen also shortens the distance between the source and the point of use, a more efficient strategy.

At this time the domestic hot water is electrically heated, although during the remodeling some initial steps were taken toward having a roof-mounted solar water heater. Two insulated copper pipes were run from the basement to the attic, which at some point would be the connection between the collectors and a storage tank. We are now planning to use a ground-mounted breadbox passive water heater instead of an active flat plate collector, which obviates the use of these pipes. But if a new owner were to opt for the latter system, the plumbing would be in place and save the work and expense of tearing up walls to conceal pipes.

Our calculations show that the combined effects of the solar and conservation improvements have cut the house's heating load about in half. The cost for the solar improvements alone was about $3000, and we estimate that from just an energy standpoint this investment will be paid back in seven to eight years. Of course, we have also created some very nice living spaces with the various south glass additions, the benefits of which really have no price tag.

As fuel costs continue to rise, the feasibility of retrofitting older houses shifts from a possibility to a necessity. We found that breathing new life into this old house at a cost of about $19 per square foot for all the improvements, solar, conservation and remodeling, was more cost-effective, less energy-intensive and provided more jobs than building a new house. This formerly derelict building has been rejuvenated to become a handsome energy-efficient residence that represents a new generation of "turned-on" houses. We'll be seeing a lot more of them in the near future.

W. Laurence Doxsey

The O'Boyle House

The O'Boyle family considered their 1950s ranch house a total disaster, and with good reason. When they bought it, it was a jumble of no less than seven separate, badly insulated additions tacked onto a two-room house, all nestling under a rotting roof with several disjunct pitches. When Paul Peters, friend, builder and designer, was hired to replace the roof, he proposed an even more extensive face-lift.

"Everything was primed for it," he recalls now. This unbeautiful house in eastern Pennsylvania happened to be beautifully sited to catch the sun, with 88 feet of south-facing wall shaded by a line of tall deciduous trees. For Paul, solar design was the answer to the house's aesthetic and energy deficiencies, but while he had long been a solar enthusiast, the O'Boyles didn't know much about the subject.

To show them the remarkable solar potential of their property, Peters produced plans for a 12-foot-deep, glass-walled addition to run the length of the south side. The addition, a unified sweep of glass, wood and stucco, was divided into several spaces. From west to east, it included a roofed outdoor patio, a solarium, a Trombe-type wall, an air-lock vestibule and a greenhouse. The O'Boyles examined the plans and became enthusiastic about the face-lift, increased energy efficiency and more living space.

The solar work was done in 1978, at a cost of $9800, and the new roof, built at the same time, cost another $7000. The O'Boyles' solar sideshow has increased their usable living space by 20 percent while decreasing the overall heating load by 35 percent. No less important, all these improvements have about doubled the value of the house. It also makes a great case for a solar retrofit as an aesthetic bonus.

All the glass used in the addition is double pane, ⅛ inch thick with a ⅜-inch airspace. The new exterior walls are 12-inch-thick concrete block with sand compacted inside. The outside of the walls is insulated with Dow-Corning Styrofoam insulation covered with standard galvanized wire mesh and stucco. The floors are 4- to 8-inch concrete slabs poured over 1 inch of Styrofoam. In the solarium, the concrete is topped with Vermont slate. In addition to under-slab insulation, the perimeter footings, extending 3 feet below grade, are also insulated with an inch of foam.

The solarium is the heart of the new living space. At 12 by 30 feet, its proportions are ample for family dining in summer. On sunny winter afternoons, it often reaches 80°F (27°C) and even on cloudy winter days it never drops below 48°F (9°C). The solarium works in part because great care was taken during construction to make it as airtight as possible, along with plenty of insulation. Getting warm air into the house only requires opening the windows that connect the solarium with a bedroom. There is a floor-level cold air return vent on the back wall of the solarium.

The Trombe wall, almost an afterthought, was built to fill a gap between the solarium and the vestibule. Three floor registers supply the Trombe wall with cool intake air, which, when heated by the sun, rises and flows in the house through the tops of dining room and kitchen windows.

Photo VI-10: The O'Boyle House

The adjoining 6 by 8-foot vestibule is an air-lock system that keeps the heat in the house despite the frequency with which assorted O'Boyles, friends, and relatives stride in and out through its insulated steel door.

The 11 by 14-foot greenhouse, used primarily for space heating, has the highest proportion of glass of all the additions. In winter, the greenhouse temperature can rise to 95°F (35°C); in summer when it's vented, it usually doesn't get above 90°F (32°C).

Tom O'Boyle considers his greenhouse the greatest success of all the additions, and so, by extension, is the room behind it. The once drafty old family room has become the focus of redecorating plans and family gatherings since the greenhouse sprouted on its south wall. It's brighter, warmer and prettier now, the favored spot to sit and watch the begonias flowering. Warm air flows in from the greenhouse through windows on the original south wall of the family room.

The impetus to redecorate the rest of the house up to the standard of the new addition has diverted funds from further steps toward energy efficiency. There are still several things that need to be done, particularly adding window insulation to the solarium and the greenhouse to reduce heat loss at night. Extra water thermal mass in both these rooms would help as well. In the solarium, six 18-inch-diameter water columns are planned, and in the greenhouse, six 55-gallon drums will be added to supplement the three 55-gallon drums already in place as the base of a worktable. Tom estimates that these measures should reduce his overall energy load by an additional 15 percent, a total reduction of 50 percent from the time when the house was an ugly duckling.

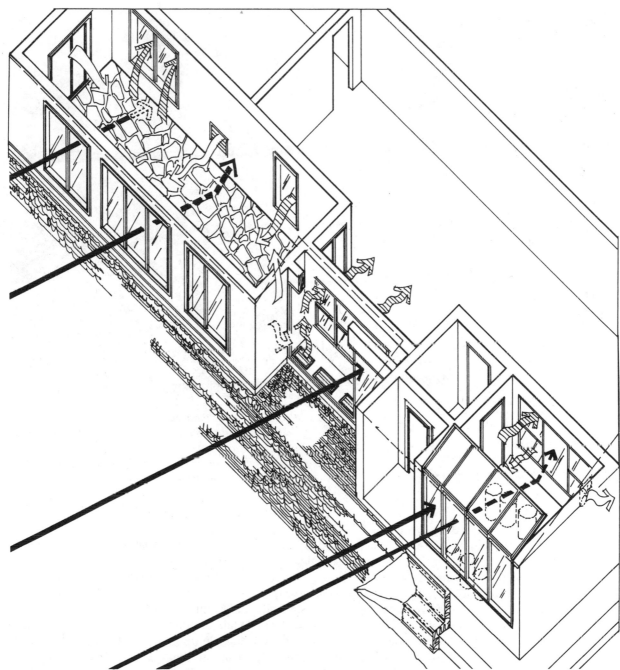

Figure VI-4: The O'Boyles moved their 1950s ranch house toward the sun by adding a 12-foot-deep, glass-walled addition along the entire length of the south wall, creating in the process a large solarium, a Trombe-type wall and a solar greenhouse, all of which pay energy dividends while enhancing the looks of the house.

Photo VI-11: The greenhouse provides a wide view to the outside. The overhead, sloped glass admits sunlight to the greenhouse interior, where it warms 55-gallon drums of water and bounces enough light around to keep the plants from leaning toward the glass.

Photo VI-12: Adding glazing outside the existing brick wall turned this section into a sort of Trombe wall. In winter, warm air passes through vents into the room behind, while in summer, the vent above the glass permits excess heat to be dumped outside.

With the house now sporting 11 rooms inside, and with four acres and several outbuildings awaiting, the O'Boyles are more than glad that their solar additions are maintenance-free and expected to remain so. There isn't all that much to go wrong. All the glass has frames of construction heart redwood that needed only an initial coat of a sealer preservative. The stucco contains a yellow sand and lime mixture that gives it a pleasant natural tint and eliminates the need to paint.

A well-designed solar system shouldn't require much work to operate it, and at the O'Boyles it all comes down to a simple tale of opening and closing windows and vents. In winter, opening the tops of the windows in the north wall of the solarium, the greenhouse, and the Trombe wall sends warm air into the rooms behind. Inlet air vents, kept open during the day, are closed at night. In summer, thermostatically controlled fans automatically exhaust heat out to the east from the Trombe wall and greenhouse, pulling in cool air from the shaded solarium. Opening the sliding glass doors at the west end of the solarium does a surprisingly effective job of letting in cooling breezes.

Photo VI-13: The large solarium is a pleasant living space year-round because it is well insulated and well shaded in summer. The floor of concrete topped with Vermont slate helps to temper the space in hot weather with the help of thermostatically controlled exhaust fans.

The house's back-up heating system consists of oil heat, and a Heatilator fireplace in the living room, which uses about two-and-a-half cords of wood a year. The thermostat is the real measure of the benefits of the new solar system. Nowadays, it is kept at 60°F (16°C) for most of the winter, while the actual house temperature is usually around 70°F (21°C). In the old days, the thermostat was set in the high 70s in winter just to keep the house at 70. And in summer, where numerous fans and a bedroom air conditioner used to puff and blow, a single portable fan now suffices.

While the solar additions are paying energy dividends, it was just as important to the O'Boyles that they also improve the looks of their house. From the inside, the warm walls and floors make life so comfortable that it's hard to imagine the reality of cold and wind outside. And, from outside, the wide expanse of glass succeeds in giving the house an open and inviting aspect that is even more appealing when the rooms are lit in the evening.

Letting the sun shine in turned the O'Boyles' unpromising rancher around. It also turned Tom O'Boyle's thinking around. "Now I want to build a totally solar home," he says. "This house is just my first step. I can hardly wait."

Marilyn Hodges

The Contributors

It took a lot of effort from a lot of people to make this book happen, and to this special group of old and new friends and associates I extend a very grateful thanks for your commitment to what was being done. Group, in two-and-a-half years we've made a good book out of the many things that were built and written, drawn and photographed, edited, checked and rechecked and finally designed into this one thing, this beautiful book. As other solar books have done, *Solarizing* proves again that people can indeed take control of their energy futures, and that solar heating systems are well within the grasp of anyone willing to find out about them. To that end, this group has created a primary tool that will be useful for many years of this country's transition into the solar age.

Other thanks are due. Carol Stoner, the executive editor of the Rodale Press Book Division, gave a tenderfoot editor a pretty free rein to make this book as good as it seems to have turned out. That in itself was an amazing opportunity, but with it she also gave much guidance and support. A heartfelt thank you is but the smallest conveyance of my gratitude for her help.

John Blackford did a lot of writing and smoothed out a lot of rough copy. His assistance was vital to molding this big lump of words into its final form. Margaret Balitas was a pillar of calm organization amidst the frenzy of getting this thing into production. She researched and wrote the very important Hardware Focus Sections, and she kept hundreds of details in order, staving off the ever-present threat of confusion and disarray. As the senior designer for this book, Kim E. Morrow had to take on a few hundred articles, illustrations and photographs and wrestle them into meaningful, readable form. The result, as you've seen it, is a truly elegant design. We gave Dolores Plikaitis what we were certain was "clean" copy, but thank goodness she was able to pick out specks, erase blotches and fill gaping holes with her exacting copy editing. Herb Wade was our assurance of technical accuracy. His sharp solar acumen eliminated dozens of bloopers that would otherwise have been immortalized in type. It was always a relief to hear that Herb thought our stuff was good. Somebody had to do a heck of a lot of typing to transform reams of editorial scribblings into pristine manuscript pages, and most of it went across Bobbie Hartranft's desk. When a piece of edited manuscript looked utterly mad, Bobbie would render it sane. When the typing load overflowed, as it often did, Dorothy Smickley helped out too. From the tenderfoot, a special thanks to all.

David Bainbridge, founder of the Passive Solar Institute in Davis, California, is the author of numerous articles, reports and books on energy, bicycling, community planning and environmental concerns. He wrote the Passive Solar Catalogs series, and he is one of the authors of Rodale's *Village Homes' Solar House Designs.*

Prior to joining the editorial staff of Rodale Press, *Margaret Balitas* was a community development agent with the Cooperative Extension Service of Pennsylvania State University. She helped to establish numerous community energy projects including educational programs in which youth designed and built various solar systems.

An architect and member of the Lehigh Valley Solar Energy Association, *Ken Baxter* has experience in the design and construction of sprayed urethane domes, earth-sheltered structures and passive solar dwellings. Currently, he is designing solar retrofits. His illustrations appear in this book.

John Blackford, an assistant editor at Rodale Press, is the principal writer of Rodale's *Build Your Harvest Kitchen.* He also is a free-lance writer specializing in alternative energy and home remodeling. Blackford lives with his family in a 200-year-old stone farmhouse they are planning to retrofit with solar.

From 1950 to 1968 *H. Nelson Blue* was a contractor in the construction business, and from 1968 to 1978 he owned a metal fabricating business. In 1978, he sold this business and moved to his present home in Carthage, North Carolina, where he and his wife Helen raise vegetables, fruit, grain, beef cattle and hogs.

John Burton founded Integral Design in Santa Rosa, California, which provides design and installation services for low-cost solar devices, primarily breadbox water heaters. Prior to founding Integral Design, Burton was a solar information coordinator for the Solar Office of the California Energy Commission.

As information coordinator for the Community Action Program Solar Project in Lancaster, Pennsylvania, *M.R. Carey* provides information on the numerous solar retrofit projects of this group. Carey is also contributing editor to the *Lancaster Independent Press.*

Joe Carter has written, photographed, researched, consulted and installed in the solar field. He is currently waging war on the Home Heating Index of his house in Emmaus, Pennsylvania.

David Chesebrough is a science and mathematics teacher at Sewickley Academy in Sewickley, Pennsylvania, and he also teaches solar energy courses at local colleges. He helped to found two nonprofit energy and environmental education organizations, and he writes a weekly energy column for the *Beaver County Times.*

Mic Curd, an architect and builder, is a partner in Shelter Design Group, Stony Run, Pennsylvania. Currently, half of that company's work is solar retrofitting, and most of the new construction is earth-sheltered. Some of Curd's designs appear in Rodale's *30 Energy-Efficient Houses* and *A Design and Construction Handbook for Energy-Saving Houses.*

Carl Doney, a professional photographer since 1973, is Rodale Press's studio manager and the principal photographer of *Rodale's New Shelter.* Doney's work appears in this book and in other Rodale books such as *The Fish-Lovers' Cookbook* and *Bread Winners.* His assignments for Rodale have taken him all over North America.

W. Laurence Doxsey, a solar designer, consultant and builder in Asheville, North Carolina, is president of Approtech, Inc., and coowner and cofounder of Sunbright Company. Doxsey, a research associate for Rodale's *Movable Insulation,* also lectures and directs workshops on solar and alternative energy.

Bob Flower, a mechanical engineer and mathematician, is an energy consultant to the Rodale Press North Street Design Department, the Organic Gardening and Farming Research Center and to architects and builders. Flower, who enjoys tackling problems in heat transfer, fluid mechanics and the subtleties of natural climate, also teaches classes on solar design.

Chris Fried, an energy consultant and mechanical engineer in Catawissa, Pennsylvania, conducts solar workshops throughout that state, often with the aid of his energy van, which carries a windmill and solar heaters. Fried now lives in a passive solar, earth-sheltered home he designed and built for less than $20,000.

Gary Gerber, a mechanical engineer, is a cofounder of Sun Light & Power Company, a solar consulting, design and construction firm in Point Richmond, California. Gerber, whose projects have appeared in national publications, also teaches a class on solar installation at the College of Marin in Kentfield, California.

T.L. Gettings has been a photographer with Rodale Press since 1971 and its director of photography since 1973. Before joining Rodale Press, he served as a Marine Corps photographer and worked as a graphic designer/photographer in Washington, D.C. His photographs appear in this book.

Bill Ginn and his wife, June LaCombe Ginn, designed and built a solar greenhouse onto their 1837 Cape Cod. He is executive director of the Maine Audubon Society. Ginn was coordinator of the Maine bottle bill campaign, and he is currently chairman of the Pesticide Control Board for the State of Maine.

John Hamel joined the Rodale Press Photography Department in 1974 and is currently the assistant director of photography and the principal photographer for *Organic Gardening.* His work appears in this book as well as in other Rodale books such as *Build It Better Yourself, Square Foot Gardening* and *Getting the Most from Your Garden.*

Michael Harris, a longtime contributor to the *New Hampshire Times,* is the author of *Heating with Wood* (Secaucus, N.J.: Citadel Press, 1980) and over 100 feature articles in major publications.

Marilyn Hodges is a free-lance writer and editor who has worked on energy-related books and magazines in England and the United States.

John Hoover, a fine arts graduate, is a free-lance illustrator and water color artist in Bethlehem, Pennsylvania. He creates illustrations for local companies and historical societies. His illustrations appear in this book.

Dennis Jaehne, who was the director of Passive Solar 1980, The Fifth National Passive Solar Conference held in Amherst, Massachusetts, is a writer and consultant involved in passive solar and community energy development. He was the director of Project SUEDE at the University of Massachusetts, a solar training and demonstration program.

Doug Kelbaugh, AIA, has built many award-winning passive solar structures including the first Trombe wall in the United States. Kelbaugh speaks and writes on passive solar architecture; he teaches at the New Jersey School of Architecture, and he practices with Kelbaugh and Lee Architects in Princeton, New Jersey.

Sheldon Klapper, an architect, is presently training programs manager at Western SUN in Portland, Oregon. A custom home builder for years, Klapper was previously training director for the New England SUEDE Project. Klapper designed the 3100-square-foot Wintergreen Cooperative Solar Greenhouse in Orange, Massachusetts.

Charles Klein, AIA, is an architect in Bethlehem, Pennsylvania, who is known for his design excellence in climatic responsive dwellings. Klein has incorporated passive solar tempering in his projects since 1973; he is currently president of the Lehigh Valley Solar Energy Association.

An associate editor of *Organic Gardening, Mike LaFavore* specializes in writing about energy topics, primarily wood heat. He also contributes regularly to *Rodale's New Shelter.* Before joining Rodale Press, he was a newspaper reporter and free-lance writer.

Frederic S. Langa began his writing career in the year of the Arab embargo and quickly specialized in residential energy. His articles have appeared in many magazines, and now he is associate editor of *Rodale's New Shelter.*

For 15 years, *Malcolm Lillywhite,* a physicist, worked in the aerospace industry specializing in thermal control of aircraft until 1971 when he founded the Domestic Technology Institute in Evergreen, Colorado. A 5000-square-foot passive solar greenhouse that he designed and built in Colorado has operated continuously since its construction in 1970.

Lawrence L. Lindsey, an architect, was strongly influenced by Victor Olgyay, Doug Kelbaugh, Harrison Fraker and others at Princeton University's School of Architecture when he was a student there in the late 1960s. In 1976 he and Fraker founded the Princeton Energy Group, a research and consulting firm.

Denny Long, who specializes in custom-built water storage tanks and movable insulation systems, designs and manufactures passive solar components for homes in central California. Long served as a technical advisor to HUD's *First Passive Solar Home Awards,* and he was a consultant to the *Passive Solar Handbook for California.*

Mitch Mandel, whose photographs appear in this book, is with the Rodale Press photography staff. He is the principal photographer for *The New Farm,* and his photos appear in many Rodale books and in *Rodale's New Shelter.* Mandel's photography assignments have taken him many places such as atop windmills and into underground homes.

Gene A. Mater, a cartoonist and illustrator, lives in Bethlehem, Pennsylvania. For seven years, his cartoon strip "Gremlin Village" appeared in college newspapers throughout the United States. Mater's illustrations appear in this book and in Rodale's *Build Your Harvest Kitchen,* and his cartoons in Rodale's *The Bicycle Touring Book.*

Bruce Melzer, a principal of Earth Integral in Davis, California, served as project manager for the retrofit demonstration and training project of the Sacramento Housing and Redevelopment Agency, and he also managed the breadbox water heater retrofit project in Indio. Previously, Melzer was a project manager for Living Systems.

Arthur C. Meyers III, Ph.D., is a solar energy physicist in the Energy Division of the Institute of Basic Applied Research in Ames, Iowa. Meyers, who has published extensively in his field, conducts research in passive solar design and education.

Tim Michels is president of Londe.Parker.Michels, Inc., an energy consulting firm in St. Louis. In addition to consulting work, this company conducts passive solar workshops throughout the Midwest. Michels wrote *Solar Energy Utilization* (New York: Van Nostrand Reinhold Co., 1979), recognized by *Library Journal* as one of the 100 best science and technical books of 1979.

A free-lance writer in Butte, Montana, *Barbara Miller* specializes in writing articles on energy for newspapers and magazines, and she also wrote the energy section of *The All Montana Catalog* (Missoula, Mont.: Montana Small Business Association, 1980). Miller is the founder of Window Arts in Butte which sells insulating window shades that are locally made.

Erika Morgan, presently supervisor of the Education Section in the Solar Applications Branch of the Tennessee Valley Authority, is a public educator specializing in solar and environmental topics. Highlights of her career as an educator include coordinating 74 energy workshops and 3 exhibitions in Maine and coordinating that state's first Sun Day.

David Morris is the director of the Institute for Local Self-Reliance, Washington, D.C. He is also the author of *Neighborhood Power* (Boston: Beacon Press, 1975) and *Power to the Cities* (San Francisco: Sierra Club Books, forthcoming).

Prior to joining Rodale Press as a senior book designer, *Kim E. Morrow* was employed as an advertising and marketing director and as a graphic artist. Other Rodale books to Morrow's design credit include *Movable Insulation* and *Good Cooking from India.* A fiber artist, Morrow is a juried member of the Lehigh Valley Crafts Association.

As executive director of the Center for Ecological Technology (CET) in Pittsfield, Massachusetts, *Ned Nisson* conducts energy workshops and provides technical assistance. Nisson, the author of *The Berkshire Energy Manual,* managed the Low Income Weatherization Program of the Berkshire Community Action Council for CET.

A senior scientist with Ecotope Group, Inc., in Seattle, *Larry Palmiter* is involved in research concerning passive solar applications and energy conservation for homes. Prior to joining Ecotope, Palmiter, who is the author of numerous technical papers, was a systems engineer with the National Center for Appropriate Technology.

Anastas Pollock, a solar designer and builder, founded Taconic Solar, Inc., in Lebanon Springs, New York. Previously, he worked as a crew chief/instructor for Project SUEDE and the Weatherization Solar Training Programs of the Center of Ecological Technology in Massachusetts. Pollock has been a carpenter/builder since 1971.

Jeff Reiss was a solar energy specialist for the Solar Office of the California Energy Commission from 1975 to 1979. Now, as the founder of Solar Future in Sacramento, he distributes solar products and promotes low-cost solar applications. He is a coauthor of the *1980 Year of the Sun Calendar.*

Amy Zaffarano Rowland is an associate editor at Rodale Press, and her current book project focuses on handcrafted house details. She assisted with Rodale's *Movable Insulation* and *A Design and Construction Handbook for Energy-Saving Houses.* As a free-lance writer, she profiles artists and craftspeople in magazines such as *Fiberarts.*

Managing editor of *Organic Gardening, Jack Ruttle* is also active in solar greenhouse and cold frame research at the Organic Gardening and Farming Research Center in Maxatawny, Pennsylvania. This experience is evident in his contributions as an editor of Rodale's *The Solar Greenhouse Book.*

Rick Schwolsky is president of Sunrise Builders in Grafton, Vermont, a partner in Sunrise Solar Services in Suffield, Connecticut, and chairman of the National Association of Solar Contractors, Washington, D.C. His column "Building It Right" appears in *Solar Age;* he also wrote *The Builder's Guide to Solar Construction* (New York: McGraw-Hill Book Co., 1981).

David Sellers is a research engineer in Rodale's Product Testing Department where he has tested, evaluated and written about such products as solar cookers, solar food driers, glazing materials and basement waterproofers.

Sally Ann Shenk's photographs appear in this book. She joined the Rodale Press Photography Department in 1977, and she is the principal photographer for *Bicycling*. In addition to her photography, Shenk has experience in graphic arts and interior design.

Bristol Stickney, a mechanical engineer, is on the technical staff of the New Mexico Solar Energy Association (NMSEA) in Santa Fe where he presently directs the research and development program. Stickney has presented technical papers at National Passive Solar Conferences, and he writes regularly for *SUNPAPER,* the bulletin of the NMSEA.

Jerry Thierolf, a commercial artist and a free-lance artist, lives in Reading, Pennsylvania. Thierolf, whose illustrations appear in this book, also has experience in industrial advertising and marketing communications.

In 1975 Professor *Irving H. Thomae* sat in on a class on solar heating systems at the Thayer School of Engineering, Dartmouth College, and he became hooked on solar. Thomae, a specialist in computer engineering, has developed equipment and procedures for electric utilities to use in load management research for residential applications.

Herb Wade is manager of the Solar Program, Division of Energy, Missouri Department of Natural Resources. Prior to that he was deputy director of the Arizona Solar Energy Research Commission. He is currently writing a book on underground homes for Rodale.

Ronald Wantoch, a chemical engineer and energy consultant in Overland Park, Kansas, has provided technical assistance for numerous solar projects, and he also provided performance calculations and energy consulting for three award-winning projects in federal solar demonstration programs.

Alex Wilson is executive director of the New England Solar Energy Association in Brattleboro, Vermont, and prior to this, he was associate director of the New Mexico Solar Energy Association (NMSEA). Wilson wrote *Thermal Storage Wall Design Manual* for NMSEA and the *Passive Solar Retrofit Training Manual for the Tennessee Valley Authority.*

Tom Wilson, an award winner in the 1978 HUD Passive Solar Residential Design Competition for a Trombe wall retrofit design, is a partner in the consulting firm Residential Energy Conservation in Upper Black Eddy, Pennsylvania. Wilson, vice-chairperson of the Mid-Atlantic Solar Energy Association (MASEA), edited MASEA's energy retrofit manual *Home Remedies* (Philadelphia, 1980).

Harry Wohlbach is a technician with the Product Testing Department at Rodale Press where he has built, installed and monitored drain-down, drain-back, thermosiphon and batch-type domestic hot water systems.

Ray Wolf is executive editor of Rodale Plans Books. He is the author of Rodale's *Solar Growing Frame, Insulating Window Shade* and *Solar Air Heater.* An avid do-it-yourselfer, Wolf retrofitted his 120-year-old farmhouse for solar and wood heating.

Index